studies in physical and theoretical chemistry 78

ELECTRON AND PROTON TRANSFER IN CHEMISTRY AND BIOLOGY

studies in physical and theoretical chemistry 78

ELECTRON AND PROTON TRANSFER IN CHEMISTRY AND BIOLOGY

Edited by

A. MÜLLER
Department of Inorganic Chemistry
University of Bielefeld, Germany

H. RATAJCZAK
Institute of Chemistry
University of Wrocław, Poland
and
Institute of Low Temperature and Structure Research
Polish Academy of Sciences
Wrocław, Poland

W. JUNGE
Faculty of Biology
University of Osnabrück, Germany

E. DIEMANN
Department of Inorganic Chemistry
University of Bielefeld, Germany

ELSEVIER
Amsterdam — London — New York — Tokyo 1992

ELSEVIER SCIENCE PUBLISHERS B.V.
Sara Burgerhartstraat 25
P.O. Box 211, 1000 AE Amsterdam, The Netherlands

ISBN: 0-444-88862-4

This book is printed on acid-free paper.

Printed in The Netherlands

studies in physical and theoretical chemistry

Recent titles in this series

11 **Electrodes of Conductive Metallic Oxides. Part A** edited by S. Trasatti
Electrodes of Conductive Metallic Oxides. Part B edited by S. Trasatti
12 **Ionic Hydration in Chemistry and Biophysics** by B.E. Conway
13 **Diffraction Studies on Non-Crystalline Substances** edited by I. Hargittai and W.J. Orville-Thomas
14 **Radiation Chemistry of Hydrocarbons** by G. Földiák
15 **Progress in Electrochemistry** edited by D.A.J. Rand, G.P. Power and I.M. Ritchie
16 **Data Processing in Chemistry** edited by Z. Hippe
17 **Molecular Vibrational-Rotational Spectra** by D. Papoušek and M.R. Aliev
18 **Steric Effects in Biomolecules** edited by G. Náray-Szabó
19 **Field Theoretical Methods in Chemical Physics** by R. Paul
20 **Vibrational Intensities in Infrared and Raman Spectroscopy** edited by W.B. Person and G. Zerbi
21 **Current Aspects of Quantum Chemistry 1981** edited by R. Carbó
22 **Spin Polarization and Magnetic Effects in Radical Reactions** edited by Yu.N. Molin
23 **Symmetries and Properties of Non-Rigid Molecules: A Comprehensive Survey** edited by J. Maruani and J. Serre
24 **Physical Chemistry of Transmembrane Ion Motions** edited by G. Spach
25 **Advances in Mössbauer Spectroscopy: Applications to Physics, Chemistry and Biology** edited by B.V. Thosar and P.K. Iyengar
26 **Aggregation Processes in Solution** edited by E. Wyn-Jones and J. Gormally
27 **Ions and Molecules in Solution** edited by N. Tanaka, H. Ohtaki and R. Tamamushi
28 **Chemical Applications of Topology and Graph Theory** edited by R.B. King
29 **Electronic and Molecular Structure of Electrode-Electrolyte Interfaces** edited by W.N. Hansen, D.M. Kolb and D.W. Lynch
30 **Fourier Transform NMR Spectroscopy (second revised edition)** by D. Shaw
31 **Hot Atom Chemistry: Recent Trends and Applications in the Physical and Life Sciences and Technology** edited by T. Matsuura
32 **Physical Chemistry of the Solid State: Applications to Metals and their Compounds** edited by P. Lacombe
33 **Inorganic Electronic Spectroscopy (second edition)** by A.B.P. Lever
34 **Electrochemistry: The Interfacing Science** edited by D.A.J. Rand and A.M. Bond
35 **Photophysics and Photochemistry above 6 eV** edited by F. Lahmani
36 **Biomolecules: Electronic Aspects** edited by C. Nagata, M. Hatano, J. Tanaka and H. Suzuki
37 **Topics in Molecular Interactions** edited by W.J. Orville-Thomas, H. Ratajczak and C.N.R. Rao
38 **The Chemical Physics of Solvation. Part A. Theory of Solvation** edited by R.R. Dogonadze, E. Kálmán, A.A. Kornyshev and J. Ulstrup
The Chemical Physics of Solvation. Part B. Spectroscopy of Solvation edited by R.R. Dogonadze, E. Kálmán, A.A. Kornyshev and J. Ulstrup
The Chemical Physics of Solvation. Part C. Solvation Phenomena in Specific Physical, Chemical, and Biological Systems edited by R.R. Dogonadze, E. Kálmán, A.A. Kornyshev and J. Ulstrup
39 **Industrial Application of Radioisotopes** edited by G. Földiák
40 **Stable Gas-in-Liquid Emulsions: Production in Natural Waters and Artificial Media** by J.S. D'Arrigo

41 **Theoretical Chemistry of Biological Systems** edited by G. Náray-Szabó
42 **Theory of Molecular Interactions** by I.G. Kaplan
43 **Fluctuations, Diffusion and Spin Relaxation** by R. Lenk
44 **The Unitary Group in Quantum Chemistry** by F.A. Matsen and R. Pauncz
45 **Laser Scattering Spectroscopy of Biological Objects** edited by J. Štěpánek, P. Anzenbacher and B. Sedláček
46 **Dynamics of Molecular Crystals** edited by J. Lascombe
47 **Kinetics of Electrochemical Metal Dissolution** by L. Kiss
48 **Fundamentals of Diffusion Bonding** edited by Y. Ishida
49 **Metallic Superlattices: Artificially Structured Materials** edited by T. Shinjo and T. Takada
50 **Photoelectrochemical Solar Cells** edited by K.S.V. Santhanam and M. Sharon
51 **Graph Theory and Topology in Chemistry** edited by R.B. King and D.H. Rouvray
52 **Intermolecular Complexes** by P. Hobza and R. Zahradník
53 **Potential Energy Hypersurfaces** by P.G. Mezey
54 **Math/Chem/Comp 1987** edited by R.C. Lacher
55 **Semiconductor Electrodes** edited by H.O. Finklea
56 **Computational Chemistry** by M.D. Johnston, Jr.
57 **Positron and Positronium Chemistry** edited by D.M. Schrader and Y.C. Jean
58 ***Ab Initio* Calculation of the Structures and Properties of Molecules** by C.E. Dykstra
59 **Physical Adsorption on Heterogeneous Solids** by M. Jaroniec and R. Madey
60 **Ignition of Solids** by V.N. Vilyunov and V.E. Zarko
61 **Nuclear Measurements in Industry** by S. Rózsa
62 **Quantum Chemistry: Basic Aspects, Actual Trends** edited by R. Carbó
63 **Math/Chem/Comp 1988** edited by A. Graovac
64 **Valence Bond Theory and Chemical Structure** edited by D.J. Klein and N. Trinajstić
65 **Structure and Reactivity in Reverse Micelles** edited by M.P. Pileni
66 **Applications of Time-Resolved Optical Spectroscopy** by V. Brückner, K.-H. Feller and U.-W. Grummt
67 **Magnetic Resonance and Related Phenomena** edited by J. Stankowski, N. Pislewski, S.K. Hoffmann and S. Idziak
68 **Atomic and Molecular Clusters** edited by E.R. Bernstein
69 **Structure and Properties of Molecular Crystals** edited by M. Pierrot
70 **Self-consistent Field: Theory and Applications** edited by R. Carbó and M. Klobukowski
71 **Modelling of Molecular Structures and Properties** edited by J.-L. Rivail
72 **Nuclear Magnetic Resonance: Principles and Theory** by R. Kitamaru
73 **Artificial Intelligence in Chemistry: Structure Elucidation and Simulation of Organic Reactions** by Z. Hippe
74 **Spectroscopy and Relaxation of Molecular Liquids** edited by D. Steele and J. Yarwood
75 **Molecular Design: Chemical Structure Generation from the Properties of Pure Organic Compounds** by A.L. Horvath
76 **Coordination and Transport Properties of Macrocyclic Compounds in Solution** by B.G. Cox and H. Schneider
77 **Computational Chemistry: Structure, Interactions and Reactivity. Part A** edited by S. Fraga
Computational Chemistry: Structure, Interactions and Reactivity. Part B edited by S. Fraga

Preface

This volume constitutes the proceedings of an international meeting held in September 19-21, 1990, at ZIF, Zentrum für Interdisziplinäre Forschung, Universität Bielefeld, Germany. The meeting was one of a continuing series organized through ZIF on topics which transcend the boundaries of the traditional academic disciplines.

The book describes various aspects of electron and proton transfer in chemistry and biology. The joint presentation was chosen for two reasons. Rapid electron and proton transfer is the basis of life on earth. It governs cellular energetics in the most primitive as well as in higher organisms, with photosynthetic and with heterotrophic lifestyles. Further, biology has become the ground where the various disciplines of science, that were rigorously diversified during the last century, are meeting again.

Students of the life sciences will always rely on knowledge and categories acquired from investigations of simpler and cleaner systems that are the object of pure physics and chemistry. However, some objects of biology have been characterized so well, at Ångstrom dimensions in space and at fractions of picoseconds in time (for photosynthesis), that a rigorous understanding of their function is feeding back to and further colouring physics and chemistry.

The book starts with a survey of physicochemical principles of electron transfer in the gas and the solid phase, with thermodynamic and photochemical driving force. Inner and outer sphere mechanisms and the coupling of electron transfer to nuclear rearrangements are reviewed. These principles are applied to construct artificial photosynthesis. This leads to biological electron transfer involving proteins with transition metal and/or organic redox centres. The tuning of the free energy profile on the reaction trajectory through the protein by single amino acids or by the larger ensemble that determines the electrostatic properties of the reaction path is one major issue. Another one is the transformation of one-electron to paired-electron steps with protection against hazardous radical intermediates. The diversity of electron transport systems is represented in some chapters with emphasis on photosynthesis, respiration and nitrogenases. In photosynthesis of green plants light driven vectorial electron transfer is coupled to protolytic reactions, with about one quarter of the useful work derived from light quanta utilized for proton pumping across a coupling membrane. That is where the interchange of electrochemical (Dm) and chemical (ATP) forms of free energy storage and transfer in cellular energetics starts. The proton is distinguished from other reactants by an extremely small diameter and the ability of tunneling at reasonable rates. This is the basis for particular polarization, solvent and isotope effects as well as for hydrogen-bonded networks that are suited to long-range proton transfer.

This is, of course, not a comprehensive treatise. As the subject developed in to many dimensions not all topics could be considered, for example those related to double exchange phenomena, solid state devices (*e.g.* molecular electronics), resonance effects in long range electron transfer and aspects of tunneling (*e.g.* the investigation of electron transfer by scanning tunelling microscopy), and picosecond kinetics.

The meeting was generously supported by ZIF, the Westfälisch-Lippische Universitätsgeselschaft, and the Fonds der Chemischen Industrie.

A. Müller
H. Ratajczak
W. Junge
E. Diemann

Bielefeld, Wrocław, Osnabrück

September, 1991

Contents

List of Contributors

Althoff G.

Lehrstuhl Biophysik
Fachbereich Biologie
Universität Osnabrück
Barbarastr. 11
D-4500 Osnabrück, Germany

Bokranz M.

Institut für Mikrobiologie
J.W. Goethe-Universität
Frankfurt am Main, Germany

Borgis D.

Laboratoire de Physique Theorique des Liquides
Université Pierre et Marie Curie, T16, 5eme etage
75252 Paris Cedex, France

Brandt U.

Universitätsklinikum Frankfurt
ZBC, Therapeutische Biochemie
Theodor-Stern-Kai 7, Haus 25B
6000 Frankfurt am Main 70, Germany

Cannon R D.

School of Chemical Sciences
University of East Anglia
Norwich NR4 7TJ, England

Droß F.

Institut für Mikrobiologie
J.W. Goethe-Universität
Frankfurt am Main, Germany

Dutton P.L.

Department of Biochemistry and Biophysics
University of Pennsylvania
Philadelphia, PA 19104, USA

Engelbrecht S.

Lehrstuhl Biophysik
Fachbereich Biologie
Universität Osnabrück
Barbarastr. 11
D-4500 Osnabrück, Germany

Gehring S.

Institut für Physikalische Chemie
Technische Hochschule Darmstadt
Petersenstr. 20
D-6100 Darmstadt, Germany

Grätzel M.

Institut de Chimie Physique
Ecole Polytechnique Fédérale de Lausanne
1015 Lausanne, Switzerland

Gutman M.

Laser Laboratory for Fast Reactions in Biology
Department of Biochemistry
Tel Aviv University
Ramat Aviv 69978, Israel

Gutmann M.

Institut für Mikrobiologie
J.W. Goethe-Universität
Frankfurt am Main, Germany

Haase W.

Institut für Physikalische Chemie
Technische Hochschule Darmstadt
Petersenstr. 20
D-6100 Darmstadt, Germany

Jahns P.

Lehrstuhl Biophysik
Fachbereich Biologie
Universität Osnabrück
Barbarastr. 11
D-4500 Osnabrück, Germany

Jeżowska-Trzebiatowska B.

Institute of Chemistry
University of Wrocław
14, Joliot-Curie St.
50-383 Wrocław, Poland

Junge W.

Lehrstuhl Biophysik
Fachbereich Biologie
Universität Osnabrück
Barbarastr. 11
D-4500 Osnabrück, Germany

Kaim W.
Institut für Anorganische Chemie
Universität Stuttgart
Pfaffenwaldring 55
D-7000 Stuttgart 80, Germany

Keske J.M.
Department of Biochemistry and Biophysics
University of Pennsylvania
Philadelphia, PA 19104, USA

Klimmek O.
Institut für Mikrobiologie
J.W. Goethe-Universität
Theodor-Stern-Kai 7
D-6000 Frankfurt am Main, Germany

Körtner C.
Institut für Mikrobiologie
J.W. Goethe-Universität
Theodor-Stern-Kai 7
D-6000 Frankfurt am Main, Germany

Krafft T.
Institut für Mikrobiologie
J.W. Goethe-Universität
Theodor-Stern-Kai 7
D-6000 Frankfurt am Main, Germany

Kröger A.
Institut für Mikrobiologie
J.W. Goethe-Universität
Theodor-Stern-Kai 7
D-6000 Frankfurt am Main, Germany

Kryachko E.
Institute for Theoretical Physics
Kiev 252130, Ukraine

Kunkely H.
Institut für Anorganische Chemie
Universität Regensburg
Universitätstraße 31
D-8400 Regensburg, Germany

Lauterbach F.
Institut für Mikrobiologie
J.W. Goethe-Universität
Theodor-Stern-Kai 7
D-6000 Frankfurt am Main, Germany

Lill H.

Lehrstuhl Biophysik
Fachbereich Biologie
Universität Osnabrück
Barbarastr. 11
D-4500 Osnabrück, Germany

Limbach H.-H.

Institut für Physikalische Chemie
Universität Freiburg i.Br.
Albertstr. 21
D-7800 Freiburg, Germany
and
Freie Universität Berlin
Fachbereich Chemie
Takustr. 3
W-1000 Berlin 33, Germany

Link T.A.

Universitätsklinikum Frankfurt
ZBC, Therapeutische Biochemie
Theodor-Stern-Kai 7, Haus 25B
6000 Frankfurt am Main 70, Germany

Lowe D.J.

AFRC IPSR Nitrogen Fixation Laboratory
University of Sussex
Brighton BN1 9RQ, England

Maróti P.

Department of Biophysics
Joszef Attila University
Szeged, Hungary

Moser Ch.C.

Department of Biochemistry and Biophysics
University of Pennsylvania
Philadelphia, PA 19104, USA

Pistorius E.K.

Lehrstuhl Zellphysiologie
Fakultät für Biologie
Universität Bielefeld
4800 Bielefeld 1, Germany

Pope M.T.

Department of Chemistry
Georgetown University
Washington DC 20057, USA

Ratajczak H.

Institute of Chemistry
University of Wrocław
50-383 Wrocław, Poland
and
Institute of Low Temperature and Structure Research
of the Polish Academy of Sciences
50-950 Wrocław, Poland

Sandorfy C.

Département de Chimie
Université de Montréal
Montréal, Québec, H3C 3J7, Canada

Schönknecht G.

Lehrstuhl Biophysik
Fachbereich Biologie
Universität Osnabrück
Barbarastr. 11
D-4500 Osnabrück, Germany

Shimoni E.

Laser Laboratory for Fast Reactions in Biology
Department of Biochemistry
Tel Aviv University
Ramat Aviv 69978, Israel

So H.

Department of Chemistry
Georgetown University
Washington DC 20057, USA

Takahashi E.

Department of Plant Biology
University of Illinois
Urbana, IL 61801, USA

Tsfadia Y.

Laser Laboratory for Fast Reactions in Biology
Department of Biochemistry
Tel Aviv University
Ramat Aviv 69978, Israel

Vogler A.

Institut für Anorganische Chemie
Universität Regensburg
Universitätstraße 31
D-8400 Regensburg, Germany

von Jagow G.

Universitätsklinikum Frankfurt
ZBC, Therapeutische Biochemie
Theodor-Stern-Kai 7, Haus 25B
6000 Frankfurt am Main 70, Germany

Warncke K.

Department of Biochemistry and Biophysics
University of Pennsylvania
Philadelphia, PA 19104, USA

Wojciechowski W.

Technical University of Wrocław
Wrocław, Poland

Wraight C.A.

Department of Plant Biology and
Department of Physiology and Biophysics
University of Illinois
Urbana, IL 61801, USA

Zundel G.

Institute of Physical Chemistry
University of Munich
Theresienstr. 41
D-8000 München 2, Germany

Electron and Proton Transfer in Chemistry and Biology, edited by A. Müller et al.
Studies in Physical and Theoretical Chemistry, Vol. 78

Mechanisms of Electron Transfer

R.D. Cannon

School of Chemical Sciences
University of East Anglia
Norwich NR4 7TJ (England)

Summary

The basic physical principles underlying electron transfer between metal ions and other well–localised redox centres are summarised. Recent experimental progress is reviewed, with particular reference to electron transfer in the gas phase – between large molecules; in solution – results of volume of activation studies; and in the solid phase – transfer between paramagnetic centres. The potential value of neutron scattering spectroscopy, as a probe of low–energy electronic transitions in mixed–valence materials, and as a method of measuring rate processes in the range 10^{10} - 10^{12} s^{-1}, is discussed.

Introduction

The subject of electron transfer in chemical systems has always been well served by reviews, and at the present time there is no shortage of surveys of both theoretical principles and experimental data (ref. 1). Rather than attempt to add another — which in any case would be impossible in a single lecture — it seemed appropriate, for an interdisciplinary Symposium, to attempt something more modest. In the first part of this lecture, a few well understood principles will be summarised, with some indication of which ones are well supported by experiment and which are not (ref. 4). In the second part, some relatively new experimental approaches will be outlined: the selection of these is admittedly personal, and perhaps arbitrary.

1. Electron Transfer and Mixed Valence

1.1 The Marcus-Hush Model

We might almost speak of the Marcus–Hush *paradigm*. The ideas have been so widely discussed that it is easy to forget that in this subject, as in all others, there is still a need to ask fundamental questions. Equation 1 symbolises an electron transfer reaction in solution, starting with separate reactants A^+ and B, where the positive sign indicates the higher of two possible valence states, and (env) denotes the ligand and solvent environment at thermal equilibrium.

$$A^+(env) + B(env) \rightarrow A(env) + B^+(env) \quad (1)$$

A general mechanism is shown in equation 2, the symbols i, p, s, f, denoting initial, precursor, successor and final states :

$$\underset{i}{A^+ + B} \underset{k_{pi}}{\overset{k_{ip}}{\rightleftarrows}} \underset{p}{A^+.B} \overset{k_{et}}{\rightarrow} \underset{s}{A.B^+} \rightarrow \underset{f}{A + B^+} \qquad (2)$$

In the 1960s considerable ingenuity was exercised in demonstrating that each of the steps in scheme 2 could be realised physically in particular systems (ref. 4). Successor complexes were detected and isolated, precursor complexes were shown to be labile intermediates (refs. 6, 7) and variants of kinetic behaviour were demonstrated, depending on the relative magnitudes of the individual rate constants k_{ip}, k_{pi}, k_{et}, *etc.* (refs. 8 and 9). Finally precursor complexes with rigid frameworks were prepared, so that intramolecular rate constants k_{et} could be measured free from other kinetic complications. Of special interest are symmetrical mixed-valence complexes of the type $[A^+\cdots X\cdots A]$. After the first of these was prepared by Creutz and Taube (ref. 10), the criteria for trapped or non–trapped valences were worked out (refs. 11, 12) and determinations of rates of internal, symmetrical electron transfer began, using various spectroscopic line–broadening measurements (ref. 1).

In solution however, the commonest case is simple second–order kinetics, with either the first or the second step rate–limiting:

$$\text{Rate} = k_2\,[A^+][B] \qquad (3)$$

$$k_2 = k_{ip}\,k_{et}/(k_{pi} + k_{ip}) \qquad (4)$$

Early theoretical treatments, following Levich and Dogonadze (ref. 13) tended to express the probability of electron transfer over a range of distances, rather than over the fixed or narrow range of distances implied by the stepwise mechanism. Thus one may write

$$k_2 = \int_{R_c}^{\infty} p(R_{AB}) \exp\{-w(R_{AB})\}\, 4\pi\, R^2_{AB}\, d\, R_{AB} \qquad (5)$$

where $p(R_{AB})$ is the probability per unit time of electron transfer over the distances in the range R_{AB} to $(R_{AB} + dR_{AB})$, $w(R_{AB})$ is the work required to bring the reactants up to this distance, and R_c is the distance of closest contact. In detailed analyses of distance–dependence, and of medium effects, this is still a useful concept (ref. 14). But mostly, the chemically intuitive form of equation 4 is preferred, and expressions have been given for K_{ip} $(= k_{ip}/k_{pi})$ for the case of an outer–sphere reaction between spherically shaped, charged reactants, allowing for ionic medium effects on the basis of Debye–Hückel theory, (refs. 15 and 16).

In comprehensive reviews of the state of the semiclassical Marcus theory, in 1982-3, Sutin (ref. 15) introduced the compact notation

$$k_{et} = \nu_n \, \kappa_n \, \kappa_{el} \tag{6}$$

where ν_n is called the nuclear frequency factor, κ_n is a nuclear rearrangement factor, and κ_{el} is the electronic transmission coefficient; and he gave expressions for K_{ip}. The factor κ_n expresses the combined effects of reorganisation energy λ and thermodynamic driving force:

$$\kappa_n = \exp(-\Delta G^{*}/RT) = (\lambda/4)(1 + \Delta G_{et}^{o}/\lambda)^2 \tag{7}$$

This is the classical Marcus–Hush expression (ref. 17), illustrated by — and derivable from (ref. 18) — a diagram of the type of Figure 1, where x is a suitable reaction coordinate.

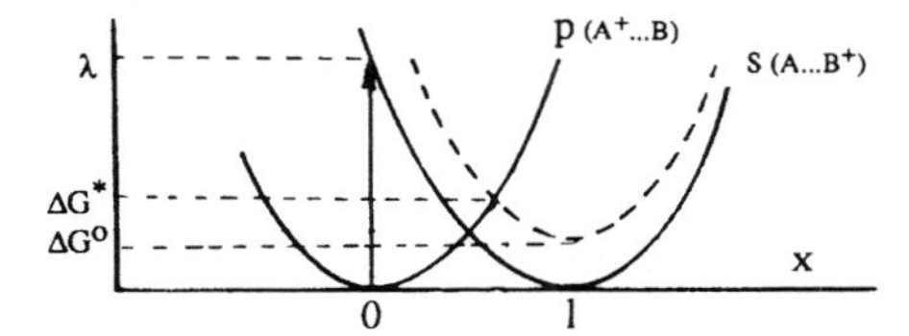

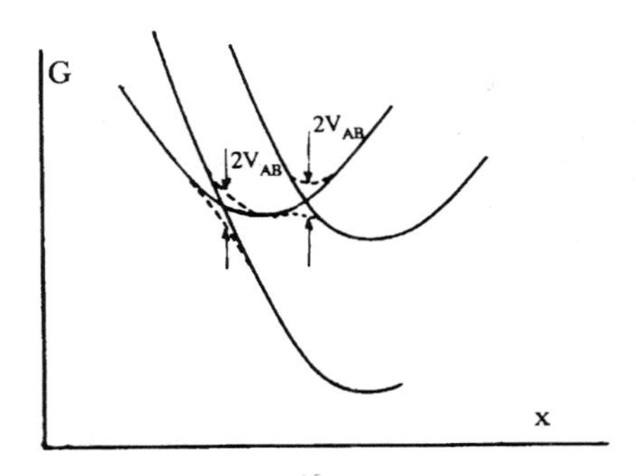

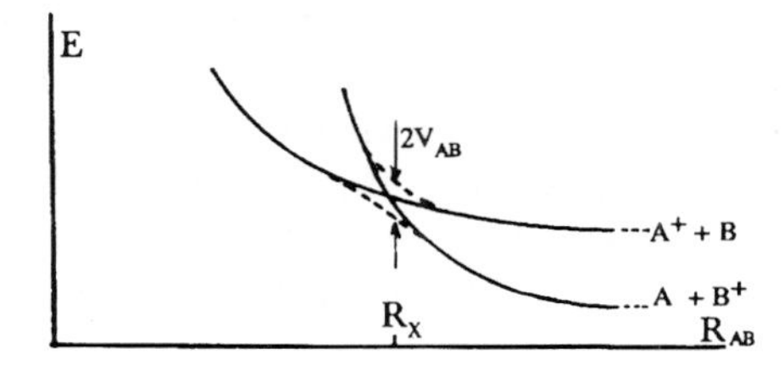

Fig. 1. Reaction profiles for electron transfer reactions. *Top to bottom:* (1) General reaction $A^+ + B \rightarrow A + B^+$, where A and B are complex ions in solution. See text, equations 1, 2 and 7. (2) Adiabatic and non–adiabatic transfer in the normal and inverted free–energy regimes. See text and equation 8. (3) Potential energy curves for the reaction $A^+ + B \rightarrow A + B^+$ in the gas phase. The energy E is the internal energy of the pseudomolecule $(A \cdots B)^+$ as a function of internuclear distance R, for each of the two electronic configurations $A^+ \cdots B$ and $A \cdots B^+$.

The essence of the model is that free energy (or internal energy if an isolated molecule is considered) is a harmonic function of x, with the same characteristic λ for precursor and successor states. Two historic treatments lead to expressions for the separate contributions to λ, due to distortion of inner–sphere bonds (ref. 19), and reorganisation of outer-sphere solvent molecules (ref. 20).

Among the tests of equation 7, the correlation of ΔG^* with ΔG^o for related reactions, and the Marcus cross-relation, are well known (ref. 5). When ΔG^o is significantly far from zero, linear free energy plots have slopes $\alpha = \partial\Delta G^*/\partial\Delta G^o =$ $= d \log k / d \log K = (1/2)(1 + \Delta G^o/ \lambda) < 0.5$, and this too has been verified in many cases (ref. 5). Non-linear dependences of rate upon ΔG^o tend to a maximum of $k_{et} =$ $= \nu_n$ or $k_2 = k_{ip}$ as the case may be, and some famous examples of this behaviour are shown in Figure 2.

The electronic transmission coefficient κ_{el} originates with the Eyring transition state theory and is the probability that the reacting system, on passing through the crossing point (Figure 1) will change from precursor to successor electronic configuration. Arguments based on the Fermi Golden Rule lead to expressions for κ_{el} in terms of V_{AB}, the electron tunnelling integral (ref. 15), equal to the resonance energy which arises when the electronic energies at the crossing point are calculated by a static first-order perturbation treatment.

Reactions with $\kappa_{el} << 1$ are termed *non-adiabatic* and in this case the rate constant k_{et} (at the limit of high temperature when all atomic motions can be modelled classically) is given by (ref. 15)

$$k_{et} = (4\pi^2/h)\,(V_{AB})^2\,(4\pi\lambda\, RT)^{-1/2}\,\kappa_n \qquad (8)$$

There is a good deal of interest in detecting this situation experimentally (ref. 25). One approach is to compare large numbers of rates against the cross-relation, and to search for anomalies due to non-cancellation of values of κ_{el} (ref. 26). Another is to compare measured and calculated entropies of activation, including a contribution R ln κ_{el} for the probability of electron transfer at the crossing point (refs. 16 and 27). Yet another is to calculate values of V_{AB} *ab initio*, and this has been done for some important reactions, such as $[Fe(H_2O)_6]^{3+/2+}$ and $[MnO_4]^{-/2-}$. Slater functions have been used for donor and acceptor orbitals and various detailed structures for the precursor complexes have been considered (refs. 28 and 29). By these and other arguments (ref.15) it has been concluded that reactions between complex ions with polypyridyl ligands, such as $[M(bpy)_3]^{3+/2+}$, are generally more adiabatic than those of aquo- or ammine-cations (for which theoretical estimates of κ_{el} are mostly in the range 10^{-1} to 10^{-3}) and that self-exchange reactions can be more adiabatic than cross reactions involving couples with dissimilar types of ligands (ref. 16). But there is still a good deal of uncertainty in this area, since assessment of the results involves calculating all the factors that contribute to rates through the equations cited above. This is clearly a difficult exercise, and the case of the important reaction $[Fe(OH_2)_6]^{3+/2+}$, for example, is still by no means settled.

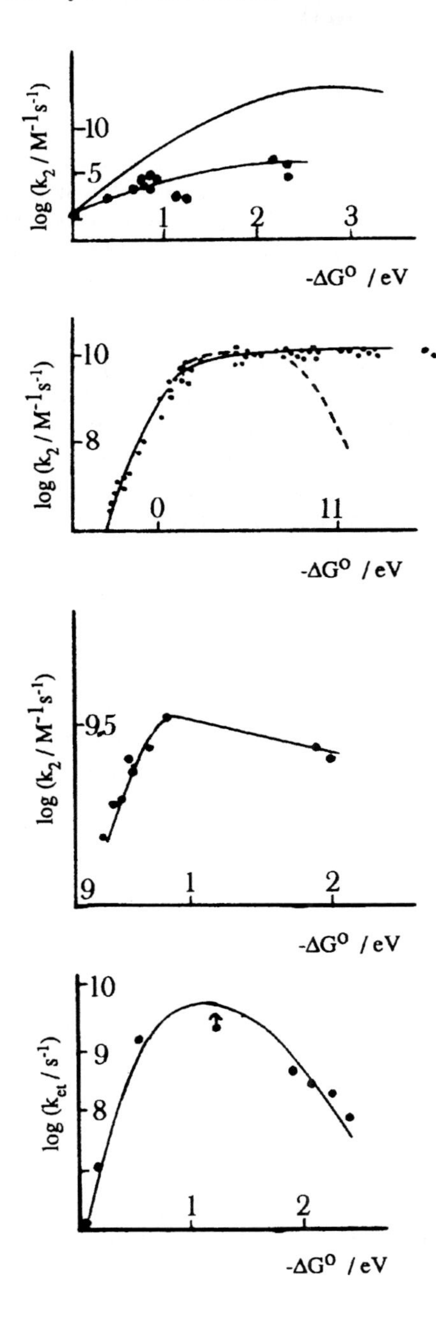

Fig. 2. Exploring the inverted region. Examples of series of electron transfer reactions which deviate from the Marcus model.*Top to bottom:* (1) Reactions $[M(OH_2)_6]^{3+}$ + $[M'L_6]^{2+}$ (M = Co, Mn, Cr *etc.*) in acidic aqueous solution. The upper curve is calculated from the Marcus equations omitting work terms (ref. 21). The observed limiting value $k_2 \cong 10^5\ M^{-1}s^{-1}$ was later shown to to be consistent with diffusion-controlled encounter of the highly charged cationic reactants. (2) Fluorescence quenching of electronically excited hydrocarbons such as anthracene, by electron donors such as amines. The dotted curve shows the predictions of equations 4 and 7 (ref. 22). The fact that the data do not follow the curve can be explained by supposing that when $\Delta G^o << 0$ for the conversion of reactants into ground-state products, other electronic pathways come into play. (3) reactions of the type $^*[Ru(bpy)_3]^{2+}$ + $[M(bpy)_3]^{3+}$ (* = excited state). The slight fall in rate at the most negative values of ΔG^o is at tributed to the Marcus inversion effect, offset by increased nuclear tunnelling (ref. 23). (4) Intramolecular electron transfer reactions $A\text{-}X\text{-}B^+ \rightarrow A^+\text{-}X\text{-}B$, A = biphenyl group, B = quinonoid group, X = steroid framework, (ref. 24).

1.2 Non-Marcus Behaviour (1) The Inverted Region

According to equation 7, when $\Delta G^o \leq - \lambda$, the slope of a plot of log k_2 against $-\Delta G^o$ becomes negative. Most experiments with separated reagants in fluid media have failed to disclose this, though as long ago as 1977, Creutz and Sutin reported some examples, shown in Figure 2 (ref. 23). Dutton and co–workers have made extensive studies of bacterial photosystems in which the electron receptor sites are systematically replaced with other molecules of different reduction potentials (ref. 30) and work by Miller and co–workers using rigid glass media has clearly shown rates $k_{et} < \nu_n$ at highly negative ΔG^o (ref. 24). More recently Closs and Miller (ref. 24) measured k_{et} in rigid molecules of the type A-X-B$^-$ (B$^-$ = reduced biphenyl residue, X = steroid framwork, A = various quinone residues) and obtained a bell–shaped plot of k_{et} *versus* $-\Delta G^o$, with a maximum ca 10^{10} s^{-1} (Figure 2).

Since this rate is well below the theoretical upper limiting value of $\nu_n = (RT/Lh) = 10^{12.8}$ s^{-1} at T = 298K, the inference is that $\kappa_{el} << 1$, and the reactions are non–adiabatic. A value for the tunnelling integral was estimated, $V_{AB} = 6.2$ cm^{-1}. Creutz and Sutin (ref. 23) have pointed out that although in the normal region, adiabaticity (large V_{AB}) favours electron transfer, in the inverted region the reverse is true (see Figure 1). For an inverted crossing however, nuclear tunnelling can be more facile, and if such tunnelling becomes effective at energies significantly different from the crossing energy, the simple factorisation of rate into activation and probability factors (equation 6) breaks down. Theoretical treatments involve consideration of the probabilities of individual transitions between vibrational levels in the precursor and successor states which are given by the Franck–Condon factors, *i.e.* the overlap integrals between the nuclear probability density functions (ref. 31). Treatments vary in the extent to which other relevant factors are modelled classically or quantum–mechanically. It is possible that the more general model of Piepho, Krausz and Schatz will eventually predominate (ref. 32); though at present the PKS predictions are available only in the form of a computer algorithm.

For highly negative ΔG^o, the descending portion of the log k_{et} ($-\Delta G^o$) plot approximates to a monotonic curve, less steep than the Marcus parabola, and the present kinetic data bear this out (Figure 2). The same theory is used for radiationless transitions and fluorescence intensities and leads to the ‘energy gap law’ according to which the probability of an internal electronic rearrangement in the case of weak coupling, such as the $T_1 \rightarrow S_0$ transition of an aromatic molecule, *decreases* as the transition energy becomes more favourable. This of course is quite well supported by experiment (ref. 33).

1.3 Non–Marcus Behaviour (2) The Low–Temperature Limit.

At sufficiently low temperatures, all these models predict a transition to temperature–independent rates. The classic demonstration of this effect is electron transfer in bacterial photosystems, first reported by DeVault and Chance in 1966. Typically these show Arrhenius behaviour down to T ≈ 150 K, and almost no further change in rate from 100 to 4.3K (ref. 34). Efforts to explain the transition from activated to activationless behaviour are still apparently controversial: good fits can be obtained from the above models assuming that essentially the same electron transfer reaction is rate-limiting at all temperatures, but other explanations have also been considered. These include phase changes, changes in inter-reactant distance, changes in protein conformation, and changes in dynamical behaviour of the protein (ref. 35). With the rapidly increasing knowledge of structures of these photosystems, clarification may be expected, and indeed this topic will be treated by others in this Symposium. Within the low temperature regime however, it is difficult to doubt that long-range, non-adiabatic electron transfer prevails.

It should be mentioned here (sadly!) that another report of activationless transfer, in a solid inorganic medium, can no longer be accepted. This was based on Mössbauer emission spectra of a cobalt(II) ferricyanide, doped with radiactive ^{57}Co. The primary process $^{57}Co \rightarrow {}^{57}Fe$ is expected to produce iron(II) which could then undergo electron transfer, $Fe^{II}NCCo^{III} \rightarrow Fe^{III}NCCo^{II}$. Rates of transfer were calculated and held to be temperature-independent below T ≈ 50K (ref. 36). Recent work by Gütlich and coworkers with improved instrumentation has failed to demonstrate any such reaction within the Mössbauer time-range (ref. 37). This is not the first time that an excellent experimental idea has had to be discarded but has nevertheless had a beneficial effect on the development of the subject. Here again, optical spectroscopy provides evidence of an analogous effect with fewer problems of interpretation. Intervalence charge transfer (see the vertical transition in Figure 1) has been discussed by Hush (refs. 38 and 39), and in several cases the temperature-variation of band intensity and width fits the prediction of a continuous transition from high to low-temperature regimes. A good example is the complex $[(NC)_5Fe(pyrazine)Fe(CN)_5]^{5-}$ which has been shown in this way to be a weakly coupled iron(III,II) mixed-valence species, (ref. 41).

1.4 Long-Range Electron Transfer

In the above models, internuclear distance R_{AB} affects rate in two main ways — *via* the reorganisation energy λ and *via* the tunnelling integral V_{AB}. Effects on λ are usually given in terms of the formula

$$\lambda \propto (1/2a_1 + 1/2a_2 - 1/R_{AB}) \qquad (9)$$

which is valid for the case of spherical reactant molecules of radii a_1 and a_2 at a relatively long distance, $R_{AB} >> (a_1 + a_2)$. For short distances, a model which treats the precursor complex as an ellipsoid has been used, at first with the simplification that the reaction centres are located at the foci of the ellipsoid (ref. 42) but now without this restriction (ref. 43). The distance dependence of λ has been tested for both optical and thermal electron transfer using bridged ruthenium(III,II) systems in the normal free-energy range (ref. 44).

The distance dependence of V_{AB}, from the *ab initio* calculations mentioned above, is exponential (refs. 15 and 45), and experimental evidence is now building up to confirm this. Experiments using reaction centres separated by glassy matrices, by proteins, and by monolayers, have been reviewed (ref. 24). The experiments of Closs and Miller, using rigid ligands, have already been mentioned. Data from intramolecular electron transfer in long–chain organic molecules, and in biological photosystems, will be reviewed by other speakers in this meeting.

1.5 Gas–Phase Analogues

This brief review of factors controlling electron transfer rates may conclude with a reminder that some of these factors can be isolated more clearly in entirely different types of experiment (ref. 5).

The fundamental resonance transfer between identical donor and acceptor sites with no vibronic coupling, has been extensively studied. Equation 10 gives the simplest possible case:

$$H^+ + H \rightarrow H + H^+ \qquad (10)$$

When the nuclei are at rest there is no reaction in any chemical sense — the system is simply a molecule H_2^+. But when the nuclei are in relative motion, real electron transfer can take place with a frequency governed by the time-dependent Schrödinger wave equation (Figure 3).

In the type of experiment in which a beam of protons is passed through a gas of randomly moving H atoms, the oscillatory character of the wave function shows up directly in a periodic dependence of electron–transfer cross section upon beam velocity.

The adiabatic/non–adiabatic distinction is well shown by reactions between unlike atoms, such as

$$Ar^+ + Na \rightarrow Ar + Na^+ \qquad (11)$$

In such a system, the resonance condition is fulfilled only at a certain internuclear distance R^*, corresponding to the crossing point in Figure 1. This leads to an optimum beam velocity for electron transfer. The effect has been verified many times. A simple rationalisation is that for electron transfer to take place, the time spent by the reactants at distances close to R^* should be of the order of the time for one half–oscillation of the wave function. More sophisticated models have been extensively discussed, and values of R^* and V_{AB} have been obtained.

The effect of inner–sphere reorganisation is neatly demonstrated by comparisons between electron transfer rate processes and vertical electron loss transitions — *i.e.* photoelectron emission. In the series of reactions

$$CH_4 + A^+ \rightarrow CH_4^+ + A \qquad (12)$$

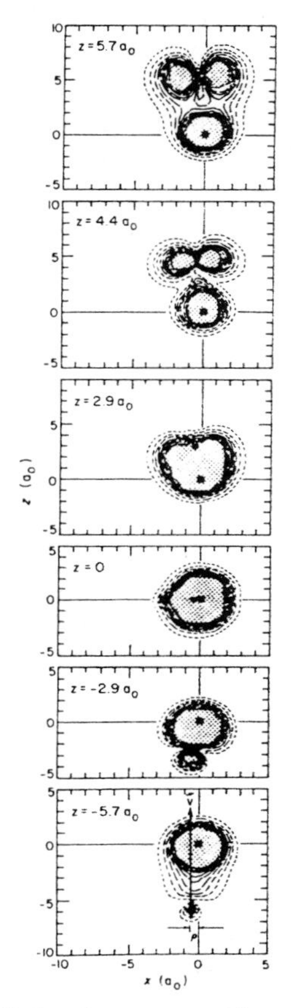

Fig. 3. An electron transfer reaction in progress. Calculated electron probability densities at different times during the reaction $He^{2+} + H \rightarrow He^{+} + H^{+}$. This calculation is for nuclei passing very closely, at kinetic energy 1keV. In frame 1 (*bottom*) the He^{2+} is moving up towards the H atom, whose electron cloud is already somewhat polarised towards the helium nucleus. In frame 3, the electron cloud is nearly spherical. As the helium nucleus moves out, in frame 6 (*top*) it can be seen to be capturing the electron in a $2p\pi$ orbital (ref. 46).

where A is a rare gas atom, rates vary in the order He > Ne < Ar ≈ Kr > Xe. They correlate with the intensities of the photoelectron spectrum of CH_4, at ionisation potentials equal to the net energy changes of the reactions. The inference is that ion–molecule reactions of this type are controlled by the Franck–Condon principle so that for example when $\Delta E = -16$ eV (A = = Ar), an electronic excited state of CH_4^+ is available with comparatively little structural change from CH_4; but when $\Delta E = -20$ eV (A = = Ne), this is not the case.

2. Recent Experimental Progress

2.1 Gas Kinetics: Large Molecules

When this field was reviewed in the late 1970s, reactions of the type just mentioned were about the most complex systems to have been studied (ref. 5). Atoms and high–energy free radicals predominated and there were no data on reactants of the type that can exist in solution. Since then, Richardson and co–workers have measured kinetics of reactions such as the following, using Fourier-transform ion cyclotron mass spectrometry:

$$Fe(C_5H_5)_2^+ + Fe(C_5H_5)_2 \rightarrow Fe(C_5H_5)_2 + Fe(C_5H_5)_2^+ \quad (13)$$

$$Ru(hfac)_3 + Ru(hfac)_3^- \rightarrow Ru(hfac)_3^- + Ru(hfac)_3 \quad (14)$$

Rate constants ($2 \cdot 10^{11}$ and $2 \cdot 10^{10}$ $M^{-1}s^{-1}$ at 375K) are some 10^5 times higher than in solution (in MeCN at 298K) and this is a direct measure, for the first time, of the effects of solvent on the rates of encounter and energies of outer–sphere reorganisation (ref. 47). At the low kinetic energies being used — unlike the molecular beam experiments referred to above — Marcus–type models can be applied. Energy surfaces have been plotted for stepwise reaction mechanisms involving precursor and successor complexes, analogous to equation 2, and the first indication of a correlation between ΔG^* and ΔG^0 has been found, in a series of reactions of SF_6^- with with different electron acceptors (ref. 48). Effects of inner–sphere reorganisation can be seen in a series of rates of self–exchange reactions. The reaction $CO^+ + CO \rightarrow CO + CO^+$ is close to the limiting collision rate. Rates of other reactions have been rationalised in terms of bond–length differences between oxidised and reduced states, and in the case of the manganocenes, a change of spin state (refs. 49, 50).

2.2 Solution Kinetics: Volume of Activation

This is a topic with a long early history and a rapid recent growth. The volume of activation of a reaction is defined by

$$\Delta V^{\neq} = - RT\, (\partial \ln k/\partial P)_T \qquad (15)$$

Higher–order dependences of ln k upon pressure are also important but will not be mentioned here. Pressure effects on inorganic reactions in solution began to be studied in the 1950s, and in 1965, Candlin and Halpern pointed out that inner–sphere electron transfer reactions should show characteristically more positive values of $\Delta V^{\neq}$ than outer-sphere reactions, owing to the release of a solvent molecule in the transition state (ref. 51). They verified this with a number of well–established examples, including Taube's (ref. 52) original reaction (17):

$$[Co(NH_3)_5Cl]^{2+} + [Cr(H_2O)_6]^{2+} \rightarrow [Co(NH_3)_5ClCr(H_2O)_5]^{4+} + H_2O \qquad (16)$$

Stranks calculated negative values of $\Delta V^{\neq}$ for several outer–sphere reactions and this too has been generally verified (ref. 53). Actually there was an error in one part of Stranks' calculation, as a result of which it seemed for a time that agreement with experiment was better than was in fact the case. Correction of the error (ref. 54) led to renewed interest (ref. 55) and to increasingly detailed calculations of different contributions to $\Delta V^{\neq}$ due to such factors as interionic repulsion, Debye–Hückel salt effects, and solvent reorganisation (contributing respectively to w_{AB}, K_{ip} and κ_n in equations 4 and 6 above). The hope now is that information can be gained on the non–adiabaticity and inter–reactant distances. The argument is that on compressing the solution, R_c is decreased, and if the decrease is proportional to contractions elsewhere, it follows that R_c is proportion al to $\rho^{-1/3}$ where ρ is the density of the medium, and thence *via* equation 15 the non–adiabaticity contribution to $\Delta V^{\neq}$ is (ref. 58b)

$$\Delta V^*_{NA} = -2\, RT\, \alpha\, R_c\, \kappa /3 \qquad (17)$$

where α is the coefficient in the distance dependence of V_{AB}, *i.e.* $V_{AB} = V^0_{AB} \exp(-\alpha R_{AB})$, and κ is the compressibility of the solvent. As an example, a recent study by Jolley, Stranks and Swaddle suggests that the uncatalysed $[Fe(H_2O)_6]^{3+/2+}$ self–exchange is adiabatic (ref. 56). As mentioned above, this reaction, one of the most–studied of all simple electron transfers, is still controversial. It has also been considered adiabatic on the basis of comparisons with other reactions *via* the Marcus cross–relation, whereas on the other hand calculations of the tunnelling integral V_{AB} favour non–adiabatic transfer. All conclusions about this reaction are highly model-dependent, and there is not even universal agreement as to an inner– or an outer–sphere mechanism (ref. 57). Other recent $\Delta V^{\neq}$ studies include the reactions $Ru(hfac)_3^{-/0}$ and $MnO_4^{-/2-}$ — concluded to be adiabatic and non–adiabatic respectively (ref. 58); and $[Mn(NCR)_6]^{3+/2+}$ in a range of solvents (ref. 59). This last has proved to be much more complicated owing to ion pairing effects. Wherland and co–workers concluded that the variation of $\Delta V^{\neq}$ 'does not correlate directly with solvent parameters such as molal volume, viscosity, dielectric constant (D), index of refraction (n), the combined solvent parameter ... $(1/n^2 - 1/D)$, dipole moment, donor or acceptor number, HOMO or LUMO energies, ... or a variety of other parameters...(ref. 59b)'.

2.3 Magnetic Interactions

Prior to 1985 all theories of electron transfer rates implicitly assumed that the transferring electron is the sole unpaired electron in the reacting system. But many systems of interest have high magnetic moments. What is the effect of the resident spins of the donor and acceptor sites ? Qualitatively the answer was given by Anderson and Hasegawa's treatment of what they termed 'double exchange' (ref. 60). Consider a complex $M_A^+...M_B^+$ in which M_A^+ and M_B^+ are identical ions with spin S represented by vectors $\mathbf{S_A}$, $\mathbf{S_B}$ respectively, and consider the effect of adding one more electron, of spin **s**. When the added electron is close to one of the two atoms, its spin will couple strongly to the local spin, giving $\mathbf{S_A} \pm \mathbf{s}$ or $\mathbf{S_B} \pm \mathbf{s}$ as the case may b e. In order for the electron to be delocalised, *i.e.* to move freely from one atom to the other, $\mathbf{S_A}$ and $\mathbf{S_B}$ should be parallel, so that in effect they are ferromagnetically coupled. Conversely if for some reason the coupling is antiferromagnetic, the travelling electron will be localised. Antiferromagnetic coupling is promoted by, among other things, covalent bonding *via* a bridging ligand (superexchange). Thus in general the magnetic character of a mixed–valence complex or electron transfer intermediate will be the outcome of two sets of opposing effects, those which promote resonance delocalisation, and those which promote superexchange. Antiferromagnetic complexes will tend to be of the localised, Robin–Day (ref. 11) Class II type, ferromagnetic the delocalised Class III. Theories which quantify this generalisation have been developed for model systems of two (ref. 61) and three (ref. 62) magnetic ions. Most mixed–valence complexes with net spin greater than 1/2 do indeed fit clearly into one or other of the two types. Early examples of the one type include dimanganese(IV,III) in $[L_2Mn(O)_2MnL_2]^{3+}$ (ref. 63), and triiron(III,III,II), in $[Fe_3O(OOCCH_3)_6L_3]$ (ref. 64), both ligand–bridged, antiferromagnetic and valence–trapped. Early examples of the other type include dinickel(II,I) in $[Ni_2(quinuclidine)_4Br_2]^+$ (ref. 65) and diruthenium(III,II) in $[Ru_2(OOCR)_4]^+$ (ref. 66), both metal–metal bonded, ferromagnetic, and delocalised. More subtle are the Fe_3S_4 centres in ferredoxins which consist of a delocalised, ferromagnetic iron(III,II) pair, antiferromagnetically coupled to the third iron(III) centre (ref. 67).

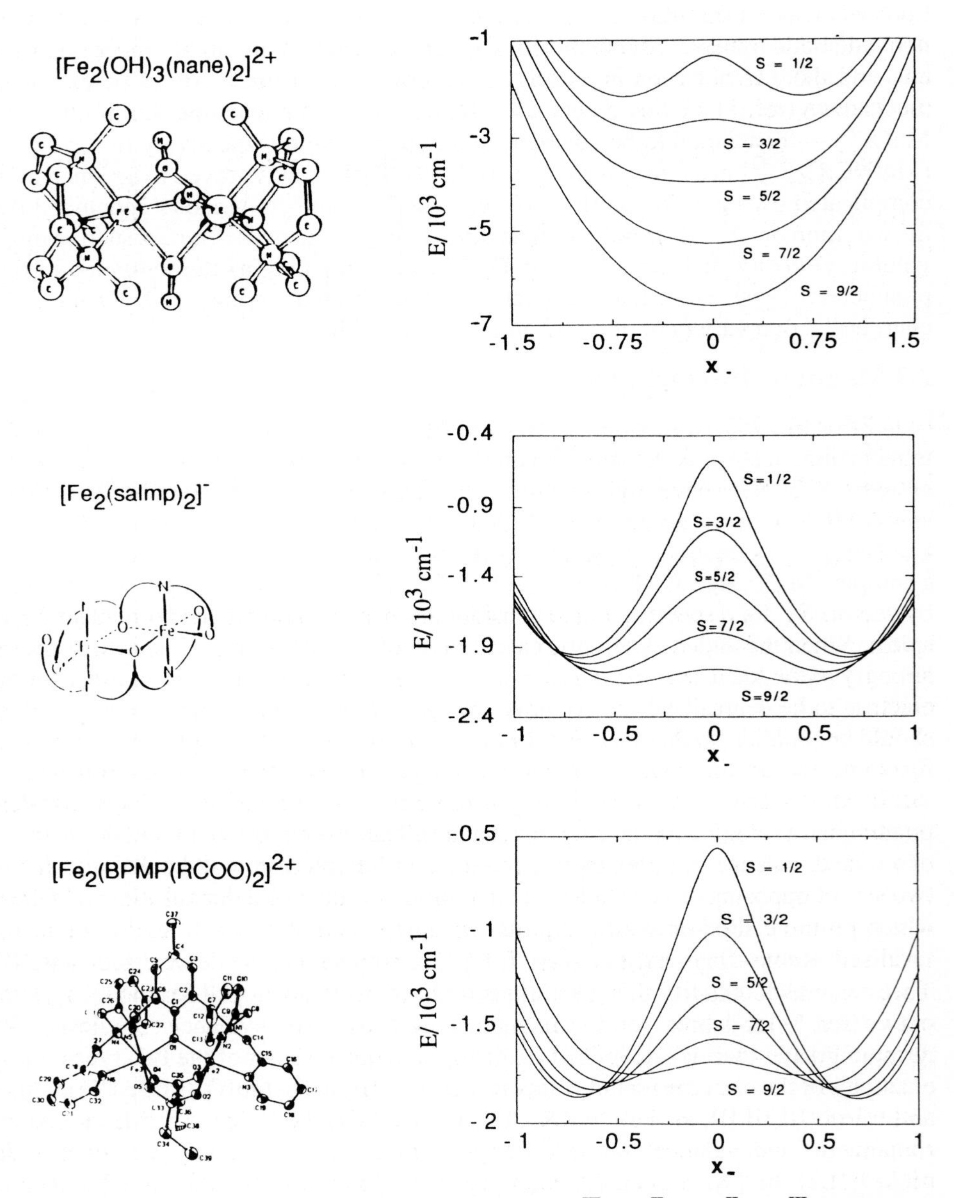

Fig. 4. Calculated potential energy curves for reactions $Fe^{III}... Fe^{II} \rightarrow Fe^{II}... Fe^{III}$ in binuclear mixed–valence iron(III,II) complexes, for different spin states, based on measured magnetic properties. Data, refs. 70 a, b; 71 and 72, calculations, ref. 69.

As was emphasised at the start, the localised and delocalised limits of mixed valency correspond to precursor and successor complexes of reaction scheme 2. Thus to account for the magnetic character of mixed–valence complexes is necessary to add in the effect of vibronic coupling (ref. 68). Using the three parameters λ, V_{AB}, and J, Blondin and Girerd have calculated the ordering of magnetic levels as a function of the reaction coordinate (ref. 69). Three examples are compared in Figure 4 .

Example 1 is strongly delocalised and ferromagnetic, S = 9/2, in the ground state but localised in an excited, ferromagnetic, S = 1/2 state. Example 2 is again ferromagnetic in the ground state, but is much more weakly localised owing to a smaller ratio of V_{AB}/ λ. Example 3 is antiferromagnetic and strongly localised in the ground state, but much less localised in the electronically excited higher–spin states. Thus the activation energy for thermally activated electron transfer will depend on whether a mechanism exists for mixing the spin states.

There are no kinetic data for electron transfer in any of these systems as yet, apart from the fact that the spectroscopic techniques which have been applied confirm delocalisation or localisation with respect to the various time scales. It will highly interesting to see actual rate measurements from appropriate lifetime–broadening measurements.

2.4 Neutron Scattering. (1) Inelastic Scattering

Until recently the main source of information on the energies and ordering of spin levels in magnetic systems was the variation of magnetic susceptibility with temperature. With improvements in instrumentation, and especially the advent of the SQUID magnetometer, this is still a major tool. Magnetic ground states are also characterised by other methods such as EPR, NMR and magnetic field-dependent Mössbauer spectroscopy. But direct transitions between spin levels in the absence of an applied magnetic field have always been difficult to observe, being spin–forbidden in optical spectra. Inelastic incoherent neutron scattering spectroscopy (IINS) avoids this limitation. Because of the spin of the neutron itself, the selection rule is $\Delta S = 0$ or ± 1. Applications of IINS to coordination compounds were pioneered by Güdel and Furrer who determined the singlet–triplet transition energy in the copper(II) acetate dimer, in good agreement with previous data (ref. 73). For polynuclear systems characterised by more than one J value, the neutron technique is invaluable. Recently the complete spectrum of levels of the trichromium(III) cluster in $[Cr_3O(OOCCD_3)_6(OD_2)_3]Cl{\cdot}5D_2O$ was determined from $S_{tot} = 1/2$ to 9/2 (a limitation of this method is that deuterated samples are required, since the ^1H nucleus has a high scattering cross-section) (ref. 74). At higher resolution, spectra of the same chromium compound showed splittings due to a slight lowering of symmetry of the metal atom triangle, and the splitting of the ground state was measured accurately for the first time (ref. 75). The first mixed–valence systems to have been studied are of the type (ref. 76) $[M_3O(OOCR)_6(py)_3]$, (M_3 = iron(III,III,II) and manganese(III,III,II) (refs. 64, 77 and

78). In the limit of low temperature, and at a moderate resolution (ca 7 cm^{-1}) the spectra of the iron complexes are well fitted by a Heisenberg spin–only model (refs. 79, 80). The coupling is strongly antiferromagnetic with J values similar to J($Fe^{III}Fe^{III}$) and J($Fe^{III}M^{II}$) in mixed–metal trimers with no mixed–valence interaction (refs. 79, 81), and this confirms that at low temperature, valence localisation is complete. However the mixed–valence systems differ from the mixed–metal in two ways. On raising the temperature the lines appear to broaden, while at low temperature and higher resolution the lowest transition splits into a complicated subset (Figure 5).

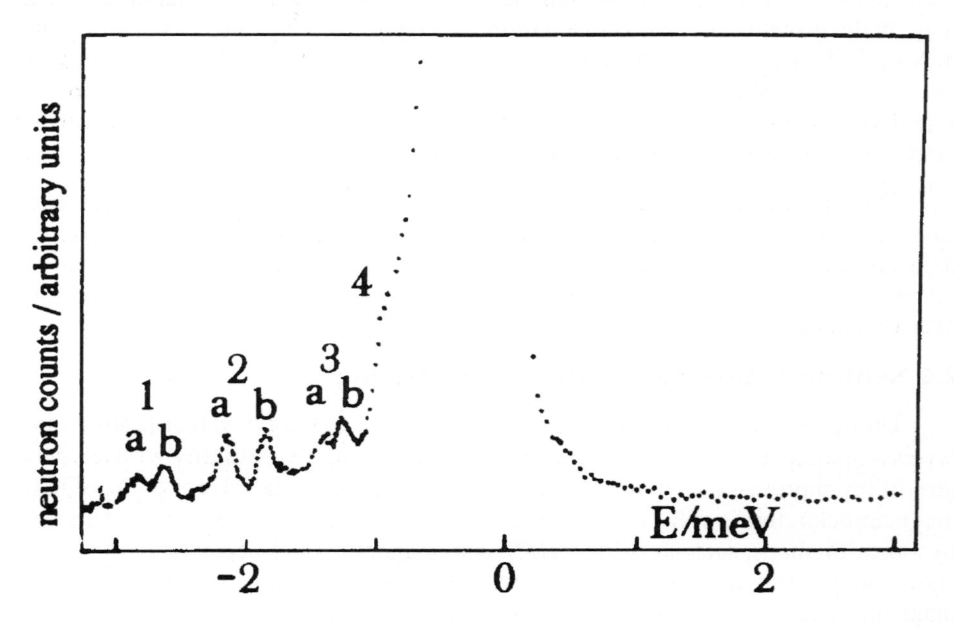

Fig. 5. Inelastic neutron scattering spectrum of $[Fe_3O(OOCCD_3)_6(C_5H_5N)_3](C_5H_5N)$ showing low–energy absorption bands. These are considered to be the fine structure of the electronic transition $|S, S_{ab}\rangle = |1, 1\rangle \rightarrow |S, S_{ab}\rangle = |0, 2\rangle$, the splittings being due to interactions betweeen ordered spins of the localised $Fe^{III}{}_2Fe^{II}$ molecules. T = 1.5K, incident neutron beam wavelength 6.0 Å. Resolution 63.26 eV. Instrument IN5, ILL, Grenoble.

This manifold is assigned to the transition $|S, S_{ab}\rangle = |1,1\rangle \rightarrow |S, S_{ab}\rangle = |0, 2\rangle$ where the two quantum numbers refer to the total spin of the trinuclear cluster, and the intermediate spin of the Fe^{III}-Fe^{III} pair, respectively (ref. 80).

We propose that the effects are due in part to magnetic ordering between clusters and that this is a consequence of the mobility of the electron in the mixed–valence system. At room temperature, both the mixed–metal $Fe^{III}{}_2Co^{II}$ and the mixed–valence $Fe^{III}{}_2Fe^{II}$ compounds crystallise in the space group R32; that is, X–Ray diffraction appears to show all three metal atoms in a cluster equivalent (ref. 82). In both cases this is clearly due to disorder. In the mixed–valence case, the valences are localised on the IR time scale, at room temperature, but delocalised on the Mössbauer time scale, and detailed Mössbauer and specific heat measurements by Hendrickson and co–workers imply that the molecules undergo rapid pseudorotation by electron transfer. Below a series of phase transitions — which have been carefully documented — electron transfer ceases (ref. 83). We suggest that electron localisation takes place in an ordered fashion, hence the detailed fine structure of the magnetic spectrum. The nature of the ordering — whether the iron(III,III,II) triangles point the same way within a domain, or whether they are related in some more complicated fashion, is still to be determined. In the mixed–metal compound however, disorder must be static, and must persist down to low temperature. It is significant that in a solid solution of the the $Fe^{III}{}_2Fe^{II}$ compound in the $Fe^{III}{}_2Co^{II}$, the fine structure of the spectrum does not appear.

These are the first IINS results on mixed–valence systems. There is clearly much more to be done. One intriguing feature of the present data is the strong intermolecular magnetic coupling. In the crystal structure the terminal pyridine ligands of neighbouring molecules are interleaved. Close contact of aromatic π–electron systems may provide an exchange pathway — and this recalls the facile electron transfer between complex ions in solution with polypyridyl ligands, mentioned above.

2.5 Neutron Scattering. (2) Quasielastic

Quasielastic incoherent neutron scattering (QINS) is due to the slight change in energy of neutrons which are scattered by moving nuclei within a sample. It is an established tool for studying processes such as hydrogen atom diffusion in metals, conformational motions in polymers, and molecular motion in liquids (ref. 84). When electron transfer is strongly coupled to vibrations, in principle QINS is a method of measuring electron transfer rates. Moreover the time–scale of QINS is of the order of 10^{10} to 10^{12} s^{-1} and this fills a gap between the other spectroscopies which have been used for lifetime–broadening measurements (*cf.* Mössbauer, 10^{-7} to 10^{-9} s^{-1}; IR, 10^{-13} s^{-1}).

The mixed–valence trimer complexes already mentioned usually crystallise with a non–coordinated solvent molecule, *i.e.* $[M_3O(\ OOCCH_3)_6L_3](L')$ with L = py; L' = py, C_6H_6, $CHCl_3$ *etc.* Hendrickson and co–workers have shown that the rotation of the molecule L' can be closely related to the internal electron transfer in the tri–metal cluster. Some typical quasielastic scattering spectra of the iron(III,III,II) compound are shown in Figure 6.

The broadening due to molecular motion sets in over the same temperature range as does the electron transfer. From the scattering cross–sections of the nuclei present, the broadening is known to be mainly due to motions of hydrogen atooms, and comparisons with selectively deuterated compounds show that in this case it is the pyridine, rather than the methyl groups which are responsible. Analysis of band width and intensity as functions of momentum transferred are consistent with a molecule of

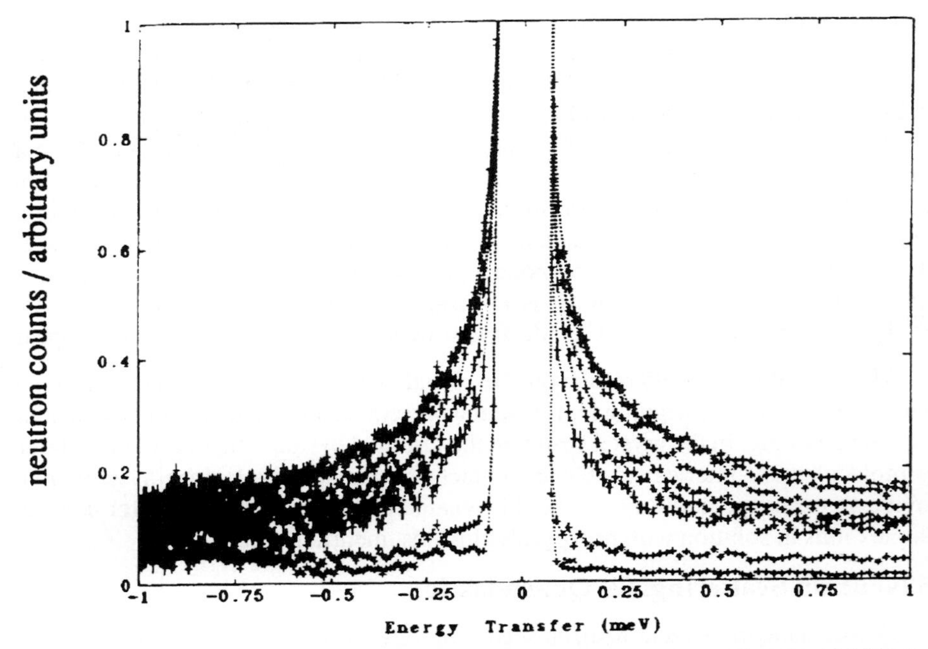

Fig. 6. Quasielastic neutron scattering spectra of $[Fe_3O(OOCCD_3)_6(C_5H_5N)_3](C_5H_5N)$, at temperatures T = 5, 100, 175, 183, 188.5, 200, 250, 300K. Instrument IN5, ILL, Grenoble. At the lowest temperature, the lowest curve shows only the elastic scattering — the curve centred at zero energy transfer, broadened by the instrumental resolution. At higher temperatures the curves are broadened by transfer of energy to and from moving parts of the molecules, especially rotation of the non–coordinated pyridines. The greatest temperature–sensitivity coincides with the gradual onset of electron delocalisation.

radius *ca* 2.1 to 2.3 , undergoing jump rotation between three equivalent sites, with a frequency of *ca.* 10^{11} s^{-1} at 200K and activation energy 1.7 kJ mol^{-1} (ref. 85 and 86).

In a single lecture it has been imposible to do justice to such a broad theme as electron transfer: what I have tried to do has been to set the scene for the detailed discussions which will follow, and to show that research in this area still has many new and interesting directions in which to go.

Acknowledgements

All of the neutron scattering work mentioned above has been done at the Institut Laue-Langevin (ILL), Grenoble, in collaboration with Dr. U. A. Jayasooriya (UEA) and Dr. R. P. White (ILL). I thank Dr. S.A. Borshch, Prof. P. Gütlich, Prof. J.–J. Girerd and Dr. G. R. Moore for useful discussions and for pre–publication copies of MSS.

References

1 Work on electron transfer, both experimental and theoretical, has been regularly reviewed in refs. 2 and 3.

2 *Inorganic Reaction Mechanisms* Chemical Society Specialist Periodical Report Series, Vols. 1 -7. The Chemical Society, London, 1971-1981

3 M.V. Twigg, (Ed.) *Mechanisms of Inorganic and Organometallic Reactions* Vols 1 - 6. Plenum Press, New York, 1983- (continuing)

4 When detailed references are not given, these will be found in ref. 5.

5 R.D. Cannon *Electron Transfer Reactions* Butterworth & Co., London 1980.

6 R.D. Cannon and J. Gardiner, J. Am. Chem. Soc., **92** (1970) 3800

7 D. Gaswick and A. Haim, J. Am. Chem. Soc., **93** (1971) 7347

8 N. Sutin, Acct. Chem. Res., **1** (1968) 225

9 A. Haim, Progr.Inorg. Chem., **30** (1983) 273

10 C. Creutz and H. Taube, J. Am. Chem. Soc., **91** (1969) 3988; C. Creutz, Progr.Inorg. Chem., **30** (1983) 1.

11 M.B. Robin and P. Day, Adv. Inorg. Chem. Radiochem, **10** (1967) 247

12 B. Mayoh and P. Day, J. Am. Chem. Soc., **94** (1972) 2885

13 R.A. Marcus, Ann. Rev. Phys. Chem., **15** (1964) 155

14 E. Waisman, G. Worry and R.A. Marcus, J. Electroanal. Chem., **82** (1977) 9

15 N. Sutin, Acct. Chem. Res., 275 (1982); Progr.Inorg. Chem., **30** (1983) 441.

16 A. Haim, Comments Inorg. Chem., **4** (1985) 113

17 ΔG^* and ΔG^0_{et} are the activation free energy and standard free energy change for the step p→s of equation 1, while $\Delta G^{\neq}$ and G^0 refer to the overall reaction.

18 T.W. Newton, J.Chem Ed., **45** (1968) 571

19 N. Sutin, Ann. Rev. Nucl. Sci., **12** (1962) 285; cf. Ref [6], p. 186

20 R.A. Marcus, J. Chem. Phys., **24** (1956) 966; cf N.S. Hush, Trans. Faraday Soc., **57** (1961) 557, and ref. 5, 194-200.

21 M.R. Hyde, R. Davies and A.G. Sykes, J. Chem. Soc., Dalton Trans., 1838 (1972).

22 D. Rehm and A. Weller, Israel J. Chem., **8** (1970) 259.

23 C. Creutz and N. Sutin, J. Am. Chem. Soc., **99** (1977) 1.

24 G.L. Closs, L.T. Calcaterra, N.J. Green, K.W. Penfield and J.R. Miller, J. Phys. Chem., **90** (1986) 3673

25 H. Taube, in: K.N. Raymond (Ed.) *Bioinorganic Chemistry–II* (Advances in Chemistry Series, 162). American Chemical Society, Washington, D.C., 1977; pp. 127-144.

26 V. Balzani, F. Scandola, G. Orlandi, N. Sabbatini and M.T. Indelli, J. Am. Chem. Soc., **103** (1981) 3370.

27 P. Bernhard, L. Helm, A. Ludi and A.E. Merbach, J. Am. Chem. Soc., **107** (1985) 312.

28 M.D. Newton, Int. J. Quantum Chem., Symp. **14** (1980) 363; ACS Symposium Series., No. 198, 255 (1982).

29 S.P. Dolin, R.R. Dogonadze and E.D. German, J. Chem. Soc. Faraday I, **73** (1977) 648. See also refs. cited in ref. 2. vol 6, p 6.

30 M.R. Gunnmer, D.E. Robertson and P.L. Dutton, J. Phys. Chem., **90** (1986) 3783.

31 (a) N.R. Kestner, J. Jortner and J. Logan. J. Phys. Chem., **78** (1974) 2148.
(b) I. Webman and N.R. Kestner, J. Chem. Phys., **77** (1982) 2387.
(c) R.P. Van Duyne and S.F. Fischer, Chem. Phys., **5** (1974) 183.
(d) S.F. Fischer and R.P. Van Duyne, Chem. Phys., **26** (1977) 183.
(e) J. Ulstrup and J. Jortner, J. Chem. Phys., **63** (1975) 4358.
(f) M.D. Newton, Int. J. Quantum Chem., Quantum Chem. Symp., **14** (1980) 363.
(g) B.S. Brunschwig, J. Logan. M. Newton and N. Sutin, J. Am. Chem. Soc., **102** (1980) 5798.
(h) J. Jortner, J. Chem. Phys., **64** (1976) 4860.
(i) R.R. Dogonadze, A.M. Kuznetsov, M.A. Vorotyntsev and M.G. Zakaroya, J. Electroanal. chem., **75** (1977) 315.
(j) R.R. Dogonadze, A.M. Kuznetsov and M.A. Vorotyntsev, Z. phys. chem., **100** (1976) 1.

32 P.N. Schatz, in D.B. Brown (Ed.) *Mixed–Valence Compounds* D.Reidel, 1980. pp 115 -150.

33 G.W. Robinson and R.P. Frosch, J. Chem. Phys., **38** (1963) 1187.

34 D.Q. DeVault, Rev. Biophys., **13** (1980) 387.

35 T. Guarr and G. McLendon, Coord. Chem., Rev., **68** (1985) 1.

36 V.P. Alekseev, V.I. Goldanskii, V.E. Prusakov, A.V. Nefedev and R.S. Stukan, JETP Lett., **16** (1972) 43.

37 M. Alflen, C. Henner, H. Spiering and P. Gütlich, in: K. Prassides (Ed.) *Mixed Valence Systems: Applications in Chemistry, Physics and Biology* [Proc. NATO Advanced Research Workshop, Heraklion, Crete, June 1990]. Kluwer Academic Publishers, Dordrecht, Netherlands (to appear 1991), p. 299.

38 N.S. Hush, Prog. Inorg. Chem., **8** (1967) 391.

39 The thermodynamic significance of the optical intervalence charge transfer energy has recently been discussed: it is shown to be equivalent to a free energy G rather than an internal energy E or enthalpy H. (ref. 40). See also ref. 16 for a critical discussion of experimental data.

40 R.A. Marcus and N. Sutin, Comments Inorg. Chem., **5** (1986) 119.

41 F. Felix and A. Ludi, Inorg Chem., **17** (1978) 1782.

42 R.D. Cannon, Chem. Phys. Lett., **49** (1977) 299.

43 B.S. Brunschwig, S. Ehrenson and N. Sutin, J. Phys. Chem., **90** (1986) 3657.

44 M.J. Powers, D.J. Salmon, R.W. Callahan and T.J. Meyer, J. Am. Chem. Soc., **98** (1976) 6731.

45 Other possible distance dependences are discussed by N.S. Hush, Coord. Chem. Rev., **64** (1985) 135.

46 T.G. Winter and N.F. Lane, Phys. Rev. A., **31** (1985) 2698.

47 D.E. Richardson, Coord. Chem. Rev., **93** (1989) 59.

48 D.E. Richardson, J. Phys. Chem., **90** (1986) 3693.

49 P. Sharpe, C.S. Christ, J.R. Eyler and D.E. Richardson, Int. J. Quantum Chem., Quantum Chem. Symp., **22** (1988) 601.

50 J.R. Eyler and D.E. Richardson, J. Am. Chem. Soc., **107** (1985) 6130.

51 J.P. Candlin and J. Halpern, Inorg. Chem., **4** (1965) 1086.

52 H.Taube *Electron Transfer between Metal Complexes – Retrospective* (Nobel Lecture, 8 December 1983). *Les Prix Nobel 1983* Almqvist and Wiksell Int., Stockholm, 1984. pp. 149 - 169.

53 D.R. Stranks, Pure Appl. Chem., **38** (1974) 303.

54 S. Wherland, Inorg. Chem., **22** (1983) 2349.

55 (a) I. Krack, P. Braun and R. Van Eldik, Physica B **139 & 140** (1986) 680
(b) I. Krack and R. Van Eldik, Inorg. Chem., **25** (1986) 1743.

56 W.H. Jolley, D.R. Stranks and T.W. Swaddle, Inorg. Chem., **29** (1990) 1948.

57 J.T. Hupp and M.J. Weaver, Inorg. Chem., **22** (1983) 2557.

58 (a) H. Doine and T.W. Swaddle, Inorg. Chem., **27** (1988) 665;
(b) T.W. Swaddle and L. Spiccia, Physica B, **139 & 140** (1986) 684;
(c) L. Spiccia and T.W. Swaddle, Inorg. Chem., **26** (1987) 2265.

59 (a) R.M. Nielson, J.P. Hunt, H.W. Dodgen and S. Wherland, Inorg. Chem., **25** (1986) 1964.
(b) M. Stabler, R.M. Nielson, W.F. Siam , J.P. Hunt, H.W. Dodgen and S. Wherland, Inorg. Chem., **27** (1988) 2893.

60 P.W.Anderson and Hasegawa, Phys. Rev., **100** (1955) 675.

61 M.I. Belinskii, B.S. Tsukerblat and N.V. Gerbeleu, Fiz. Tverd. Tela, **25** (1983) 869, (Sov. Phys. Solid State, **25** (1983) 497);
B.S. Tsukerblat and M.I. Belinskii, Fiz. Tverd. Tela (Sov. Phys. Solid State), **25** (1983) 2024;
M.I. Belinskii and B.S. Tsukerblat, Fiz. Tverd. Tela, **25** (1983) 758, (Sov. Phys. Solid State, **26** (1984) 458);
M.I. Belinskii and B.S. Tsukerblat, Khim. Fiz. **4** (1985) 606, (Sov. J. Chem. Phys., **4** (1987) 982);
M.I. Belinskii, V.Ya. Gamurar and B.S. Tsukerblat, Phys. Stat. Sol., **135** (1986) 189, 555;
M.I. Belinskii, V.Ya Gamurar and B.S. Tsukerblat, Fiz. Tverd. Tela, **29** (1987) 208, (Sov. Phys. Solid State, **29** (1987) 116).

62 M.I. Belinskii, Fiz. Tverd. Tela, **27** (1985) 1761, (Sov. Phys. Solid State, **27** (1985) 1057).

63 P.M. Plaksin, R.C. Stoufer, M. Mathew and G.J. Palenik, J. Am. Chem. Soc., **94** (1972) 2121.

64 (a) R.D. Cannon, L. Montri, D.B. Brown, K.M. Marshall and C.M. Elliott, J.Am.Chem.Soc., **106** (1984) 2591;
(b) L. Meesuk, U.A. Jayasooriya and R.D. Cannon, J.Am.Chem.Soc., **109** (1987) 2009.

65 L. Sacconi, C. Mealli and D. Gatteschi, Inorg. Chem. **13** (1974) 1985.

66 V.M. Miskowski, T.M. Loehr and H.B. Gray, Inorg. Chem., **26** (1987) 1098.

67 V. Papaefthymiou, J.–J. Girerd, I. Moura, J.J.G. Moura and E. Münck, J. Am. Chem. Soc., **109** (1987) 4703.

68 S.A. Borshch, I.N. Kotov and I.B. Bersuker, Teor. Eksp. Khim., **20** (1984) 675, (Soviet Physics JETP, **20** (1984) 633);
Chem. Phys. Lett., **111** (1984) 264;
Khim. Fiz. **3** (1985) 667, (Sov. J. Chem. Phys., **3** (1985) 1009);
S.A. Borshch, Dokl. Akad. Nauk SSSR, **280** (1985) 652, (Doklady Phys. Chem., **280** (1985) 59);
S.A. Borshch, Zh. Strukt. Khim., **28** (1987) 36.

69 G. Blondin and J.–J. Girerd, Chem. Rev., **90** (1990) 1359.

70 B.S. Snyder, G.S. Patterson, A.J. Abrahamson and R.H. Holm, J. Am. Chem. Soc., **111** (1989) 5214;
K.K. Surerus, E. Münck, B.S. Snyder and R.H. Holm, J. Am. Chem. Soc., **111** (1989) 5501.

71 S. Drüeke, P. Chaudhuri, K. Pohl, K. Wieghardt, X.–Q. Ding, E. Bill, A. Sawaryn, A.X. Trautwein, H. Winkler and S.J. Gurman, J. Chem. Soc. Chem. Commun., 1989, 59.

72 A.S. Borovik, V. Papaefthymiou, L.F. Taylor, O.P. Anderson and L. Que, J. Am. Chem. Soc., **111** (1989) 6183.

73 H.U. Güdel, A. Stebler and A. Furrer, Inorg. Chem., **18** (1979) 1021.

74 U.A. Jayasooriya, R.D. Cannon, R.P. White and G.J. Kearley, J. Am. Chem. Soc., to be submitted.

75 U.A. Jayasooriya, R.D. Cannon, R.P. White and G.J. Kearley, Angew. Chem., Int. Ed. English, **28** (1989) 930.

76 R.D. Cannon and R.P. White, Prog. Inorg. Chem., **36** (1988) 195.

77 R.P. White, L.M. Wilson, D.J.Williamson, G.R. Moore, U.A. Jayasooriya and R.D. Cannon, Spectrochim.Acta, **46A** (1990) 917.

78 L. Meesuk, R.P. White, B.G. Templeton, U.A. Jayasooriya and R.D. Cannon, Inorg. Chem. **29** (1990) 2389.

79 R.P. White, J.O. Al–Basseet, R.D. Cannon, G.J Kearley and U.A. Jayasooriya, Physica B, **156 &157** (1989) 367.

80 U.A. Jayasooriya, R.D. Cannon, R.D. White and G.J. Kearley, J. Am. Chem. Soc., to be submitted.

81 A.B. Blake, A. Yavari, W. Hatfield and C.N. Sethulekshmi, J. Chem. Soc., Dalton Trans., 1985 2509.

82 S.E. Woehler, R.J. Wittebort, S.M. Oh, T. Kambara, D.N. Hendrickson, D. Inniss and C.E. Strouse, J. Am. Chem. Soc., **109** (1987) 1063.

83 M. Sorai, K. Kaji, D.N. Hendrickson and S.M. Oh, J. Am. Chem. Soc., **108** (1986) 702.

84 M. Bée *Quasielastic Neutron Scattering* Adam Hilger, Bristol, 1988.

85 R.D. Cannon, U.A. Jayasooriya and R.P. White *Inelastic Neutron Scattering Studies of Mixed–Valency Compounds* in: K. Prassides (Ed.) *Mixed Valency Systems: Applications in Chemistry, Physics and Biology* [Proc. NATO Advanced Research Workshop, Heraklion, Crete, June 1990]. Kluwer, Academic Publishers, Dordrecht, Netherlands (to appear 1991), p. 283.

86 R.D. Cannon, U.A. Jayasooriya, S. ArapKoske, R.P. White and J.H. Williams, J. Am. Chem. Soc., in press.

Electron and Proton Transfer in Chemistry and Biology, edited by A. Müller et al.
Studies in Physical and Theoretical Chemistry, Vol. 78
1992 Elsevier Science Publishers B.V.

Electron Transfer Effect in Chemical Compounds

B. Jeżowska-Trzebiatowska and W. Wojciechowski*

Institute of Chemistry, University of Wrocław
Wrocław (Poland)

*Technical University of Wrocław
Wrocław (Poland)

Summary

Effect of the electron transfer on the physicochemical properties, in particular, on the magnetic and spectroscopic properties of coordination compounds with d-electron ions is presented in this work.

The influence of the electron transfer on the catalytic properties of compounds has been also examinated with respect to the mechanism of bond formation between the molecular gases and coordination compounds as well as metal monocrystals.

In most of the chemical and physical processes we come across electron transfer effects. In dissociation of water, for example, an electron from the hydrogen atom in neutral molecule is transferred to the oxygen atom which results in the formation of the OH^- ion.

Electron transfer is responsible for many phenomena occurring in the living organisms, *e.g.*, the process of the oxygen uptake by hemoglobin or the oxidation - reduction reactions which proceed in these organisms. The majority of these processes leads to the formation of the metal complexes with ligands occurring in the organisms. In many cases, such as the formation of the complex compounds, this induces the stabilization of the oxidation state of some metal ions.

Electron transfer plays an important part in the catalytic processes, in the formation of the so-called chemisorption bonding (adsorption of small molecules on the surface of metal monocrystals), in the processes of the energy transfer outside the system (laser) and in many other processes.

Moreover, changes in the electron density within the given molecule in comparison with the electron density distribution for the free atoms or ions may also be considered as the electron transfer processes. Therefore, it can be stated that the formation of the chemical bonding of any type, including hydrogen bonding, is related to the electron transfer. Electron transfer effects are also very important for the magnetic properties of compounds.

Transition of the electron within one molecule, *e.g.*, in the d^n - ion complex compound of the O_h symmetry (in the d^7 - d^4 electron configuration), from the t_{2g} to e_g orbitals, causes the changes of the magnetic properties of this molecule depending on the value of the crystal field parameter. The high - spin compound converts into the low - spin state. The reverse process is also possible. This type of the electron transition proceeding within the given ion yields the luminescence spectrum being used in the quantum electronics. These transitions have been observed also in the case of some solids, *e.g.*, in the compounds consisting of d-electron metals and oxygen, in which, according to the theory of the exchange interaction, the electron transfer from the anion orbitals to the cation orbitals leads to the occurrence of anti- or ferromagnetism. This type of phenomenon has been also observed in the complex compounds of d-electron ions.

In the experimental and theoretical studies performed by Jeżowska - Trzebiatowska and coworkers, and later by continuers (ref. 1) it has been revealed that the formation of the oxygen bridge between two central magnetic ions within the one molecule leads to the occurrence of a diamagnetism. In such case we deal with the so-called intramolecular antiferromagnetism. The source of this effect has been explained by Jeżowska - Trzebiatowska *et al.* (ref. 1) on the basis of the molecular orbital theory. Complex compounds are very interesting examples of the electron transfer process. This is due to the fact that in such compounds there are various active centers related with one (or more than one) central ion and the ligands. Centers arising from the ligands have the donor properties, whereas those originating from the metals have acceptor properties. The character of these centers is variable and depends on the type of bonding which is related to the electronic structure of the central ions, electron density distribution and the energy of the ligand orbitals as well as the difference in the electronegativity between metal ions and the ligand donor atoms or ions. The presence of the active centers in the complex compounds affords possibilities for the occurrence of various types of interactions, among them also the exchange interactions, which leads to the appearance of the non-typical magnetic properties.

Electron transfer is also connected with the problem of the energy transfer. For example, in the compounds of the type: $MMnX_3$ (M = K, Rb, Cs, NH_4; X = Cl, Br, J) doped with Nd^{3+} or Eu^{3+} ions, the excitation of the Mn^{2+} ion is accompanied by the increase in intensity of the luminescence line arising from the given lanthanide ion. The energy from terms of the Mn^{2+} ion is transferred to the Ln^{3+} ion terms (ref. 2). In this process some conditions have to be fulfilled, in particular, the values of the energy of the corresponding terms in both these metal ions should be similar. This type of the energy transfer and such relationships have also been found in many other systems, among them, in various kinds of glasses containing ions of the d- or f-electron elements (Jeżowska - Trzebiatowska *et al.* (ref. 3)).

Electron transfer process leads sometimes to the so-called mixed - valence compounds, *e.g.*, well-known binuclear cobalt complex compound with O_2^- and O_2^{2-} bridges which will be discussed in a further part of this work. Other examples are the monocompounds of Eu^{2+} with oxygen, sulphur, selenium and tellurium where the action of the external agents (*e.g.* the change of a pressure) is followed by the electron transition to the conduction band, eventually leading to the change of the formal oxidation state of the lanthanide ion (ref. 4). Electron transfer phenomenon also occurs in the high -temperature superconductors. In the so - called 1, 2 and 3 superconductors depending on the oxygen content, the formal oxidation state of the copper ions changes from +1 to +3.Superconductivity in these systems is most probably caused by the electron transfer in the planes in which the copper ions and some of the oxygen atoms are located (ref. 5).

In order to understand the electron transfer in the complex compounds it is necessary to elucidate the role of the bridging bond between the metal atoms. The non-metal bridges between the meta atoms are excellent mediating agents in the exchange of electrons. The best in this respect are oxygen bridges. This specific significance of the oxygen bridges has been found by us (ref. 1) in the systems, in which the bridge is acting between two magnetic atoms (possessing magnetic moment). For example, two Re(IV) ions of the d^3 -electron configuration and with the magnetic moment equal to about 3.8 MB show diamagnetism in dimes, whereas their oxidation state remains the same. This is caused by the formation of the compound $K_4Re_2Cl_{10}$ with the metal - oxygen - metal bond.

A series of this type of compounds has been obtained with rhenium, ruthenium, osmium and other elements (ref. 6). This unusual phenomenon has been explained on the basis of the molecular orbital theory (ref. 1).

Let us consider now the influence of the electron transfer processes on the magnetic properties of the coordination compounds of d-electron ions. The example of the binuclear complex compound of copper (II) is examined in the first place. When the Cu(II) ion is in the field of the axial or rhombic symmetry, then the ground state of each copper ion is nondegenerated. In copper (II) acetate, in the ground state, the electron is on the $d_{x^2-y^2}$ orbital. The other orbital, d_{z^2}, may take part in the coupling of two copper ions. Thus, there are four orbitals, two from each ion, which are responsible for the exchange interaction. The remaining orbitals of the metal ions also change their energy, but do not have any effect on the interacting electrons. This consideration requires introduction of the two types of overlap between orbitals of copper ions, namely (d_{z^2}–d_{z^2}) and ($d_{x^2-y^2}$ – $d_{x^2-y^2}$). These two types of orbitals determine the splitting of the molecular orbitals formed in the dimmer which contain electrons originating from the $d_{x^2-y^2}$ orbitals and determine the magnitude of the exchange integrals.

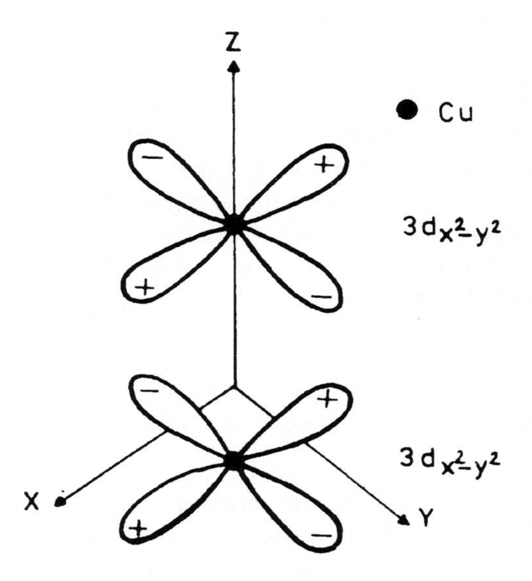

It has been shown experimentally (ref. 7), that the δ interaction have not much significance in electron spin-coupling. The structure of copper (II) acetate as well as the structures of other dimmers of copper, which result from the coupling of two paramagnetic metal centers, afford possibilities for the magnetic superexchange interaction between copper ions through the system of four carboxylic bridges (Cu-O-C-O-Cu). Superexchange is related with the overlap of $2p_\pi(O)$ - $2p_\pi(C)$ - $2p_\pi(O)$ atomic orbitals of the mesomeric system ^-O - CH = O/O = CH - O^-. Such a delocalization of the π type molecular orbitals on these three π nuclei yields a low-energy path for the electron migration (electron hole).

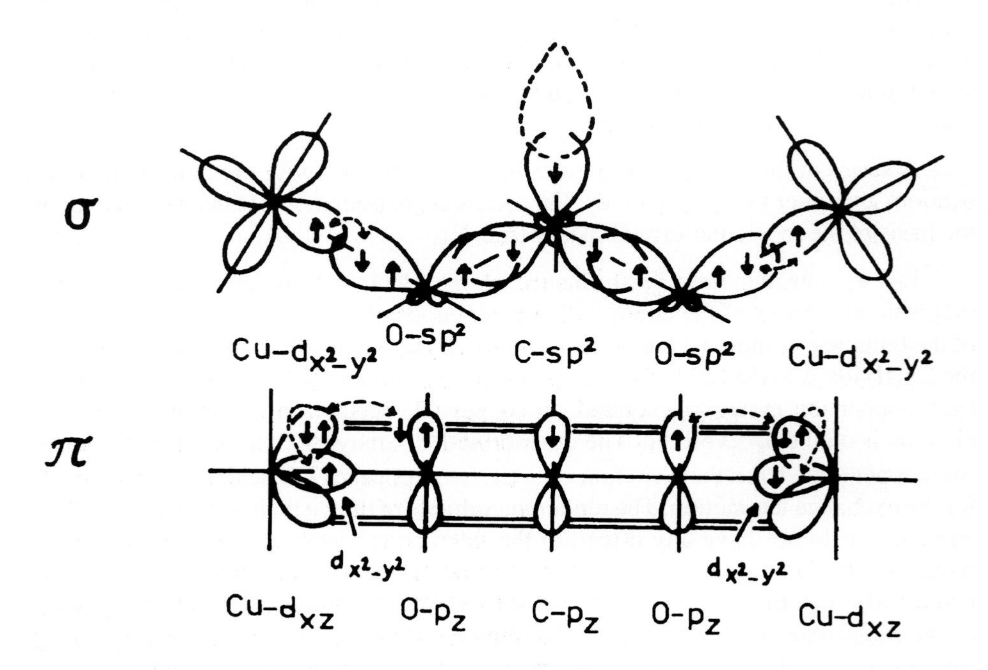

If the examined structure consists of the infinite chain of the copper centers, then the approximation of the experimental magnetic data (dependence of magnetic susceptibility on the temperature) can be performed on the basis of the following models of exchange interaction:

- Ising model for the one-dimensional spin system of S=1/2
- Heisenberg model for the one-dimensional N+1 spin system, and S=1/2 in zero field.

In the case of copper (II) compounds of such a structure, similar results have been obtained from the measurements of the electric conductivity in the solid state. For copper (II) phthalocyanine these measurements indicate the hopping mechanism of conductivity (ref. 8). Unfortunately, these measurements have been performed only on the polycrystalline samples. For a few copper compounds these measurements have been done also on monocrystals and the one-dimensional type of conductivity was found (ref.). Very characteristic for copper (II) phthalocyanine is a small energy gap (about 0.5 eV). Addition of some soot (physical mixture) which contains free radicals, being found in the EPR measurements, to this chemical compound, induces the appearance of the metal conductivity at the specified ratio of these components (ref. 9). Similar effect has been obtained from the exposure of this compound with the X-radiation (ref. 10). Complex compounds are particularly interesting objects of studying electron transfer since this process is facilitated by the bonding of various atoms or group of atoms having different character and electron density in one molecule. These atoms are connected with each other by various kinds of interactions and the most important among them are chemical bonds.

One of the phenomena of electron transport is activation of of the molecular oxygen occurring in the natural systems and in their synthetic analogs the complex compounds (ref. 11). The processes of transport and activation of the molecular oxygen by the natural systems have been known for many years. Bioinorganic chemists and biophysicists were involved in the study of the structure and dynamics of the systems engaging oxygen in metabolic processes.

Many papers have been reported so far, however, a lot of problems concerning the electron mechanism of the oxygen uptake and the activation of O_2 as well as the structure of M - O_2 arrangement still require explanation. Our studies on the models of dioxygen carriers and enzymatic systems resulted in the detection and characterization of new synthetic systems responsible for the O_2 and CO_2 uptake. Among them are Co(II) complexes with bidentate Schiff bases (BSB) which are the substrata for the formation of these systems. These are tetracoordinated, high spin, tetrahedral complexes, which in the presence of N-organic base (B) under the oxygen-free conditions are transformed into the "active" form of the complex being able to activate and transport molecular oxygen. This "active" form is the pentacoordinated complex of the structure of tetragonal pyramid which results from the flattening of the $Co(BSB)_2$

complex and the attachment of the N-base. The formation of this complex was confirmed by the spectrophotometric titration with the N-base which yielded a linear dependence for the relation: log $[(A_0\text{-}A)/A]$ *vs.* logB with a slope β equal to 1.

At higher concentration one can observe the bend line. This is caused by the appearance of the subsequent equilibrium involving the hexacoordinated, non-active complex unable for the oxygen transport.

The fact that the B values are positive, evidences that the decrease of electron density on the cobalt ion favors the coordination of the axial ligand making complex a stronger Lewis acid. Oxygen uptake by the model system is accompanied by the violent change of the color of the solution from dark-yellow to brown-red. The complex with the coordinated O_2 molecule is a monomer like in the case of the hem O_2 carriers.

The EPR spectra are characteristic for the monomeric cobalt complexes with the unpaired electron on the O_2 molecule. This eight line hyperfine structure results from interaction of the unpaired electron of which the spin density is localized mainly on the π^* orbital of O_2, with the nuclear spin I=7/2 of ^{59}Co. In the natural carriers and enzymes the axial ligands are very significant in determining properties of the prosthetic groups. We have stated (ref. 11) that the increase in basicity of these ligands being connected with the increase in the σ -donation is responsible for increasing thermodynamic stability of the complex with the O_2 molecule. Simultaneously, for the π-acceptor ligands the decreasing stability of the oxygenated complex has been observed.

The active form of the dioxygen carrier is the high-spin complex of tetragonal pyramid structure. Its symmetry should be related with the electronic configuration $(d_{xz})^2$ $(d_{yz})^2$ ρ $(d_{xy})^1(d_z)^1(d_{x^2-y^2})^1$. The relative energy of these orbitals is determined by the σ and π donor-acceptor properties of the axial and equatorial ligands. The extremely high values of the stability constant of hexacoordinated complexes have been found for such ligands as py and 4-Mepy. This fact can be interpreted as the influence of the π-acceptor properties of these ligands. Empty π orbitals of the pyridine nitrogen are able to overlap with the filled $\pi(d_{yz},d_{xz})$ metal orbitals and this causes the reduction of the repulsive interaction between the sixth ligand and the metal through the decrease of the electron density on the cobalt ion. Thus, the d_{xz} and d_{yz} orbitals become more stable and the d_{z^2} orbital becomes slightly less stable for the axial ligands of low basicity.When the O_2 molecule with two half-filled π^* orbitals approaches the complex, one electron (π^*) from oxygen one electron from the d_{z^2} orbital metal (*i.e.* the orbital best fitting in symmetry and energy) are forming the σ binding. Because of the symmetry of the π^* orbitals with respect to the O-O axis the O_2 molecule is bound at the angle of 120°. The second unpaired electron on the π^* orbital of oxygen interacts with suitably filled $\pi(d_{yz}, d_{xz})$ orbital and forms the π - back bonding stabilizing the M - O_2 core.

The schematic order of the molecular orbitals should be as follows (ref. 12):

$$(\pi_{yz},\pi_y)^2(d_{xz})^2(d_{xy})^2(d_z^2,\pi_u)^2(\pi_{yz},\pi_v)^1(d_{x^2-y^2})^0(\sigma_{z^2},\pi_h)^0$$

The π-back interaction contributes to the formation of the anisotropy in the EPR spectra and induces a polarizing factor of the σ bonding.

These examples elucidate the exchange interactions which accompany almost all phenomena occurring in systems differing in the electron density. The occurrence of the exchange interactions and consequently, the movement of electrons characterizes the formation of the chemical bond. In the compound under the discussion there are π - acceptor axial ligands which induce the π- back interaction. Such an interaction occurs simultaneously with the polarization of the δ electron pair. The π - acceptor effect decreases the electron density on the d_π metal orbitals (d_{xy}, d_{yz}). In such a way their stabilization through the reduction of the π -back bonding causes a reduction of the $Co\text{-}O_2$ bonding.

Specificity of the complex compounds in an electron transport makes them very significant in the homogeneous as well as the heterogeneous catalysis. The above mentioned tetrahedral complexes capable of transport and oxygen uptake have the catalytic properties. Therefore the complexes of Co with the Schiff bases, $Co(BSB)_2$, can act as the catalysts in the oxidation of hydrazine and its aryl- and aldehyde-derivatives. The main products of this process are N_2, H_2O and the respective hydro-carbons. It should be emphasized that in this very complicated process oxygen is bound with the cobalt center and becomes converted into the formal group O_2^-. The electrons in the resulting unstable complex undergo a slow rearrangement "-redox process" through the metal center and simultaneously the hydrogen atom is being released. In the first stage the coordinated subtract undergoes oxidation to the di-imino group. The significance of the complex not only consist in the activation of a substrate through its coordination but also in the electron transfer in the catalytic processes of reduction and oxidation. This is the reason that such a system is similar to some of the enzymatic systems.

We shall discuss now the processes of the electron transfer in the binuclear cobalt compounds with the O_2^{2-} and O_2^- bridges because in these compounds there is also the activation of a small molecule, *i.e.* oxygen molecule. It results from the X-ray studies that each of the complex ions (ref. 13-17):

$$[(NH_3)_5CoO_2Co(NH_3)_5]^{4+}, \quad [(NH_3)_5CoO_2Co(NH_3)_5]^{5+},$$
$$[(CN)_5CoO_2Co(CN)_5]^{6-}, \quad \text{and} \quad [(CN)_5CoO_2Co(CN)_5]^{5-},$$

has an approximate C_{2h} symmetry. The two cobalt ions lie in the same plane as well as a junctive oxygen bridge.

The calculations performed for the oxygen bridge in the Co complex with ammonia, the charge of 5+, yielded the following electron configuration (ref. 18-20):

$$[2a_g(\sigma,\pi)^b]^2[1a_u(\pi)^b]^2[3a_g(\pi,\sigma)^b]^2$$
$$[2b_u(\pi,\sigma)^*]^2[4b_g(\pi)^*]^1[12b_u(\sigma,\pi)^*]^0.$$

For the oxygen bridge in the complex ion $[(CN)_5CoO_2Co(CN)_5]^{5-}$ a similar electronic structure has been obtained:

$$[1a_u(\pi)^b]^2[2b_u(\sigma,\pi)^b]^2[2a_g(\pi,\sigma)^b]^2$$
$$[12a_g(\sigma,\pi)^*]^2[9b_g(\pi)^*]^1[22b_u(\pi,\sigma)^*]^0.$$

In these ions the electronic structure of the oxygen bridge is very similar to that of the free oxygen molecule being under the influence of the ligand field.

The antibonding molecular orbitals of the oxygen bridge, $b_u^*(a_g^*)$ and b_g^* are degenerated just as the π_y^* and π_z^* orbitals for the free oxygen molecule.

In the case of the oxygen bridge one electron is located on the antibonding orbital $b_g\pi^*$ (which is empty in a free oxygen molecule) and this indicates the presence of the superoxide O_2^- ligand. This conclusion has been confirmed by the results from the electron paramagnetic resonance studies (ref. 21) and the X-ray data (ref. 13-17).

For the oxygen bridges in the ion complexes $[(NH_3)_5CoO_2Co(NH_3)_5]^{4+}$ and $[(CN_5)\ CoO_2Co(CN)_5]^{6-}$ the resulting electronic structure is very similar to that in the superoxide - bridges:

$$[b_u(\sigma,\pi)^b]^2[a_u(\pi)^b]^2[2a_g(\pi,\sigma)^b]^2$$
$$[a_g(\sigma,\pi)^*]^2[b_g(\pi)^*]^2[b_u(\pi,\sigma)^*]^0.$$

In this case the antibonding molecular orbital of the oxygen bridge $b_g(\pi)^*$ the oxygen bridge is the peroxide ligand O_2^{2-}. Since the molecular orbital $b_g(\pi)^*$ is filled with the second electron, the O-O bond lengths increases and approaches the value close to the corresponding one in BaO_2 and H_2O_2.

From the comparison of the electronic structures of the complex ions investigated it can be stated that in all these complexes both cobalt atoms are equivalent and change of the charge on the complex ion is caused by the altering charge on the oxygen bridge. Calculations performing for the complex ion (refs. 18-20) $[(NH_3)_4CoO_2NH_2Co(NH_3)_4]^{4+}$ indicate that the molecular orbitals mainly consist of the d - cobalt and p - oxygen atomic orbitals and the series of the molecular orbitals is as follows:

$$O\ [2a_1(\sigma^b) << 1b_1(\pi)^b << 3a_1(\pi,\sigma)^b] << Co\ [d_\pi(1a_2, 2a_2, 5b_2, 3b_1, 8a_1, 5b_1)] <<$$
$$O\ [6b_2(\sigma^*)] << Co\ [d_\sigma(7b_2, 9a_1)] << O\ [4a_2(\pi^*)] << Co\ [d_\sigma(8b_2, 10a_1)] <<$$
$$O\ [11b_2(\pi,\sigma)^*].$$

In this case the valence electron of the complex ion is localized on the d_δ orbital of the cobalt ion $[7b_2(d_{x^2-y^2})]$. The oxygen bridge has the following electronic structure:

$$[2a_1(\sigma^b)]^2[1b_1(\pi^b)]^2[3a_1(\pi,s)^b]^2[6b_2(\delta^*)]^2[4a_2(\pi^*)]^0[11b_2(\pi,\sigma)]^0.$$

The superoxide bridge has a greater acceptor-donor character in the $[(NH_3)_4CoO_2Co(NH_3)_4]^{4+}$ ion than in the $[(NH_3)_5CoO_2Co(NH_3)_5]^{5+}$ ion. This is proved by the enlarged contribution of the d_{yz} orbitals of the cobalt ion in the antibonding orbital of the superoxide bridge, as compared to the corresponding contributions in the analogous molecular orbitals of the uni-bridged complexes. This is also supported by the presence of filled orbitals of the oxygen bridge in empty σ orbitals of cobalt ions.

Electron transfer in the investigated compounds is particularly well observed in their electronic spectra. Brągiel and coworkers (ref. 22) have interpreted the electronic spectra of these complex ions on the basis of calculations with the CNDO method.

It follows from the data presented that in ref. 2 mechanism of the electron transition one can distinguish the transfer inside the bridge as well as the electron transfer from the O_2^{2-} or O_2^- bridge to the cobalt ion and *vice versa.*

Electron transfer from the O_2^- bridge to the Co ion occurs only in $[Co_2(\mu - O_2)(NH_3)_{10}]^{5+}$ cation and is related with the compensation of charges on the central ions.

Process of activation of small molecules has been studied also in the case of adsorption of small molecules on the surface of monocrystals of metals (ref. 23). This involves formation of so-called chemisorption bond that is also the electron transfer. The energy of such bond is comparable with the energy of a typical ionic bond and depends on the degree of surface coating and its structure (it also depends on which wall the adsorption occurs).

The energy of the chemisorption bond is comparable to that of the M - Co bond in transition metal carbonyls. On the basis of the results obtained, the series illustrating the changes in the chemisorption bond energy can be defined as follows:

$$N > O > H > Co > NO > N$$

Calculations performed with various quantum-chemical methods indicate that the energy of this chemisorption bond depends on the mutual position of the Fermi level in metal as well as on the HOMO/LUMO orbital energies of the system being formed. This energy proportional to the overlap integral, that is, to the change in the electron density distribution.

References

1 B. Jeżowska-Trzebiatowska, H. Kozłowski and L. Natkaniec, Bull. Acad. Polon. Sci. **19** (1971) 115.

2 M. Pawłowska, W. Stręk and T. Trabjerg, Phys. Stat. Sol.(a) **124** (1991) K63.

3 B. Jeżowska-Trzebiatowska, E. Łukowiak, W. Stręk, A. Buczkowski, S. Paleta, J. Radojewski and J. Sarzyński, Sol. Energy, Mat. **13** (1986) 267.

4 W. Suski in: J.Z. Damm and J. Klamut (Eds.) *Fizyka i chemia ciała stałego* Ossolineum, Wrocław, 1987, p. 41, (in Polish).

5 R. Horyń, J. Klamut, T. Kopeć and A. Zalewski in: J.Z. Damm and J. Klamut (Eds.) *Fizyka i chemia ciała stałego* Ossolineum, Wroclaw, 1987, p.231, (in Polish).

6 B. Jeżowska-Trzebiatowska, Coord. Chem. Rev. **26** (1968) 255.

7 E. Sinn and H. Ewald, Inorg. Chem. **8** (1969) 537.

8 W. Wacławek and M. Ząbkowska-Wacławek, Phys. Stat. Sol. **108** (1988) 331.

9 M. Ząbkowska-Wacławek and W.Wojciechowski, Phys. Stat. Sol. (a) **118** (1990) K87.

10 S. Turkmani and W. Wojciechowski, (Mat. Sci.), in press.

11 P. Sobota, B. Jeżowska-Trzebiatowska, Coord. Chem. Rev. **6** (1978) 71.

12 L. Natkaniec, M. Rudolf and B. Jeżowska-Trzebiatowska, Theor. Chim. Acta **28** (1973) 193.

13 W.P. Schaefer and R.E. Marsh, Acta Cryst. **21** (1966) 735.

14 W.P. Schaefer, Inorg. Chem. **7** (1968) 725.

15 F.R. Fronczek, W.P. Schaefer and R.E. Marsh, Acta Cryst. **B30** (1974) 117.

16 F.R. Fronczek and W.P. Schaefer, Inorg. Chim. Acta **9** (1974) 143.

17 F.R. Fronczek, W.P. Schaefer and R.E. Marsch, Inorg. Chem. **14** (1975) 611.

18 I. Hyla-Kryspin, L. Natkaniec and B. Jeżowska-Trzebiatowska, Bull. Acad. Polon. Sci. **25** (1977) 193.

19 I. Hyla-Kryspin, L. Natkaniec, B. Jeżowska-Trzebiatowska, Chem. Phys. Letters **35** (1975) 311.

20 I. Hyla-Kryspin, L. Natkaniec and B. Jeżowska-Trzebiatowska, Bull. Acad. Polon. Sci. **26** (1978) 985.

21 B. Jeżowska-Trzebiatowska, J. Mroziński and W.Wojciechowski, J. Prakt. Chem. **34** (1966) 57.

22 P. Bragiel, W. Wojciechowski and M. Czerwiński, Koord. Khim. **14** (1988) 1665.

23 W. Wojciechowski and P. Bragiel, Mat. Sci. **15** (1988) 33.

Electron and Proton Transfer in Chemistry and Biology, edited by A. Müller et al.
Studies in Physical and Theoretical Chemistry, Vol. 78

Light Induced Electron Transfer of Metal Complexes

A. Vogler and H. Kunkely

Institut für Anorganische Chemie
Universität Regensburg
Universitätsstraße 31
D-8400 Regensburg (Germany)

Summary

Light induced electron transfer of metal complexes has been studied extensively during the last decade. This interest was stimulated by attempts to develop an artificial photosynthesis for the conversion and chemical storage of solar energy. Even if this goal has not yet been achieved photochemical redox processes of coordination compounds are now much better understood. In this review the various possibilities of photoinduced electron transfer are discussed and illustrated by selected examples. A distinction is made between intra– and intermolecular electron transfer which may occur as a direct optical transition or by an excited state electron transfer mechanism.

Introduction

The study of photochemical electron transfer processes involving coordination compounds (refs. 1-17) has become an important research subject during the last 10 years. These investigations were initiated, at least partially, by the desire to create an artificial photosynthetic system for the conversion and chemical storage of solar energy (refs. 18-20). It is well known that natural photosynthesis requires a light induced electron transfer as the basic process employing chlorophyll as the key compound. In order to imitate nature we have to improve our knowledge of photoredox processes. Coordination compounds, particularly those of transition metals, are excellent candidates for such studies. Metal complexes are generally redox active. Their lowest electronically excited states are frequently luminescent under ambient conditions. These properties facilitate the investigation of photochemical electron transfer reactions.

This review will illustrate the various possibilities of photoinduced electron transfer involving metal complexes. More recent developments and the interest of the authors are emphasized. A systematic approach requires a classification. As guide lines we applied two criteria. Intra– and intermolecular electron transfer are distinguished. In addition, electron transfer may occur by a direct optical transition or after an initial internal electronic excitation of the reductant or oxidant.

Finally, the extent of electronic coupling (refs. 8, 13 and 21-23) between the electron donor and acceptor is important. If the coupling is weak a "whole" electron changes its location. In the case of strong coupling which involves some electron delocalization between donor and acceptor only a fraction of an electron may be shifted.

In many cases the primary light induced electron transfer is followed by a rapid back electron transfer. Although a permanent photochemical change does then not occur such processes can be studied by time–resolved spectroscopy. However, under suitable conditions the charge separation is followed by efficient secondary reactions which compete successfully with the charge recombination. Consequently, stable photoproducts will be formed. In this review the latter situation is emphasized.

Intramolecular Electron Transfer

Direct Optical Charge Transfer

The majority of intramolecular photoredox processes of metal complexes which have been reported take place upon direct optical charge transfer (CT) excitation (refs. 1-11). Since metal complexes consist of metal centers and ligands they represent also the redox sites. The electronic interaction between these redox sites induces the occurance of optical CT transitions which give rise to absorption bands in the electronic spectrum (ref. 24). Light absorption into these CT bands is associated with an electron transfer which may be followed by secondary reactions yielding stable photoproducts. Optical CT transitions are classified according to the redox sites (refs. 24 and 25): Ligand to Metal (LM), Metal to Ligand (ML), Metal to Metal (MM), Ligand to Ligand (LL), and Intraligand (IL). While LMCT and MLCT are the classical CT transitions MMCT, LLCT and ILCT are of more recent interest.

LMCT

LMCT absorptions appear at low energies if the ligand is reducing and the metal oxidizing (ref. 24). Most Fe^{III} complexes are characterized by long–wavelength LMCT bands. Such absorptions cause also the colors of d^0 oxometallates such as CrO_4^{2-} and MnO_4^-. By definition a LMCT transition involves the reduction of the metal and oxidation of a ligand. In suitable cases the reduced metal is stabilized by secondary processes before a charge recombination regenerates the starting complex:

$$[Co^{III}(NH_3)_5X]^{2+} \xrightarrow{H_2O} Co^{2+}aq. + 5NH_3 + X \qquad \text{(refs. 1 and 2)}$$

$$[Cr^{I}(CO)_5I] \xrightarrow{CH_3CN} [Cr^{0}(CO)_5CH_3CN] + I \qquad \text{(ref. 26)}$$

The oxidized ligand radicals X (halogen) undergo then further reactions.

MLCT

MLCT bands appear at long wavelength if the metal is reducing and a ligand provides empty orbitals at low energies (ref. 24). Complexes such as $Fe^{II}(CN)_6^{4-}$ and $Ru^{II}(bipy)_3^{2+}$ (bipy = 2,2'–bipyridyl) are typical cases. Since the metal–ligand bonding does usually not change very much MLCT excitation induces rarely intramolecular photoreactions. For the same reason MLCT states are the most prominent luminescing excited states of coordination compounds. Nevertheless, MLCT excited complexes may undergo efficient intermolecular reactions such as excited state electron transfer (see below). A special type of excited state electron transfer is the generation of solvated electrons (refs. 1, 2 and 11):

$$[Fe^{III}(CN)_6]^{4-} \rightarrow [FeIII(CN)_6]^{3-} + e^-_{solv.}$$

In the MLCT state a ligand–localized excess electron is transferred to the solvent. The solvent itself or another electron scavenger such as N_2O can be reduced irreversibly. As an alternative the electron–rich ligand of MLCT states is susceptible to an electrophilic attack. The addition of protons to coordinated carbynes illustrates this type of excited state reactivity:

$$[Os^0(CPh)(CO)(PPh_3)_2Cl]+HCl \rightarrow [Os^{II}(CHPh)(CO)(PPh_3)_2Cl_2] \qquad \text{(ref. 27)}$$

MLCT excitation is associated with a shift of electron density from the metal to the carbyne ligand. In a limiting description the equilibrated excited state can be viewed as an Os^{II} complex which contains a deprotonated carbene ligand. A stabilization takes place by the addition of a proton to the coordinated carbene anion. The Os^{II} takes up a Cl^- ligand to complete its octahedral coordination.

MMCT

Since MMCT requires at least two metal centers it occurs only in bi- or poly-nuclear complexes which contain reducing and oxidizing metals (ref. 24). Both metals may be linked by a bridging ligand which mediates the electronic interaction. Typical examples are mixed–valence compounds which contain the same metal in two different oxidation states (refs. 13 and 21-23). The color of Prussian Blue which consists of Fe^{II}–CN–Fe^{III} units is caused by a MMCT absorption involving an electronic transition from Fe^{II} to Fe^{III}. The MMCT excitation of suitable heteronuclear complexes induces a photoredox reaction (refs. 4,8):

$$[(NH_3)_5Co^{III}NCRu^{II}(CN)_5]^- \xrightarrow{H_2O} Co^{2+}aq. + 5NH_3 + [Ru^{III}(CN)_6]^{3-} \qquad \text{(ref. 28)}$$

Light absorption by the Ru^{II} to Co^{III} MMCT band creates the kinetically labile Co^{II} ammine complex. The decay of this complex competes successfully with back electron transfer. The metal–metal coupling in the Co^{III}/Ru^{II} compound is rather weak.

The absorption spectrum of the binuclear complex is composed of the spectra of its mononuclear components and the additional MMCT band (λ_{max}=375nm).

MMCT bands appear also in the spectra of polynuclear complexes which contain direct, but polar metal–metal bonds between metals of different electronegativity (refs. 29-31). Of course, the metal–metal coupling is strong in these cases and electron delocalization is only limited by the different energies of the overlapping metal orbitals. This type of polar metal-metal bond occurs in the complex $(Ph_3P)Au^I$—$Co^{-I}(CO)_4$ (ref. 29). The absorption spectrum exhibits a low–energy Co^{-I} to Au^I MMCT band. The MMCT transition involves the promotion of a σ^b(M-M) electron to a σ^*(M-M) orbital. Consequently, the MMCT ($\sigma^b \rightarrow \sigma^*$) excitation leads to a photoredox reaction which is associated with a homolytic cleavage of the polar metal–metal bond (ref. 29):

$$[(PPh_3)Au^I - Co^{-I}(CO)_4] \rightarrow Au^0 + PPh_3 + \frac{1}{2}[Co_2(CO)_8]$$

Metallic gold and cobalt carbonyl are the final products of the photolysis.

LLCT

Optical LLCT is only of recent interest (ref. 25). Such absorptions appear if one ligand is reducing and another oxidizing (L_{red}-M-L_{ox}). The metal only mediates and modifies the electronic ligand–ligand interaction. LLCT bands appear in the spectra of complexes such as $Be(bipy)X_2$ with X^-=halide or alkyl anions. The color of these compounds is caused by X^- to bipy LLCT absorptions. LLCT bands are also exhibited by ligand–based mixed–valence complexes. An interesting family of such compounds contains 1,2–diimine (*e.g.* bipy) donor and 1,2–ethylenedithiolate acceptor ligands (ref. 25):

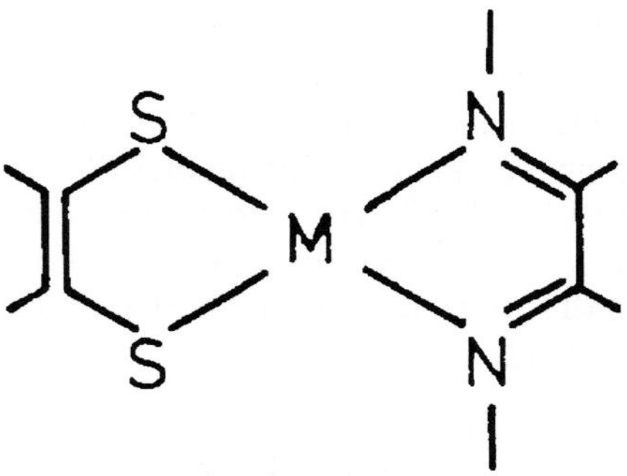

The basic electronic structure of both ligands is very similar but differs in their redox states by two electrons. The absorption spectrum of square planar d^8 complexes with M^{II}=Ni^{II}, Pd^{II}, and Pt^{II} is characterized by an intense dithiolate to diimine CT band which is strongly solvatochromic.

If M^{II} is Zn^{2+} the complexes are tetrahedral (ref. 32):

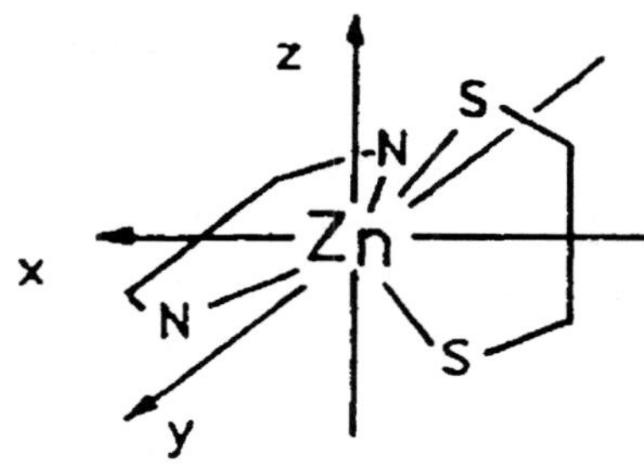

Due to the orthogonal orientation of the planes of both ligands the LLCT transition is symmetry–forbidden. The corresponding absorption is now of very low intensity. It has been suggested that these Zn^{II} complexes are good inorganic models for an efficient light–induced charge separation. The LLCT state has much in common with the so–called twisted intramolecular charge transfer (TICT) states of certain organic molecules (ref. 33).

ILCT

A ligand itself may consist of a reducing and oxidizing part (ref. 25). The spectrum of such a complex as well as that of the free ligand can then exhibit a ILCT band. Such an absorption was identified for the complex (ref. 34):

ILCT involves the transition from sulfur lone pairs to the diimine moiety.

Excited State Electron Transfer

In distinction to the direct optical charge transfer an intramolecular photoredox process may also occur by an excited state electron transfer. An internally excited chromophoric group of a coordination compound can undergo an electron transfer to or from another part of the same molecule (ref. 8). This process is usually facilitated by large driving forces. Compared to ground states the reducing and oxidizing strength of excited states is increased by the amount of the excitation energy.

In many cases the excited state electron transfer is followed by a rapid back electron transfer which regenerates the starting complex. A net photolysis is thus not observed. Permanent photoproducts are only formed if the primary electron transfer yields labile species which undergo fast secondary reactions. Co^{III} ammine complexes are well suited as acceptors for irreversible electron transfer because Co^{II} ammines are not stable in aqueous solution but decay rapidly ($k \sim 10^6 s^{-1}$) to Co^{2+}aq. (ref. 35). Aromatic molecules may be used as excited state electron donors. They can be attached to Co^{III} ammines *via* a carboxylic group which coordinates to the metal (ref. 8).

A variety of complexes of the type $[Co^{III}(NH_3)_5O_2CR]^{2+}$ with *e.g.* R=1–,2–naphthalene, 4–stilbene, 9–anthracene, 4–biphenyl was studied (refs. 8 and 36). The interaction of donor and acceptor is certainly of the inner–sphere type. However, the electronic coupling is apparently weak. The donor and acceptor occur as independent chromophores of the complex. A LMCT band involving direct optical charge transfer from the aromatic group to Co^{III} does also not appear.

Light absorption by the aromatic substituent is associated with a complete fluorescence quenching indicating an efficient electron transfer from the excited $\pi\pi^*$ singlet state of R to Co^{III} (refs. 8 and 36) (* denotes an excited state):

$$[(NH_3)_5Co^{III}O_2CR]^{2+} \xrightarrow{h\nu} [(NH_3)_5Co^{III}O_2CR^*]^{2+}$$

$$[(NH_3)_5Co^{III}O_2CR^*]^{2+} \rightarrow [(NH_3)_5Co^{II}O_2CR^{\oplus}]^{2+}$$

The electron transfer is certainly favored by a large driving force. The excited aromatic groups are very strong reductants ($E^0 < -2V$) while Co^{III} ammines are weakly oxidizing ($E^0 \sim +0.1V$). The product formation is determined by the competition between back electron transfer and decay of the Co^{II} ammine complex:

$$[(NH_3)_5Co^{II}O_2CR^{\oplus}]^{2+} \rightarrow [(NH_3)_5Co^{III}O_2CR]^{2+}$$

$$[(NH_3)_5Co^{II}O_2CR^{\oplus}]^{2+} \rightarrow 5NH_3 + Co^{2+}\text{aq.} + \text{oxidized carboxylate}$$

The quantum yield of Co^{2+} formation was very much dependent on R. A simple correlation was not apparent.

An interesting extension of this study included complexes of the type (ref. 37):

$$[(NH_3)_5Co^{III}O_2C(\text{—}CH_2\text{—})_nNHCO\text{–2–naphthyl}]^{2+} \quad n=1 \text{ to } 5$$

The excited state electron transfer within these complexes is still an intramolecular process but not by an inner–sphere mechanism since the methylene groups of the peptide linkage are electronically insulating. As a consequence the excited state electron transfer is not as efficient as that discussed above. This is indicated by the observation that the quenching of the naphthalene donor is not any more complete. Interestingly, the rate of excited state electron transfer increased from $k=4.9 \cdot 10^9\ s^{-1}$ for n=1 to $5.6 \cdot 10^9$ (n=2), $6.6 \cdot 10^9$ (n=3) and $9.2 \cdot 10^9$ (n=4) and dropped then to $6.0 \cdot 10^9\ s^{-1}$

for n=5 (ref. 38). The rate of back electron transfer is much slower: $k=1.07{\cdot}10^7\ s^{-1}$ (n=1), $1.26{\cdot}10^7$ (n=2), $2.71{\cdot}10^7$ (n=3), $4.02{\cdot}10^7$ (n=4), and $2.30{\cdot}10^7$ (n=5) (ref. 40). However, this transfer rate reaches also a maximum at n=4.

Generally, outer sphere electron transfer becomes slower with a larger distance between donor and acceptor. Our observation suggests that the actual distance between the naphthyl group and Co^{III} decreases with an increasing chain length of the peptide from n=1 to 4. It is assumed that donor and acceptor come to a closer approach by an appropriate bending of the flexible peptide bridge if n grows from 1 to 4. This approach may be favored by hydrogen bonding between coordinated ammonia of the $Co(NH_3)_5$ moiety and the carbonyl group of the peptide linkage (ref. 37):

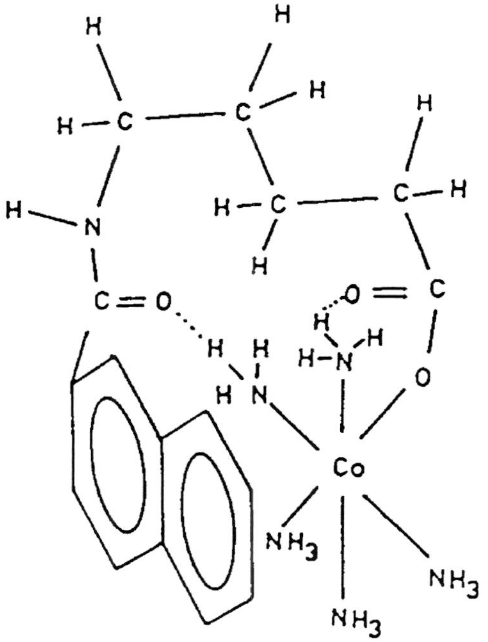

At n=5 electron transfer slows down in both directions. The donor–acceptor distance may now increase by an extension of the peptide chain.

Intermolecular Electron Transfer

While the intramolecular coupling between an electron donor and acceptor may vary considerably (ref. 8) the intermolecular interaction is generally much weaker. Nevertheless, light induced outer sphere electron transfer may take place by a direct optical CT transition or by a bimolecular encounter of an excited donor (or acceptor) with a ground state acceptor (or donor).

Direct Optical Charge Transfer

An optical intermolecular or outer sphere (OS) CT transition can occur if a reducing and an oxidizing molecule or ion are in close contact which provides some orbital overlap between donor and acceptor. This close contact is frequently facilitated by the electrostatic attraction within an ion pair (refs. 4-7, 9-11 and 17). But also neutral molecules may be close enough in suitable cases, particularly at high concentrations in solution or in the solid state (ref. 17).

In analogy to intramolecular CT transitions (see above) optical OS CT can be classified according to the predominant localization of the donor and acceptor orbitals at the metal or ligand. Under appropriate conditions OS MLCT, LMCT, MMCT, and LLCT absorption bands will be observed (ref. 17).

The ion pair $[Rh^{III}(bipy)_3]^{3+}$ $[Ru^{II}(CN)_6]^{4-}$ provides an example of OS MLCT (ref. 41). The electronic spectrum of the aqueous ion pair displays a new absorption which is assigned to a CT transition from Ru^{II} of the cyano complex anion to the bipy ligands of the rhodium complex cation. This OS MLCT band appears at much shorter wavelength (λ_{max} = 379 nm) compared to the corresponding inner sphere MLCT absorption of $Ru^{II}(bipy)_3^{2+}$ (λ_{max} = 448 nm) (ref. 24). This shift seems to reflect the larger distance between Ru^{II} and bipy in the ion pair.

The aqueous ion pair $[Co^{III}(NH_3)_6]^{3+}$ $[Ru^{II}(CN)_6]^{4-}$ shows a Ru^{II} to Co^{III} MMCT band at λ_{max} = 344 nm (refs. 17 and 42). It is quite interesting that this absorption appears at λ_{max} = 375 nm for the binuclear complex $[(NH_3)_5Co^{III}NCRu^{II}(CN)_5]^-$ (ref. 28). Again, the decreasing distance between the redox sites decreases also the energy of the CT transition. Of course, this conclusion is based on the assumption the electronic coupling is weak in both cases. Upon MMCT excitation the ion pair undergoes a photoredox decomposition quite analogous to that of the binuclear complex (see above) (refs. 17, 42 and 43):

$$[Co^{III}(NH_3)_6]^{3+}[Ru^{II}(CN)_6]^{4-} \rightarrow Co^{2+}aq + 6NH_3 + Ru^{III}(CN)_6]^{3-}$$

A large number of organometallic salts consist of an oxidizing metal carbonyl or metallocenium cation and a reducing metal carbonyl anion (ref. 17). It has been shown quite recently that the intense colors of these salts originate from OS MMCT absorptions (refs. 17 and 44). For example, the ion pair $[Co^{I}(CO)_3(PPh_3)_2]^+[Co^{-I}(CO)_4]^-$ displays an Co^{-I} to Co^{I} MMCT band at λ_{max} = 386 nm in acetone (ref. 45). The MMCT excitation creates a radical pair in the first step. These radicals are labile and undergo ligand dissociation and exchange reactions before they finally to form a metal–metal bond (ref. 45):

$$[Co^{I}(CO)_3(PPh_3)_2]^+[Co^{-I}(CO)_4]^- \rightarrow [Co^{0}(CO)_3(PPh_3)_2][Co^{0}(CO)_4]$$

$$[Co^{0}(CO)_3(PPh_3)_2][Co^{0}(CO)_4] \rightarrow [(PPh_3)(CO)_3Co^{0}\text{–}Co^{0}(CO)_3(PPh_3)]\text{+}CO$$

It is now recognized that light–induced electron transfer plays an important role in organometallic chemistry (refs. 17 and 44).

Excited State Electron Transfer

As pointed out above any electronically excited molecule is a much stronger reductant and oxidant than the same molecule in its ground state. Accordingly, intermolecular excited state electron transfer may take place while it does not occur in the ground state owing to thermodynamic limitations. However, the diffusional encounter must take place before the excited molecule is deactivated to the ground state. Generally, only the lowest excited states are long–lived enough to undergo a bimolecular electron exchange. If the molecule is luminescent the emission quenching by an acceptor or donor can be used to obtain kinetic data on the excited state electron transfer.

The first example of a transition metal complex which participates in excited state electron transfer was reported by Gafney and Adamson in 1972 (ref. 46):

$$[Ru^{II}(bipy)_3{}^{2+}]^* + [Co^{III}(NH_3)_6]^{3+} \rightarrow [Ru^{III}(bipy)_3]^{3+} + Co^{2+}aq. + 6NH_3$$

Since then numerous studies of intermolecular excited state electron transfer were carried out. The majority of investigations involved $Ru(bipy)_3{}^{2+}$ and related complexes as excited donors or acceptors (refs. 9, 10, 12, 47 and 48). The ruthenium complex (refs. 47 and 48) offers many advantages for such studies. It can be reversibly reduced and oxidized. The lowest excited state which is a MLCT triplet undergoes an efficient and relatively slow (τ~1μs) phosphorescence under ambient conditions. The excited state redox potentials for the couples $Ru(bipy)_3{}^{2+*/3+}$ and $Ru(bipy)^{3+/2+*}$ are E^0 = = -0.86 V and E^0 = +0.84 V. In aqueous solution the complex is thermally and photochemically quite stable.

However, all attempts to utilize $Ru(bipy)_3{}^{2+}$ as sensitizer for the conversion and chemical storage of solar energy did not lead to any practical applications. The light induced charge separation yields primary redox products with a large potential difference which favors a rapid electron transfer. The absorbed light is then simply converted to heat. All attempts to slow down the charge recombination and to couple the primary electron transfer step to secondary reactions yielding kinetically stable products with a high energy content were not yet successful. Most efforts were directed toward the photochemical splitting of water. While first results were rather promising the final goal has not yet been achieved. Nevertheless, as a spin–off these investigations were beneficial for a better understanding of electron transfer reactions. The theories developed by Marcus, Hush, and others could be tested and extended (ref. 49).

At this point we will briefly discuss an example of intermolecular excited state electron transfer to illustrate the mechanism. In this case a comparison with a corresponding intramolecular electron transfer is quite interesting and informative. Naphthalene carboxylate cannot only be used as sensitizer for intramolecular (see above) but also for intermolecular excited state electron transfer to Co^{III} ammine complexes such as $Co^{III}(NH_3)_5O_2CCH_3^{2+}$. The sequence of events may be described by the following simplified scheme (NC^-=1–naphthalenecarboxylate) (ref. 50):

$$NCH^* + [Co^{III}(NH_3)_5O_2CCH_3]^{2+} \rightarrow NCH^+ + [Co^{II}(NH_3)_5O_2CCH_3]^+$$

$$NCH^+ + [Co^{II}(NH_3)_5O_2CCH_3]^+ \rightarrow NCH + [Co^{III}(NH_3)_5O_2CCH_3]^{2+}$$

$$NCH^+ + [Co^{II}(NH_3)_5O_2CCH_3]^+ \rightarrow NCH^+ + Co^{2+}aq + 5NH_3 + CH_3COO^-$$

Electron transfer from the excited $\pi\pi^*$ singlet of NCH to the Co^{III} complex is associated with the fluorescence quenching of NCH and the irreversible formation of Co^{2+}. The analysis of the Stern–Volmer plot (Fig. 1) shows that the excited state electron transfer is very fast ($k=3.6\cdot109\ M^{-1}\ s^{-1}$) (ref. 38) and takes place with an efficiency of almost unity.

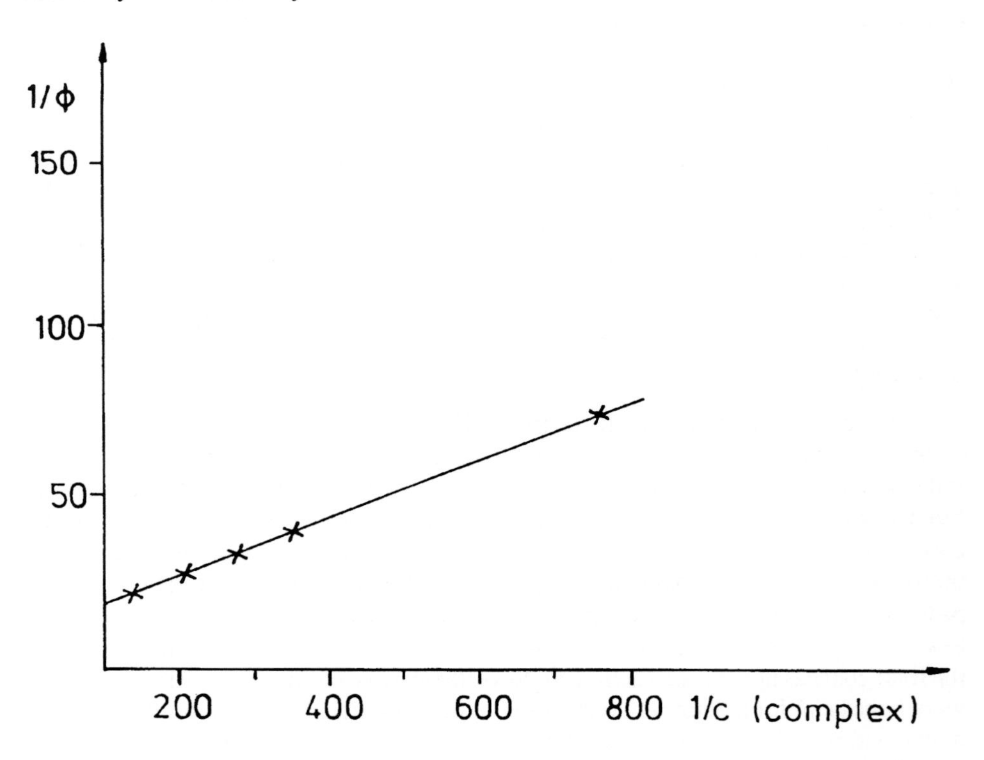

Fig. 1. Excited state electron transfer from 1–naphthalinecarboxylic acid ($8\cdot10^{-4}$M) to $Co(NH_3)_5$ acetate^{2+} (c in M) in a mixture of CH_3OH/H_2O (1:1) and $HClO_4$ (0.01 M); Stern–Volmer plot of the reciprocal quantum yield (ϕ) of Co^{2+} formation *vs.* the reciprocal concentration of $Co(NH_3)_5$ acetate^{2+}.

This is a consequence of the large driving force. In its first excited singlet state naphthalene is strongly reducing ($E_{1/2}$ = -2.25 V *vs.* SCE) while Co^{III} ammine complexes are weak oxidants ($E^0 \sim 0.1$ V) (ref. 37). On the contrary, the back electron transfer is much slower ($k=4\cdot10^6 M^{-1}s^{-1}$) (ref. 40) and reduces the limiting quantum yield for Co^{2+} production to $\phi = 0.2$.

The intramolecular excited state electron transfer in $[Co^{III}(NH_3)_5NC]^{2+}$ must be also rather fast and efficient since the fluorescence of the coordinated NC^- is completely quenched. However, the quantum yield of Co^{2+} formation ($\phi = 0.005$) is now much lower. If all other processes occur with the same rates for inter– and intramolecular electron transfer the charge recombination in $Co(NH_3)_5NC^{2+}$ is much faster ($k = = 2\cdot10^8 s^{-1}$) than that for the bimolecular reaction (see above). At moderate driving forces the electron transfer is apparently facilitated by an inner–sphere mechanism.

The systems which were studied for applications in solar energy conversion suffer by an efficient energy–wasting charge recombination. In order to slow down this process various approaches were suggested (ref. 51). In natural systems electron transfer across an interface which may be a cell membrane seems to be important. In an attempt to imitate nature many studies were devoted to heterogeneous excited state electron transfer (ref. 51). Microheterogeneous environments were created by the application of micelles, vesicles, microemulsions and similar systems. A very simple model for a membrane is formed by the interface between two immiscible solvents. If the electron donor is soluble only in one phase and the acceptor in the other solvent an excited state electron transfer may take place across the interface. We explored this possibility and dissolved naphthalene in hexadecane and $Co(NH_3)_6^{3+}$ in water (0.1 M HCl) (ref. 52). In a simple pyrex beaker both solutions formed two immiscible layers. A vertical light beam passed the layer and was absorbed by the naphthalene in the top layer near the interface. As a result $Co(NH_3)_6^{3+}$ was reduced in the aqueous layer. The excited naphthalene apparently transferred an electron to $Co(NH_3)_6^{3+}$ across the interface of both immiscible solvents. A decrease of concentration of $Co(NH_3)_6^{3+}$ led also to a drop of the quantum yield of Co^{2+} formation. A nearly linear Stern–Volmer relationship was obtained when $1/[Co(NH_3)_6^{3+}]$ was plotted *versus* $1/\phi_{rel.}(Co^{2+})$. However, naphthalene is soluble in water to a small extent. Accordingly, Co^{2+} was not only generated by heterogeneous electron transfer across the interface but also by homogeneous electron transfer in the aqueous phase. Since we were not able to separate both processes the mechanistic implications of these observations are not yet clear.

Acknowledgment

Support of this research by the Deutsche Forschungsgemeinschaft and the Fonds der Chemischen Industrie is gratefully acknowledged.

References

1 V. Balzani and V. Carassiti *Photochemistry of Coordination Compounds* Academic Press, New York, 1970.

2 A.W. Adamson and P.D. Fleischauer (Eds.) *Concepts of Inorganic Photochemistry* Wiley, New York, 1975.

3 G.L. Geoffroy and M.S. Wrighton *Organometallic Photochemistry* Academic Press, New York, 1979.

4 A. Vogler, A.H. Osman and H. Kunkely, Coord., Chem. Rev., **64** (1985) 159.

5 H. Hennig, D. Rehorek and R.D. Archer, Coord. Chem. Rev., **61** (1985) 1.

6 V. Balzani, N. Sabbatini and F. Scandola, Chem. Rev., **86** (1986) 319.

7 H. Hennig, D. Rehorek and R. Billing, Comments Inorg. Chem., **8** (1988) 163.

8 A. Vogler, in: M.A. Fox and M. Chanon (Eds.) *Photoinduced Electron Transfer* Elsevier, Amsterdam, 1988, part D, p. 179.

9 N. Serpone, in: M.A. Fox and M. Chanon (Eds.) *Photoinduced Electron Transfer* Elsevier, Amsterdam, 1988, part D, p. 47.

10 V. Balzani and F. Scandola, in: M.A. Fox and M. Chanon (Eds.) *Photoinduced Electron Transfer* Elsevier, Amsterdam, 1988, part D, p. 148.

11 C. Gianotti, S. Gaspard and P. Krausz, in: M.A. Fox and M. Chanon (Eds.) *Photoinduced Electron Transfer* Elsevier, Amsterdam, 1988, part D, p. 200.

12 V. Balzani, F. Bolletta, M.T. Gandolfi and M. Maestri, Top. Curr. Chem., **75** (1978) 1.

13 T.J. Meyer, Acc. Chem. Res., **11** (1978) 94.

14 N. Sutin and C. Creutz, Pure Appl. Chem., **52** (1980) 2717.

15 T.J. Meyer, Progr. Inorg. Chem., **30** (1983) 389.

16 N. Sutin and C. Creutz, J. Chem. Ed., **60** (1983) 809.

17 A. Vogler and H. Kunkely, Top. Curr. Chem., **158** (1990) 1.

18 M. Grätzel, in: M.A. Fox and M. Chanon (Eds.) *Photoinduced Electron Transfer* Elsevier, Amsterdam, 1988, part D, p. 394.

19 T.J. Meyer, Acc. Chem. Res., **22** (1989) 163.

20 M. Gratzel *Heterogeneous Photochemical Electron Transfer* CRC Press, Boca Raton, 1989.

21 D. Brown (Ed.) *Mixed–Valance Compounds* Reidel, Dordrecht, 1980.

22 C. Creutz, Prog. Inorg. Chem., **30** (1980) 1.

23 N.S. Hush, Progr. Inorg. Chem., (1967) 391.

24 A.B.P. Lever *Inorganic Electronic Spectroscopy* Elsevier, Amsterdam, 1984.

25 A. Vogler and H. Kunkely, Comments Inorg. Chem., **9** (1990) 201.

26 H. Kunkely and A. Vogler, J. Organomet. Chem., **378** (1989) C15.

27 A. Vogler, J. Kisslinger and W.R. Roper, Z. Naturforsch., **38b** (1983) 1506.

28 A. Vogler and H. Kunkely, Ber. Bunsenges. Phys. Chem., **79** (1975) 83.

29 A. Vogler and H. Kunkely, Z. Naturforsch., **44b** (1989) 132.

30 A. Vogler and H. Kunkely, J. Organomet. Chem., **355** (1988) 1.

31 H. Kunkely, G. Stochel and A. Vogler, Z. Naturforsch., **44b** (1988) 145.

32 R. Benedix, H. Hennig, H. Kunkely and A. Vogler, Chem. Phys. Letters, **175** (1990) 483.

33 W. Rettig, Angew. Chem. Int. Ed. Engl., **25** (1986) 971.

34 A. Vogler and H. Kunkely, Inorg. Chim. Acta, **54** (1981) L273.

35 M. Simic and J. Lilie, J. Am. Chem. Soc., **96** (1974) 291.

36 A. Kern, Ph. D. thesis, University of Regensburg, 1978.

37 A.H. Osman and A. Vogler, in: H. Yersin and A. Vogler (Eds.) *Photochemistry and Photophysics of Coordination Compounds* Springer, 1987, p. 197.

38 The calculations were based on the assumption that the fluorescence of the free naphthalene chromophore decays with $k = 10^8\ s^{-1}$. This is an average for many naphthalene derivatives (ref. 39).

39 I.B. Berlman *Handbook of Fluorescence Spectra of Aromatic Molecules* Academic Press, New York, 1971.

40 For the calculations we assumed that Co^{II} ammine complexes decay with $k = 10^6 s^{-1}$. This is, however, only a lower limit (ref. 35).

41 A. Vogler and H. Kunkely, Inorg. Chem., **26** (1987) 1819.

42 A.H. Osman, Ph. D. thesis, University of Regensburg, 1987.

43 A. Vogler and J. Kisslinger, Angew. Chem. Int. Ed. Engl., **21** (1982) 77.

44 T.M. Bockmann and J.K. Kochi, J. Am. Chem. Soc., **111** (1989) 4669.

45 A. Vogler and H. Kunkely, Organometallics, **7** (1988) 1449.

46 H.D. Gafney and A.W. Adamson, J. Am. Chem. Soc., **94** (1972) 8238.

47 K. Kalyanasundaram, Coord. Chem. Rev., **46** (1982) 159.

48 A. Juris, V. Balzani, F. Barigelletti, S. Campagna, P. Belser and A. von Zelewsky, Coord. Chem. Rev., **84** (1988) 85.

49 V. Balzani (Ed.) *Supramolecular Photochemistry* D. Reidel Publishing Company, Dordrecht, 1987.

50 S. Schäffl and A. Vogler, unpublished results.

51 M. Grätzel in ref. 49, p. 394.

52 A. Vogler and P.C. Ford, unpublished results.

Electron and Proton Transfer in Chemistry and Biology, edited by A. Müller et al.
Studies in Physical and Theoretical Chemistry, Vol. 78

Coordinative Aspects of Single Electron Exchange between Main Group or Transition Metal Compounds and Unsaturated Organic Substrates

Wolfgang Kaim

Institut für Anorganische Chemie
Universität Stuttgart
Pfaffenwaldring 55
D-7000 Stuttgart 80 (Germany)

Summary

Several examples are presented for inner sphere and outer sphere electron transfer in the reactions between electron rich organometallic main group or transition element compounds and electron deficient unsaturated substrates. Emphasis is placed on various steric and electronic aspects of (multiple) coordination between metal centers and the substrate as a requirement or consequence of intra– and intermolecular single electron exchange. The examples include thermal and photoinduced electron transfer processes which result in various new stable dia– and paramagnetic follow–up products.

Introduction

Single electron transfer is now becoming widely recognized as a fundamental and sometimes separable step of reaction sequences in various areas of chemistry, (refs. 1-4). This article focuses on single electron transfer processes between coordinated organic acceptor ligands and electron rich (organo)metal fragments or parts thereof, *e.g.* the metal center proper or other co–ligands.

Organometallic complexes with their electronically flexible ligands (→ delocalization, covalent bonding) and often open coordination shells (ref. 5) are ideal candidates for inner sphere electron transfer reactivity. The term "inner sphere" was originally derived by Taube, (ref. 6), for reactions like (1) in classical coordination chemistry in order to rationalize facilitated electron transfer between two metal centers which can share a common bridging ligand (1), thus implying the existence of a three–component precursor complex M—(μ-L)—M'.

$$\mathrm{M{-}(L)} + \mathrm{M'} \longrightarrow \underset{\text{precursor complex}}{\mathrm{M{-}(\mu\text{-}L){-}M'}} \longrightarrow \mathrm{M^-} + \mathrm{[(L){-}M']^+} \qquad (1)$$

e.g. $M = (NH_3)_5Co^{3+}$, $L = Cl^-$, $M' = Cr(aq)^{2+}$

In what can be called (ref. 7) an "inverse inner sphere" situation (2), one may formulate an analogous three component precursor complex L-M-L' or R_nM-Su (6) between the coordinatively unsaturated (organo)metal fragment L-M = R_nM and the bonded organic acceptor substrate L' = Su; the precursor complex should be poised for the then intramolecular electron transfer which can be triggered thermally, photolytically or chemically, *e.g.*, by protonation, solvent or "salt" effects (cf. below).

$$\begin{array}{ccc} L{-}M + L' \longrightarrow & L{-}M{-}L' \longrightarrow & L^{+\bullet} + (M{-}L')^{-\bullet} \\ & \text{precursor complex} & \downarrow \\ & & (L{-}M)^{+\bullet} + (L')^{-\bullet} \text{ or} \\ & & L^{\bullet} + (M{-}L')^{\bullet} \end{array} \qquad (2)$$

L : organometallic (C–bonded) ligand; L' = acceptor ligand

This kind of intra–complex electron transfer is very likely a fundamental elementary step in reaction sequences involving metal catalysis; many oxidative addition/reductive elimination processes and reactions between coordinated ligands (ref. 8) may contain such elementary steps (ref. 5).

As a spin–off from the postulate of a classical inner sphere precursor complex M-L-M' followed (ref. 6) the synthesis and recognition of mixed–valence dimers M^{n+}-L-$M^{(n+1)+}$ such as the Creutz–Taube ion (3A) or its organometallic analogue (3B) (ref. 9) as attractive probes for metal–to–metal charge transfer (MMCT, excited state situation) and electron (de)localization (ground state phenomenon). A similar problem for the "inverse–type" situation (2) has been recognized in the form of the singly reduced or MLCT–excited states of complexes ML_3 (4), *e.g.* M = Ru^{2+}, L = 2,2'–bipyridine (ref. 10).

A: $[(H_3N)_5Ru\text{-}N\bigcirc N\text{-}Ru(NH_3)_5]^{5+}$

(3)

B: $[mer\text{-}(OC)_3(P^iPr_3)_2W\text{-}N\bigcirc N\text{-}W(P^iPr_3)_2(CO)_3]^{+}$

$$ML_3 \quad + \quad e^- \quad \longrightarrow \quad M(L)_2(L^{-\bullet})$$

$$ML_3 \quad + \quad h\nu \quad \underset{MLCT}{\longrightarrow} \quad *[M^{+}(L)_2(L^{-\bullet})] \qquad (4)$$

In a recent comprehensive article, (ref. 2), Kochi has elaborated on the different conceptual approaches towards thermal and photoinduced electron transfer, including the inner sphere/outer sphere formalism. It is the purpose of this article to present evidence for the

- possibility to find and study isolable but still controlled electron transfer–reactive precursor complexes of inverse inner sphere electron transfer reactions (2), the
- coordinative requirements and consequences of strongly inner sphere enhanced electron transfer, the
- role of paramagnetic radical products resulting from spin decoupling, the
- effects of apparently innocent solvent molecules and of additional ions in the medium, the
- chemical and spectroscopic criteria for less obvious electron transfer, and the
- opportunities leading to interesting new chemistry and new products from such processes.

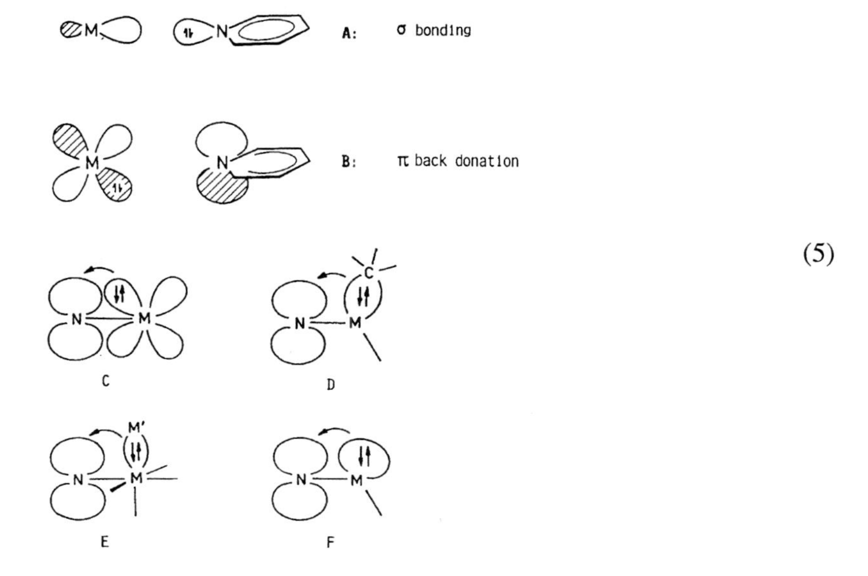

(5)

While the electron rich metal fragments described here range from main group alkyls to organometallic and related inorganic transition metal species, the substrates are invariably unsaturated acceptor molecules with lone pair–containing heteroatoms for metal coordination. This latter restriction turned out to be useful in terms of geometrical restraint, the σ/π separation (5A,B) facilitating simplified rationalizations within well established concepts of bonding in organometallic coordination chemistry (ref. 5).

Single electron transfer to accessible (energy match, symmetry) π^* orbitals of coordinated substrates may occur from filled d orbitals (5C), metal–element σ bonds (5D,E) or unshared electron pairs (5F). Consequences of electron removal include simple one–electron oxidation (first row transition metal centers, see below), substitutional activation (bond cleavage, refs. 11 and 13) and rehybridization.

Main Group Organometallics

Although many synthetically useful "carbanion" reactions of organolithium, –magnesium, –aluminum compounds R_nM may proceed *via* organometal–to–substrate single electron transfer, (refs. 1, 7 and 12) there have been only few cases where the majority of the expected intermediates could be detected or even isolated (ref. 13).

In the prototypical sequence (6), (ref. 7), the organometallic donor R_nM and the acceptor Su may form a precursor complex R_nM-Su which should be distinguished by a low–energy charge transfer excitation energy HOMO(R_nM) $\rightarrow$ LUMO(Su): Irradiation into that absorption band or thermal activation may lead to electron exchange (ref. 2) and a pair of oppositely charged radical ions will result if both components were neutral and diamagnetic at the beginning (for electron transfer between an ion pair, cf. sequence 19).

$$R^{\bullet} + (R_{n-1}M^{+})(Su^{-\bullet})$$

$$\uparrow$$

$$^{*}[(R_nM^{\delta+})--(Su^{\delta-})] \qquad\qquad R_nM^{+\bullet} + Su^{-\bullet}$$

charge transfer ↑ hν (from R_nM--Su to the excited state; excited state ↘ to radical-ion pair) — escape, dissociation ↑ (from radical-ion pair to $R_nM^{+\bullet} + Su^{-\bullet}$)

$$R_nM + Su \longrightarrow \underset{\text{charge transfer complex}}{R_nM--Su} \longrightarrow \underset{\text{radical-ion pair}}{(R_nM^{+\bullet})--(Su^{-\bullet})} \tag{6}$$

cage, collapse ↓

$$R_{n-1}M-Su-R$$

$$\downarrow$$

$$R_{n-1}M^{+} + Su-R^{-}$$

The reactivity of the primary radical (ion) pair after electron transfer is generally quite high. Bond cleavage may occur between the substrate and the metal fragment or within the metal fragment alone (2) or within the reduced substrate. In principle, this electron transfer–activated entity, existing perhaps only very briefly inside a solvent cage, has then the alternatives

i) to revert to the starting material, dissipating the energy as heat or emitted light (emission, cf. electron transfer induced chemiluminescence (ref. 14)),

ii) to recombine or "collapse" within the solvent cage in a different, rearranged form, leading to a new diamagnetic product complex, or

iii) to dissociate, *e.g.*, along a homolytically cleaved metal–carbon bond and let the radical products "escape" into solution where they may react further or be detected spectroscopically.

If products according to alternative iii) cannot be detected at all, the identification of a single electron transfer pathway in a reaction becomes very hard. Indeed, there is a fuzzy area (ref. 15) between true single electron transfer (SET) and "normal", "polar" electron pair shift mechanisms which depends, among other things, on the time scale considered. There may well be SET and conventional pathways coexisting for a given reaction.

The bond between an alkyl or aryl carbon center and an electropositive metal is very susceptible towards cleavage by electron transfer because, as an electron rich "carbanionic" σ bond, its propensity is to lose an electron ($\sigma^2 \rightarrow \sigma^1$) which normally results in bond cleavage (refs. 1, 2). The term "carbanion", although still frequently used in organic synthetic chemistry, is misleading in so far as the "carbanion"–stabilizing metal is an indispensable corollary and often that part of the $M^{\delta+}$–$C^{\delta-}$ bond that is more active in terms of synthetic selectivity (ref. 16).

An Example for Chelate–Stabilization of the Precursor Complex and for Chelate–Enhancement of Electron Transfer Reactivity

The new example (7), (ref. 13), offers several advantages over related, previously studied systems, (refs. 1, 7), including

- variable stability, even isolability and therefore structural and spectroscopic characterization of precursor complexes,
- the possibility of thermal and photolytical activation,
- the complete characterization of different paramagnetic products and a synthetically attractive dichotomy of diamagnetic products.
- In addition, the reaction (7) involves relatively simple components, diorganozinc compounds R_2Zn and symmetrical hetero-1,4-dienes such as R'N=CH-CH=NR', which can be treated theoretically with some confidence (ref. 13d).

Organozinc compounds have recently enjoyed a renaissance as stereoselective reductants and alkylating agents in organic synthesis (refs. 13c, 17). Diorganozinc compounds are not only among the earliest, (ref. 5a), but also among the most simple organometallics, they exist as linear non–associated molecules in hydrocarbon solutions and in the gas phase (ref. 18). The metal atom is coordinatively unsaturated, however, and it can add *e.g.* two n donor centers. Such additional coordination to the closed shell (d^{10}) metal center lowers the symmetry and thus creates a dipole moment between the electropositive metal center and the "carbanionic" organic groups, (ref. 17), enhancing corresponding reactivity. If the (π-)unsaturated substrate contains n donor center(s), the electrostatic effect of (σ-)coordinative binding (5) of a principally electropositive metal center is transformed into some amount of electron withdrawal from the π system as well (σ/π coupling as second order effect), (ref. 19), making it more susceptible for reduction not only by external electrons but also by a "carbanionic", already metal–coordinated alkyl or aryl group. A synergistic tendency is thus produced towards the reaction between two coordinated ligands at an electron transfer–innocent metal with no strongly preferred coordination geometry, *i.e.*, in the absence of ligand field effects. The synergism mentioned is apparent experimentally as a lowering of the oxidation potential of the organometallic fragment due to ligand n donation and geometry change (*e.g.*, bending of a linear molecule) and as facilitated reduction of the unsaturated ligand due to coordination of the electropositive metal. Energetically, this means a narrowing of the HOMO-LUMO gap with the energy required coming from the formation of coordinative bonds.

In an initially formalistic approach, the process described may then be separated into a single–electron transfer step in which the substrate forms an anion radical by electron uptake into the π^* LUMO whereas the metal–carbon σ bond, the complex HOMO, loses an electron. The latter leads to formation of a neutral carbon radical (which can then dimerize, disproportionate, or abstract a proton from another molecule) and an organometal fragment as a solvent– or anion radical–coordinated entity.

Evidence for such an electron transfer pathway has been gained from an extensive study, (ref. 13d), of reaction (7) using diorganozinc compounds and chelating 1,4-diazabutadienes (dab). The synthetically useful reactions, (ref. 13c), of diorganozinc with such non–aromatic α–diimine compounds and related conjugated heterodienes (stereospecific β lactame synthesis!, ref. 20), yield various amounts of dia– and paramagnetic products (7). The ESR–distinguishable paramagnetic products in the reaction of 1,4-di-tert.-butyl-1,4-diazabutadiene **1** and R_2Zn **2**, beyond the precursor complex **3** *via* an assumed trigonal pyramidal transition state **4** are the neutral complex **7** and the anion radical complex **8** which may be formed by a "carbanion transfer"–equilibrium (7). Diamagnetic products include, besides the C–C–dimerized neutral radical **7·7** (equilibrium), (ref. 13b), the N– and the C–alkylated complexes **5** and **6**.

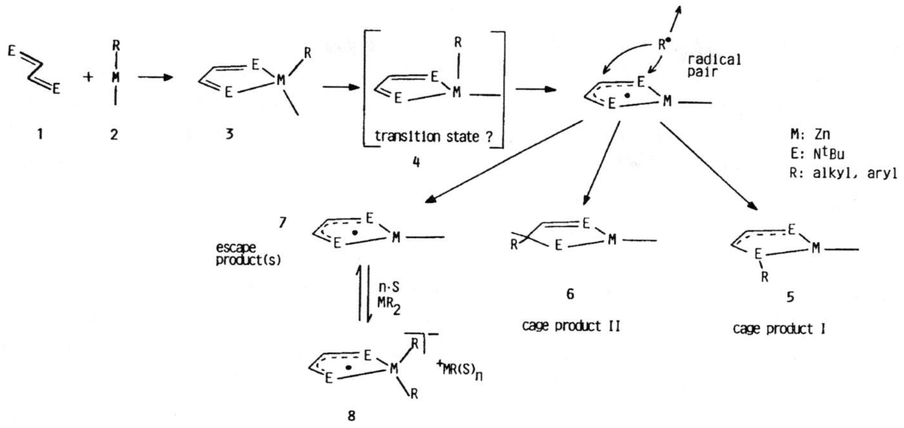

1 2 3 4 5 6 7 8

transition state ?

radical pair

escape product(s)

cage product II

cage product I

n·S
MR_2

$^+MR(S)_n$

M: Zn
E: N^tBu
R: alkyl, aryl

(7)

The reaction is presumed (ref. 13d) to proceed as follows: The open chain diimine **1** in its s–trans conformation reacts with diorganozinc compounds **2** to yield an often isolable chelate precursor complex **3** with tetracoordinate metal and s–cis conformation of the heterodiene. The structure analysis, (ref. 13d), of a Me_2Zn(dab) complex shows that this complex contains a planar five membered chelate ring and is thus particularly activated for thermal or photoinduced electron transfer: Pseudopotential calculations, (ref. 13d), reveal the antibonding combination of Zn—C σ bonds as the HOMO and the π^*_1 orbital of the heterodiene as the LUMO (which is occupied by a single electron in the persistent anion radical complex **8**).

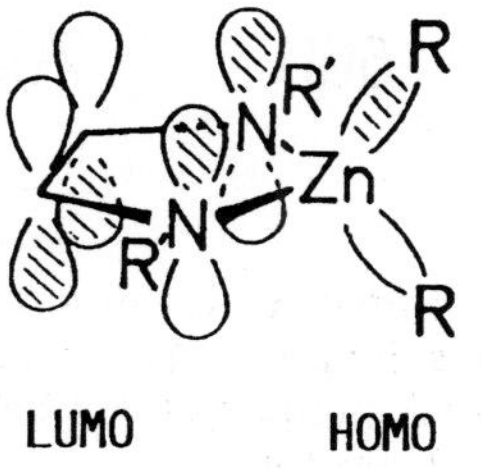

(8)

In the rigid chelate situation of the precursor complex **3** with no rotational isomerism possible, the HOMO and LUMO can almost optimally overlap in a symmetry–allowed way (8) which explains the susceptibility of this seemingly inert type of complex, composed out of neither very electron poor acceptors ($E_{red} \approx -1.8$ V *vs.* SCE) nor very electron rich organometallics ($E_{ox} \approx +1.0$ V *vs.* SCE, there are also electron–transfer reactive magnesium and aluminium analogues, refs. 7, 13c, 21 and 22) to intra–complex electron transfer:

Irradiation into the fairly intense ($\varepsilon > 1000\ M^{-1}cm^{-1}$) long–wavelength charge transfer band (HOMO→LUMO excitation) readily produces radical species (Fig. 1) and a characteristic spectrum of diamagnetic products, especially C–alkylated species and dimerized neutral radical complexes which, as loose cage and escape products, are believed to result from a less tightly bound intermediate.

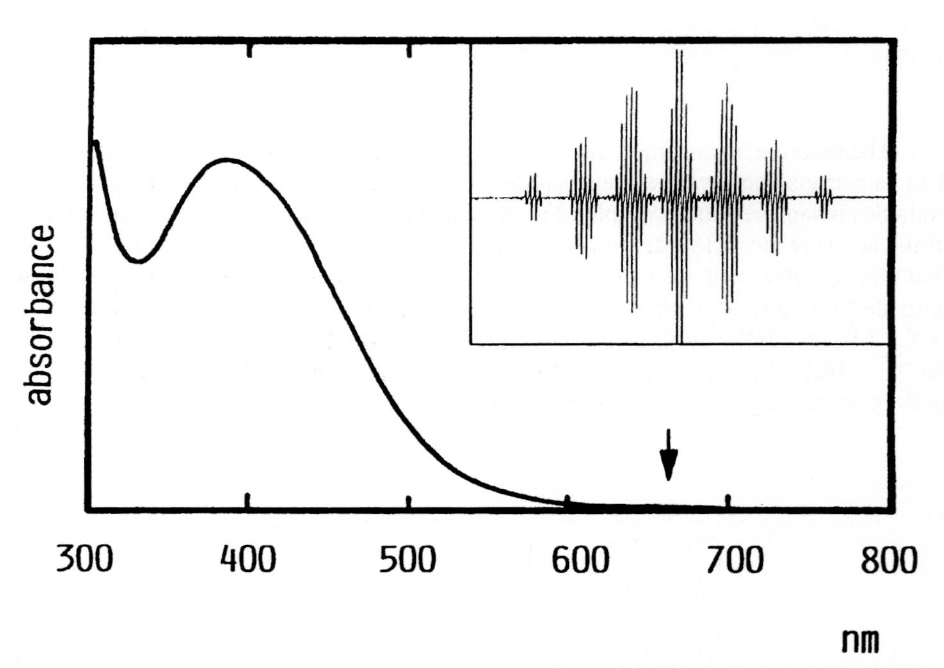

Fig. 1. Absorption spectrum of the dimethylzinc complex of 1,4-di-tert.-butyl-1,4-diazabutadiene in benzene with a charge transfer band maximum at 388 nm ($\varepsilon > 1000\ M^{-1}cm^{-1}$). Irradiation into the long wavelength tail (665 nm) produces the ESR–detectable neutral radical complex (insert spectrum).

In agreement with the notion of a metal–to–(dab–)ligand or "carbanion"-to-(dab-)ligand charge transfer, pseudopotential calculations of the triplet excited state have shown a significant lengthening of the metal carbon bond, accompanied by a decreased carbon/metal/carbon angle (ref. 13d). The photoprocess may thus be designated as a ligand–to–ligand charge transfer (LLCT) of a d^{10} metal complex; few such examples and results as concerns photoreactivity have been described (ref. 23).

The thermal reaction requires widely varying amounts of activation energy: The (tert.–Bu)$_2$Zn complex of 1,4–di–tert.– butyl–1,4–diazabutadiene reacts even at 200 K, the Me_2Zn complex converts slowly at ambient temperatures, and the complexes of $(Me_3SiH_2C)_2Zn$ and diarylzinc are thermally stable even at elevated temperatures, (refs. 13c-e, 22). While increasing reactivity parallels decreasing HOMO–LUMO differences in the series Me, Et, iso–Pr, tert.–Bu, as is also evident from strongly bathochromically shifted charge transfer energies,(ref. 13c), the other more complicated substituents indicate that there are additional steric reasons that can strongly influence the rate of the reaction. Both types of more special ligands, the trimethylsilylmethyl and the aryl (phenyl, tolyl, mesityl) groups have a strong conformational preference for a position with their main plane perpendicular to a planar entity to which they are bonded. It thus can be assumed that the thermal pathway is a tighter one than the photochemical alternative, it may even be a pathway where the distinction between inner–sphere electron transfer and "concerted" carbanion transfer gets blurred (ref. 15). However, even the thermal reactions, if they do occur, are accompanied by small amounts (2%), (ref. 13b,c), of obvious "escape" products, *i.e.* radical complexes and their dimers. On the other hand, those complexes which are not reactive thermally still produce some radical products on irradiation (ref. 13d).

The transition state geometry of the initially distorted tetrahedral metal center, (ref. 13d), for **3** on the way to a trigonal planar three coordinate species in the relaxed primary radical product **7** is assumed to be trigonal pyramidal with a long bond to the apex (7). Such an assumption does not seem unreasonable; there are structural precedents in other activated tetracoordinate main group organometallic species such as the tetramethylstannane radical cation Me_4Sn^{+}, (9), (refs. 2, 24) and in tetracoordinate copper centers activated for electron transfer ($Cu^{I/II}$: biological "blue" copper centers in proteins, (10), (ref. 25)), or metal–to–ligand ($Cu^{I}\rightarrow L$) charge transfer luminescence in the solid state of dinuclear complexes (11, distances in pm),(ref. 26).

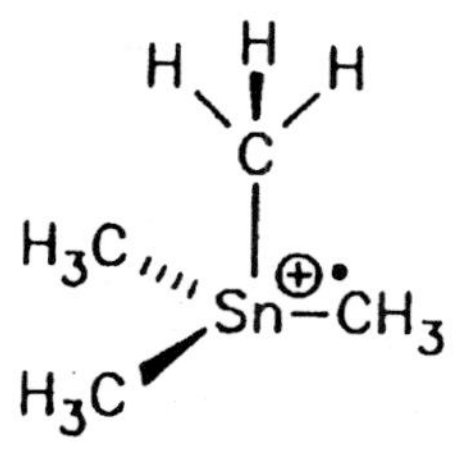

(9)

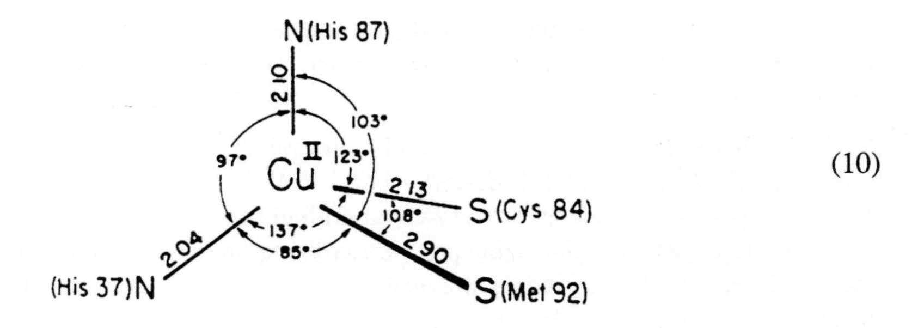

(10)

P = PPh_3

$]^{2+}$

$(BF_4^-)_2$

(11)

An approximately trigonal pyramidal transition state with an elongated bond to the pyramid apex would

- ensure an even better overlap than depicted in (8) between one metal–carbon σ bond to be cleaved and the π^* acceptor orbital,
- explain the activation barrier for aryl– and particularly mesityl–zinc as compared to tert.–butyl–zinc complexes to undergo intra–complex reactions because of the difficulty associated with their standing orthogonal to a plane (steric hindrance, 12).

tert.-butyl mesityl

(12)

Attractive as this hypothesis may be, at the present level of calculations, (ref. 13d), the question remains open as to what extent a single electron exchange occurs before or during the necessary breaking of one metal–carbon bond. A more detailed theoretical analysis of possible reaction pathways and different transition states will have to answer such questions, in any case, the two diamagnetic products **5** and **6** (7) were calculated to be close in total energy (ref. 13d).

The mechanism (7) invites a comparison with general metal catalysis. Just as in homogeneous (organo)metal catalysis of reactions between coordinated ligands, (refs. 5, 8), the function of even a non–redox active main group metal center in promoting electron transfer can be divided, for simplicity, into two separate parts.

$$\mathrm{L} + \mathrm{Su} \xrightarrow[\;]{\mathrm{M}\;\; E_a(c)} \underset{\text{precursor complex}}{\mathrm{L{-}M{-}Su}} \xrightarrow{E_a(r)_{cat}} \underset{\text{successor complex}}{\mathrm{L}\cdots\mathrm{M}\cdots\mathrm{Su}} \xrightarrow{E_a(d)\;\; \nearrow \mathrm{M}} \mathrm{L{-}Su} \qquad (13)$$

$$\mathrm{L} + \mathrm{Su} \xrightarrow{E_a(r)} \mathrm{L{-}Su}$$

M: metal center of a coordinatively unsaturated complex (with co–ligands)

The first role of the metal is to bring the two reactants together by coordination of both substrates as bound ligands for a sufficient length of time. This function of the metal is directed against the random diffusion in dilute solution where the probability for an encounter of the two substrates in the right orientation and with the right energy may be very low. On a more detailed level, the metal can — as a template — bind the two substrates in such a mutual spatial orientation that the orbital overlap necessary for electron transfer from one part of the complex to the other is strongly favoured, and that the desired reaction can thus occur with little activation energy, *i.e.* with little geometrical (translational, rotational) reorientation (→chelate effect). Optimal is a situation in which the high energy precursor complex (energy from formation of coordinative bonds as driving force !) assumes a strained geometry which is already close to the transition state of the desired reaction (entatic state of the catalyst/substrate system) (ref. 25). The geometric constraints at the metal center may be different according to the reaction to be catalyzed; while some processes require a considerable amount of flexibility at the coordination site (d^{10} metal ions with low coordination numbers and no crystal field stabilization), other reactions need more rigid coordination geometries *e.g.* in order to ensure stereoselectivity. The role of seemingly remote coligands and substituents cannot be underestimated in this respect.

The second function of the metal is, of course, to electronically activate the substrate(s), *e.g. via* σ acceptor and n or π donor interactions, as outlined above for example (7). This aspect may be studied by separately treating the metal center with either one of the reacting ligands or — in favourable cases — by studying the precursor complex itself; according to the entatic state theory of catalysis (ref. 25), the electronic activation should also work towards the transition state geometry — this time affecting particularly bond lengths and angles of coordinated ligands.

The example (7) has some significance beyond organometallic chemistry since not only α–diimines, including the less reactive aromatic derivatives such as 2,2'–bipyridine, (ref. 7) but also synthetically as well as biochemically interesting other electron transfer active heterodienes (14) such as α–iminocarbonyls (→pterins and flavin coenzymes, (refs. 27, 28)), α–dicarbonyls, (refs. 28, 29) (→dehydroascorbate, cofactor PQQ, and other ortho–quinones, (ref. 21)), or α–dithiones (→the postulated molybdopterin cofactor, 15, (ref. 30)) exhibit the structural and electronic advantage for effective intra–complex electron transfer after metal coordination and fixation in a sterically favourable five–membered chelate ring.

(14)

X,Y = NR: α–diimines

X,Y = O: α–dicarbonyl compounds, incl. {o–quinones, o–semiquinones, catecholates}

X,Y = S: α–dithiolenes

X =O, Y = NR: α–iminoketones

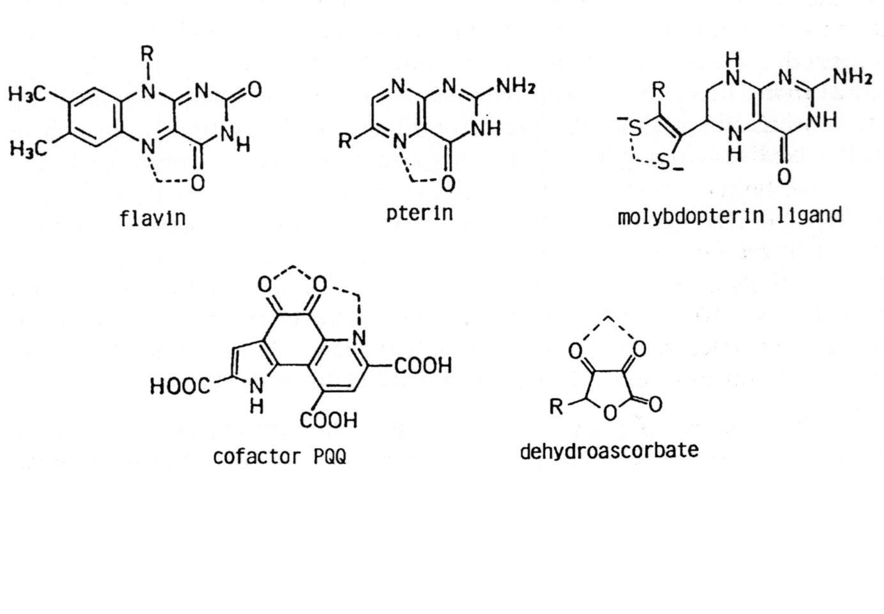

flavin pterin molybdopterin ligand

(15)

cofactor PQQ dehydroascorbate

Facilitated Electron Transfer and Enhanced Radical Product Stability through Non-Chelate Double Metal Coordination

While the susceptibility for intra–complex electron transfer and therefore the inner sphere reaction rate is enhanced by excellent overlap between HOMO and LUMO (8) even if the frontier orbital energy difference within the complex is still relatively large, some other unexpected stabilities of radical complex products as obtained from electron transfer reactions between metal fragments and unsaturated organic substrates can be due to multiple coordination and electrostatic stabilization at electronically coupled binding sites.

Reduced substrates, for instance, are able to lower their excess charge and therefore electrostatic disadvantage by coordinating to quite innocent cationic or neutral metal species. The role of multiple effective charge compensation can be quite striking as shown in the following: Organolithium compounds react with pyrazine to form 2–phenyl–1,2–dihydropyrazine as cage product (ref. 31) and (with excess lithium organyl) the persistent, ESR–detectable pyrazine radical anion–dilithium triple ion (ref. 32) as escape product (16), (refs. 7 and 33).

$$\text{pyrazine} \xrightarrow[+\,Li^{\bullet},\ +\,LiBPh_4]{+\,3\,LiR,\ -\,R^{\bullet}\ \text{SET (escape)}} [Li^{+}\,pz^{\bullet -}\,Li^{+}] \equiv [Li\text{–}pz^{\bullet +}\text{–}Li]\ \ [LiR_2]^{-}\ \text{or}\ [BPh_4]^{-} \tag{16}$$

$$\text{pyrazine} \xrightarrow{+\,LiR\ \text{(cage)}} \text{1-Li-2-R-1,2-dihydropyrazine}$$

The stability of the latter is evident from cyclic voltammetry of pyrazine in the presence of Li^+ (17 and Fig. 2): Whereas weak coordination of Li^+ to the neutral molecule facilitates reduction by only 0.2 V peak potential difference, the reduced substrate gets reoxidized in the reverse scan at a much more positive potential, virtually in the region of quinone or benzophenone reduction, (ref. 34), thus producing a potential shift of more than 1 V, (ref. 35).

$$\begin{array}{ccccc} pz + 2\,Li^{+} & \rightleftharpoons & [pz(Li^{+})] + Li^{+} & \rightleftharpoons & [pz(Li^{+})_2] \\ \big\downarrow E_{pc} & & & & \big\uparrow E_{pa} \\ pz^{\bullet -} + 2\,Li^{+} & \rightleftharpoons & [pz^{\bullet -}(Li^{+})] + Li^{+} & \rightleftharpoons & [pz^{\bullet -}(Li^{+})_2] \end{array} \tag{17}$$

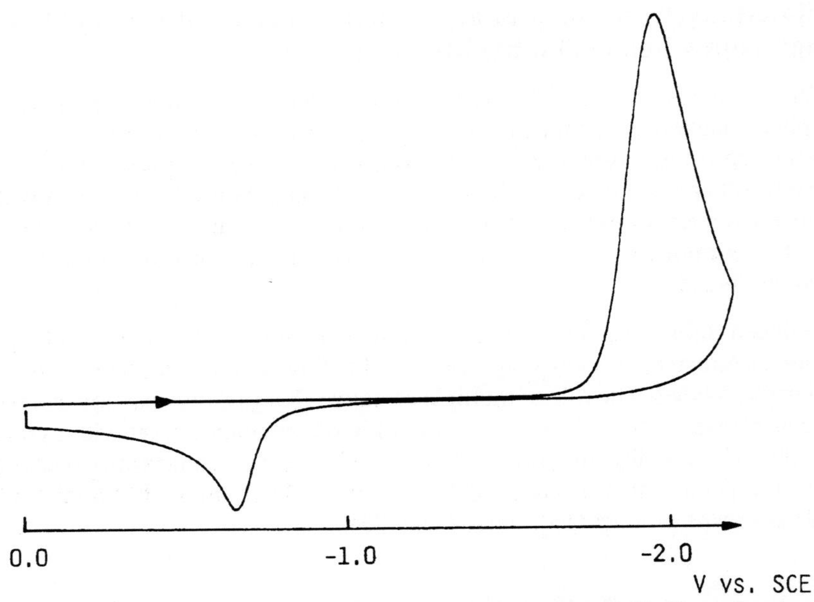

Fig. 2. Typical cyclic voltammogram of pyrazine in the presence of alkali metal cations (here: tenfold molar excess of Li^+ in DMF).

Similar effects were noted before for quinones and azo compounds (ref. 36) but were never found to be that large: a result of high electron densities at the two conjugatively coupled nitrogen coordination sites in the reduced (7π) pyrazine molecule (refs. 33, 37). The important role of the charge itself is obvious from the fact that the effects for K^+ and Na^+ are only slightly smaller than for Li^+ (ref. 35) whereas coordination of H^+ produces a shift of more than 2 V, (ref. 38). The unexpected formation of radical complexes in reaction (16) constitutes a rather striking example for the "salt effect" in electron transfer reactions,(ref. 39); here the alkali metal cation is a product of the reaction.

Brief Discussion of Salt Effects, Proton, Base, and Solvent Coordination in Electron Transfer Reactions

Extrapolating the salt effect one step further, one arrives at the more familiar effects of protonation and deprotonation upon electron transfer, (ref. 40) or, more general, redox reactions. High proton concentrations can enormously facilitate reductions of H^+–coordinating (*i.e.* basic) molecules such as 2,2'–bipyridine, especially after multiple coordination at coupled sites; conversely, one electron–reduced bases such as bipyridine anion radicals can have extremely high basicities (18), (ref. 41).

On the other side, oxidized molecules such as amine or alkylbenzene cation radicals are often very acidic (ref. 42). Addition of a base can thus facilitate electron transfer reactivity in the direction of oxidation of deprotonated or base–attacked species; a very familiar example is the formation of semiquinones from hydroquinones, base and an oxidant such as air.

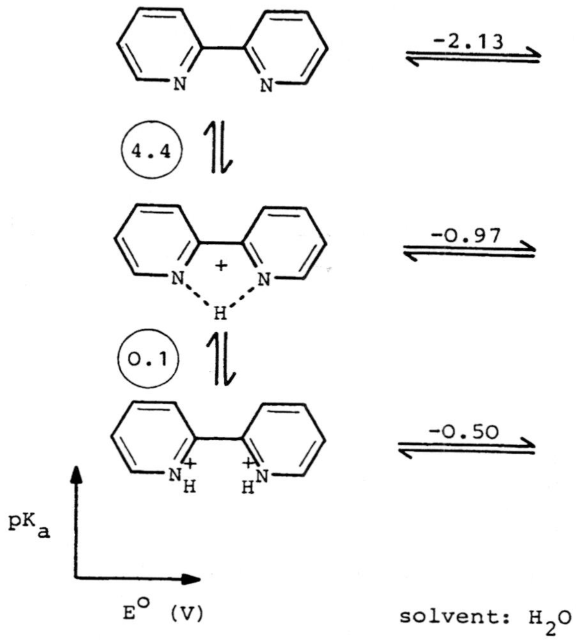

-2.13 -0.97 -0.50

4.4   0.1   24   8.0

pK_a

E^o (V)   solvent: H_2O

(18)

All the above effects involve some kind of complexation of at least one of the electron transfer substrates by an added reagent (cation or anion of a salt, H^+ or OH^- in an acid/base process, Lewis acid/base species). Such a coordination can in turn be influenced by the solvent (as can, of course, the primary encounter between metal fragment and substrate, (refs. 1-4)); a very elucidating example from transition metal chemistry, (ref. 43), involving the Lewis base effect of a solvent is shown in (19).

$$[(CO)_5W{=}C{=}N{=}CR_2]^+AlBr_4^- \xrightarrow[-AlBr_3\cdot THF]{+THF} [(CO)_5W{=}C{=}N{=}CR_2]^+Br^- \xrightarrow{SET} \{(CO)_5W{-}C{\equiv}N{-}\dot{C}R_2,\ Br^\cdot\}_{\text{radical pair}}$$

$$\{(CO)_5W{-}C{\equiv}N{-}\dot{C}R_2,\ Br^\cdot\} \xrightarrow[M=W]{\text{cage reaction}} trans\text{-}Br(CO)_4W{\equiv}C{-}N(CR_2) + CO$$

$$\{(CO)_5W{-}C{\equiv}N{-}\dot{C}R_2,\ Br^\cdot\} \xrightarrow[-Br\cdot]{\text{escape reaction}} [(CO)_5W{-}C{\equiv}N{-}\dot{C}R_2] \quad R = \text{mesityl} \tag{19}$$

$$\xrightarrow[M=W]{R=\text{Aryl} \neq \text{Mesityl}} (CO)_5W{-}C{\equiv}N{-}CR_2{-}CR_2{-}N{\equiv}C{-}W(CO)_5 + \text{polymers}$$

Although organometallic compounds generally act as electron donors in electron transfer processes, cationic systems such as $^{+}SiR_3$ or $^{+}MnCp(CO)_2$ may function as acceptors. For instance, the pentacarbonyltungsten(0) cation with an isonitrile ligand is stable with the hardly nucleophilic $AlBr_4^-$ counter–anion in dichloromethane solution. Addition of the Lewis–basic solvent tetrahydrofuran (THF) triggers a truly remarkable reaction sequence (19), (ref. 43): THF forms a very stable complex with the strong Lewis acid $AlBr_3$, leaving bromide instead of tetrabromoaluminate as counterion for the organometallic cation. Yet, the bromide ion is a sufficiently strong reductant in dichloromethane solution to bring about electron exchange with the organometallic cation, resulting in an intermediate pair of neutral radicals within the solvent cage. In the cage (recombination or "collapse") process the bromine radical substitutes the (trans) carbon monoxide ligand (19), while the "escaped" organometallic free radical can either dimerize or, if sterically protected, crystallize as a stable species out of the solution (ref. 43).

Outer Sphere Electron Transfer Involving a Non–Coordinating Main Group Organometallic Donor ?

While the organozinc example presented above (7) clearly represents a case of strongly chelate coordination–enhanced and MO situation–facilitated inner–sphere electron transfer, the sterically protected but extremely electron rich 1,4–bis– (trialkylsilyl)–1,4–dihydropyrazines such as **9**, (refs. 37, 46) are candidates rather for an outer-sphere type of electron exchange of an organometallic donor.

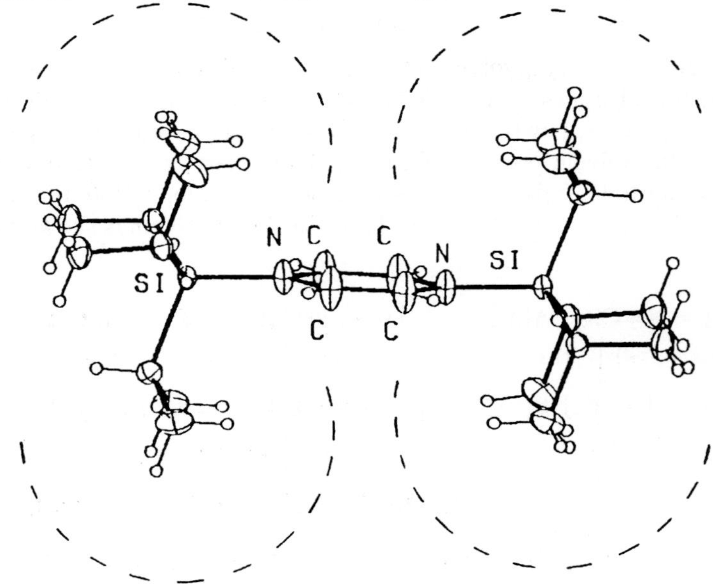

Fig. 3. Molecular structure of 1,4–bis(triisopropylsilyl)–1,4–dihydropyrazine **9** with steric shielding of Si–N bonds and of the 8 π electron ring.

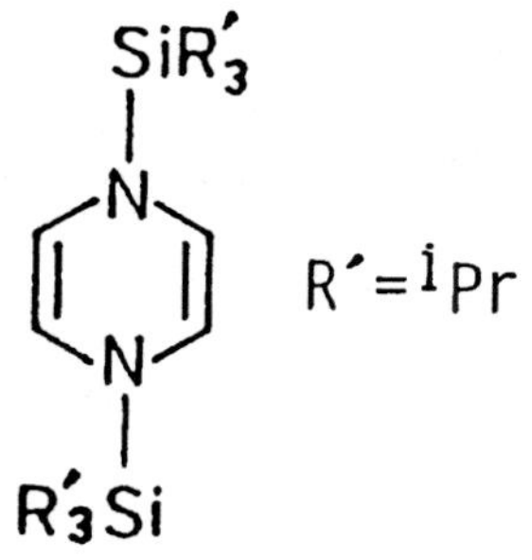

SiR'_3 / R'_3Si — **9** — $R' = {}^{i}Pr$ (20)

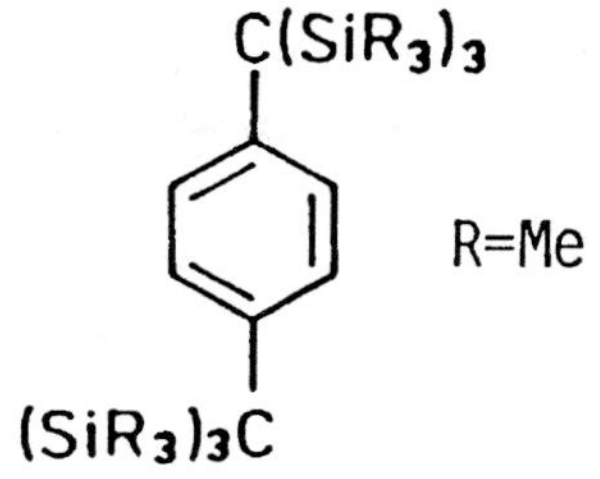

$C(SiR_3)_3$ / $(SiR_3)_3C$ — **10** — R=Me (21)

Steric protection of the sensitive but rearrangement–safe Si—N bonds (Fig. 3) has the consequence that simple electron exchange does occur but only with rather electron deficient acceptors such as tetracyanoethene (TCNE) whose reduction potential lies above the oxidation potential of the organometallic donor, (refs. 3 and 47). Since both reactants undergo relatively little geometry change on electron transfer, (refs. 46-49), and should therefore exhibit rather small amounts of reorganisation energy as required by the Marcus equation for outer–sphere electron transfer, (ref. 3), there is no "coordinative match" necessary between the two components so that the globular donor (Fig. 3) and a flat acceptor like TCNE can undergo rapid electron exchange yielding the fairly stable but not crystalline pair of oppositely charged radical ions (refs. 46 and 47). Fast electron exchange in this case did not allow us to observe a coloured charge–transfer precursor complex; however, if an even more voluminous, similarly globular organometallic donor such as **10** (21), (ref. 50), is not sufficiently electron rich, (ref. 51), for outer sphere electron transfer according to the Marcus equation, then the naively estimated charge transfer absorption of an expected precursor complex is not observed at all (ref. 52).

Scheme (22) summarizes the spectrum of reactivity between electron rich organosilicon donors with the prototypical acceptor tetracyanoethene TCNE (ref. 46). According to the mechanistic sequence presented in (6), the reactivity varies from the formation of a charge transfer complex (sometimes very weakly bound, (ref. 2), because of steric hindrance (ref. 52)) to single electron transfer with sometimes persistent but mostly dissociatively labile, (refs. 50, 53) organosilicon cation radicals.

single electron transfer, outer sphere process, radical-ion pair

+ TCNE$^{\cdot -}$

no interaction (steric hindrance)

TCNE

(22)

oxidative desilylation, formation of TCNE$^{\cdot -}$ as intermediate

coloured charge transfer (CT) complexes

n = 1−6

Introducing Redox–Active Transition Metal Centers: Opportunities and Complications

In contrast to main group element organometallics with non–redox–active metal centers, transition metal fragments contain a metal center which can itself — *via* the d orbitals (5) — undergo (single) electron transfer. This is particularly true for the first row transition metals, (ref. 5), creating further opportunities for electron transfer reactivity but also additional complications in defining and understanding electronic structures. Particularly suitable for the study of electron transfer (not necessarily for synthetic chemistry) are those coordinatively unsaturated transition metal fragments which are inert in two neighbouring oxidation states of the metal. While ruthenium(II/III) ammines have long been used in that function for classical inorganic chemistry in aqueous solutions, (ref. 6), the manganese(I/II) fragments $(C_5R_5)(CO)_2Mn^{0/+}$ can play a similar role for organometallic chemistry in nonaqueous media (ref. 45b). Continuing with the well documented electron transfer reactivity of TCNE, (refs. 1, 46-48, 54 and 55), a few quite remarkable examples for possible effects of redox–active transition metal fragments will be discussed.

TCNE cannot only accept one or two electrons fairly easily, (ref. 48), it is also at the same time an olefin, forming complexes with η^2-(C=C-) bonded metal centers, (ref. 55), and a polynitrile, using one or more nitrogen lone pairs of the nitrile groups for metal coordination (refs. 56 and 57). In the reaction with a few prototypical electron rich 16 valence electron metal carbonyl fragments (stabilized as solvates or phosphite complexes with 18 valence electrons), very variable rates of formation, modes of coordination, and electronic structures were observed (23), (refs. 54 and 58).

$$\mathrm{TCNE} + \begin{cases} (CO)_5W(THF) \longrightarrow (CO)_5W(\eta^2\text{-}TCNE) & k_2 = 0.043\ M^{-1}s^{-1} \\ (CO)_5Cr(THF) \longrightarrow (CO)_5Cr(\eta^2\text{-}TCNE) & k_2 = 0.39\ M^{-1}s^{-1} \\ (C_6Me_6)(CO)_2Cr(PX_3) \longrightarrow (C_6Me_6)(CO)_2Cr(\eta^1\text{-}TCNE) & k_2 = 1.59\ M^{-1}s^{-1} \\ (C_5H_4Me)(CO)_2Mn(THF) \longrightarrow (C_5H_4Me)(CO)_2Mn(\eta^1\text{-}TCNE) & k_2 > 14000\ M^{-1}s^{-1} \end{cases} \qquad (23)$$

$X = OCH_3$

The following dichotomy emerged, (refs. 54 and 58):
Pentacarbonylchromium and –tungsten fragments form typical olefin–type π complexes with TCNE in relatively slow THF/TCNE exchange reactions. Although intense metal–to–ligand charge transfer bands in the near–infrared region indicate a small HOMO–LUMO gap, a complete organometal–to–(TCNE) ligand electron transfer has not yet occurred. Significantly, the pentacarbonylmetal complexes are reduced at the same potential as the free ligand TCNE, suggesting that the polarization of the TCNE ligand by the organometallic electrophile is just about compensated by metal–to–ligand π back donation in the ground state. It is quite intriguing that the electrochemically reversible reduction of the π(C=C)–bonded tungsten complex leads to an ESR signal which indicates an unsymmetrical, presumably σ(C≡N)–type of coordination (refs. 54a and 59). While there are several nitrile–coordinated complexes of $TCNE^{-\cdot}$, (refs. 57 and 60), no π–bonded radical complex of $TCNE^{-\cdot}$ has ever been reported. This remarkable fact is presumably due to the greater need for π back–donation in the π bonded situation; $TCNE^{-\cdot}$ has its additional unpaired electron in that very orbital which is needed for π back–donation: the olefin π^* orbital. The nitrile lone pair, on the other hand, should be capable of binding a metal center even in the absence of back–donation.

The two more electron rich carbonyl and carbocycle containing metal fragments (23) react as THF or phosphite coordinated species very rapidly with TCNE in what is believed to be an electron transfer catalyzed process, (ref. 54), based on the redox–activity of the metal center (refs. 61 and 62):

Single electron exchange can occur between TCNE and the solvated species as an equilibrium reaction in a similar fashion as described for the electron rich organosilicon compound **9**; (ref. 46), the redox potentials, (refs. 46 and 61), and presumed reorganization energies (structural changes), (refs. 57 and 63), are such that even an outer–sphere process should be feasible. The oxidized organometallic solvate has then become a true 17 valence electron (VE) species and as such is now extremely labile with regard to its most weakly bonded (solvent) ligand (ref. 61). The phosphite ligand in the complex $(\eta^6\text{–}C_6R_6)(CO)_2Cr[P(OMe)_3]$ is, of course, less labile than a THF molecule would be; however, THF solvates of the $(C_6R_6)(CO)_2Cr$ fragment are not available in fully converted form such as the solvates $(C_5R_5)(CO)_2Mn(THF)$. The high lability of the 17 valence electron complex allows it to undergo rapid substitution of THF or phosphite by TCNE (or TCNQ), and a catalytic cycle is closed because the cationic (17 VE) TCNE–complex can, in a homogeneous electron transfer step, oxidize and therefore labilize the 18 VE starting material and yield the neutral TCNE-substituted product (24), (ref. 54). While even the substitution of the phosphite by TCNE is rather efficient, the THF complex of $(C_5R_5)(CO)_2Mn$ is substituted extremely fast, by at least five orders of magnitude more rapidly than the pentacarbonylmetal species mentioned above. Since the one–electron oxidation equivalents necessary for cycle (24) come from a preceding equilibrium reaction of the starting materials and not from an electrode or an external oxidant, this type of reaction mechanism was referred to as electron transfer autocatalytic, (ref. 54).

TCNX + R_nML

$-TCNX^{\bullet-}$ ⇅ Start

$(R_nML)^{\bullet+}$

$R_nM(TCNX)$ ← electron transfer ← R_nML TCNX → substitution → L   (24)

$[R_nM(TCNX)]^{\bullet+}$

TCNX: TCNE or TCNQ

L: oxidatively labilized ligand (THF or PR_3)

There are, of course, many other criteria applicable to the eventually formed complex which should indicate that an electron transfer according to a one–electron oxidative substitution has occurred.

Structural data of $(C_5R_5)(CO)_2Mn(TCNE)$, (R = H, ref. 57), have not only confirmed the nitrile coordination postulated previously, (ref. 54), but also the one electron–reduced state of the TCNE ligand. Similarly, C≡N, C≡O, and C=C vibrational stretching frequencies point to $TCNE^{\cdot-}$ and Mn^{II} as most likely oxidation states in that compound (ref. 54). Substitutional effects at the carbocyclic rings suggest that the long wavelength absorption bands are due to (reduced) ligand–to–(oxidized) metal charge transfer transitions, and the reduction potentials of the complexes lie more negative than that of free TCNE itself, confirming that electron(s) have already been transferred to the ligand from the metal center. The absence of an ESR signal from the singly reduced form at room temperature is in agreement with a singly occupied MO centered on the metal; on the other hand, some of the corresponding TCNQ complexes dissociate into paramagnetic pairs of radical ions in highly dielectric solvents (ref. 54).

However, the chemically the most valuable consequence of intra–complex electron transfer again concerns coordination: The increased electron density in the TCNE ligand as resulting from ground state metal–to–ligand electron transfer activates this polydentate ligand towards full coordinative saturation, *i.e.*, all cyano groups of (reduced) TCNE can now bind an organometallic fragment to yield the first neutral tetranuclear TCNE complexes with a remarkable electronic structure (25), (ref. 64).

NC CN NC CN + THF—Mn CO CO fast -THF TCNE [Mn]

+ 3 Mn CO CO

[Mn] = Mn CO CO

(25)

Concluding Remarks

Direct coordination between metal centers and substrates plays an important role in many areas of chemical electron transfer because not only electronic activation but also favourable mutual orientation are effected by forming coordinative bonds. Biological electron transfer between or inside proteins, (ref. 65), on the other hand, involves the protein as a large sophisticated scaffold for organic or inorganic redox centers, thereby separating the problem of orientation from the activation by immediate ligation or electrostatic effects. Only now have chemists begun to create similarly modular structures for energy– and electron–transfer purposes in what has become known as "supramolecular" chemistry (ref. 66).

Acknowledgements

Several of the results presented were obtained in cooperation with Professors H. Preuß and H. Stoll (Universität Stuttgart), Dr. H.–D. Hausen and Dr. Baumgarten (Universität Stuttgart), Mr. M. Kaupp (Universität Erlangen), Prof. G. van Koten and Mr. E. Wissing (Utrecht University) and Prof. H. Fischer (Universität Konstanz). I also wish to thank my coworkers, Mr. T. Stahl, Dr. C. Vogler, Dr. C. Bessenbacher, Dr. B. Olbrich–Deussner, Dr. B. Schwederski and Dr. R. Gross–Lannert for their essential contributions to the results described here. Generous support from Volkswagen–Stiftung (Program "Organometallic Reagents for Organic Synthesis"), Deutsche Forschungsgemeinschaft and NATO is gratefully acknowledged.

References

1 J.K. Kochi *Organometallic Mechanisms and Catalysis* Academic Press, New York, 1978.

2 J.K. Kochi, Angew. Chem., **100** (1988) 1331; Angew. Chem. Int. Ed. Engl., **27** (1988) 1227.

3 L. Eberson *Electron Transfer Reactions in Organic Chemistry* Springer–Verlag, Berlin, 1987.

4 M.A. Fox and M. Chanon (Eds.) *Photoinduced Electron Transfer* Elsevier, Amsterdam, 1988.

5 (a) C. Elschenbroich and A. Salzer *Organometallchemie* (2. Aufl.), Teubner, Stuttgart, 1988.
(b) J.P. Collman and L.S. Hegedus *Principles and Applications of Organotransition Metal Chemistry* University Science Books, Mill Valley, 1980.

6 H. Taube, Angew. Chem., **96** (1984) 315; Angew. Chem. Int. Ed. Engl., **23** (1984) 329.

7 W. Kaim, Acc. Chem. Res., **18** (1985) 160.

8 P.S. Braterman (Ed.) *Reactions of Coordinated Ligands* Plenum, New York, 1989.

9 (a) W. Bruns and W. Kaim, J. Organomet. Chem., **390** (1990) C45.

(b) W. Bruns and W. Kaim in *Mixed Valency Systems – Applications in Chemistry, Physics and Biology* (Ed.: K. Prassides), Kluwer Academic Publishers, Dordrecht, in press.

10 (a) M.K. DeArmond and M.L. Myrick, Acc. Chem. Res., **22** (1989) 364.

(b) W. Kaim, S. Ernst and V. Kasack, J. Am. Chem. Soc., **112** (1990) 173.

11 (a) B. Olbrich–Deussner and W. Kaim, J. Organomet. Chem., **340** (1988) 71.

(b) W. Kaim, B. Olbrich–Deussner, R. Gross, S. Ernst, S. Kohlmann and C. Bessenbacher in *Importance of Paramagnetic Organometallic Species in Activation, Selectivity and Catalysis* (Ed.: M. Chanon), Kluwer Academic Publishers, Dordrecht, 1989, p. 283.

(c) B. Olbrich-Deussner and W. Kaim, J. Organomet. Chem., **361** (1989) 335.

(d) W. Kaim and B. Olbrich–Deussner in *Organometallic Radical Processes* (Ed.: W. C. Trogler), Elsevier, Amsterdam, 1990, p. 173.

12 E.C. Ashby, Acc. Chem. Res., **21** (1988) 414.
Cf. also M. Newcomb and D.P. Curran, Acc. Chem. Res., **21** (1988) 206.

13 (a) J.T.B.H. Jastrzebski, J.M. Klerks, G. van Koten and K. Vrieze, J. Organomet. Chem., **210** (1981) C49.

(b) G. van Koten, J.T.B.H. Jastrzebski and K. Vrieze, J. Organomet. Chem., **250** (1983) 49.

(c) G. van Koten in A. de Meijere, H. tom Dieck (Eds.), *Organometallics in Organic Synthesis* Springer–Verlag, Berlin, 1988, p 277.

(d) M. Kaupp, H. Stoll, H. Preuss, W. Kaim, T. Stahl, G. van Koten, E. Wissing, W.J.J. Smeets and A.L. Spek, J. Am. Chem. Soc., in print.

(e) G. van Koten, private communication.

14 A. Weller and K. Zachariasse in M. Cormier, D.M. Hercules, J. Lee (Eds.) *Chemiluminescence and Bioluminescence* Plenum, New York, 1973.

15 A. Pross, Acc. Chem. Res., **18** (1985) 212.

16 E. Buncel and T. Durst (Eds.) *Comprehensive Carbanion Chemistry* Elsevier, Amsterdam, 1987.
Cf. also G. Boche, Top. Curr. Chem., **146** (1988) 3.

17 (a) A.P. Kozikowski, T. Konoike and A. Ritter, Carbohyd. Res., **171** (1987) 109.

(b) M. Kitamura, S. Okada, S. Suga and R. Noyori, J. Am. Chem. Soc., **111** (1989) 4028.

18 A. Almenningen, T.U. Helgaker, A. Haaland and S. Samdal, Acta Chem. Scand., **A36** (1982) 159.
Cf. also: A. Haaland, Angew. Chem., **101** (1989) 1017;
Angew. Chem. Int. Ed. Engl., **28** (1989) 992.

19 W. Kaim, Coord. Chem. Rev., **76** (1987) 187.

20 M.R.P. van Vliet, J.T.B.H. Jastrzebski, W.J. Klaver, K. Goubitz and G. van Koten, Recl. Trav. Chim. Pays–Bas, **106** (1987) 132.

21 (a) G.A. Razuvaev, G.A. Abakumov, E.S. Klimov, E.N. Gladyshev and P. Y. Bayushkin, Izv. Akad. Nauk SSSR, Ser. Khim., **1977**, 1128.

(b) M. Okubo, Y. Fukuyama, M. Sato, K. Matsuo, T. Kitahara and M. Nakashima, J. Phys. Org. Chem., **3** (1990) 379.

22 T. Stahl and W. Kaim, unpublished results.

23 C. Kutal, Coord. Chem. Rev., **99** (1990) 213.

24 B.W. Walther, F. Williams, W. Lau and J.K. Kochi, Organometallics, **2** (1983) 688.

25 R.J.P. Williams, J. Mol. Catal., Review Issue, **1986** 1.

26 C. Vogler, H.-D. Hausen, W. Kaim, S. Kohlmann, H.E.A. Kramer and J. Rieker, Angew. Chem. 101 (1989) 1734; Angew. Chem. Int. Ed. Engl. 28 (1989) 1659.

27 (a) S. Bhattacharya, S.R. Boone and C.G. Pierpont, J. Am. Chem. Soc., **112** (1990) 4561.

(b) C. Bessenbacher, W. Kaim and C. Vogler, Inorg. Chem., **28** (1989) 4645.

28 M.J. Clarke, Comments Inorg. Chem., **3** (1984) 133.

29 B. Schwederski, V. Kasack, W. Kaim, E. Roth and J. Jordanov, Angew. Chem., **102** (1990) 74; Angew. Chem. Int. Ed. Engl., **29** (1990) 78.

30 S.P. Kramer, J.L. Johnson, A.A. Ribeiro, D.S. Millington and K.V. Rajagopalan, J. Biol. Chem., **262** (1987) 16357.

31 R.E. van der Stoel and H.C. van der Plas, Recl. Trav. Chim. Pays-Bas, **97** (1978) 116.

32 S.A. Al–Baldawi and T.E. Gough, Can. J. Chem., **48** (1970) 2798.

33 W. Kaim, Angew. Chem., **95** (1983) 201; Angew. Chem. Int. Ed. Engl., **22** (1983) 171.

34 H. Yamataka, N. Fujimura, Y. Kawafuji and T. Hanafusa, J. Am. Chem. Soc., **109** (1987) 4305.

35 B. Olbrich–Deussner and W. Kaim, unpublished results.

36 (a) T. Nagaoka, S. Okazaki and T. Fujinaga, J. Electroanal. Chem., **133** (1982) 89.

(b) D.A. Gustowski, V.J. Gatto, A. Kaifer, L. Echegoyen, R.E. Godt and G.W. Gokel, J. Chem. Soc. Chem. Commun., **1984**, 923.

37 W. Kaim, Rev. Chem. Intermed., **8** (1987) 247.

38 J. Swartz and F.C. Anson, J. Electroanal. Chem., **114** (1980) 117 and Inorg. Chem., **20** (1981) 2250.

39 Cf. B. Goodson and G.B. Schuster, Tetrahedron Lett., **27** (1986) 3123.

40 Cf. T. Caronna, S. Morrocchi and B.M. Vittimberga, J. Am. Chem. Soc., **108** (1986) 2205.

41 C.V. Krishnan, C. Creutz, H.A. Schwarz and N. Sutin, J. Am. Chem. Soc., **105** (1983) 5617.

42 F.D. Lewis, Acc. Chem. Res., **19** (1986) 401.

43 F. Seitz, H. Fischer, J. Riede, T. Schöttle and W. Kaim, Angew. Chem., **98** (1986) 753; Angew. Chem. Int. Ed. Engl., **25** (1986) 744.

44 J.C. Giordan, J.H. Moore, J.A. Tossell and W. Kaim, J. Am. Chem. Soc., **107** (1985) 5600.

45 (a) R. Gross and W. Kaim, J. Chem. Soc., Faraday Trans., **1** 83 (1987) 3549.

(b) W. Kaim and R. Gross, Comments Inorg. Chem., **7** (1988) 269.

46 J. Baumgarten, C. Bessenbacher, W. Kaim and T. Stahl, J. Am. Chem. Soc., **111** (1989) 2126 and 5017.

47 W. Kaim, Angew. Chem., **96** (1984) 609; Angew. Chem. Int. Ed. Engl., **23** (1984) 613.

48 D.A. Dixon and J.S. Miller, J. Am. Chem. Soc., **109** (1987) 3656.

49 H.–D. Hausen, A. Schulz and W. Kaim, Chem. Ber., **121** (1988) 2059.

50 H. Bock and W. Kaim, Acc. Chem. Res., **15** (1982) 9.

51 H. Bock and U. Lechner–Knoblauch, J. Organomet. Chem., **294** (1985) 295.

52 H. Bock, W. Kaim and H.E. Rohwer, Chem. Ber., **111** (1978) 3573.

53 H.–D. Hausen, C. Bessenbacher and W. Kaim, Z. Naturforsch., **43b** (1988) 1087.

54 (a) B. Olbrich–Deussner, R. Gross and W. Kaim, J. Organomet. Chem., **366** (1989) 155.
(b) B. Olbrich–Deussner, W. Kaim and R. Gross–Lannert, Inorg. Chem., **28** (1989) 3113.

55 W.H. Baddley, Inorg. Chim. Acta Rev., **2** (1968) 7.

56 W. Beck, R. Schlodder, K.H. Lechler, J. Organomet. Chem. 54 (1973) 303.

57 H. Braunwarth, G. Huttner and L. Zsolnai, J. Organomet. Chem., **372** (1989) C23.

58 B. Schwederski, B. Olbrich–Deussner and W. Kaim, to be submitted.

59 T. Roth, B. Olbrich–Deussner and W. Kaim, unpublished results.

60 (a) M.F. Rettig and R.M. Wing, Inorg. Chem., **8** (1969) 2685.
(b) P.J. Krusic, H. Stoklasa, L.E. Manzer and P. Meakin, J. Am. Chem., Soc. **97** (1975) 667.

61 J.K. Kochi, J. Organomet. Chem., **300** (1986) 139.

62 K.L. Amos and N.G. Connelly, J. Organomet. Chem., **194** (1980) C57.

63 Cf. N.G. Connelly, M.J. Freeman, A.G. Orpen, A.R. Sheehan, J.B. Sheridan and D.A. Sweigart, J.Chem. Soc., Dalton Trans., **1985**, 1019.

64 (a) R. Gross and W. Kaim, Angew. Chem., **99** (1987) 257; Angew. Chem. Int. Ed. Engl., **26** (1987) 251.
(b) R. Gross-Lannert, W. Kaim and B. Olbrich–Deussner, Inorg. Chem., **29** (1990) 5046.

65 R.J.P. Williams, Adv. Chem. Ser., **226** (1990) 3.

66 (a) J.-M. Lehn, Angew. Chem., **100** (1988) 91; Angew. Chem. Int. Ed. Engl., **27** (1988) 89.
(b) F. Vögtle *Supramolekulare Chemie* Teubner, Stuttgart, 1989.

Electron and Proton Transfer in Chemistry and Biology, edited by A. Müller et al.
Studies in Physical and Theoretical Chemistry, Vol. 78

Electron Hopping and Delocalization in Mixed–Valence Metal–Oxygen Clusters

Hyunsoo So and Michael T. Pope

Department of Chemistry
Georgetown University
Washington DC 20057 (USA)

Summary

The behavior of a single unpaired electron, formally a d^1 metal center, in mixed–valence polynuclear oxometalate anions have been probed by electron spin resonance (ESR) and by nuclear magnetic resonance (NMR) spectroscopies. The discrete metal–oxygen clusters are excellent models for extended oxide lattices and the partially reduced species therefore are potential molecular analogs of semiconducting and "metallic" oxides. In the mixed–valence polyanions the unpaired electron may be trapped on a single metal atom, it may undergo intramolecular "hopping" between two metal atoms, or it may be delocalized (on the ESR time scale) over two or more metal atoms. Temperature–dependent ESR spectra, observed when the electron is hopping, can be analyzed and simulated using modified Bloch equations, and the rate constants for the intramolecular electron transfer are determined. When electron hopping is very fast, the ESR spectra are temperature–independent, and only the lower limit of the rate constant can be estimated. Activation energies controlling the hopping rates for 3d, 4d, and 5d electrons in polyvanadates, polymolybdates, and polytungstates respectively, are influenced by the magnitude of the bond angle in the metal–oxygen–metal bridge, and by protonation of the bridging oxygen atom. In most cases the electron becomes trapped on one metal atom at very low temperature, but partial delocalization by tunneling can be detected from ESR superhyperfine interactions. The NMR spectra of polymolybdates and polytungstates with <u>two</u> d^1 metal centers demonstrate that both electrons undergo rapid hopping at ambient temperatures but remain strongly spin–coupled.

I. Introduction

Interatomic electron transfer within molecular frameworks or rigid lattices is a critical elementary step underlying the phenomena of conductivity and superconductivity. Unlike the situation with electron transfer in solution, the rates of which are frequently limited by molecular diffusion or by ligand substitution, the electron–donor and –acceptor atoms in solids occupy fixed positions relative to one another, and are most commonly the same element in different oxidation states. A well-studied "molecular" example of such a mixed–valence compound is the Creutz-Taube ion $[(NH_3)_5Ru(NC_4H_4N)Ru(NH_3)_5]^{5+}$, which formally contains Ru^{II} and Ru^{III}, although the actual electronic structure has been a matter of considerable debate (ref. 1).

In the present article, we discuss the electron transfer characteristics of mixed–valence species that contain two or more metal atoms bridged by the ubiquitous and important ligand, oxide ion. Such complexes provide models for electron transfer and electron delocalization in extended oxide lattices.

A. Mixed–Valence Compounds

Systematic investigation of mixed valency can be said to date from the publication of two review articles published in 1967 (refs. 2 and 3). The classification of mixed–valence compounds proposed by Robin and Day (ref. 2) has proved to be convenient, and has been widely adopted. In brief, three classes of mixed–valence compounds are distinguished by the similarity or difference between the sites containing the two different oxidation states, and by the extent of electronic interaction between them. If we represent the two oxidation states as (a) and (b), and the two sites as A and B, then we can imagine two limiting situations, corresponding to the pure valence bond configurations, {A(a)B(b)} and {A(b)B(a)}.

At one extreme are Robin–Day **class I** compounds in which sites A and B are crystallographically distinguishable by bond length, coordination number, *etc.* The energies of the two valence configurations are then very different, and one of them is an accurate description of the ground state. An example of such a compound is Sb_2O_4, the structure of which contains $Sb^{III}O_3$ trigonal pyramids and $Sb^{V}O_6$ octahedra. The properties of class I mixed–valence compounds are therefore usually a simple superposition of those of the individual valence states, and electron delocalization between A and B is, to first order, zero. The other extreme case (**class III**) occurs when A and B are indistinguishable and the valence states are completely averaged, as for example in $[Nb_6Cl_{12}]^{2+}$, or the cubic tungsten bronzes, Na_xWO_3.

Class II species comprise the continuum of cases that are intermediate between class I and class III. They are often conveniently defined in terms of a perturbation of class I, *i.e.* by describing the ground state as a linear combination of the two extreme configurations {A(a)B(b)} and {A(b)B(a)} with an appropriate valence delocalization coefficient, α.

$$\phi = (1-\alpha^2)^{1/2}\{A(a)B(b)\} + \alpha\,\{A(b)B(a)\} \qquad (1)$$

Class II behavior is expected when A and B are, in principle (ref. 4), crystallographically distinguishable, but are nevertheless similar. With compounds of this type the possibility exists for intramolecular adiabatic electron transfer between A and B, resulting in exchange of valences. The electronic spectra of class II compounds also exhibit an intervalence charge transfer absorption band resulting from the transition between the two valence configurations, in addition to the expected absorption bands characteristic of the individual valence states.

B. Electron Transfer Models

Since 1967 some theoretical work on mixed–valence systems has appeared, but it was not until 1978 that the full vibronic problem for a two–site one–electron model was solved by Piepho, Krausz, and Schatz (PKS), (ref. 5). According to the PKS model, a mixed–valence system is characterized by three or four parameters: (1) the vibronic (or electron-phonon) coupling parameter λ , (2) the electronic coupling parameter ε, (3) the frequency of the single effective vibrational mode ν, and, in the unsymmetric case $A \neq B$, (4) the difference in zero–point energy, W, between the two coupled states of the mixed–valence dimer.

It has been shown that Robin–Day class I, II, and III systems correspond respectively to the cases $|\varepsilon| << (\lambda^2 + W)$, $|\varepsilon| \lesssim (\lambda^2 + W)$, and $|\varepsilon| > (\lambda^2 + W)$. The potential energy surfaces for symmetrical binuclear (*i.e.*, A = B) class I, II, and III systems are shown in Figure 1.

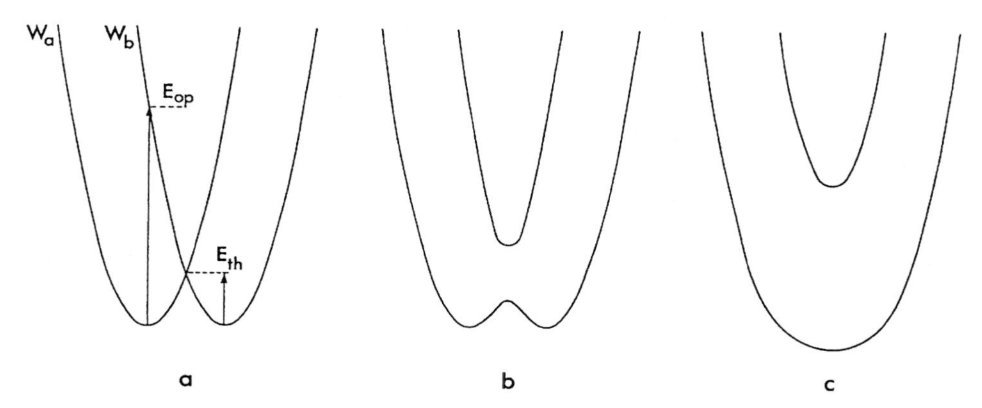

Fig. 1. Potential energy surfaces of a symmetrical binuclear mixed–valence species in terms of a vibrational coordinate, q. The minima correspond to $q = \pm\lambda$. E_{op}, intervalence charge transfer energy; E_{th}, the height of the energy barrier for thermal electron transfer, see Eq. (2). (a) $\varepsilon = 0, \lambda \neq 0$ (class I); (b) $|\varepsilon| < \lambda^2$ (class II); (c) $|\varepsilon| > \lambda^2$ (class III).

While class III has no potential barrier in the lower surface, class II has a small potential barrier, E_{th}, the height of which is given by

$$E_{th}/h\nu = 1/2\ \lambda^2 - |\varepsilon| + \varepsilon^2/2\lambda^2;\ |\varepsilon| \leq \lambda^2 \qquad (2)$$

The PKS model has been widely used to explain various experimental data of mixed–valence systems. However, more recently Ondrechen *et al.* have shown that the interpretation of experimental data may be considerably different if the bridging ligand is included as a third site (ref. 6). However, the PKS model, because of its simplicity, seems still very useful for a qualitative explanation of various properties of mixed–valence systems.

C. Experimental Methods

Attempts to define the electronic structures, and in particular the degree of valence electron delocalization, and the rates of electron transfer in class II mixed–valence compounds, have involved the application of a battery of spectroscopic probes with different time–scales, *e.g.* from *ca.* 10^{-3} s (NMR) to ca 10^{-16} s (X–ray photoelectron spectroscopy). These often provide a series of "snapshots" of the system from which approximate electron transfer rates may be estimated. The energy, intensity, and contour of the optical intervalence absorption band, (resulting from the vertical transition E_{op} in Figure 1) may be used, (ref. 5), to deduce values for ε and λ, and hence E_{th} and the electron transfer rate. However, direct observation of electron transfer over a wide temperature range is best provided by electron spin resonance (ESR) measurements when these are feasible. As we shall see, in favorable cases it is possible to distinguish between electron hopping and ground state electron delocalization by this technique, and to provide unambiguous activation energies.

Mixed–valence Cu^{II}/Cu^{I} compounds provide simple systems in which electron hopping between the two copper atoms can be detected clearly by ESR spectroscopy. If the unpaired electron is trapped on a copper atom, the solution ESR spectrum consists of four hyperfine lines from ^{63}Cu (I = 3/2, 69.1%) and ^{65}Cu (I = 3/2, 30.9%), (ref. 7). When the electron is hopping rapidly between two equivalent copper atoms, a seven–line spectrum is expected. Such seven–line spectra have been observed for a binuclear mixed–valence compound by Gagné *et al.* (ref. 8), and for related compounds by Hendrickson *et al.* (ref. 9); see Figure 2.

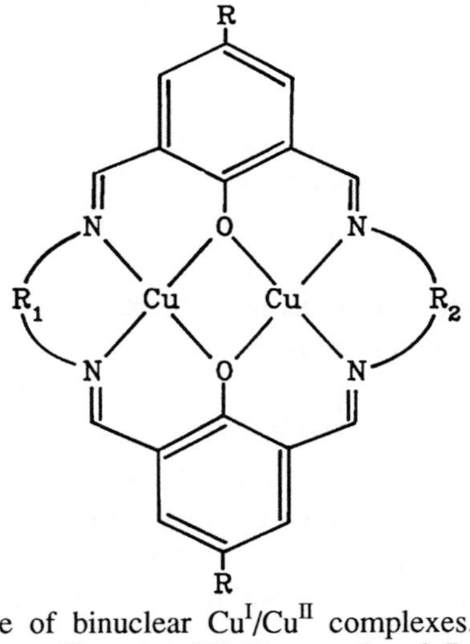

Fig. 2. Molecular structure of binuclear Cu^{I}/Cu^{II} complexes. Compound I: R=methyl, $R_1 = R_2$ = propylene. Compound II: R = tert–butyl, $R_1 = R_2$ = butylene.

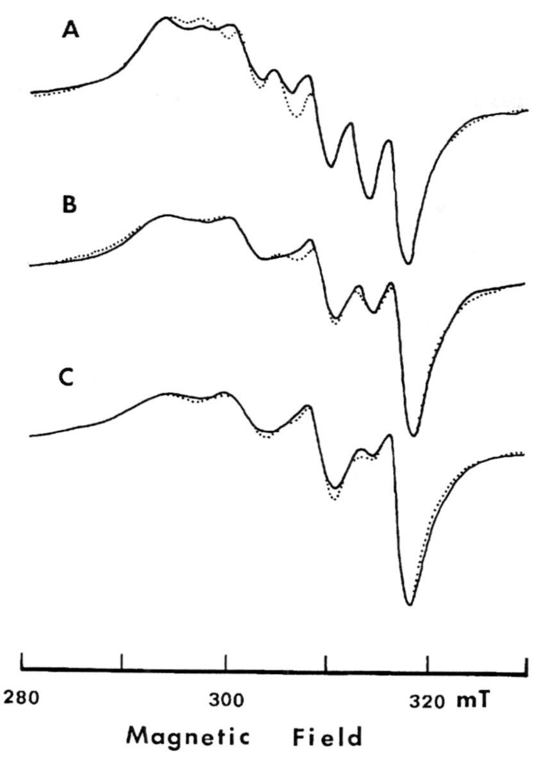

Fig. 3. Measured (——) and simulated (····) ESR spectra of Compound II in acetone at (A) 306, (B) 275, and (C) 261 K. Measured spectra are from ref. 9. The rate constants used in simulation are (A) 1.51, (B) 0.92, and (C) 0.72 in units of 10^9 s^{-1}; see ref 10.

As shown in Figure 3, the number of hyperfine lines in the ESR spectra of these complexes changes from seven to four as the temperature is lowered.

The temperature–variation of the spectrum can be simulated using a hopping electron model (ref. 10). The general approach is to calculate the positions of the hyperfine lines expected when the unpaired electron is trapped on one of the copper atoms, then to calculate the spectrum for each combination of the nuclear spin states using modified Bloch equations originally derived to describe the NMR line shape for a jumping spin (ref. 11), and finally to add up the spectra for all possible combinations of the nuclear spin states.

The rate constants calculated for the intramolecular electron transfer for compounds I and II in Figure 2 fall in the range (0.72 - 61) x 10^9 s^{-1}, and an activation energy, determined by fitting the rate constants to the Arrhenius equation, was found to be 2.6 kcal/mol (910 cm^{-1}).

It is generally accepted that the thermally activated electron hopping is preceded by geometrical adjustments to equalize the environments of both metal sites (ref. 12). For the binuclear copper complexes, it was suggested that a structure intermediate between square planar for Cu^{II} and square planar with slight tetrahedral distortion for Cu^{I} might be involved (ref. 7). (This intermediate structure corresponds to the top of the energy barrier in Figure 1(b).) Compounds I and II which have the same group for R_1 and R_2 in Figure 2 exhibit seven–line ESR spectra, indicating that this intermediate structure is attained most easily for these compounds. Only four–line spectra were observed for those compounds having two quite different groups (*e.g.* propylene and butylene) for R_1 and R_2 (ref. 9). For these unsymmetric systems the difference in zero–point energies (represented by W) also contributes to the activation energy. Thus both class I and class II systems (on the ESR time scale) can be generated by varying R_1 and R_2 groups in these binuclear copper(I,II) compounds.

II. Metal–Oxygen Cluster Anions

Several examples of the very large class of polyoxometal cluster anions ("hetero-polyanions" (ref. 13) contain the trinuclear structural motif (Figure 4), which may be approximately described as three MO_6 octahedra sharing edges.

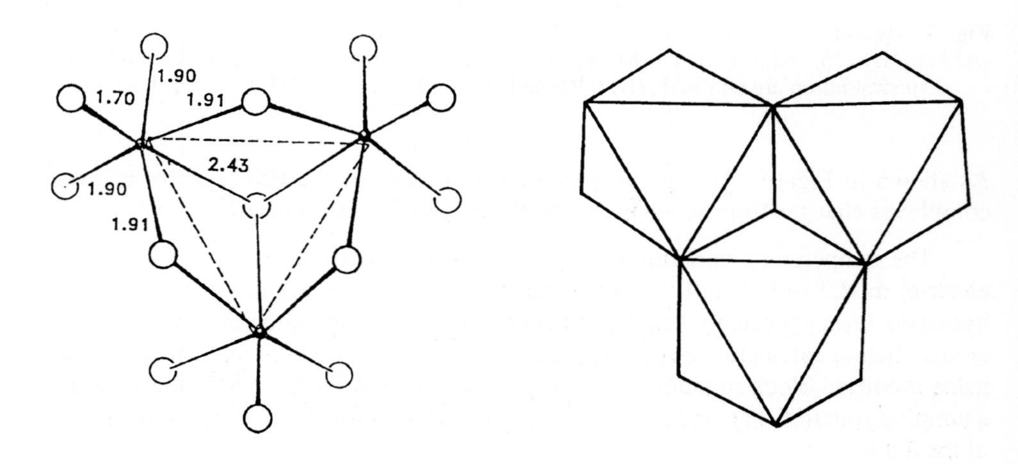

Fig. 4. The W_3O_{13} edge–shared group in α–$[PW_{12}O_{40}]^{3-}$ in bond and polyhedral representations. The bond lengths in units of Å are given.

Each octahedron actually experiences a pronounced tetragonal distortion since the metal atoms are displaced towards the terminal (unshared) oxygen atoms as shown in the Figure. The trinuclear group of Figure 4 is clearly seen in the three complete structures $\{M_6O_{19}\}$, $\{(XO_4)(M_{12}O_{36})\}$ ("Keggin"), and $\{(XO_4)_2(M_{18}O_{54})\}$ ("Dawson") shown in Figure 5.

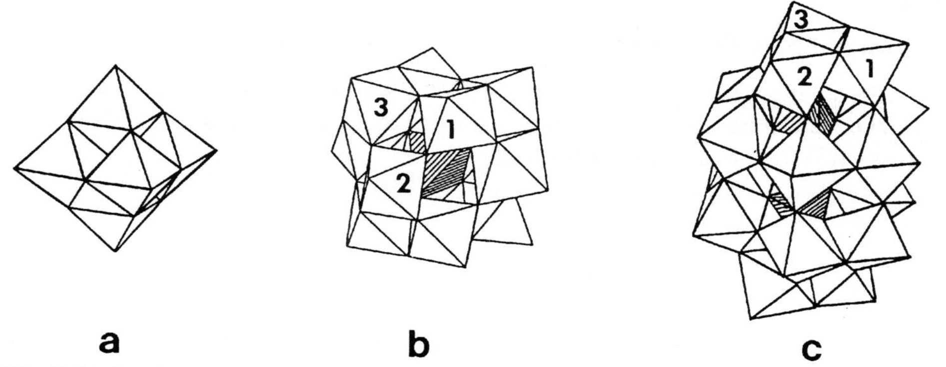

Fig. 5. (a) $\{M_6O_{19}\}$–strucuture, (b) the Keggin anion, and (c) the Dawson anion in polyhedral representations. Anions having two or three vanadium atoms in the numbered octahedra are discussed in this paper.

Such structures are formed with M = Mo^{VI}, W^{VI}, and mixtures of these elements with V^{V}. The central tetrahedral core units (XO_4) of the Keggin and Dawson structures are frequently $[PO_4]^{3-}$.

An important property of the polyoxoanion structures shown in Figure 5 is that they may be reduced stepwise to isostructural mixed–valence anions that contain (formally) one or more Mo^{V}, W^{V}, or V^{IV} (d^1) centers (refs. 13 and 14). The reduced anions have been shown in most cases to be class II mixed–valence species, and they provide excellent candidates for examining electron delocalization and hopping pathways in metal oxide structures, via variable temperature ESR spectroscopy.

A. Vanadium(IV)/Vanadium(V) Systems

Mixed–valence V^{IV}/V^{V} compounds are also convenient systems for studying the intramolecular electron transfer involving a single unpaired electron. Heteropolyanions containing two or three adjacent vanadium atoms, when reduced to a mixed–valence state, provide excellent systems for ESR study. Di– and trivanadium–substituted Keggin and Dawson anions have been studied (see Figure 5), and a variety of multi–line ESR spectra have been observed (refs. 15 and 16). The spectrum varies with

1. the mode of junction between the two VO_6 octahedra (corner–shared, $\angle$VOV *ca* 150°, or edge–shared, $\angle$VOV *ca* 120°).

2. the number of adjacent vanadium atoms.
3. protonation of the bridging oxygen atoms.

In the following discussion the electron mobility in mixed–valence heteropolyanions and in V_2O_5 containing oxygen vacancies or lithium impurities (refs. 17 and 18) will be described.

1. Corner–Shared Divanadium–Substituted Keggin Anions

The solution ESR spectrum of α-1,2-$[XV^{IV}VW_{10}O_{40}]^{n-}$ (XV_2, X = P or Si), where two VO_6 octahedra are corner–shared, consists of 15 hyperfine lines with relative intensity of 1,2,3...8...3,2,1 (Figure 6(A)).

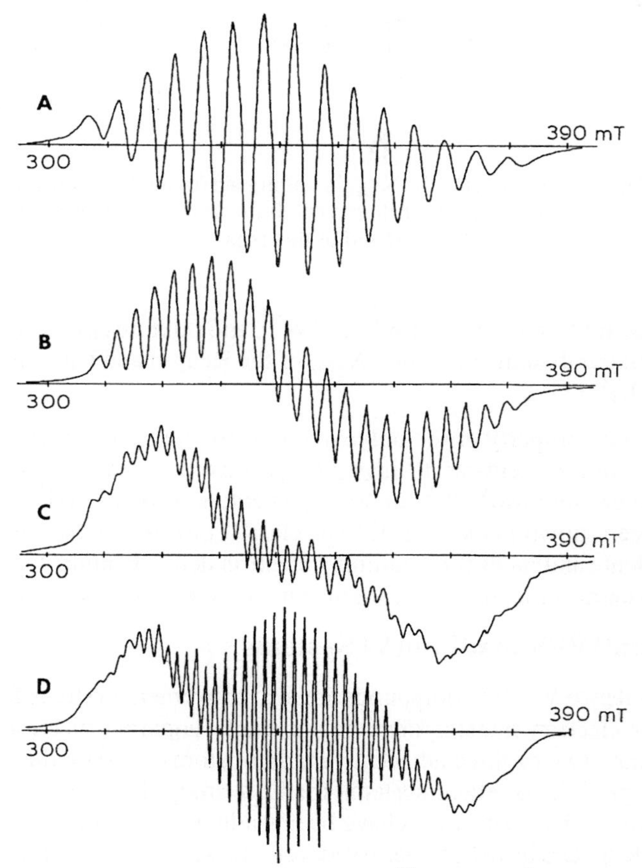

Fig. 6. Solution ESR spectra of (A) α-1,2-$[PV^{IV}VW_{10}O_{40}]^{6-}$ at 336 K, (B) 1,2,3-$[P_2V^{IV}V_2W_{15}O_{62}]^{10-}$ at 336 K, (C) 1,2,3-$[HP_2V^{IV}V_2W_{15}O_{62}]^{9-}$ at 353 K, and (D) α-1,2,3 –$[HSiV^{IV}V_2W_9O_{40}]^{7-}$ at 357 K.

This spectrum indicates that the unpaired electron is interacting equally with the two vanadium nuclei (^{51}V, I = 7/2, 100%). The unpaired electron can interact with two vanadium nuclei by a thermally activated hopping process or by ground state delocalization. The solution spectrum was simulated using a hopping electron model (ref. 19). However, since this type of spectrum can also be simulated using a delocalized electron model, it is not possible to determine whether the electron is completely delocalized in the ground state or hopping very fast on the basis of the solution spectrum. The polycrystalline ESR spectrum of the same compound exhibits temperature–dependence that is characteristic of electron hopping (ref. 20); see Figure 7.

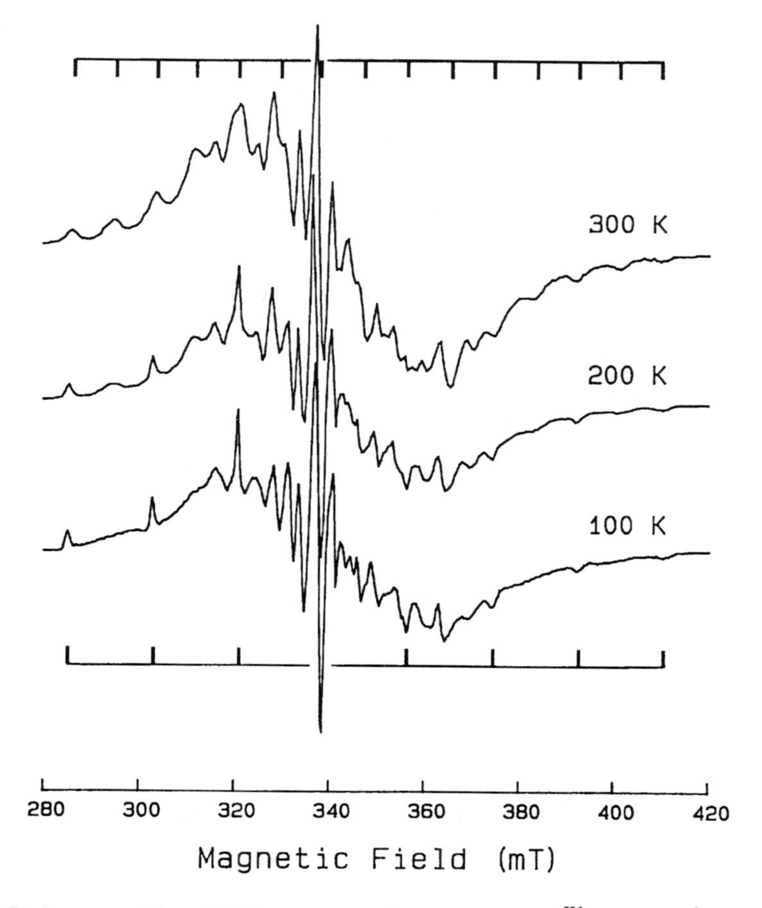

Fig. 7. Polycrystalline ESR spectra of α–1,2,3–$[HPV^{IV}V_2W_9O_{40}]^{6-}$ doped into $K_6[HSiV_3W_9O_{40}]$. The spectra are the same as those of α-1,2-$[PV^{IV}VW_{10}O_{40}]^{6-}$, a pure sample of which is difficult to prepare. Positions of the fifteen parallel lines for the 300 K spectrum and the eight parallel lines for the 100 K spectrum are shown.

At room temperature, 15 parallel lines are clearly seen. As the temperature is lowered every second parallel line broadens gradually. The spectrum at 100 K consists of eight sharp parallel lines and some very broad lines between them. The gradual change of the parallel lines from fifteen to eight could be simulated for the temperature range 300-150 K, using modified Bloch equations as for the copper complexes (ref. 21). The resulting computed rate constants were $7.0 \times 10^9\ s^{-1}$ (300 K) and $2.0 \times 10^9\ s^{-1}$ (150 K), and the activation energy was 0.77 kcal/mol (270 cm^{-1}).

2. Corner–Shared Trivanadium–Substituted Keggin Anions

The solution ESR spectrum of α-1,2,3-$[HXV^{IV}V2W_9O_{40}]^{n-}$ (HXV_3, X = P or Si) consists of 43 lines, (Figure 6(D), refs. 15 and 23). The frozen solution spectrum with 57 parallel lines and some additional perpendicular lines (Figure 8) is temperature-independent between 20 and 180 K (refs. 22 and 23), indicating that the electron is delocalized over three vanadium atoms on the ESR time scale.

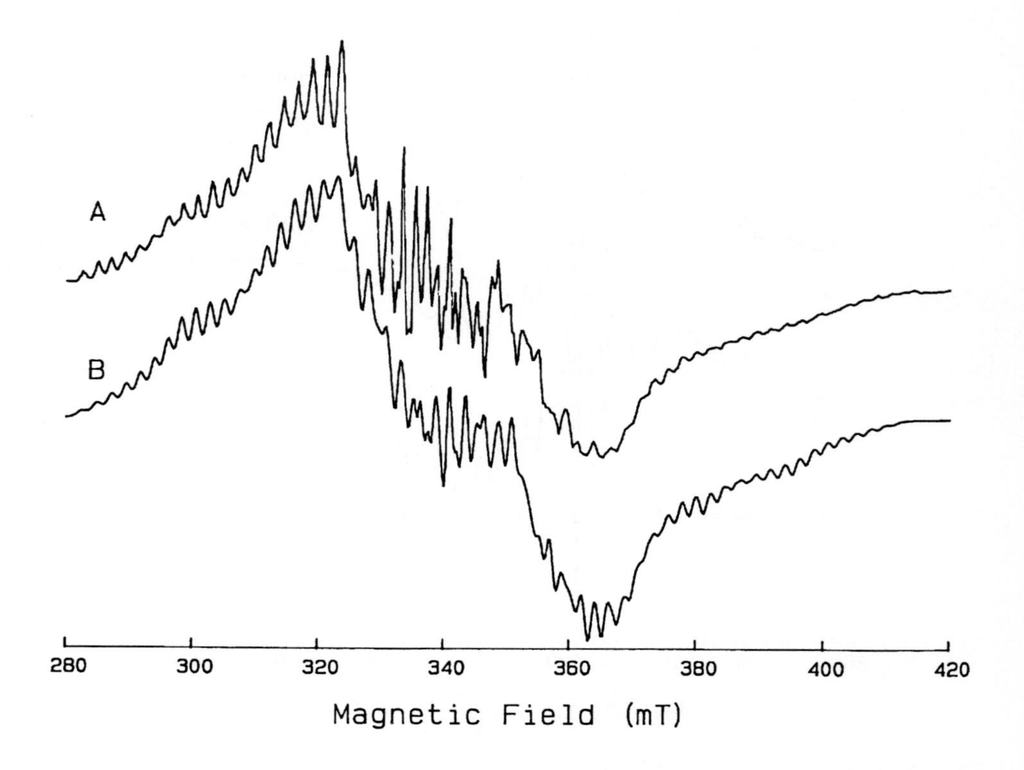

Fig. 8. (A) Measured and (B) simulated frozen solution spectra of α-1,2,3-$[HPV^{IV}V_2W_9O_{40}]^{6-}$ The parameters used in simulation are $g_{\parallel} = 1.940$, $g_{\perp} = 1.966$, $A_{\parallel} = 0.0125\ cm^{-1}$, $A_{\perp} = 0.0040\ cm^{-1}$ for V_1; $A_{\parallel} = 0.00255\ cm^{-1}$, $A_{\perp} = 0.0010\ cm^{-1}$ for V_2 and V_3. $A_{\parallel}$ represents the hyperfine coupling constant along each V = O vector.

The solution spectrum was simulated using two different hyperfine coupling constants: $a_1 = 0.0064\ cm^{-1}$ for one vanadium atom (V_1) and a_2 (a_3) = $0.0015\ cm^{-1}$ for the other two vanadium atoms (V_2 and V_3) (ref. 23). The ratio of the two hyperfine coupling constants (*ca* 4) produces a 43-line pattern. Inequivalence of the vanadium atoms is believed to originate in the protonation of one bridging oxygen atom (see Section 5 below).

The frozen solution spectrum could be simulated reasonably by using two different sets of anisotropic hyperfine coupling constants for V_1 and V_2 (V_3) (ref. 23). The hyperfine coupling constants are such that when the magnetic field is along the V_1=O vector, the hyperfine splitting by V_2 and V_3 is about one sixth of the splitting by V_1. This ratio gives 57 parallel lines. The isotropic hyperfine coupling constants calculated from the parameters for the polycrystalline spectrum (see the caption of Figure 8) are quite similar to those measured for the solution spectrum at 350 K, indicating that the unpaired electron densities on three vanadium atoms are independent of temperature.

Thus all ESR data of this trivanadium–substituted species are consistent with electron delocalization on the ESR time scale. The question arises as to whether this compound is class III, or class II with a very low activation energy. If the electron transfer between the two metal sites is faster than the vibrational time scale(10^{-13} s), the complex would be a genuine class III system. Since the polycrystalline ESR spectrum of this species is expected to be temperature–independent when the rate constant is greater than $10^{11}\ s^{-1}$, it is not possible to answer the above question unambiguously.

It is interesting to determine if the unpaired electron in this system could be described by a Hückel–type molecular orbital. The unpaired electron on the V^{4+} ion occupies the $3d_{xy}$ orbital which has π symmetry with respect to the V–O(V) bond. So we need to consider only the $3d_{xy}$ orbitals of the vanadium atoms and the p_π orbitals of the bridging oxygen atoms. If the bridging oxygen atom is not protonated, the low energy molecular orbitals which can accomodate the unpaired electron are two degenerate anti–bonding orbitals (ref. 24), which are essentially

$\Phi_1 = 2\phi_1 - \phi_2 - \phi_3$ and

$\Phi_2 = \phi_2 - \phi_3$

where ϕ_i represents d_{xy} orbitals. Protonation of the oxygen atom that bridges V_2 and V_3 breaks this degeneracy, and the unpaired electron occupies a molecular orbital related to Φ_1. If there is no interaction between V_2 and V_3, this orbital has the following form

$$\Phi' = \sqrt{2}\phi_1 - \phi_2 - \phi_3.$$

Thus the ratio of the unpaired electron densities on V_1 and V_2 should be 2 - 4 depending upon the extent of interaction between V_2 and V_3. The isotropic hyperfine coupling constant is a measure of the unpaired electron density, and the observed ratio, $a_1/a_2 = = 4$, lies in the expected range. If a more realistic molecular orbital which can explain the ESR data is found, it will provide a deeper insight about the electronic structure of this compound.

Assuming that HXV_3 is class II (hopping) rather than class III, why should the activation energy be lower than that of XV_2? The vibronic coupling parameter λ is likely to be similar for these two species, and the difference probably stems from differences in the electronic coupling. The bond length of W–O (*ca* 1.90 Å) is greater than that of V–O (*ca* 1.80 Å). So when some tungsten atoms in a Keggin anion are replaced by vanadium atoms, that porton of the anion should be contracted. Model building suggests that the trivanadium cluster in HXV_3 can be contracted relatively easily, while the V–O bonds in XV_2 cannot be contracted without considerable distortion in the framework of the anion. Thus a shorter V–O bond length in HXV_3, which corresponds to a larger electronic coupling parameter , will lower the activation energy.

3. Edge–Shared Di– and Trivanadium–Substituted Species

ESR spectra were measured for three different edge-shared species: the divanadium–substituted Dawson anion 1,2-$[P_2V^{IV}VW_{16}O_{62}]^{9-}$ (P_2V_2), and unprotonated and monoprotonated trivanadium–substituted Dawson anions 1,2,3-$[P_2V^{IV}V_2W_{15}O_{62}]^{10-}$ (P_2V_3) and 1,2,3-$[HP_2V^{IV}V_2W_{15}O_{62}]^{9-}$ (HP_2V_3). The frozen solution spectra of all these species consist of eight broad lines with some additional features, indicating that the unpaired electron is trapped on a single vanadium atom (ref. 16).

However, the solution spectrum is quite distinct for each species; see Figure 6. The solution spectra of P_2V_2 and P_2V_3 consist of 15 and 22 lines, indicating that the unpaired electron interacts equally with two and three vanadium atoms respectively. The 36–line solution spectrum of HP_2V_3 reveals that the electron interacts extensively with two kinds of vanadium atoms (ref. 16).

Thus the unpaired electron in each of these compounds is trapped on a single vanadium atom at low temperatures, but undergoes rapid hopping among the vanadium atoms at room temperature. All solution spectra were simulated using the modified Bloch equations (ref. 19). A single rate constant was used to simulate the 15– and 22–line spectra. Two different rate constants were needed to simulate the 36–line spectrum: k_1 for the transition from V_1 to V_2 (V_3) and k_2 for the reverse transition. If the ratio of these two rate constants k_1/k_2 is approximately 3, a 36-line pattern is obtained. Electron transfer between the two vanadium atoms connected by the protonated bridging oxygen atom is assumed to be effectively prevented.

The broad lines of the frozen solution spectrum were attributed to partial delocalization of the unpaired electron to other vanadium atoms (ref. 16). In fact, some superhyperfine structure is observed on certain lines, suggesting that the broad lines consist of many overlapping superhyperfine lines. The frozen solution spectrum of HP_2V_3 could be simulated by assuming that the unpaired electron localized on V_1 has superhyperfine interaction with the other two equivalent vanadium atoms; see Figure 9. The linewidth used in the simulation was 0.95 mT (9.5 G), which was the same as that for HXV_3.

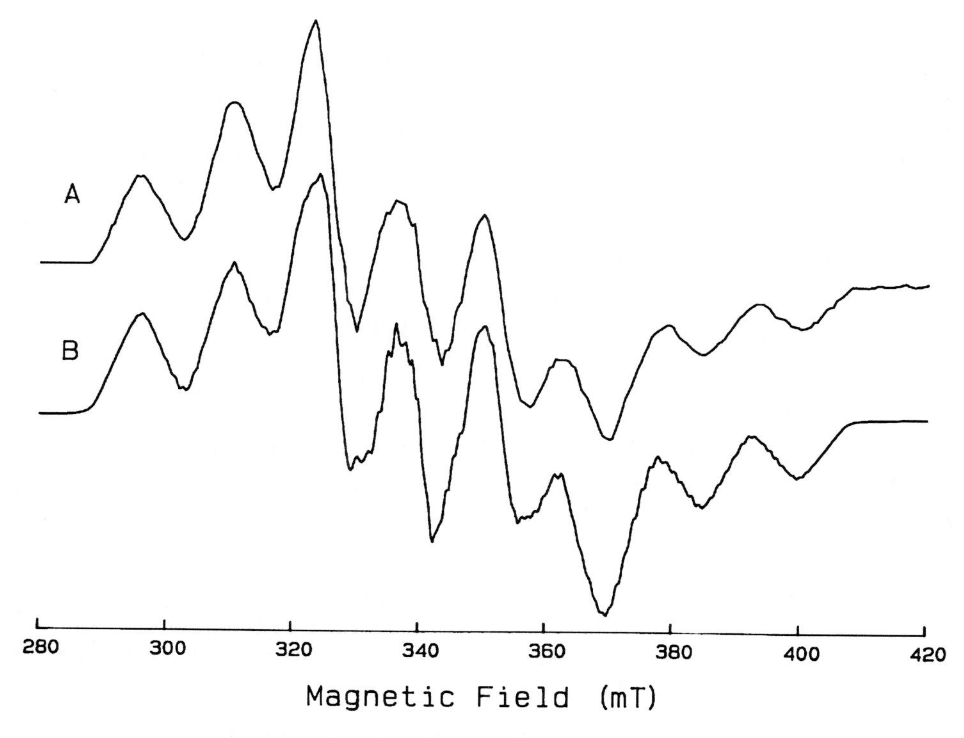

Fig. 9. (A) Measured and (B) simulated frozen solution spectra of 1,2,3-$[HP_2V^{IV}V_2W_{15}O_{62}]^{9-}$. The parameters used in simulation are $g_x = 1.972$, $g_y = 1.957$, $g_z = 1.942$, $A_{\parallel} = 0.0134\ cm^{-1}$, $A_{\perp} = 0.0045\ cm^{-1}$ for V_1; $A_{\parallel} = 0.0015\ cm^{-1}$, $A_{\perp} = 0.0008\ cm^{-1}$ for V_2 and V_3. $A_{\parallel}$ represents the hyperfine coupling constant along each V = O vector.

The ratio of the isotropic hyperfine coupling constants for the frozen solution spectrum at 150 K, $a_1/a_2 = 7.5$, is much larger than that for the solution spectrum at 350 K, $a_1/a_2 = 3$, if the solution spectrum is interpreted using a delocalized electron model. This result suggests that the electron is partially delocalized on to V_2 and V_3 by tunneling at low temperatures, and that electron hopping makes the electron spend more time on V_2 and V_3 at higher temperatures.

The edge–shared species have higher activation energies than the corner–shared species. The major difference between these two species is the V–O–V angle. It has been suggested that a large V–O–V angle in the corner–shared species increases the d–p–d π interaction, thus lowering the activation energy (ref. 16). More quantitative work is needed to confirm this explanation.

4. Vanadium Pentoxide

Sanchez *et al.* have studied the ESR spectra of single crystals of V_2O_5 containing oxygen vacancies (ref. 17) or lithium impurities (ref. 18). The V_2O_5 single crystal containing lithium impurities shows a 29–line spectrum (when the magnetic field is parallel to the V=O vectors) at low temperatures. This indicates that the unpaired electron associated with the impurity is delocalized over four equivalent vanadium atoms. As the temperature is raised, the lines start to broaden at 100 K, and a single broad line is observed at 140 K. Its width then decreases with temperature up to 210 K, above which line–broadening is again observed.

The broadening and subsequent narrowing of the spectrum at 100 - 210 K is due to the electron hopping to other sites. When the hopping is slow, a lifetime broadening is observed. As the hopping rate increases, the electron interacts with increasingly more vanadium nuclei. Since the hopping electron is more likely to feel an average magnetic field near the center of the spectrum as it interacts with more vanadium nuclei, the narrowing of the spectrum occurs. The increase of the linewidth at higher temperatures may be attributed to a mechanism causing spin–lattice relaxation. The activation energy for the electron hopping determined by linewidth analysis was 1.73 kcal/mol (600 cm^{-1}).

Vanadium pentoxide having oxygen vacancies shows a 15–line spectrum below 150 K, indicating that the electron associated with the vacancy is delocalized over two vanadium atoms. The temperature–dependence of the spectrum was interpreted in the same way as for V_2O_5 with lithium impurities.

We have seen several different behaviors of the unpaired electron in mixed–valence compounds containing V^{IV} and V^{V}. The electron can be delocalized over several vanadium atoms on the ESR time scale, as in HXV_3 and in V_2O_5 with impurities, it can hop between two vanadium atoms as in XV_2, or it can be trapped on a single vanadium atom with partial delocalization to other vanadium atoms as in the edge–shared heteropoly species. In general, compounds having corner–shared VO_6 octahedra have little or no activation energy, while those having edge–shared VO_6 octahedra have relatively high activation energies. The distorted VO_6 octahedra in V_2O_5 are joined by corner–sharing (ref. 25), and the electron delocalization over two or four vanadium atoms is consistent with the above generalization.

5. Effect of Protonation on Electron Transfer

Protonation of a bridging oxygen atom in a heteropolyanion has a very significant effect on the intramolecular electron transfer. As was mentioned above, the unprotonated edge–shared species P_2V_3 shows a 22–line solution ESR spectrum (ref. 16), indicating that the unpaired electron spends equal time on each vanadium atom. But the spectrum of the monoprotonated species HP_2V_3 consists of 36–lines, which can be simulated by assuming that electron–hopping through the protonated oxygen atom is completely prevented.

A similar effect is observed for the corner–shared species. The solution spectrum of the monoprotonated species HPV_3 at pH 5.5 consists of 43 lines. At pH 3.6, superimposed on the 43–line component is a 15–line component, which may be attributed to the diprotonated species H_2PV_3 (ref. 26). If the electron transfer through the two protonated oxygen atoms in H_2PV_3 is prevented, the electron can interact with only two vanadium atoms producing a 15–line spectrum.

It seems that protonation of a bridging oxygen atom destroys the d_π - p_π overlap between the vanadium and oxygen atoms which is needed for both the electron hopping and ground state delocalization. Some X–ray data are available showing that protonation of the bridging oxygen atom increases the metal–oxygen bond length (refs. 27 and 28).

Protonation of a bridging oxygen atom may also be partially responsible for the low activation energy in HXV_3. When a V–O bond is elongated by protonation, other V–O bonds originating in the same vanadium atom may be shortened. A short V–O bond will increase the d_π-p_π overlap, lowering the activation energy.

The effect of protonation is also seen for the multi-electron reduction products of $HSiV_3$. Its two–electron reduction product is diamagnetic, indicating that the electron spins are coupled. But the three–electron reduction product $[H_3SiV^{IV}{}_3W_9O_{40}]^{7-}$ has three unpaired electrons, whose spins are not coupled because all three bridging oxygen atoms are protonated (refs. 15 and 22).

B. Molybdenum(V)/Molybdenum(VI) Systems

Electron transfer in mixed valence polyions containing Mo^V and Mo^{VI} has also been studied by ESR spectroscopy (ref. 29). The very low temperature spectra of $[Mo^VMo_5O_{19}]^{3-}$, $[XMo^VMo_{11}O_{40}]^{n-}$ (X = P, As, Si, Ge), β–1,2,3–$[SiMo^VMo_2W_9O_{40}]^{5-}$, α–1,2,3–$[PMo^VMo_2W_9O_{40}]^{4-}$ exhibit hyperfine lines from ^{95}Mo (I=5/2, 15.7%), and ^{97}Mo (I = 5/2, 9.6%), the splitting of which indicates that the unpaired electron is trapped on a single molybdenum atom with partial delocalization to other molybdenum atoms.

As the temperature is raised, the lines broaden and the hyperfine structure disappears at 45 - 120 K; see Table 1.

Table 1.

Thermal activation energies for some Mo^{V}–Mo^{VI} systems

Compound	E_{th}, kcal/mol (cm^{-1})		T_d[a], K
$[Mo_6O_{19}]^{3-}$	3.57	(1250)	117
β–$[SiMo_3W_9O_{40}]^{5-}$	1.04	(360)	60
α–$[PMo_{12}O_{40}]^{4-}$	0.81	(280)	40

[a] temperature at which the hyperfine structure disappears.

The disappearance of the hyperfine lines and broadening of the central line can be explained in terms of the electron hopping between the molybdenum atoms. Thermal activation energies given in Table 1 were determined from an analysis of the ESR linewidth at 4 - 200 K (ref. 29).

The activation energy is higher for Mo_6O_{19} than for $SiMo_3$ and PMo_{12}. The hexamolybdate has only edge–shared MoO_6 octahedra, while $SiMo_3$ has corner–shared octahedra, and PMo_{12} has both corner–shared and edge–shared octahedra. The lower activation energy for $SiMo_3$ and PMo_{12} was attributed to the corner–sharing mode of junction. This result is in agreement with the observation on mixed–valence compounds containing V^{IV} and V^{V}. The activation energy for PMo_{12} is comparable to that of XV_2.

C. Degree of Ground State Delocalization in Mo Systems

For the mixed–valence polyions containing Mo^{V} and Mo^{VI} no multi–line spectrum showing directly the interaction of the unpaired electron with more than one molybdenum nuclei has been observed. This may be due to the fact that most work was done on natural abundance molybdenum compounds containing 75% ^{96}Mo (I = 0). But even some samples enriched with ^{95}Mo (I = 5/2) did not generate multi–line spectra (ref. 22).

The extent of ground state delocalization (which originates in tunneling for the most compounds discussed here) can be estimated by analyzing the ESR parameters. The appropriate expressions (ref. 30) for a d^1 system with the unpaired electron in a d_{xy} orbital are

$$A_{\parallel} = P[-4/7\alpha^2 - \kappa + g_{\parallel} - 2.0023 + 3/7(g_{\perp} - 2.0023)]$$

$$A_{\perp} = P[2/7\alpha^2 - \kappa + 11/14(g_{\perp} - 2.0023)]$$

where P is 2.0023 $\beta_e g_N \beta_N \langle r^{-3} \rangle$, α is the coefficient of the d_{xy} orbital in the molecular orbital, and κ is the isotropic Fermi contact term. Both α and κ are measures of the unpaired electron density on the molybdenum atom, but κ seems to be more sensitive to the electron delocalization.

Listed in Table 2 are κ values for various molybdenum compounds including some mixed–valence compounds. The largest value is found for $[Mo^VW_5O_{19}]^{3-}$ (MoW_5), where the MoO_6 octahedron is joined to four WO_6 octahedra by edge-sharing. When an edge–shared WO_6 octahedron is replaced by a corner–shared WO_6 or edge– or corner–shared MoO_6 octahedron, the value of κ is reduced. Thus using MoW_5 as the reference we may express κ as follows:

$$\kappa = \kappa_r - \sum d_i$$

where d_i represents the delocalization parameter which depends on the kind of metal (W or Mo) joined to the central atom *via* the oxygen atom and the mode of junction (edge–shared or corner–shared).

Table 2.

Fermi contact terms for polyions containing Mo^V

Compound	$g_{\parallel}$	$g_{\perp}$	$A_{\parallel}{}^a$	$A_{\perp}{}^a$	$\kappa_{exp}{}^b$	κ_{cal}	ref.
$[MoW_5O_{19}]^{3-}$	1.917	1.924	76.3	35.7	0.813[c]	0.813	29
α-$[PMoW_{11}O_{40}]^{4-}$	1.913	1.939	73.1	34.3	0.787[c]	0.787	31
α-$[SiMoW_{11}O_{40}]^{5-}$	1.914	1.931	73.7	32.1	0.759	0.787	32
$[Mo_6O_{19}]^{3-}$	1.916	1.930	72.0	30.2	0.725[c]	0.725	d
α-$[PMo_2W_{10}O_{40}]^{4-}$	1.913	1.949	70.3	28.6	0.724[c]	0.724	22
		1.934		32.1			
α-$[PMo_3W_9O_{40}]^{4-}$	1.935	1.953	57.5	28.2	0.632	0.661	22
		1.944					
β-$[SiMo_3W_9O_{40}]^{5-}$	1.921	1.950	56.9	28.0	0.617	0.661	29
α-$[PMo_{12}O_{40}]^{4-}$	1.938	1.949	54.9	24.7	0.576	0.617	29
α-$[SiMo_{12}O_{40}]^{5-}$	1.931	1.944	59.0	29.0	0.647	0.617	29

[a] In units of 10^{-4} cm^{-1}
[b] P = 0.0055 cm^{-1} was used.
[c] Values used to determine the delocalization parameters d_i
[d] M. Che, M. Fournier and J.P. Launay, J. Chem. Phys., **71** (1979) 1954.

Since κ varies according to the central atom of the Keggin anion (29), the d_i values depend upon the choice of the compounds. If we use the Keggin anions having phosphorus as the central atom, we get the following values:

d(W,c) = 0.013, d(Mo,e) = 0.022, d(Mo,c) = 0.063

where c and e represent corner–shared and edge–shared modes, respectively. The calculated values are given in Table 2.

It is noted that the following relations hold:
for the same mode of junction: d(Mo) > d(W)
for the same metal: d(corner–shared) > d(edge–shared).

Similar relations are derived for the compounds where the unpaired electron is localized mainly on a vanadium atom; see Table 3.

Table 3.

Fermi contact terms for polyions containing V^{IV}

Compound	$g_{\parallel}$	$g_{\perp}$	$A_{\parallel}$[a]	$A_{\perp}$[a]	κ[b]	ref.
$[VW_5O_{19}]^{4-}$	1.949	1.969	167.0	61.0	0.713	c
$[PVW_{11}O_{40}]^{5-}$	1.910	1.966	167.0	59.3	0.689	d
$[PVMo_{11}O_{40}]^{5-}$	1.939	1.974	151.2	53.4	0.632	31
$[HP_2V_3W_{15}O_{62}]^{9-}$	1.942	1.957	134.0	45.0	0.538	e
		1.972				
$[HPV_3W_9O_{40}]^{6-}$	1.940	1.966	125.0	40.0	0.489	23

[a] In units of 10^{-4} cm^{-1}
[b] P = 0.0128 cm^{-1} was used.
[c] H. So, C.M. Flynn Jr. and M.T. Pope, J. Inorg. Nucl. Chem., **36** (1974) 329.
[d] D.P. Smith, H. So, J. Bender and M.T. Pope, Inorg. Chem., **12** (1973) 685.
[e] This work.

d(V) > d(Mo) > d(W) for the same mode of junction
d(corner–shared) > d(edge–shared) for the same metal

The greater degree of electron delocalization in the molybdovanadates than in the tungstovanadates was pointed out before (ref. 31). The sequence d(V) > d(Mo) > d(W) is consistent with the optical electronegativities of these metals (ref. 32).

As was already pointed out by Sanchez *et al.*, (ref. 29), ground state delocalization is easier through the corner–sharing mode of junction than the edge–sharing one.

D. Tungsten(V)/Tungsten(VI) Systems

Electron transfer in mixed–valence polyions containing W^{V} and W^{VI} has been studied by ESR spectroscopy (ref. 33). The frozen solution spectrum of $[W_6O_{19}]^{3-}$ at 10 K consists of an anisotropic central line from those isotopes with I = 0 (86.74%) and weak hyperfine lines from ^{183}W (I = 1/2, 13.26%). The hyperfine lines disappear around 25 K, but the spectrum remains anisotropic. It then becomes isotropic at 40 K, and a continuous broadening of the line is observed up to 200 K.

The well–resolved hyperfine structure in the low temperature spectrum indicates that the electron is trapped on a single tungsten atom. The broadening of the line was attributed to electron hopping, and an activation energy of 1.27 kcal/mol (440 cm^{-1}) was obtained by an analysis of the linewidth.

No hyperfine structure was observed at 10 K for other compounds such as $[SiW_{12}O_{40}]^{5-}$ and $[As_2W_{18}O_{62}]^{7-}$, but the temperature dependence of the spectra for these compounds was similar to that of $[W_6O_{19}]^{3-}$, and the electron behavior in these compounds was interpreted in a similar way.

III. NMR Study of Electron Hopping in Heteropolyanions

We have seen that most of the polyanions containing V^{IV}/V^{V}, Mo^{V}/Mo^{VI}, or W^{V}/W^{VI} are class II mixed–valence species. The unpaired electron is trapped on a single metal atom at low temperatures, but undergoes hopping among several or all metal atoms in the polyanion as the temperature is raised.

The effect of the electron hopping can be studied, in some heteropolyanions, by NMR spectroscopy. The one–electron reduced phosphorus–centered Keggin anions α–$[PW_{12}O_{40}]^{4-}$ and α–$[PMo_{12}O_{40}]^{4-}$ yield narrow ^{31}P NMR lines, and the ^{17}O NMR spectra of the same anions consist of four sharp lines consistent with a structure of full T_d symmetry; see Figure 10. These results demonstrate that the electron is hopping so rapidly that it is essentally delocalized over the whole anion at room temperature on the NMR time scale (ref. 34).

Four–line ^{17}O NMR spectra were also observed for the two–electron reduced Keggin anions (refs. 35 and 36), and these results also demonstrate complete electron delocalization. Using the Evans NMR susceptibility method to measure the diamagnetic susceptibilities of oxidized and two–electron reduction products of these compounds, Kozik *et al.* found that the two–electron reduction products were more diamagnetic than their oxidized parents (ref. 37). The excess diamagnetism was attributed to a ring current of the paired electrons circulating in a loop in a plane perpendicular to the external magnetic field.

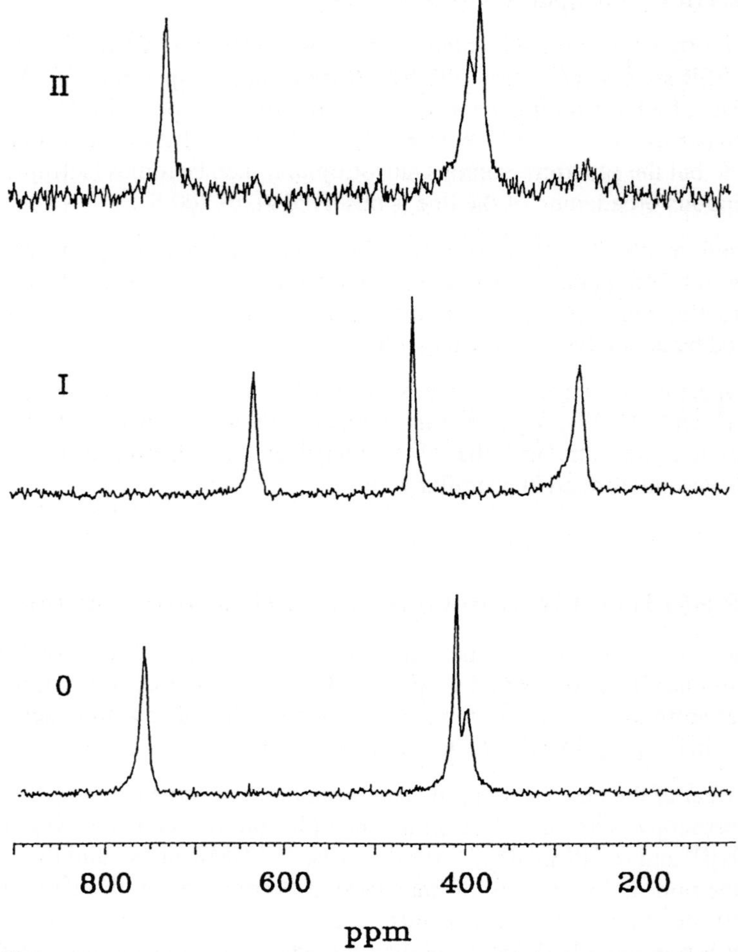

Fig. 10. ^{17}O NMR spectra of aqueous solutions of $H_4SiW_{12}O_{40}$(**0**), $H_5SiW_{12}O_{40}$(**I**), and $H_6SiW_{12}O_{40}$ (**II**). Chemical shifts relative to external H_2O. The resonance for the internal oxygen (25 - 31 ppm) is not shown.

IV. Summary and Conclusions

If the ESR spectrum of a mixed–valence compound is temperature–dependent, a hopping electron model can be used to analyze the spectrum and to determine the rate constant for the intramolecular electron transfer. Some mixed–valence Cu^I/Cu^{II} and V^{IV}/V^V compounds having rate constants on the order of 10^9 s^{-1} exhibit such temperature–dependent spectra.

When a multi–line ESR spectrum, showing that the unpaired electron interacts extensively with more than one metal atom, is temperature–independent, it can be analyzed using a delocalized electron model. In this case the electron is either delocalized or hopping very fast, and only the lower limit of the rate constant can be estimated.

The frozen solution ESR spectra of some compounds reveal that the unpaired electron is trapped mainly on a single metal atom. But some lines show superhyperfine structure, which may be attributed to tunneling of the unpaired electron to other metal atoms. The interaction of the unpaired electron with these metal atoms becomes more extensive at higher temperatures as the hopping process is turned on.

ESR and NMR spectra of polyanions containing V^{IV}/V^{V}, Mo^{V}/Mo^{VI}, or W^{V}/W^{VI} reveal that the unpaired electron is quite mobile. At very low temperatures the unpaired electron is usually trapped on a single metal atom, but can be partially delocalized by tunneling. This effect is seen as superhyperfine structure in the frozen solution spectrum of HP_2V_3 and as reduced Fermi contact terms for the compounds listed in Tables 2 and 3. As the temperature is raised, the electron undergoes hopping among the metal atoms. At room temperature the rate constants for the intramolecular electron transfer are probably greater than 10^9 s^{-1} for most of these compounds, producing multi–line ESR spectra for those containing V^{IV}/V^{V} and generating sharp ^{31}P and ^{17}O NMR lines for one–electron–reduced Keggin anions.

The activation energies and delocalization parameters suggest that the electron transfer (both through and over the energy barrier) is easier through the corner–sharing mode of junction than the edge–sharing one. But at high temperatures the electron hopping seems quite efficient through both modes of junctions.

A considerable amount of experimental data has been collected for the electron transfer in mixed–valence metal–oxygen clusters. A coherent picture of the electronic structure which can explain all these data has not yet been developed, and remains as a challenging problem.

Acknowledgments

HS thanks Sogang University for a sabbatical leave, during which this article was prepared; also the Korean Ministry of Education and the Korea Science Foundation for research support. MTP thanks the National Science Foundation, the Petroleum Research Fund, administered by the American Chemical Society, and ARCO Chemical Corporation for research support.

References

1 M.J. Ondrechen, J. Ko and L. Zhang, J. Am. Chem. Soc. **109** (1987) 1672.

2 M.B. Robin and P. Day, Adv. Inorg. Chem. Radiochem. **10** (1967) 247.

3 N.S. Hush, Prog. Inorg. Chem. **8** (1967) 391.

4 In many cases high pseudosymmetry imposed by the space group results in crystallographic disorder and prevents distinction of A and B.

5 (a) S.B. Piepho, E.R. Krausz and P.N. Schatz, J. Am. Chem. Soc. **100** (1978) 2996.
(b) K.Y. Wong, P.N. Schatz and S.B. Piepho, J. Am. Chem. Soc. **101** (1979) 2793.

6 (a) L.J. Root and M.J. Ondrechen, Chem. Phys. Lett. **93** (1982) 421.
(b) M.J. Ondrechen, J. Ko and L.J. Root, J. Phys. Chem. **88** (1984) 5919.
(c) J. Ko and M.J. Ondrechen, Chem. Phys. Lett. **112** (1984) 507.

7 ^{63}Cu and ^{65}Cu have similar nuclear moments, and their hyperfine lines overlap when the lines are broad.

8 R.R. Gagné, C.A. Koval, T.J. Smith and M.C. Cimolino, J. Am. Chem. Soc. **101** (1979) 4571.

9 D.N. Hendrickson and R.C. Long, J. Am. Chem. Soc. **105** (1983) 1513.

10 H. So, Bull. Korean Chem. Soc. **8** (1987) 111.

11 (a) H.S. Gutowsky, D.W. McCall and C.P. Slichter, J. Chem. Phys. **21** (1953) 279.
(b) H.S. Gutowsky and A. Saika, J. Chem. Phys. **21** (1953) 1688.

12 F. Basolo and R.G. Pearson *Mechanism of Inorganic Reactions* 2nd ed., Wiley, New York, 1968, p. 454.

13 M.T. Pope *Heteropoly and Isopoly Oxometalates* Springer-Verlag, New York, 1983.

14 M.T. Pope, in: *Mixed–Valence Compounds* NATO-ASI Ser. C, D.B. Brown (Ed.), Reidel, Dordrecht, 1980, p. 365.

15 M.M. Mossoba, C.J. O'Connor, M.T. Pope, E. Sinn, G. Hervé and A. Tézé, J. Am. Chem. Soc. **102** (1980) 6864.

16 S.P. Harmalker, M.A. Leparulo and M.T. Pope, J. Am. Chem. Soc. **105** (1983) 4286.

17 C. Sanchez, M. Henry, J.C. Grene and J. Livage, J. Phys. C: Solid State Phys. **15** (1982) 7133.

18 C. Sanchez, M. Henry, R. Morineau and M.C. Leroy, phys. stat. sol. **122** (1984) 175.

19 H. So, C.W. Lee and D. Lee, Bull. Korean Chem. Soc. **8** (1987) 384.

20 C.W. Lee and H. So, Bull. Korean Chem. Soc. **7** (1986) 318.

21 H. So and M.T. Pope, unpublished results. The overall shape of the experimental spectrum at low temperatures deviates from that of the simulated one. The experimental spectrum seems to have an extra broad component.

22 M.M. Mossoba, Ph.D. Thesis, Georgetown University, 1980.

23 C.W. Lee and H. So, Bull. Korean Chem. Soc. **11** (1990) 115.

24 J.P. Launay and F. Babonneau, Chemical Physics **67** (1982) 295.

25 The V = O vectors in V_2O_5 are parallel and ∠VOV is 210°.

26 C.W. Lee, H. So and K.R. Lee, Bull. Korean Chem. Soc. **7** (1986) 108.

27 J.N. Barrows, G.B. Jameson and M.T. Pope, J. Am. Chem. Soc. **107** (1985) 1771.

28 S.H. Wasfi, W. Kwak, M.T. Pope, K.M. Barkigia, R.J. Butcher and C.O. Quicksall, J. Am. Chem. Soc. **100** (1978) 7786.

29 C. Sanchez, J. Livage, J.P. Launay, M. Fournier and Y. Jeannin, J. Am. Chem. Soc. **104** (1982) 3194.

30 See, for example, B.R. McGarvey, Transition Metal Chem. **3** (1966) 89.

31 J.J. Altenau, M.T. Pope, R.A. Prados and H. So, Inorg. Chem. **14** (1975) 417.

32 H. So and M.T. Pope, Inorg. Chem. **13** (1974) 831.

33 C. Sanchez, J. Livage, J.P. Launay and M. Fournier, J. Am. Chem. Soc. **105** (1983) 6817.

34 K. Piepgrass, J.N. Barrows and M.T. Pope, J. Chem. Soc., Chem. Commun. (1989) 10.

35 L.P. Kazansky, M.A. Fedotov, I.V. Potapova and V.I. Spitsyn, Dokl. Chem., Engl. Trans. **244** (1979) 36.

36 M. Kozik, C.F. Hammer and L.C.W. Baker, J. Am. Chem. Soc. **108** (1986) 2748.

37 M. Kozik, N. Casan–Pastor, C.F. Hammer and L.C.W. Baker, J. Am. Chem. Soc. **110** (1988) 7697.

Electron and Proton Transfer in Chemistry and Biology, edited by A. Müller et al.
Studies in Physical and Theoretical Chemistry, Vol. 78

Electron Transfer in Semiconducting Colloids and Membranes, Applications in Artificial Photosynthesis

Michael Grätzel

Institut de Chimie Physique
Ecole Polytechnique Fédérale de Lausanne
1015 Lausanne (Switzerland)

Summary

A new molecular photovoltaic system for solar light harvesting and conversion to electricity will be discussed. It is based on the spectral sensitization of a thin ceramic membrane by suitable transition metal complexes. The film consists of nanometer-sized colloidal titanium dioxide particles sintered together to allow for charge carrier transport. When derivatized with a suitable chromophore these membranes give extraordinary efficiencies for the conversion of incident photons into electric current, exceeding 90% for certain transition metal complexes within the wavelength range of their absorption band. The present paper discusses the underlying physical principles of these astonishing findings. Exploiting this discovery, we have developed a new type of photovoltaic device whose overall lights to electric energy conversion yield is 12% in diffuse day light and 7% under direct (AM1) solar irradiation.

Introduction

In a conventional p-n-junction photovoltaic cell, made *e.g.* from silicon, the semiconductor assumes two roles at the same time: it harvests the incident sunlight and conducts the charge carriers produced under light excitation. In order to function with a good efficiency the photons have to be absorbed close to the p-n interface. Electron-hole pairs produced away from the junction must diffuse to the p-n contact were the local electrostatic field separates the charges. In order to avoid charge carrier recombination during the diffusion the concentration of defects in the solid must be small. This imposes severe requirements on the purity of the semiconductor material rendering solid state devices of the conventional type very expensive.

Molecular photovoltaic systems separate the function of light absorption and carrier transport. The light harvesting is carried out by a sensitizer which initiates electron transfer events leading to charge separation. This renders unnecessary the use of expensive solid state components in the system. While simple from the conceptual point of view the practical implementation of such devices must overcome formidable obstacles if the goal is to develop molecular systems that convert sunlight to electricity with an efficiency comparable to that of silicon photovoltaic devices. The approach taken by us will now be outlined in more detail. We begin by giving a very brief account of some basic principles employed by green plants, algae and cyanobacteria to harvest and convert solar energy which have inspired us in the choice of our strategies.

Principles of Light Energy Harvesting and Conversion in Green Plants

The light reaction in plants is called photophosphorylation. It involves the reduction of an electron carrier, *i.e.* nicotinamide adenine dinucleotide phosphate (NADP), by water to produce NADPH and oxygen. This photoinduced redox reaction is coupled to the generation of adenosine triphosphate (ATP) from adenosine diphosphate:

$$2H_2O + 2NADP + 3ADP + 3P \rightarrow 2NADPH + H^+ + 3ATP + O_2 \qquad (1)$$

The electron transfer process takes place in the thylakoid membranes located in the interior part of the chloroplasts of plant cells. The photosynthetic unit assembled in these membranes is composed of antenna pigments and a reaction center consisting of two photosystems. The absorption of light causes electrons to be ejected from chlorophyll reaction centers and then passed between various electron-transferring mediators (M) of the photosystems. In that way, a chain of redox reactions is induced by light. Chlorophyll and the participating mediators, for example pheophytin, ferrodoxin, cytochromes, and various quinones, are spatially ordered in such a way that the electron transfer takes place directionally ("vectorially") from the inner to the outer section of the photosynthetic membrane, Figure 1.

The spatial arrangement of the mediators is known from the recent successful isolation of the reaction center protein of *Rhodopseudomonas viridis* (ref. 1). This transmembrane electron flow is completed within a few hundred microseconds, resulting in light-induced charge separation. The negative charge is then located on the outer side of the membrane, and is ultimately used for the reduction of the $NADP^+$. (The NADPH thereby produced is employed in the dark reaction, the Calvin cycle, for the fixation of carbon dioxide). The positive charges remaining on the inner side of the membrane in the form of chlorophyll cations serve for the oxidation of water to oxygen. The overall reaction, despite its complexities in detail corresponds to the simple equation above.

In summary, the strategy employed by nature to accomplish this thermodynamically uphill chemical conversion is to use a molecular absorber and not a solid state device such a silicon p-n junction. The key element to achieve light induced charge separation is the presence of a membrane which serves to organize spatially the electron transfer mediators in an optimal fashion. This allows for the vectorial displacement of photogenerated electrons from the inside to the outside of the thylakoid vesicles. The conversion efficiency achieved in the primary charge separation step of plants' photosynthesis is rather high, *i.e. ca* 12%. Unfortunately, the overall conversion yield drops to at most 3-4% for top efficiency crops under optimal conditions of biomass generations.

The Molecular Machine of Chloroplasts

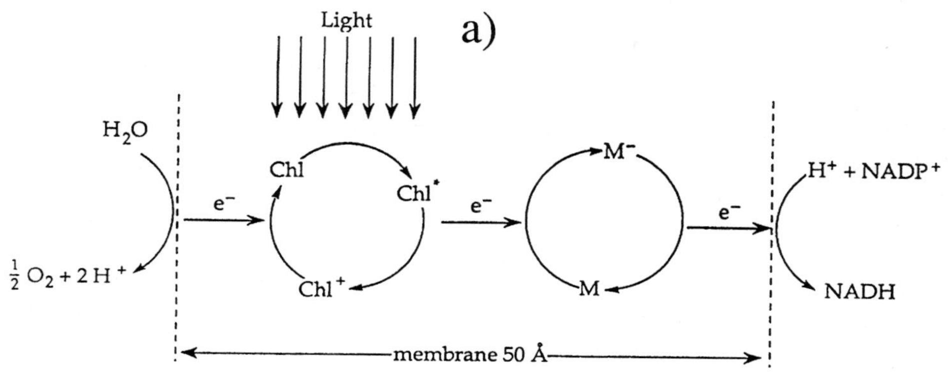

Chl = chlorophyll
M = electron relay (mediator)

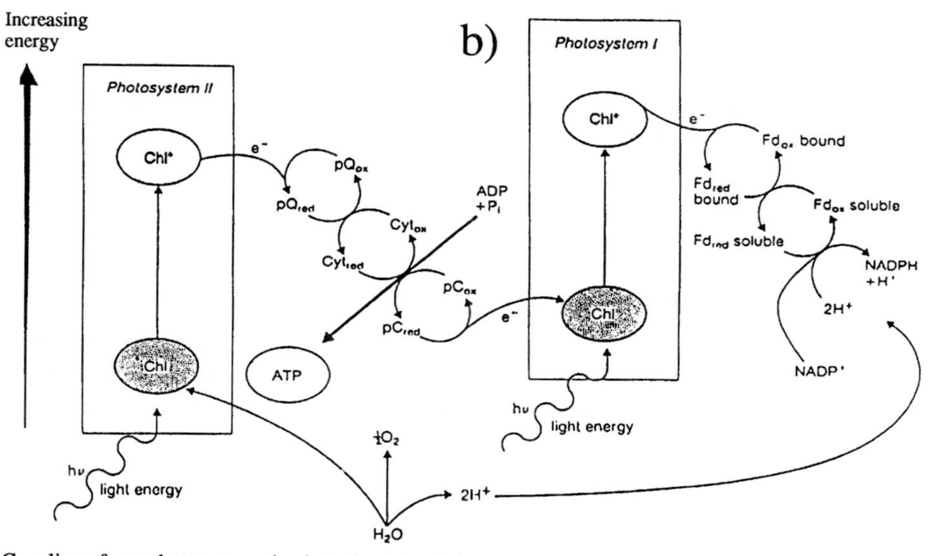

Coupling of two photosystems in photophosphorylation
Chl = chlorophyll; Chl* = electronically excited chlorophyll; pQ = plastoquinone; Cyt = Cytochrome; pC = plastocyanine; Fd = ferrodoxine. The subscripts 'red' and 'ox' refer to reduced and oxidized form, respectively.

Fig. 1. Light energy harvesting and light-induced charge separation in photosynthesis: a) the light driven molecular electron pump operation in green leaves, b) a more detailed presentation of the redox events occurring in the photosynthetic reaction centers of green plants.

Molecular Photovoltaics and Artificial Photosynthesis

Our newly developed photovoltaic cell replicates the most important principles of its prototype, photosynthesis. In any case, the components of the synthetic system must be selected to satisfy the high stability requirements encountered in principal applications. A Photovoltaic system must remain serviceable for 20 years without significant loss of performance. In living systems this stability is less significant, since unstable components are continuously renewed. Because chlorophyll, and likewise the lipid membrane, are labile *in vitro*, they cannot be adopted directly. in artificial photosynthesis chlorophyll is therefore substituted by a more stable sensitizer molecule (S). One of the most remarkable achievements of research in inorganic chemistry during the last two decades has been the development of a great variety of transition metal complexes (ref. 2) mainly of the elements osmium and ruthenium which are exceptionally stable and display good absorption in the visible. We have submitted some of these sensitizers to long time illuminations where they sustained as much as 10^7 redox cycles under light without noticeable decomposition. The redox potentials of these complexes can be adjusted to the desired value by suitable choice of the ligands and their substituents.

The role of the sensitizer is the same as that of chlorophyll: it must absorb the incident sunlight and exploit the light energy to induce a vectorial electron transfer reaction. In place of the biological lipid membrane, a titanium dioxide ceramic membrane is employed.

Titanium dioxide is a semiconductor, which does not absorb visible light because of its large (about 3eV) band gap. It is a harmless environment-friendly material, remarkable for its very high stability. It occurs in nature as ilmenite, and is used in quantity as a white pigment and as an additive in toothpaste. World annual production is of the order of a million tons, at a price of about US$1/kg. Since the membrane is about 5 microns thick, about 10 grm. of titanium dioxide is used per square meter of solar collector surface, representing an investment of only 1 cent per square meter.

The role of the titanium dioxide film is to provide a support for the sensitizer which must be applied to the surface of the membrane as a monomolecular layer. Furthermore, the conduction band of the titanium dioxide accepts the electrons from the electronically excited sensitizer. The electron injected into the conduction band travels very rapidly across the membrane. Its diffusion is at least 10^4 times faster than that of a charged ion in solution. The time required for crossing a TiO_2 membrane, say five micrometer thick, is only about 2 microseconds. During migration the electrons maintain their high electrochemical potential which is equal to the Fermi level of the semiconductor. Thus, the principal function of the TiO_2, apart for supporting the sensitizer, is that of charge collection and conduction. The advantage of using a semiconductor membrane rather that a biological one as employed by natural photosynthesis, is that such an inorganic membrane or film is more stable and allows extremely fast trans-membrane electron movement. The charge transfer across the photosynthetic membrane is less rapid since it takes about 100μs to displace the electron across to the 50Å thick thylakoid layer. Moreover, nature has to sacrifice about half of the absorbed photon energy to drive the transmembrane redox process at such a rate. In the case of the semiconductor membrane, the price to pay for the rapid vectorial charge displacement is small. It corresponds to about 0.2-0.3V of voltage drop required to establish the electrical field in the space charge layer at the semiconductor/electrolyte junction.

It is important to note that minority carriers, i.e. holes in the case of an n-type conductor such as TiO_2, do not participate in the photoconversion process. This is a great advantage in comparison to conventional photovoltaic devices, where, without exception, the generation and transport of minority carriers is required. The performance characteristics of the conventional device are strongly influenced by the minority carrier diffusion length, which is very sensitive to the presence of imperfections and impurities in the semiconductor lattice. Our cell operates entirely on majority carriers whose transport is not subjected to these limitations and hence will be much less sensitive to lattice defects.

As in natural photosynthesis, in the new photovoltaic device sunlight sets in action a molecular electron pump, whose principle is schematically represented in Fig. 2.

ARTIFICIAL PHOTOSYNTHESIS

⇒ The sensitizer S replaces chlorophyll
⇒ The semiconducting membrane replaces the biological membrane

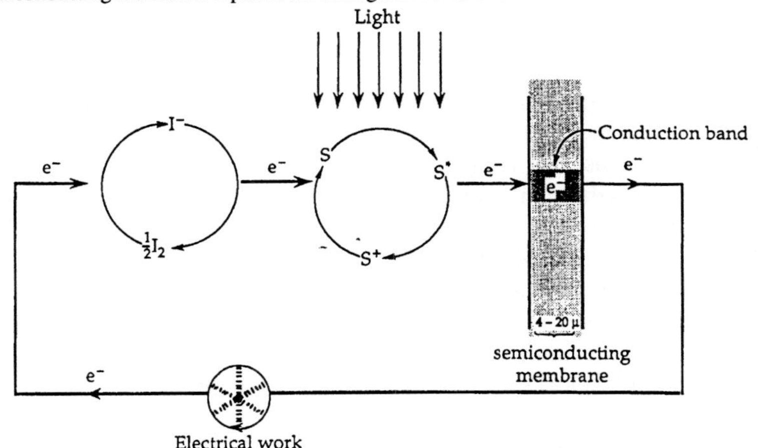

The new molecular machine constitutes an electron pump, driven by sunlight.

Fig. 2. The principles of the artificial leaf: the chlorophyll is replaced by a transition metal sensitizer while the phospholipid membrane is exchanged for a ceramic semiconducting membrane made of TiO_2. As in photosynthesis, the new solar converter constitutes a molecular electron pump driven by sunlight.

The sensitizer (S) is bound as a monomolecular coating on the surface of the titanium dioxide membrane. It absorbs the incident solar rays, and is thereby raised to the electronically excited state S^*. From this state it injects an electron into the conduction band of the titanium dioxide. The conduction band electrons then cross the membrane and are directed through a charge collector into external current circuit where electrical work is done. The electrons are then returned to the cell through a counter electrode. Between this counter electrode and the titanium dioxide membrane is an electrolyte containing a redox, *i.e.* iodine and iodide. This redox electrolyte allows for the transport of electrical charge between the two electrodes. The electrons reduce iodine to iodide ions which diffuse from the counter electrode to the titanium dioxide membrane, where they regenerate the sensitizer by electron transfer to the sensitizer cations, while simultaneously the iodide is oxidised back to iodine. The redox cycle leading to the conversion of light into electrical current is thereby closed. In direct sunshine each sensitizer molecule follows this cycle about twenty times per second. The molecular machine runs therefore at 1200 rpm.

Light Harvesting by Monomolecular Layers

For the absorption of solar rays by sensitizers attached as monolayers to the surface of a titanium dioxide membrane, there is a fundamental problem of the limited light capture cross section of the dye molecule. The cross section σ is related to the decadic molar extinction coefficient $\varepsilon(\lambda)$ by the formula:

$$\sigma(\lambda) = \varepsilon(l) \times 1000/N_A, \text{ where } N_A = 6 \times 10^{23} \quad (2)$$

Typical values for the decadic extinction coefficient of transition metal complexes lie between 10^4 and 5×10^5 $M^{-1}cm^{-1}$. This implies a light capture cross section between 0.16 and 0,8Å^2. In contrast, the sensitizer molecule occupies an area of about 100Å^2 on the surface of the membrane. It is clear from this comparison that the surface are requirement of the sensitizer is at least 125 times larger than its light capture cross section. That signifies that a monomolecular layer of the sensitizer on a smooth surface absorbs less that 1% of the incident light in the wavelength range of maximum absorption. One could naturally think, then, of depositing several molecular layers of sensitizer on the semiconductor membrane in order to increase the light absorption. This would, however, be a mistaken tactic, since the outer dye layers would act only as a light filter, with no contribution to electrical current generation. The application of a monomolecular layer of sensitizer is therefore unavoidable.

A successful strategy to solve the problem of light absorption through such extremely this molecular layers is found in the application of textured titanium dioxide membranes. It is possible using the sol-gel method to produce transparent membranes consisting of colloidal titanium dioxide particles with diameters of 10-20 nm. The electronic contact between the particles is produced by a brief sintering at about 500°C (ref. 3). A microporous structure with a very high effective surface area is thereby formed. For example the effective surface of a 5 micron thick film of such a colloidal structure is at least 300 times greater than that of a smooth membrane. On the geometric projection of such a surface a sensitizer concentration of $\Gamma = 3x10^{16}$ molecules cm^{-2} is reached when colloidal membranes are used. The optical density

$$OD(\lambda) = \Gamma \times \sigma(\lambda) \tag{3}$$

calculated for this coating level and a light capture cross section per sensitizer molecule of $0.5Å^2 = 0.5 \times 10^{-16}$ cm^2, is 1.5. The light harvesting efficiency of the device LHE(λ) is then given by:

$$LHE(\lambda) = 1\text{-}10^{-\Gamma \times \sigma(\lambda)} = 1 - 10^{OD(\lambda)} \tag{4}$$

With an optical density of 1.5, 97% of the light is absorbed by the membrane covered by a monolayer of sensitizer.

The Quantum Yield of Charge Injection

The quantum yield of charge injection (ϕ_{inj}) is the fraction of the absorbed photons which are converted into electrons injected in the conduction band. Charge injection from electronically excited sensitizer into the conduction band of the semiconductor is in competition with other radiative or radiationless deactivation channels. Taking the sum of the rate constants of these nonproductive channels together as k_{eff} results in:

$$(\phi_{inj}) = k_{inj} / (k_{eff} + k_{inj}) \tag{5}$$

One should remain aware that the deactivation of the electronically excited state of the sensitizer is generally very rapid. Typical k_{eff} values lie in the range from 10^6 to10^{10} sec^{-1}. To achieve a good quantum yield the rate constant for charge injection should be at least 100 times higher than k_{eff}. That means that injection rates up to 10^{12} sec^{-1} must be attained. In fact, in recent years sensitizers have been developed that satisfy these requirements. These dyes should incorporate functional groups ("interlocking groups") as for example carboxylates or chelating groups, which besides bonding to the titanium dioxide surface, also effect as enhanced electronic coupling oft he sensitizer with the conduction band of the semiconductor.

Very promising results have, so far, been obtained with ruthenium complexes where at least one of the ligands was 4,4'-dicarboxy-2,2'-bipyridyl. The two carboxylates groups serve to attach the Ru complex to the surface of the TiO_2 and to establish good electronic coupling between the π^* orbital of the electronically excited complex and the 3d wavefunction manifold of the TiO_2 film. The substitution of the bipyridyl with the carboxylate groups lowers also the energy of the π^* orbital of the ligand. Since the electronic transition is of MLCT (metal to ligand charge transfer) character, this serves to chanel the excitation energy into the right ligand, that is the one from which electron injection into the semiconductor takes place. With molecules like these the quantum yield of charge injection generally exceeds 90% (refs. 3 and 4).

Light-induced Charge Separation and Current Yield

As the last step of the conversion of light into electrical current, a complete charge separation must be achieved. On thermodynamic grounds, the preferred process for the electron injected into the conduction band of the titanium dioxide membrane is the back reaction with the sensitizer cation, Fig. 3.

PHOTO-INITIATED ELECTRON TRANSFER CYCLE

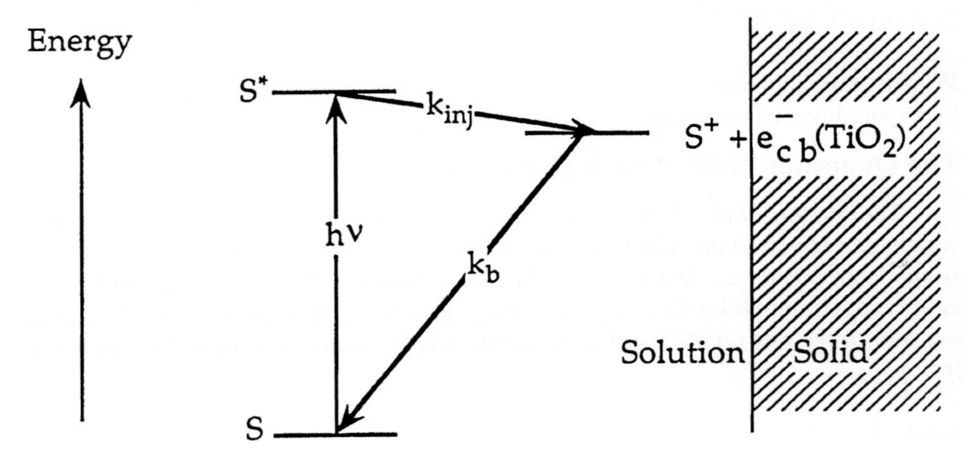

k_{inj} rate constant for charge injection [s^{-1}]
k_b rate constant for recombination

Fig. 3. Photoinduced charge separation on the surface of titanium dioxide, k_{inj} and k_b represent the rate constants for electron injection and recombination, respectively.

Naturally, this reaction is undesirable, since instead of electrical current it merely generates heat.

For the characterisation of the recombination rate an important kinetic parameter is the rate constant k_b. It is of great interest to develop sensitizer systems for which the value of k_{inj} is high and that of k_b low. Fortunately, for the transition metal complexes we use, the ratio k_{inj}/k_b is often greater than 10^3, which significantly facilitates the charge separation. The reason for this behaviour is that the molecular orbitals involved in the back reaction overlap less favorably with the wavefunction of the conduction band electron than those involved in the forward process. For example for our Ru-complexes bound to the titanium dioxide membrane, the injecting orbital is the π^* wavefunction of the carboxylated bipyridyl ligand since the excited state of this sensitizer has a metal to ligand charge transfer character. The carboxylate groups interact directly with the surface Ti(IV) ions resulting in good electronic coupling of the π^* wavefunction with the 3d orbital manifold of the conduction band of the TiO_2. As a result the electron injection from the excited sensitizer into the semiconductor membrane is as extremely rapid process. By contrast, the back reaction of the electrons with the oxidized ruthenium complex involves a d-orbital localized on the ruthenium metal whose electronic overlap with the TiO_2 conduction band is small. The spatial contraction oft he wavefunction upon oxidation of the Ru(II) to the Ru(III) state further reduces this electronic coupling and this explains the large difference between the forward and backward electron transfer rates.

Of great significance for the inhibition of charge recombination is the existence of an electric field in the titanium dioxide membrane. This field is established spontaneously by electron flow from the semiconductor to the redox electrolyte when these are brought into contact. This charge transfer is driven by the difference between Fermi level in the titanium dioxide and the redox potential of the solution. It will stop once the electrochemical potentials on both sides of the junction are equal. The depletion of majority charge carriers leads to the establishment within the titanium dioxide of a space charge layer (ref. 5). Within this layer the conduction band is curved towards the interior of the semiconductor. On injection of electrons from an excited sensitizer, the band bending acts to draw them from the surface to the interior of the membrane. This effect in addition depresses the rate recombination further, by a factor of the order of 1000 (ref. 3). Expressing by η_e the proportion of photoinduced electrons which avoid recombination and pass into the external current circuit, the monochromatic current yield is given by:

$$\eta_i(\lambda) = LHE(\lambda) \times \phi_{inj} \times \eta_e \qquad (6)$$

This current yield expresses the ratio of the measured electric current to the incident photon flux for a given wavelength. By development of appropriate sensitizers and systematic improvement of the electronic properties of the titanium dioxide membrane, systems are now available for which all three factors in eqn. (6) are close to unity. Thereupon, within the wavelength range of the sensitizer absorption band a quantitative conversion of incident photons to electrons is obtained.

A graph which presents the monochromatic current output as a function of the wavelength of the incident light is called the "photocurrent action spectrum". An action spectrum of this type for two trinuclear ruthenium complexes as sensitizers appears as Fig. 4 (ref. 6). The dotted line 1' applies to complex1 and is derived from curve 1 by correction for light losses in the conducting glass which serves as substrate for the titanium dioxide film. It establishes the very high efficiency of current generation, exceeding 75% over a wide range of wavelengths with these complexes.

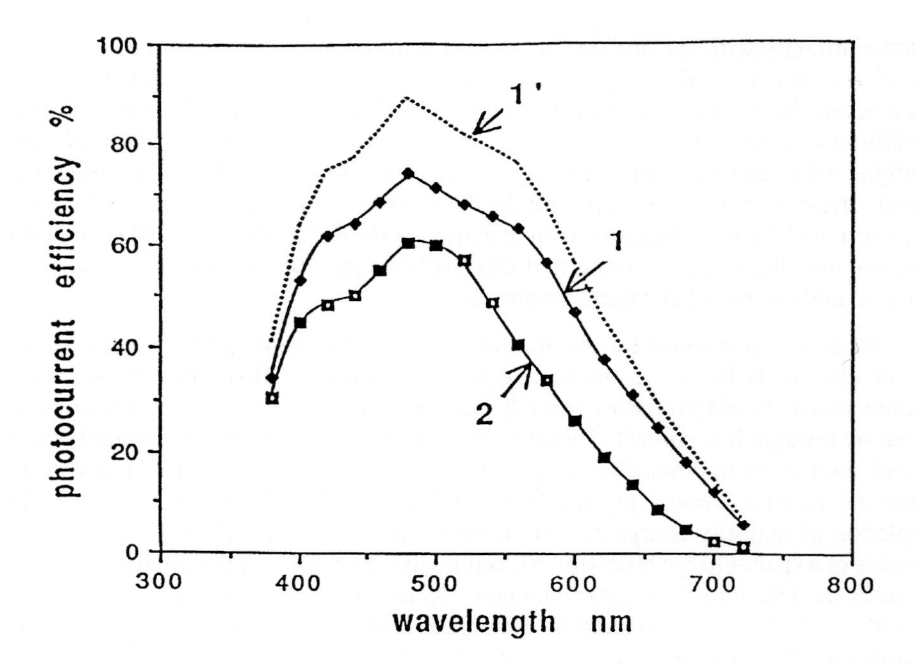

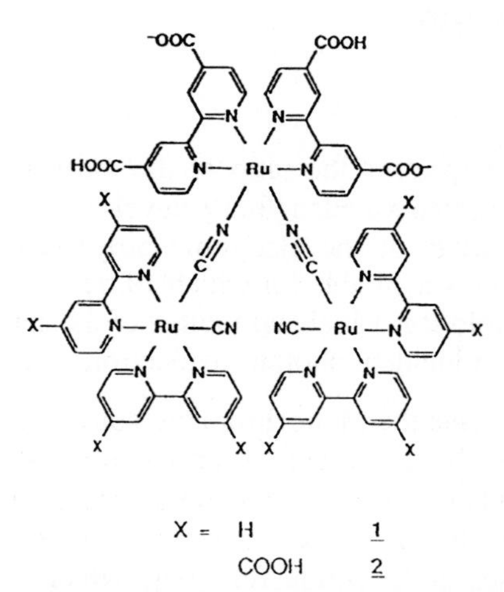

Fig. 4. Photocurrent action spectrum of titanium dioxide films with the trinuclear sensitizers 1 and 2 observe in a thin film cell with lithium iodide/iodine solution in ethanol as electrolyte. The incident photon to current conversion efficiency is plotted as a function of the wavelength oft he exciting light. The curve dashed curve 1' was calculated from the experimentally observed curve 1 by correcting for the light absorption in the conducting glass used as support for the TiO_2 film.

Cell Voltage and Overall Conversion Efficiency

The photovoltage of our cell represents the difference between the Fermi level of titanium dioxide under illumination and the redox potential of the electrolyte, Fig. 5.

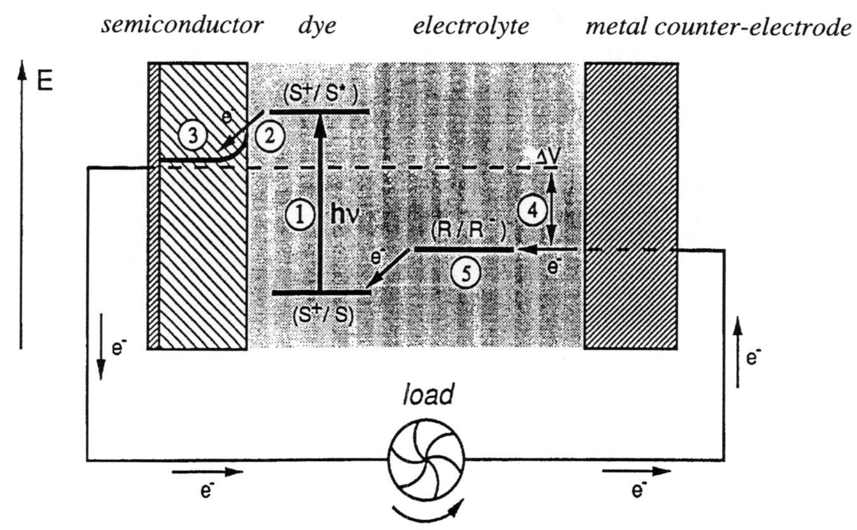

Fig. 5. Schematic representation of the principle of the new photovoltaic cell to indicate the electron energy level in the different phases. The cell voltage observed under illumination corresponds to the difference in the quasi-Fermi level of the TiO_2 under illumination and the electrochemical potential of the electrolyte. The latter is equal to the Nernst potential of the redox couple (R/R^-) used to mediate charge transfer between the electrodes.

Using an alcoholic iodine-iodide solution under full sunlight an open-circuit cell voltage of 0.7 to 1.0 V can be measured. Under a 1000-fold lower intensity the cell voltage is about 200 mV lower, a relative change of cell voltage of only 20-30%. For the conventional silicon cell, the cell voltage decreases by a factor of 3 for a comparable change of light intensity. This shows that the photovoltage of our cells is significantly less sensitive to light intensity variations than in conventional photovoltaic devices.

The overall efficiency (η_{global}) of the photovoltaic cell easily be calculated from the integral photocurrent density (i_{ph}), the open-circuit photovoltage (V_{oc}), the fill factor of the cell (ff) and the intensity of the incident light (I_s).

$$\eta_{global} = i_{ph} \times V_{oc} \times ff/I_s \tag{7}$$

The integral photocurrent density is given in turn by the overlap integral of the solar spectral emission $I_s(\lambda)$ and the monochromatic current yield.

$$i_{ph} = \int_0^{\infty} I_s(\lambda)\eta_l(\lambda)d\lambda \qquad (8)$$

For example, for an AMI distribution of the solar spectral emission (overall intensity 88.92 mW/cm^2) the integral photocurrent density for action spectrum 1 in Fig. 4 can be calculated as 11.09 mA/cm^2. Using the average value of V_{0c}, 0.85V, and for a fill factor of 0.7, the predicted overall efficiency for the cell has the the value of 7.45%. This prediction was tested in the laboratory with small cells (area 1.5 cm^2) under simulated AMI light. The measured conversion efficiency was 7% in good agreement with the expectations.

Development and Testing of the First Cell Module

Meanwhile, the development and testing of the first cell module for practical applications has begun. The assembly of the module is presented in Fig. 6.

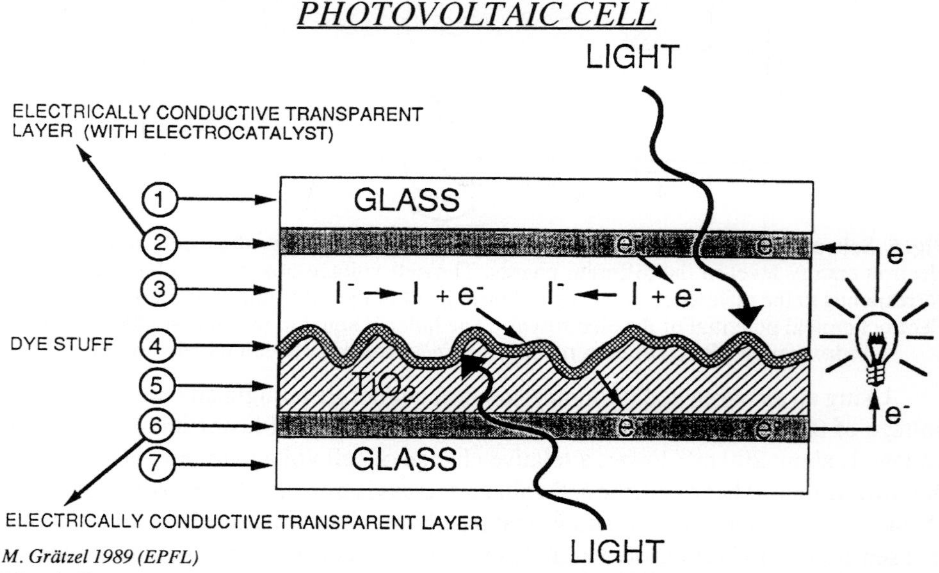

Fig. 6. Construction of a cell module with transparent glass electrodes. ① and ⑦: glass sheets; ② and ⑥: transparent conductive layer of fluorine doped tin oxide; ⑤: colloidal TiO_2 membrane (thickness ca 5 microns); ③: electrolyte containing the I_2/I^- redox couple.

The cell consists of two glass plates ① and ⑦, which are coated with a thin electrically conducting tin oxide layer. The colloidal titanium dioxide film deposited on one plate by the sol-gel procedure is notable for its high roughness factor and functions as a light trap. Visible light is absorbed by a monomolecular layer of an appropriate transition metal complex ④, which functions as sensitizer. On illumination this injects an electron into the titanium dioxide conduction band. With such a system,

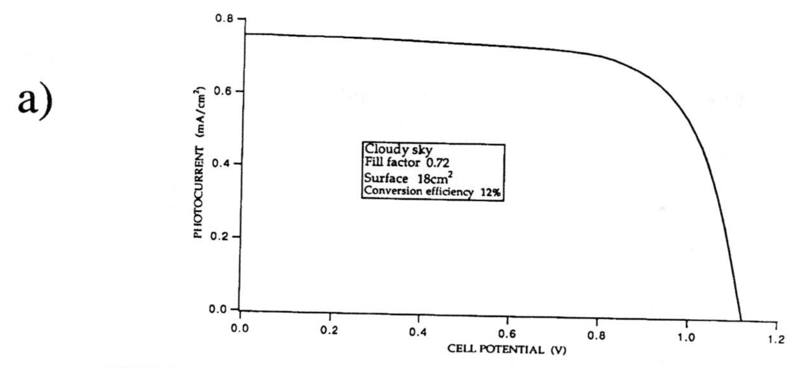

TWO CELL MODULES CONNECTED IN SERIES

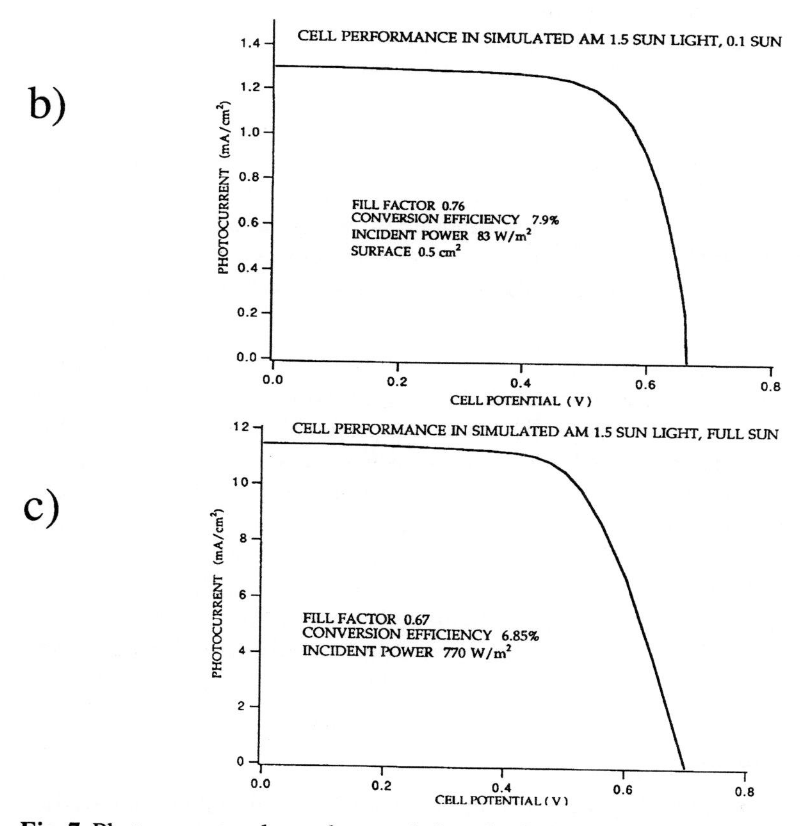

Fig. 7. Photocurrent-voltage characteristics of cells based on dye sensitized titanium dioxide films. a) cell exposed to diffuse day light (cloudy sky) in Lausanne, Switzerland. b) and c) cell exposed to simulates sun light with AM1.5 spectral distribution.

it is possible to convert 80% of the incident photons in the wavelength range of the sensitizer absorption maximum into electric current. the electrons pass over the collector electrode ⑥ into the external current where they do work. They are then returned to the cell via the counter electrode ②. The sensitizer film is separated from the counter electrode by the electrolyte ③. In the electrolyte there is a redox system, for example iodine/iodide whose function is to transport electrons from the counter electrode to the sensitizer layer, which had been positively charged as a result of the electron injection. The area of this module is between 20 and 100 cm^2.

In Fig. 7 we show the current voltage characteristics of our cell for different weather conditions. Curve a) was obtained under cloudy sky where the incident light intensity was 3 W/m^2. The overall conversion efficiency was 12% indicating that under diffuse day light conditions, the new photovoltaic device is remarkably efficient. In fact, the conversion yield under these conditions is significantly higher that that obtained with conventional single crystal silicon cells. Figure 7 b) and c) refer to direct solar radiation with AM 1.5 spectral distribution. The contribution of infrared radiation in AM 1.5 light is increased with respect to diffuse solar light and this reduces the overall efficiency. The conversion yields under one tenth and full sunlight are 7.9 and 6.85%, respectively. Note the excellent fill factors obtained with the new photovoltaic device. At low light intensity these exceed 70% which is far better than the values obtained with crystalline and amorphous silicon under the same conditions. This result is very surprising in view of the disordered structure oft he semiconductor membrane giving rise to defects such as lattice vacancies or dislocations and grain boundaries. Fortunately, the light induced charge separation process proceeds very efficiently at the surface of the colloidal membrane despite of the presence of such defect states.

In summary, then, it can be stated that in the present stage of development the overall efficiency of the module in am 1.5 sunlight is around 7%. The yield should be further increased in the near future. For this, the principal line of attack is the improvement of the spectral matching of the sensitizer absorption and the solar emission. In diffuse light or under a cloudy sky the efficiency of the cell is surprisingly high, already largely exceeding that of normal silicon cells. These new photovoltaic cells should be attractive in particular for decentralized applications, *i.e.* for domestic use where they could make an important contribution to the supply of electrical energy. Furthermore, the excellent performance of these cells under diffuse daylight would predestinate them for utilisation in countries where cloudy weather conditions prevail. Their price is expected to be ten times lower than that of the conventional silicon device.

Acknowledgement

This work was supported by a grant from the Swiss National Engineering Office (OFEN).

References

1 J. Deisenhofer, H. Michel and R. Huber, Trends Biochem. Sci., **10** (1985) 243.

2 A. Juris, V. Balzani, F. Barigiletti, S. Campagna, P. Belzer and A.V. Zelewski, Coord. Chem. Rev., **84** (1988) 85.

3 B. O'Regan, J. Moser, M. Anderson and M. Grätzel, J. Phys. Chem., **94** (1990) 8720.

4 N. Vlachopoulos, P. Liska, J. Augustynski and M. Grätzel, J. Am. Chem. Soc., **110** (1988) 1216.

5 H. Gerischer, Electrochim. Acta, **35** (1990) 1677.

6 M.K. Nazeeruddin, P. Liska, J. Moser, N. Vlachopoulos and M. Grätzel, Helv. Chim. Acta, **73** (1990) 1788.

Electron and Proton Transfer in Chemistry and Biology, edited by A. Müller et al.
Studies in Physical and Theoretical Chemistry, Vol. 78

Electron Transfer in Photosynthetic Reaction Centers

Christopher C. Moser, Jonathan M. Keske, Kurt Warncke and
P. Leslie Dutton

Department of Biochemistry and Biophysics
University of Pennsylvania,
Philadelphia, PA 19104 (USA)

Summary

The photosynthetic reaction center protein catalyzes charge separation across a biological membrane through a photon–initiated series of electron transfer reactions between bound redox cofactors which range over twelve orders of magnitude in time. Several of these intraprotein electron transfer reactions have been examined in reaction centers from two bacterial species as a function of redox cofactor distances, reaction free energy and chemical structure. The results suggest that the role of the protein in the reaction center design is primarily to act as a scaffolding which establishes the distances between the redox cofactor binding sites. The distance set overwhelmingly dominates the determination of the electron transfer rate, because the protein medium appears to be electronically homogeneous over a wide distance range. Comparison with results from chemical systems suggest that the rate–distance dependence in the RC protein is biologically unremarkable. The relatively small dependence of the rates observed for three intraprotein reactions on free energy suggests a secondary role for free energy in modulation of the rates. In addition, rates of electron transfer for several reactions are not strongly dependent on the chemical class of the cofactor. The data admit the possibility that the reorganization energies of all but the fast and small distance chlorin reactions are similar and can be described by a continuum form of a Marcus–type rate expression. A simple, unified view of the factors important for control of electron transfers in dense media emerges from this treatment.

Introduction

The photosynthetic reaction center proteins excel at stable charge separations over long distances with a remarkable near unit quantum efficiency (ref. 1) that continues to stimulate emulation by synthetic chemists in artificial systems (ref. 2). In our analysis of the factors which govern photosynthetic electron transfers, we unite the inter–cofactor distance information available from the x–ray crystallographically–determined reaction center protein (RC) structures with extensive data sets for the dependence of the electron transfer rate constants on reaction free energy, temperature, and cofactor and protein structure. The broad experimental data set explored here provides us with an opportunity, so far unique in chemistry and biology, to clarify the relative importance of distance, reaction free energy and system structure in controlling the rates of non–adiabatic electron transfer reactions in dense media, not only in the RC, but in chemical systems as well. The structural arrangement of the redox centers in the RC from the bacterium *Rhodopseudomonas viridis* as revealed by x–ray diffraction (ref. 3) is shown in Figure 1.

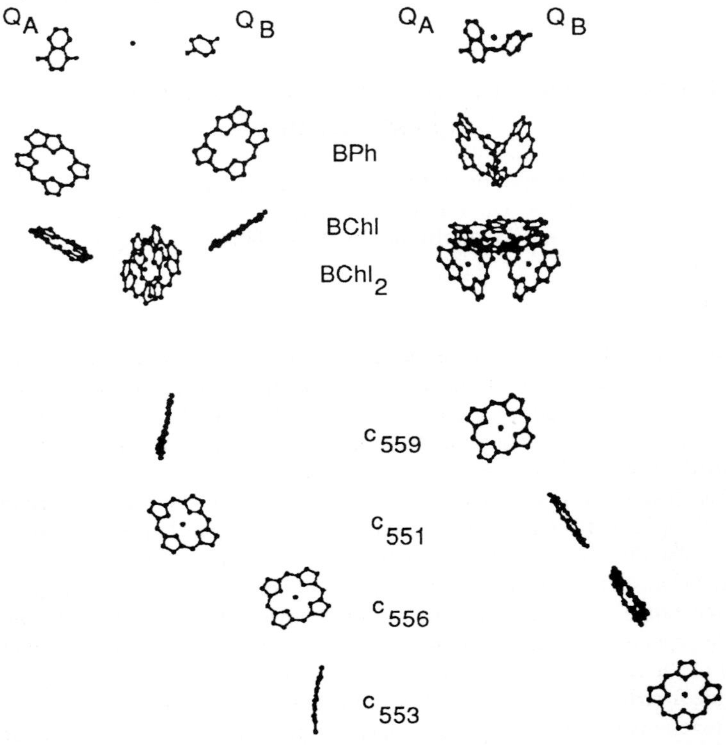

Fig.1. Structure of the redox cofactors of *Rp. viridis* as revealed by x–ray diffraction. The two views are rotated by 90 degrees about the approximate 2 fold symmetry axis. Figure based on a graphic supplied by M. R. Gunner. The functional L side of the protein is shown on the left of the left hand drawing.

The structure of the RC from *Rhodobacter sphaeroides* has also been solved, (ref. 4). The chlorins and quinones that are held by the membrane–intrinsic portion of the RC display a near two fold rotational symmetry about an axis that is perpendicular to the membrane plane. The c–type hemes are held in a linear arrangement by the cytochrome polypeptide, which extends beyond the membrane into the aqueous medium. Upon excitation of the bacteriochlorophyll dimer ($[BChl]_2$) only one of the two bacteriochlorophyll monomers (BChl) and bacteriopheophytins (BPh), those on the so–called L side, are normally involved in the picosecond charge separation. In this report we provisionally accept evidence which suggests that two distinct electron transfers in this sequence (ref. 5) lead to the formation of reduced BPh (BPh^-). The

BChl and BPh on the M side are apparently not directly involved in the native electron transfer sequence (ref. 6). The quinone acceptors, QA and QB, are reduced in turn on a 200 picosecond and 10-100 microsecond time scale, respectively. Electron transfer from the cytochrome c hemes to oxidized $BChl_2$ ($[BChl]_2^+$) takes place in the 0.1 to a few millisecond time range, depending on the species. If this electron transfer is prevented, either by prior oxidation of the hemes, or in the case of purified RC from *Rb. sphaeroides*, by the absence of the cytochrome, the various non–physiological, charge recombination reactions between BPh^-, or the reduced quinones Q_A^- and Q_B^-, and $[BChl]_2^+$ can be observed. Thus, many electron transfer processes, which occur over a wide range of distance and time, are available for study in the RC. Values for the rate constants, reaction free energies and distances for these native electron transfers are assembled in Table I.

Table 1

	Electron Transfer	Species	Log k (s^{-1})	r (Å)	-ΔG (meV)	ref
1a.	$BChl_2^* \rightarrow BChl$	sph	11.5	5.3	0	5,24-27
		vir	11.5	5.5	0	5,28,29
1b.	$BChl^- \rightarrow BPh$	sph	12.0	4.3	160	
		vir	12.2	4.8	160	
2.	$BPh^- \rightarrow Q_A$	sph	9.6	11.0	700	30,31
		vir	9.6	9.6	445	32
3.	$Q_A^- \rightarrow Q_B$	sph	3.7	14.5	70	33,34
		vir	5.1	13.5	120	35,36
Back Reactions						
4.	$BPh^- \rightarrow BChl_2^+$	sph	7.7	10.1	1200	17,24,37,38
		vir	7.7	9.5	1000	32,36
5.	$QA^- \rightarrow BChl_2^+$	sph	0.9	23.4	520	39,40
		vir	1.9	22.4	635	36
Cytochrome Reactions						
6.	c551→c556	vir	10.5	8.1	300	41-43
7.	c551→c558	vir	7.6	6.9	360	
8.	$c558 \rightarrow BChl_2^+$	vir	6.6	12.3	120	

Our evaluation of the hierarchy of factors involved in controlling these electron transfers starts with an assessment of the dependence of electron transfer rate (here reported as the logarithm of the rate constant) on the reaction free energy. The most extensive rate–free energy dependencies have been accumulated for the BPh^- to Q_A and Q_A^- to $[BChl]_2^+$ electron transfers in the RC from *Rb. sphaeroides*. Reconstitution of the Q_A site with a variety of synthetic quinones, following removal of the native ubiquinone cofactor, has been utilized to vary the reaction free energies over a roughly 600 meV range, (ref. 7). The rate–free energy dependencies have been studied from room temperature to 14K, (ref. 7). We find that these data sets can be used to obtain a working, "generic" rate–free energy relation by fitting the Q_A site data to a suitable theoretical expression for the electron transfer rate, and that this fitted relationship can then be sensibly applied to extend the rate–free energy dependences derived from other, less complete data sets.

Theoretical treatments of the electron transfer rate (rate constant, k_{et}) generally refer to Fermi's Golden Rule:

$$k_{et} = 4\pi^2/h\, |V(r)|^2\, FC \tag{1}$$

where V(r) is the distance–dependent electronic coupling of the reactant and product states and the Franck–Condon factor (FC) describes the extent of vibrational wave function overlap between the reactant and product states, which includes the dependence of rate on the nuclear geometries of the reactant and product (refs. 8,9). One of the more successful physical models used to derive expressions for the FC factors for overlap of the reactant and product potential surfaces has been provided by Marcus (ref. 10). Figure 2 illustrates this type of model, which in its simple form considers the energy surfaces of the reactant and product states to be parabolic (harmonic oscillators) along the reaction coordinate (nuclear displacement). An exact quantum mechanical expression for FC factors for the involvement of harmonic nuclear motions in transition from reactant to product states, has been described in the appendix of the excellent review by Marcus and Sutin (ref. 10), (their Eqn. 72):

$$k = 2\frac{\pi}{\hbar^2\omega}\,|V(r)|^2\, e^{-S(2\bar{n}+1)} \left(\frac{\bar{n}+1}{\bar{n}}\right)^{P/2} I_P[2S\sqrt{\bar{n}(\bar{n}+1)}] \tag{2}$$

Here, $P = \Delta G/h\nu$, where ΔG is the free energy of electron transfer between the lowest energy reactant and product states while n is considered to be the "mean frequency" of the harmonic oscillator coupled to electron transfer. The parameter S is given by $\lambda/h\nu$, where λ is the reorganization energy, which as shown in Figure 2, corresponds to the energy of the reactant state in the product geometry. Ip() is the modified Bessel function, and n includes the temperature dependence ($n = [\exp(h\nu/k_bT)-1]^{-1}$, where k_b is the Boltzman constant). An expression, based on Equation 2, has previously been shown to provide a satisfactory fit to the rather shallow free energy dependence

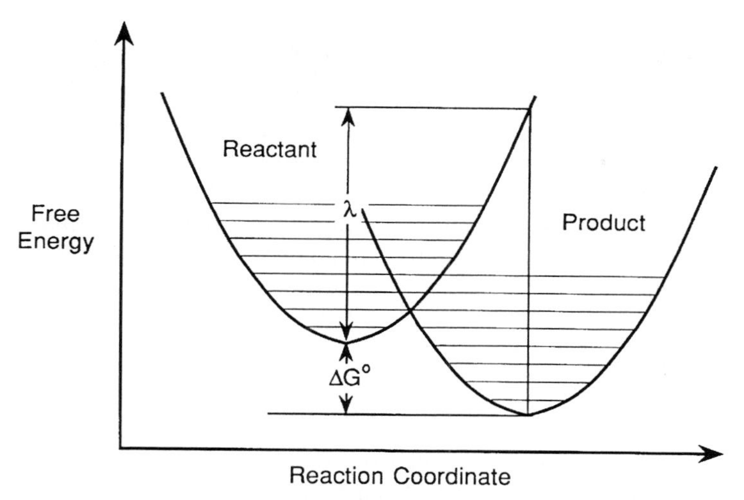

Fig. 2. Intersecting harmonic potential energy surfaces of reactant and product illustrating a simple version of a Marcus model of electron transfer. Parabolic surfaces represent approximate maxima or reactant and product wavefunctions along the reaction (nuclear) coordinate for each vibrational energy level. The intersection point represents a region of greatest vibrational wavefunction overlap and greatest electron transfer rate. Vibrational levels above or the below the intersection region have smaller but non–zero overlaps

displayed by the quinone–chlorin data sets if a combination of two vibrational modes, one of high (~1600 cm^{-1}) and one of intermediate (120 cm^{-1}) frequency, is used (ref. 7). We show in this work that the form of several rate–free energy dependencies can be satisfactorily fit using Equation 2 and an appropriate interpretation of "mean frequency".

Equation 2 specifies that the rate of the fixed–distance reaction will be at a maximum when the reorganization energy matches the free energy of the reaction. Under this condition, the rate is dependent primarily on V(r), and is therefore modulated by separation distance and the attenuation of V(r) due to the properties of the intervening protein matrix. An understanding of the effects of these contributions to the rate–distance dependence in the RC can be obtained by comparing the maximal rates for electron transfers occurring over different, known distances. We can increase the size of rate–distance data set by fitting the "generic" rate–free energy relation obtained above to incomplete data sets. The rate–distance relation allows us to ask several critical questions: Is there a general dependence of the rate on distance? What is the role of the intervening protein medium—is it a passive insulator, or do particular amino acids enhance electronic coupling between the redox centers? Broader perspective on the role of the medium is gained by comparing the RC rate–distance dependencies with those exhibited in chemical systems in disparate media, for which donor–acceptor distance and the rate maximum are known.

The results of our analysis reveal a simple and satisfying picture of the determinants of electron transfer rates in the RC. Determination of the electron transfer rate is dominated by the inter-cofactor distance. The reaction free energy plays a secondary role. The intrinsic structure of the system appears to have even less importance. Rather, the protein appears to exist primarily as a scaffolding to establish the distance of electron transfer through a homogeneous medium. These findings provide a basis for understanding the origin of the quantum and thermodynamic efficiencies displayed by the RC, and allow us to predict the operational features of any integral–membrane electron transfer protein.

Results and Discussion

Development of the Rate–Free Energy Relation

BPh^- to Q_A Electron Transfer. The photosynthetic bacterium *Rp. viridis* provides a means of exploring the effect of reaction free energies on rate that is independent of cofactor substitution. The RC from this species exhibits only a 40% amino acid homology with the RC from *Rb. sphaeroides*. *Rp. viridis* has b–type BChl and BPh and menaquinone as Q_A (*vs.* a–type BChl and BPh and ubiquinone in *Rb. sphaeroides*). The lower BPh^- and higher Q_A^- free energy levels relative to *Rb. sphaeroides* create a smaller free energy gap, while the inter-cofactor distance in the two RC are similar to a first approximation, as shown in Table I.

Figure 3 shows that the rate–free energy relation for BPh^- to Q_A electron transfer rates in the *Rp. viridis* RC, obtained from quinone substitution experiments, appears to extend the trend established for *Rb. sphaeroides* (ref. 7). The solid line represents a fit of the *Rb. sphaeroides* data to Equation 2, based on parameters described in Gunner and Dutton (ref. 7). We will subsequently test the success of this relation in accounting for rate–free energy data for other RC electron transfer reactions. The approximate continuity of the data for the BPh^- to Q_A reactions confirms that the differences in the structures of the native cofactors in the two species are not critical in the determination of the rate of electron transfer. Further, in view of the rather low amino acid homology between the two RC, these results begin to suggest the absence of significant fine–tuning of the electronic coupling across the protein medium separating the redox centers. While it can readily be argued that the natural amino acid substitutions between species will avoid residues critical to the electron transfer rate, mutagenesis directed at the reaction domains has resulted in only relatively modest changes (less than 5–fold) in rate, when changes due to losses in cofactor binding affinity or RC assembly are accounted for (ref. 11).

The dependence of rate on cofactor structure has been more stringently tested for the Q_A site–mediated reactions in *Rb. sphaeroides* (ref. 12). Figure 3 also shows that non–quinone, or "exotic", cofactors, such as 1,3-dinitrobenzene and substituted 9-fluorenones, function at the Q_A site with electron transfer rates comparable with quinone

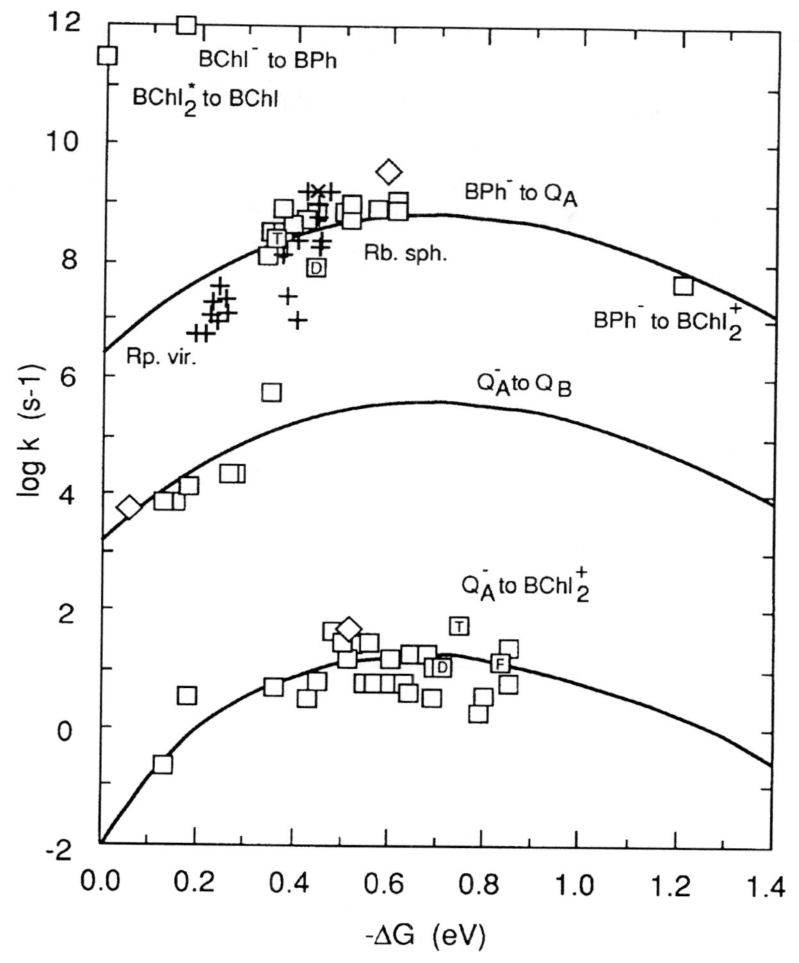

Fig. 3. Logarithm of the rate of electron transfer in the RC as a function of free energy of electron transfer (consult text for references). Those involving the *Rb. sphaeroides* Q_A site include rates for the native quinone (diamond) and modified quinones (squares). Non–quinonoid compounds are represented by lettered squares--F, tetrafluorofluorenone; D, dinitrobenzene and T, trinitrofluorenone. Data from *Rp. viridis* is represented as crosses. Curves represent the continuum form of the Marcus and Sutin rate expression (equation 2) with a mean frequency of 70 meV and a reorganization energy of 700 meV. The top curves are for data at 298 K, while the bottom curve is for data at 35 K.

replacements of similar free energies, even at temperatures down to 35 K (ref.12). This result more seriously calls into question attempts to correlate the vibrational frequency parameters, used in fitting expressions such as Equation 2 to the experimental results, with cofactor motions specific to the native quinonoid structure, such as those associated with the carbonyl group or quinone ring skeletal vibrations.

Q_A^- to $BChl_2^+$ Electron Transfer. The Q_A^- to $BChl_2^+$ back reaction provides an opportunity to examine the rate–free energy dependence of quinone–chlorin electron transfer at a considerably longer time and distance scale. We find the same form of rate–free energy relation that was successful in fitting results for the BPh^- to Q_A reaction also describes the recombination results, when uniformly slowed (*i.e.*, lowered vertically, *en bloc*) as shown in Figure 3. Once again, exotic cofactors display rate–free energy dependences comparable with the quinone cofactors (ref. 12), which demonstrates that specific features of the native, quinonoid structure do not significantly influence rate determination.

Development of the Rate–Distance Relation

Quinone–Chlorin Reactions. For both the BPh^- to Q_A and the Q_A^- to $[BChl_2]^+$ reactions, the rate with the native quinone is close to the maximum rate, as shown by the data in Figure 3, which according to the treatment leading to Equation 2, indicates that the free energy and reorganization energy for the native reactions is nearly identical. The comparable free energy and reorganization energy values calculated using RC structure–based molecular dynamics simulations concur with these results (ref. 13). The maximal rates for these reactions are plotted as a function of the edge–to–edge inter–cofactor distance obtained from the crystallographic RC structures in Figure 4.

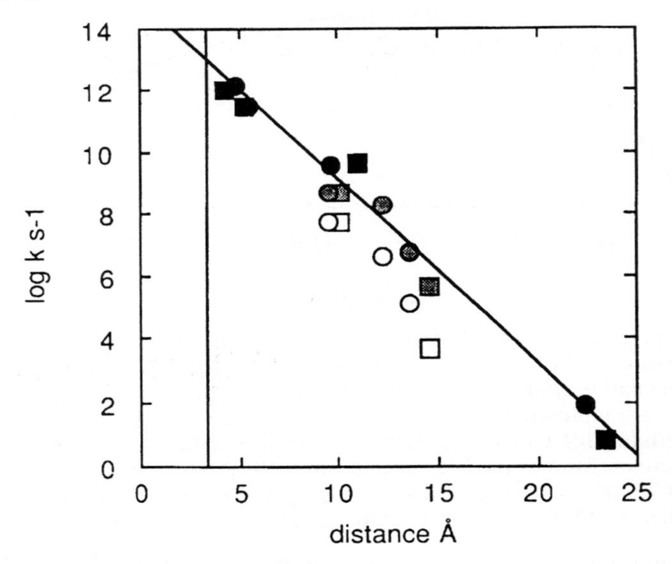

Fig. 4. Logarithm of electron transfer rate *vs.* edge carbon to edge carbon distances for reaction centers. Van der Waals contact occurs at 3.6 . Data for *Rb. sphaeroides* is represented by squares; that for *Rp. viridis* is represented by circles. Filled and open symbols represent native rates; shaded symbols represent free energy optimized rates. Consult Table 1 for references.

Figure 4 shows that a nearly linear relation exists between the maximal reaction rate and distance, which extends over 8 orders of magnitude in rate. A linear relationship is expected if the electronic wavefunction overlap decays exponentially with distance outside of the reactant molecules in a uniform manner (*i.e.*, similar attenuation factor) for all reactions (ref. 9).

Q_A^- to Q_B Electron Transfer. We now expand the rate–distance data set by using the fitted rate–free energy relation obtained from the quinone–chlorin reactions to estimate the maximum rate of reactions for which the rate–free energy data are less complete. As shown in Figure 3, the rates of the Q_A^- to Q_B electron transfer in *Rb. sphaeroides*, which occurs in the intermediate, microsecond time domain, are found to exhibit a dependence on free energy (ref. 14) (varied by quinone substitution) that is roughly comparable in form to the relations displayed by the other quinone–mediated reactions. Although this data set has not been extended far enough to reach an obvious maximum rate, it is clear that the rate of the native Q_A^- to Q_B electron transfer in *Rb. sphaeroides* is not optimal. We therefore estimate the optimum electron transfer rate by uniformly slowing the rate–free energy relation originally applied to the BPh^- to Q_A reaction. The estimated maximum rate obtained in this way is raised by two orders of magnitude from the native value, which places the point near the linear relationship shown in Figure 4.

BPh^- to $[BChl]_2^+$ Electron Transfer. Even limited rate–free energy dependencies for the other electron transfers in the reaction center are unavailable at present. This is because cofactor substitution at the chlorin binding sites presents a special set of problems. However, through use of the rate–free energy relation, and some additional assumptions, it is possible to estimate the maximum rate for these reactions. For example, the back reaction from BPh^- to $BChl_2^+$, which takes place over nearly the same distance as the BPh^- to Q_A reaction (see Table I), exhibits a rate which falls on the rate–free energy relation for this reaction, as shown in Figure 3. If it is assumed that the reorganization energy of the BPh^- to $[BChl]_2^+$ reaction is similar to that of the BPh^- to Q_A electron transfer, as these results suggest, then a uniformly slowed rate–free energy relation which includes this data point can be used to estimate the optimum rate for this recombination reaction. Remarkably, the estimated maximum rate places the data point for the BPh^- to $[BChl]_2^+$ reaction close to the line of Figure 4.

Electron Transfers Involving $[BChl_2]^*$, BChl and BPh. If we use recent measurements suggesting a two step electron transfer for the very fast rates of electron transfer between $[BChl_2]^*$, BChl and BPh (ref. 5), these reactions also fit neatly onto the linear relationship shown in Figure 4, while a single step electron transfer mechanism between $[BChl_2]^*$ and BPh generates a much larger distance for comparable rates. In order to reconcile the position of these data on the plot in Figure 4, we must consider smaller reorganization energies for these reactions. Because of the relaxation of the dielectric around the redox centers during electron transfer appears to represent a significant contribution to the total reorganization energy (refs. 10 and 49), this energy may well be less at fast time scales, where experiments suggest the relaxation is incomplete (refs. 16, 17). Similarly, at short distances of electron transfer dielectric reorganization can be less extensive. Under these conditions, reorganization energies closer to 200 meV may be expected. Molecular dynamics simulations based on the atomic coordinates of the RC have provided similar reorganization energies (ref. 18).

Electron Transfers Involving the Cytochrome c Hemes. The reorganization energies of the reactions involving the cytochrome c hemes may differ from the other reactions, especially because the effective polarity of the environment surrounding the redox centers may be higher in the cytochrome c polypeptide. The reaction free energies are given in Table I. If we choose to consider reorganization energies comparable with those of the best–characterized RC reactions, in order to obtain an estimate of the optimum rate, we find that the rate for electron transfer from cytochrome c_{558} to $BChl_2^+$ falls on the linear log rate *vs.* distance relation in Figure 4. Notice that even a dramatic drop in reaction free energy of 480 meV from a putative optimum value of 700 meV decreases the rate by only about 1.6 orders of magnitude, which is equivalent to 2.6 angstrom on this plot. The rates of electron transfers between the other cytochrome c hemes (c_{556}, c_{551} and c_{559}) have not be determined independently, but the range of optimized rates for these reactions falls close to the linear relationship, as shown in Figure 5.

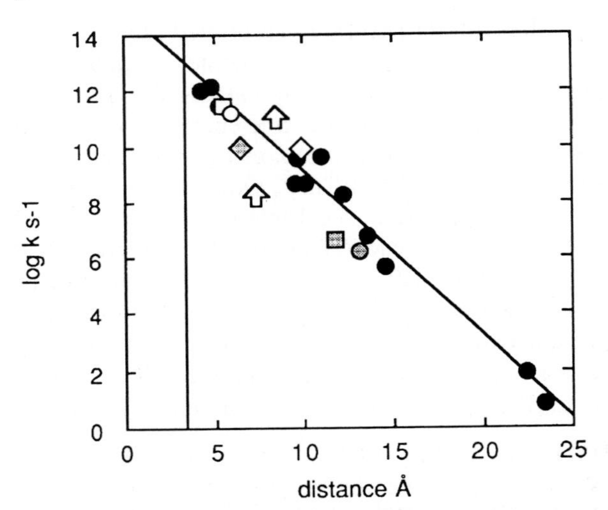

Fig 5. Free energy optimized rates for reaction center reactions as described in figure 4 (filled circles). For the partially–determined *Rp. viridis* rates for the cyt c_{551} to c_{556} and cyt c_{551} to c_{559} electron transfers, arrows indicate the calculated minimum rate from Shopes and Wraight (ref. 36). Also shown are optimized rates for various synthetic systems-- open square: Wasielewski *et al.*, porphyrin-quinone (ref. 44); open circle: Fox *et al.*, Ir--organic (ref. 45); open diamond: Joran *et al*, Porphyrin--Q (ref. 46); shaded square: Meade *et al*, Ru-cyt c (ref. 47); shaded circle: Cowan *et al.*, Ru--myoglobin (ref. 48); shaded diamond: Closs *et al.*, Biphenyl--organic (ref. 49).

The linearity of the rate–distance relationship shown in Fig. 4, suggests that the attenuation of V(r) as a function of distance is uniform for all of the reactions. Therefore, the protein acts as a relatively homogeneous medium between the different redox centers. Is the uniform attenuation of V(r) with distance a unique property of the RC protein matrix? To answer this question, we compare the rate–distance relation derived for the RC with data obtained in other dense media.

Chemical and Semi–Synthetic Systems

Fig. 5 shows the optimum rates of electron transfer observed in a number of systems, chemical (organic–organic, or organic–transition metal complex, bridged donor-acceptor systems) and semi–synthetic (metallo–proteins to which a transition metal complex has been peripherally attached), obtained from data sets for which the inter–redox center distance is fixed and its value known, and in which some measure of the free energy optimum has been performed (see Figure legend for data sources). The results from the chemical systems cluster about the linear rate–distance relation defined by the biological, RC reactions. At Van der Waals contact of donor and acceptor edge carbon atoms (roughly 3.6 Å), the intercept of the rate *vs.* distance curve is 1.3×10^{13} s^{-1}. This is a typical vibrational frequency for nuclear motion along the reaction coordinate (ref. 10), and signifies a transition to a reaction that is adiabatic. The slope of 1.7 Å per decade is in the middle of a range of other measurements made on more limited data sets (ref. 15). The results shown in Figure 5 suggest that the RC medium is biologically unremarkable in that many media, independent of their specific, chemical features, share a common distance modulation of the electronic overlap between the redox centers. In contrast, the reorganization energies displayed in the chemical systems differ significantly from the apparently "generic" relation within the RC.

The Choice of the Reorganization Energy and "Mean Frequency"

Our findings regarding the free energy and structure–dependence of the rates allow us to make several generalizations about the nature of the reorganizational contributions associated with electron transfers in the RC. These generalizations prove valuable towards estimating the reorganization energy for protein–mediated electron transfer reactions for which the reorganization energy is unknown. The observation that the rate–free energy relation shown in Figure 3, which implies a total reorganization energy of about 700 meV and a "mean frequency" of about 70 meV, provides a satisfactory fit to data obtained for three reactions which occur across different reaction domains in the *Rb. sphaeroides* RC, and include an extension to *Rp. viridis*, suggests that the reorganizational properties of the protein environment for most reactions are uniform.

The total reorganization energy for any reaction will have contributions due to both internal (redox cofactor) and external (medium) rearrangements. The correspondence of the rates for electron transfers of comparable free energy mediated at the QA site by exotic and quinone cofactors suggests that the internal cofactor reorganizational properties are common to all multicarbon cofactors. Calculations of the internal reorganization energies yield values of 100 to 200 meV (refs. 19, 20). We recognize that distance dependence of the external reorganization energies can become conspicuous for distances as short as 5 Å. In such cases we recommend estimating the solvent reorganization energy by means of a simple distance–dependent, dielectric continuum expression for the reorganization energy, such as described by Marcus (ref. 10). Application of this relation shows that an external reorganization energy at distances of 10 Å or more are reduced to about 200 meV for distances around 5 Å.

The view of the protein matrix of the RC emerging from this analysis is that of a framework which provides a homogeneous dielectric medium for the reactions. We suggest that the relatively "structureless" nature of Equation 2, which permits vibrational coupling to a multi–frequency continuum, provides the most simple, and physically realistic, way to incorporate this view into a quantitative description of the observed rate–free energy dependence. Specifically, we believe that the "mean frequency" is best interpreted as the reorganization energy–weighted average of a distribution of contributing frequencies. Furthermore, all modes of less than Boltzmann thermal energy contribute at the effective frequency of k_bT/h. For example, for the quinone-chlorin reactions, a mean frequency of about 70 meV and a reorganization energy of about 700 meV, together with the distance–dependent value for V(r) from Figure 4, appears to provide an adequate description of the observed rate–free energy dependence. Furthermore, by retaining the mean frequency, but reducing the reorganization energy, a relatively flat rate–free energy relation is predicted, which appears to match the observed dependence for the fast chlorin reactions. This mild dependence of rate on free energy for the chlorin reactions stands in contrast to the behavior predicted by the classical Marcus expressions (ref. 10), and is consistent with our recent observations of electric field effects on oriented RC multilayers (ref. 21) and observations on RC randomly embedded in a polymer (ref. 22).

Conclusions

Our analysis of the data for electron transfers in the RC strongly suggests that inter–cofactor separation distance is overwhelmingly dominant in determining electron transfer rates. The results show that this is true because the exponential attenuation of rate with distance is uniform for all reactions, regardless of distance or position in the protein. We suggest that the role of the protein in the reaction center, and perhaps in other electron transfer proteins, is to act foremost as a scaffolding that sets distances between redox centers, and that the particular arrangement of cofactor distances in the RC is specialized toward the ultimate goal of transferring an electron across a distance equal to the width of a biological membrane. Protein control of electron transfer rates through manipulation of the reaction free energy, which is exercised by establishing the relative binding strengths of the oxidized and reduced cofactor species, is of secondary importance. The "intrinsic" structure of the system appears to play a minor role in influencing control of electron transfer rates. Specific cofactor and protein structural elements do not appear significant in providing coordinates which specifically facilitate the reactions. Indeed, the reorganization energy may be essentially predetermined by the dielectric properties of the protein medium and a roughly uniform contribution from cofactor-associated, internal reorganization.

The view of electron transfer which emerges from this analysis provides a powerful means for predicting the operational features of any integral membrane, electron transfer protein, by using the log rate–distance relation (Figure 4), the form of the rate–free energy dependence (Figure 3), and the simple continuum approximation for estimating the reorganization energy. Maximum efficiency will be achieved with the largest separation distance that is compatible with high yield in the face of back reactions. In the case of the RC, except for the inherent decay rate of the excited $[BChl]_2$ light trap, these back reactions will also be distance dependent. The relative rates of the forward and back reactions can only be improved to a limited extent by selecting relatively high free energies of around -500 meV for any edge to edge electron transfer of greater than 6 Å. Probably the most critical design influence on the efficiency of reaction centers is found in the choice of the number of redox centers, which has the profound effect of lowering the edge to edge distances and accelerating rates relative to the primary light trap.

References

1 P.L. Dutton, in: L.A. Staehelin and C.T. Arnten (Eds.) *Encyclopedia of Plant Physiology* Springer-Verlag, Berlin, 1986, Vol. 19, pp. 197-237.

2 D. Gust and T.A. Moore, Science **244** (1989) 35.
D. Gust, T.A. Moore, A.L. Moore, S.-J. Lee, E. Bittersmann, D.K. Luttrull, A.A. Rehms, J.M. DeGranziano, X.C. Ma, Gao, F., Belford, R.E. and Trier, T.T., Science **248** (1990) 199.

3 H. Michel, O. Epp and J. Deisenhofer, EMBO J. **5** (1986) 2446.

4 J.P. Allen, G. Feher, T.O. Yeates, H. Komiya and D.C. Rees, Proc. Natl. Acad. Sci. USA, **84** (1987) 5730.

5 W. Holzapfel, U. Finkele, W. Kaiser, D. Oesterhelt, H. Scheer, H.U. Stilz and W. Zinth, Chem. Phys. Lett. **160** (1989) 1.
W. Zinth, personal communication.

6 D.M. Tiede, D.E. Budil, J. Tang, O. El-Kabbani, J.R. Norris, C.-H. Chang and M. Schiffer, in: J. Breton and A. Vermeglio (Eds.), *The Photosynthetic Bacterial Reaction Center, Structure and Dynamics* pp. 13-20, Plenum Press, New York, 1988.

7 M.R. Gunner and P.L. Dutton, J. Amer. Chem. Soc. **111** (1989) 3400.

8 R.A. Marcus, J. Chem. Phys. **24** (1956) 966.

9 D. Devault, Quart. Rev. Biophys. **13** (1980) 387.

10 R.A. Marcus and N. Sutin, Biochim. Biophys. Acta **811** (1985) 265.

11 For recent review see W.J. Coleman and D.C. Yuovan, Ann. Rev. Biophys. Biophys. Chem. **19** (1990) 333.

12 K. Warncke and P.L. Dutton, in: M.E. Michel-Beyerle (Ed.) *Structure and Function of Bacterial Reaction Centers* Proceedings of the Feldafing II workshop, Feldafing, FRG, 24-26 March, Springer-Verlag, Heidelberg, 1990.

13 A. Warshel, Z.T. Chu and W.W. Parson, Science **246** (1989) 112.; W.W. Parson, Z.T. Chu and A. Warshel, Biochim. Biophys. Acta **1017** (1990) 251.

14 K.M. Giangiacomo and P.L. Dutton, Proc. Natl. Acad. Sci., USA **86** (1989) 2658.

15 For example, S.L. Mayo, W.R. Ellis, Jr., R.J. Crutchley and H.B. Gray, Science **233** (1986) 948.

16 C. Kirmaier and D. Holten, (1988) FEBS Lett.

17 N.W.T. Woodbury and W.W. Parson, Biochim. Biophys. Acta **767** (1984) 345.

18 A. Warshel, Z.T. Chu and W.W. Parson, Science **246** (1989) 112.

19 W.W. Parson, Z.T. Chu and A. Warshel, Biochim. Biophys. Acta **1017** (1990) 251; and others.

20 J.R. Miller, J.V. Beitz and R.K. Huddleston, J. Am. Chem. Soc **106** (1984) 5057.

21 C.C. Moser, R. Sension, S. Repenic, P.L. Dutton and R.A. Hochstrasser, in preparation.

22 D.J. Lockhart, C. Kirmaier, D. Holten and S.G. Boxer, (J. Phys. Chem.), in press.

23 Edge carbon to edge carbon distances for *Rb. sphaeroides* were derived from X–ray derived crystallographic coordinates provided by the laboratories of D.C. Rees, J. Allen and G. Feher. The distances for the *Rp. viridis* structure were obtained from the Brookhaven Data Base.

24 C.C. Schenck, R.E. Blankenship and W.W. Parson, Biochim. Biophys. Acta **680** (1982) 44.

25 V.A. Shuvalov, A.V. Klevanik, A.V. Shardov, Y.A. Matveetz and P.G. Kryukov FEBS Letters **91** (1978) 135.

26 R. Haberkorn, M.E. Michel-Beyerle and R.A. Marcus, Proc. Natl. Acad. Sci. USA **76** (1979) 44.

27 C. Kirmaier, D. Holten and W.W. Parson, Biochim. Biophys. Acta **810** (1985) 33.

28 J.K.H. Horber, W. Gobel, A. Ogrodnik, M.E. Michel-Beyerle and E.W. Knapp, in: M.E. Michel-Beyerle (Ed.) *Antennas and Reaction Center* Springer-Verlag, Berlin pp. 292-297.

29 W. Zinth, E.W. Knapp, S.F. Fischer, W. Kaiser, J. Deisenhofer and H. Michel Chem. Phys. Lett. **119** (1985) 1.

30 M.G. Rockley, M.W. Windsor, R.J. Cogdell and W.W. Parson, Proc. Natl. Acad. Sci. USA. **72** (1975) 2251.

31 K.J. Kaufmann, P.L. Dutton, T.L. Netzel, J.S. Leigh and P.M. Rentzepis, Science **188** (1975) 1301.

32 D. Holten, M.W. Windsor, W.W. Parson and J.P. Thornber, Biochim. Biophys. Acta **501** (1978) 112.

33 C.A. Wraight, Biochim. Biophys. Acta **548** (1979) 309.

34 A. Vermeglio, in: B.L. Trumpower (Ed.) *Functions of Quinones in Energy Conserving Systems* Acad. Press , New York, pp.169-180, 1982.

35 R.P. Carithers and W.W. Parson, Biochim. Biophys. Acta **387** (1975) 194.

36 R.J. Shopes and C.A. Wraight, Biochim. Biophys. Acta **806** (1985) 348.

37 W.W. Parson, R.K. Clayton and R.J. Cogdell, Biochim. Biophys. Acta **387** (1975) 265.

38 C.E.D. Chidsey, C. Kirmaier, D. Holten and S.G. Boxer, Biochim. Biophys. Acta **766** (1985) 424.

39 J.D. McElroy, D.C. Mauzerall and G. Feher, Biochim. Biophys. Acta **333** (1974) 261.

40 B.J. Hales, Biophys. J. **16** (1976) 471.

41 S.M. Dracheva, L.A. Drachev, A.A. Konstantinov, A.Y. Semenov, V.O. Skulachev, A.M. Aruntjunjan, V.A. Shuvalov and A.M. Zaberezhnaya, Eur. J. Biochem. **171** (1988) 252.

42 R.J. Shopes, L.M.A. Levine, D. Holten and C.A. Wraight, Photosynth. Res. **12** (1987) 165.

43 R.C. Prince, J.S. Leigh and P.L. Dutton, Biochim. Biophys. Acta **440** (1976) 622.

44 M.R. Wasielewski, M.P. Niemczyk, W.A. Svec and E.B. Pewitt, J. Am. Chem. Soc. **107** (1985) 1080.

45 L.S. Fox, M. Kozik, J.R. Winkler and H.B. Gray, Science **247** (1990) 1069.

46 A.D. Joran, B.A. Leleand, P.M. Felker, A.H. Zewail, J.J. Hopfield and P.B. Dervan, Nature **327** (1987) 508.

47 T.J. Meade, H.B. Gray and J.R. Winkler, J. Am. Chem. Soc. **111** (1989) 4353.

48 J.A. Cowan, R.K. Upmacis, D.N. Beratan, J.N. Onuchic and H.B. Gray, Ann. N.Y. Acad. Sci. **550** (1988) 68.

49 G.L. Closs and J.R. Miller, Science **240** (1988) 440.

Electron and Proton Transfer in Chemistry and Biology, edited by A. Müller et al.
Studies in Physical and Theoretical Chemistry, Vol. 78

Exchange Interaction in Electron Transfer Proteins and their Model Compounds

Wolfgang Haase and Stefan Gehring

Institut für Physikalische Chemie
Technische Hochschule Darmstadt
Petersenstr. 20
D-6100 Darmstadt (Germany)

Summary

Metalloproteins with magnetically coupled metal centres are essential for many biological processes, *e.g.* the electron transfer. In this article some basic ideas of exchange interaction are given followed by a summary of bridging ligands mediating exchange coupling. Special attention is given to magnetic susceptibility measurements on protein samples. Two examples, hemocyanin and purple acid phosphatase, are presented for handling and interpretation of $\chi(T)$–data. Distance dependences in exchange coupled systems, *e.g.* in photosynthetic processes, are presented. Localized mixed valence states and resonance delocalization (double exchange) in model compounds are compared. Finally, magneto–structural correlations show that the exchange coupling is influenced by geometric and electronic parameters in a characteristic way.

1. Introduction

In biomolecules a noteworthy amount of their active sites is formed by multimetal units (refs. 1-3). Among these metals, 3d–elements mainly, predominantly antiferromagnetic exchange coupling occurs. Examples for the O_2–binding proteins are hemerythrin with two iron centres, bridged by one oxygen and two carboxylato groups (ref. 4), or hemocyanin (refs. 2 and 5) with a dimeric copper unit. The nitrogenase with a FeMo-cofactor (ref. 6), is one example from the group of oxidoreductases, whereas the acid phosphatases (ref. 7) with iron in their active site are examples for hydrolases; the active site structure of methane monooxygenase (ref. 8) seems to be similar to that in acid phosphatase. In urease the active site is formed by a dinuclear nickel–centre (ref. 9).

Within the group of electron carriers the iron sulphur containing ferredoxins (ref. 10) and the high potential iron sulfur proteins (HiPIP; (ref. 11)) are very interesting systems. Other metal containing electron carriers are cytochromes (ref. 12) with iron and zinc and blue copper proteins (ref. 13), also the ascorbate-oxidase (ref. 14) with a monomeric and a trinuclear copper unit.

An example with strong interest is the structure and function of the polynuclear manganese cluster in the photosystem II (ref. 15). Also some natural products are governed by multimetal centres, *e.g.* ferritin (ref. 16).

The examples given show that in some biomolecules as mentioned before multimetal centres are obviously prefered realized. A point with a considerable actuality is the function and the contribution of the exchange coupling within these multimetal centres to the relevant biochemical processes. To find answers to these questions the magnetic properties of such biomolecules give information. Further data can be obtained from model compounds which simulate some relevant properties of the biomolecules. It is clear in general that the catalytic activity of a biomolecule depends not only on the structure and function of the multimetal centre if this is the active unit but also on the structure and dynamic of the protein at least in the surrounding of the active site.

Another point of interest is the role of the superexchange mechanism in photosynthetic reaction processes (refs. 17-20). One model to explain the experimental data is the description of the electron transfer as an one step mechanism mediating the primary charge separation through a superexchange electronic interaction. Indeed, the photosynthetic reaction centre is a good example for the role of superexchange between singlet and triplet states not necessarily involving metal centres.

The scope of this contribution is to describe magnetic exchange effects in biomolecules and relevant model compounds from experimental and theoretical points of view. The role of the exchange coupling in electron transfer reactions and in photosynthetic reaction processes is part of this contribution.

2. Fundamental Description

The electrostatic interaction between unpaired electrons can be described in a general form by a phenomenological spin hamiltonian of the Heisenberg type (ref. 21)

$$\hat{H} = -2J\,\hat{S}_1\hat{S}_2 \tag{1}$$

with $\hat{S}_1$ and $\hat{S}_2$ as spin operators of the different centres. In this definiton -J means antiferromagnetic spin coupling (AF) with $2J = \Delta E$ as the energy difference between the ground and the excited spin state, +J means ferromagnetic spin coupling (F). Fig. 1 shows the diagram for the simplest possible case, two interacting electrons localized on different centres, *e.g.* in a Cu(II)–dimer with $S_1=S_2=1/2$.

The antiferromagnetic spin coupling having as an intrinsic principle the lowering of the spin multiplicitiy is actually the favoured coupling effect. Practically all of the natural products including proteins and most of the synthetic compounds show antiferromagnetic spin coupling. Approaches how to realize ferromagnetic spin coupling are known, *e.g.* by using the principle of orthogonality of magnetic orbitals (refs. 22, 23).

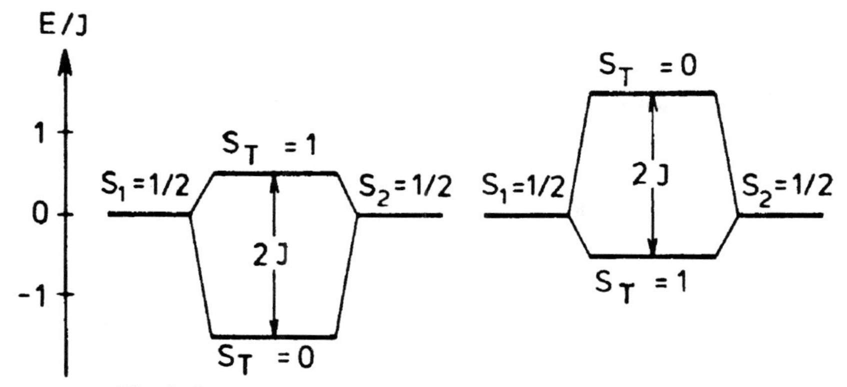

Fig. 1. Spin energy splitting in a dimeric $S_1=S_2=1/2$ system.

In general, the AF coupling is much stronger than the F one.

For spin centres with $S_i>1/2$ as in Fe_2S_2–clusters the spin coupling model can also be proved. Gibson *et al.*, (ref. 24), proposed an adequate description in terms of such coupling. Using this model the consequence in comparison to the simple case shown in Fig. 1 is a more complicated single ion behaviour and the increasing number of spin states, characterized by a more or less complicated energy ladder (refs. 25-27).

For n spin centres equation (1) becomes

$$\hat{H} = -2 \sum_{j>i=1}^{n} J_{ij} \hat{S}_i \hat{S}_j \tag{2}$$

which is the most used formula in case of isotropic exchange interaction. The number of degenerated quantum states, $N=(2S_i+1)^n$, and in a special case that of degenerated energy levels, increases rapidly with an increase of the number of centres n and of the single ion spins S_i.

As an example the energy splitting diagram for a $Fe(III)_3$–system with D_{3h}–symmetry is given in Fig. 2. It results from eq. (2) with J as isotropic exchange coupling constant between the three Fe(III) centres.

As mentioned before, a strong coupling effect is mostly realized in the antiferromagnetic coupling type. Strong coupling energy effects are in the magnitude of vibronic processes.

The spin–spin coupling can be interpreted as a *direct exchange* process between electrons on different centres either as a *through bond* effect *via* σ and π chemical bonds as in the H_2 molecule or as a *through space interaction*. The last one is equivalent to a weak chemical bond, whereas the former is equivalent to or part of a strong

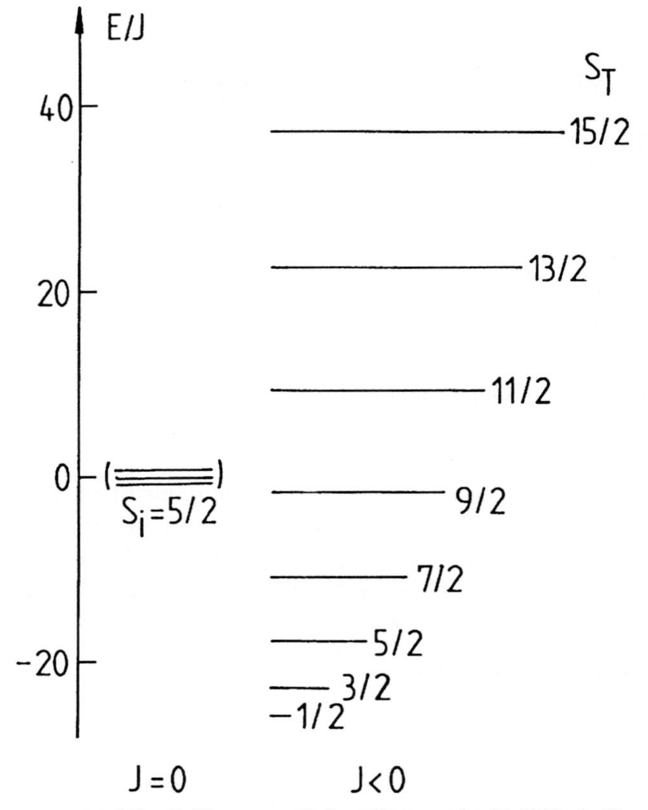

Fig. 2. Energy splitting diagram for $Fe(III)_3$ in D_{3h}–symmetry

chemical bond. Clearly, this effect is strongly distance dependent, *e.g.* in an exponential form, and becomes practically zero at long distances.

There is another way to couple spins between different centres, the so–called *superexchange* mechanism (ref. 28) through diamagnetic bridging ligands. Most of the exchange coupled magnetic diluted systems show this superexchange interaction on the molecular level. The distance dependence especially for long range exchange interactions is still open for discussion. To understand this, there are some approaches and some experimental results. Generally, the pathway of the superexchange interaction *via* various diamagnetic groups can be different even in case of multi–atom bridges. In Table 1 some bridging groups for superexchange interaction are summarized. These can be as well as closed shell atoms/groups – including also d-elements as *e.g.* CrO_4^{2-}groups – or hydrogen bond systems. The largest distance with remarkable strong exchange interaction is found in a terephthalato bridged Cu(II) dimer with Cu–Cu = 11.25 Å and 2J = -140 cm^{-1} (ref. 29).

One can imagine that the pathways for the superexchange interaction may be similar to those for some electron transfer processes, particulary for long range interactions. Therefore, understanding the electron exchange coupling gives also information about electron transfer processes.

TABLE 1.

Bridging ligands in exchange coupled systems.

Only one recent reference is given for unusual bridges, for the other ones see review articles (refs. 22 and 25-27).

$$M_1\text{—}L_n\text{—}M_2,$$
spin S_1 spin S_2

L=

O—H···O (ref. 30)

N O F S Cl Br I

O_2^{2-} N_3^- CN^- RO^- RS^- OCN^- SCN^-

$R\text{-}CO_2^-$ CO_3^{2-} NO_2^- NO_3^- $RO\text{—}PO_3^{2-}$ $(RO)_2\text{—}PO_2^-$ PO_4^{3-}

SO_3^{2-} SO_4^{2-} CrO_4^{2-} MnO_4^{2-} WO_4^{2-} AsO_4^{3-} VO_4^{3-} (refs. 31-33)

(ref. 34)

(ref. 35) (ref. 29)

(ref. 36) (ref. 37)

3. Magnetic Susceptibility Measurements on Metallo–Biomolecules

3.1 Magnetic Susceptibility Equation

The magnetic susceptibility following the Van Vleck equation in the form

$$\chi_M = \frac{N_A g^2 \mu_B^2}{3kT} \frac{\sum_n \omega_n S'_n(S'_n+1)\exp(-E_n/kT)}{\sum_n \omega_n \exp(-E_n/kT)} + N\alpha \tag{3}$$

with

S'_n	spin of state n
E_n	Energy of the spin states
ω_n	degeneracy of spin state n
$N\alpha$	temperature independent paramagnetism

allows to calculate the energy spin states E_n as function of the exchange coupling constants.

The susceptibility of any spin state is calculated using the Curie law

$$\chi_M(S'_n) = \frac{N_A g^2 \mu_B^2}{3kT} S'_n(S'_n+1) \tag{4}$$

Especially for biomolecules when using formula (3) a contribution to the bulk susceptibility arising from uncoupled centres (so–called paramagnetic impurity) has to be taken into account.

The above formulas show that for T>0 K the magnetic susceptibility reflects only an averaged property of all involved spin states, their relative energies and their respective g values. No specific information about a certain spin state can be obtained. To get more details the magnetic susceptibility data should be discussed together with EPR, ENDOR, optical, and Mössbauer data.

3.2 Protein Samples

For model compounds the experimental methods to determine J–values are wellknown in the literature (see *e.g.* (ref. 38)). Exchange coupled metalloproteins however are very diluted systems and the experimental procedure for evaluating magnetic data cannot be generalized. Some important points should be stressed here:

- The paramagnetic spin density is so small that mostly the overall contribution is diamagnetic. This means $|\chi_{dia}(\text{enzyme, buffer})| >> |\chi_{para}(\text{metal ions})|$.
- There is always a remarkable amount of uncoupled metal ions or some vacancies in the "lattices" or the molecules. Measurements on *e.g.* urease (ref. 9) revealed amounts of uncoupled $Ni(II)^{HS}$–centres up to 21 %.
- The buffer system is able to change the molecular conformation in a specific way.
- An important question is the procedure of isolation and purification of the metalloprotein, and the former treatment of the sample. In some cases the catalytic activity becomes zero in a lyophylized enzyme. Also some effects can arise from freezing the watery solution.
- Determinations of enzyme concentrations, usually *via* UV/VIS spectroscopy, are connected with comparably high uncertainties.
- Due to the weakness of the magnetic effects characterizing the exchange coupling a careful sample handling and exact correction for sample holders, buffer *etc.* are necessary.
- Use of deuterated buffer solutions for magnetization studies avoids effects from the slowly relaxing proton nuclei (ref. 39).

As it was shown by Day *et al.*, (ref. 39), a Faraday balance and a SQUID susceptometer have about the same sensitivity. In some cases a high sensitive Faraday system may provide a better resolution and an easier handling than a commercial SQUID susceptometer. For field dependent magnetization measurements usually SQUIDs are applied. The highest resolution for magnetic susceptibility data of metalloproteins is expected from self–constructed SQUIDs.

3.3 Magnetic Susceptibility Measurements on Hemocyanin and Purple Acid Phosphatase as Examples for Protein Studies

The problems related to the determination of exchange coupling parameters in metalloproteins are demonstrated in the following for two systems: hemocyanin, Hc (ref. 5) with a dinuclear copper site and purple acid phosphatase from bovine spleen (PAP) with a dinuclear iron site (refs. 7, 41 and 42).

Using the Faraday method for determination of the $\chi(T)$–data a signal $\Delta m \sim \chi$ is recorded. Fig. 3 shows the experimental result (ref. 43) for oxy–Hc (Cu(II)/Cu(II)) as a Δm *vs.* $1/T$ plot. To estimate the J–value some $\Delta m(1/T)$ functions with different J's are simulated. Within the resolution of the microbalance an uncoupled or weakly coupled behaviour can be excluded, however, the exact strength of the antiferromagnetic coupling cannot be determined. Therefore a limit of $|2J| \geq 100$ cm^{-1} for oxy–Hc

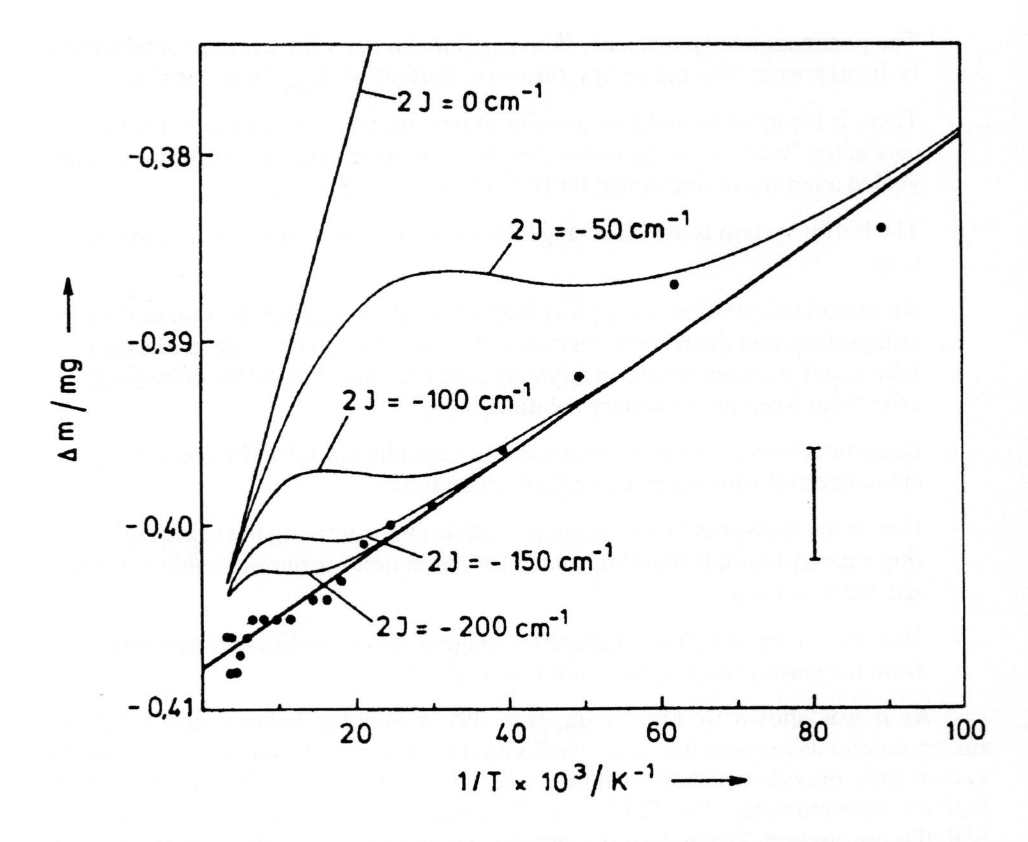

Fig. 3. Δm *vs* 1/T data with simulated functions to estimate the exchange coupling in oxy–hemocyanin, $\Delta(\Delta m) = \pm 0.003$ µg.

is given as final result related to the used microbalance's resolution of $\Delta(\Delta m) = \pm 3\mu g$. Using water as standard this Faraday system can reproduce $\chi(H_2O)$ within 1%. It is obvious that this J–value can be precised using an increased sensitivity together with an higher field (*e.g.* $\Delta(\Delta m) = \pm 0.1$ µg, $H_{max} = 8$ T as now installed in our group).

In the active site of purple acid phosphatase antiferromagnetic spin coupling between the iron centres was found (ref. 41). Uncertainty exists about the bridging ligands and their structure. The different forms are shown in Fig. 4.

The reduced system PAP_{red} contains a Fe(III)/Fe(II)–unit, the oxidized PAP_{ox} a Fe(III)/Fe(III)–unit. These forms can be easily distinguished by UV/VIS spectroscopy, also from their phosphate containing forms.

The reported coupling constants (ref. 7) differ due to the problems mentioned in section 3.2, the recently determined data are $|J| \geq 150\ cm^{-1}$ (AF coupling) for PAP_{ox}, obtained from a lyophylized sample (ref. 41) and $J=-7\ cm^{-1}$ for PAP_{red} ((ref. 41), EPR–method).

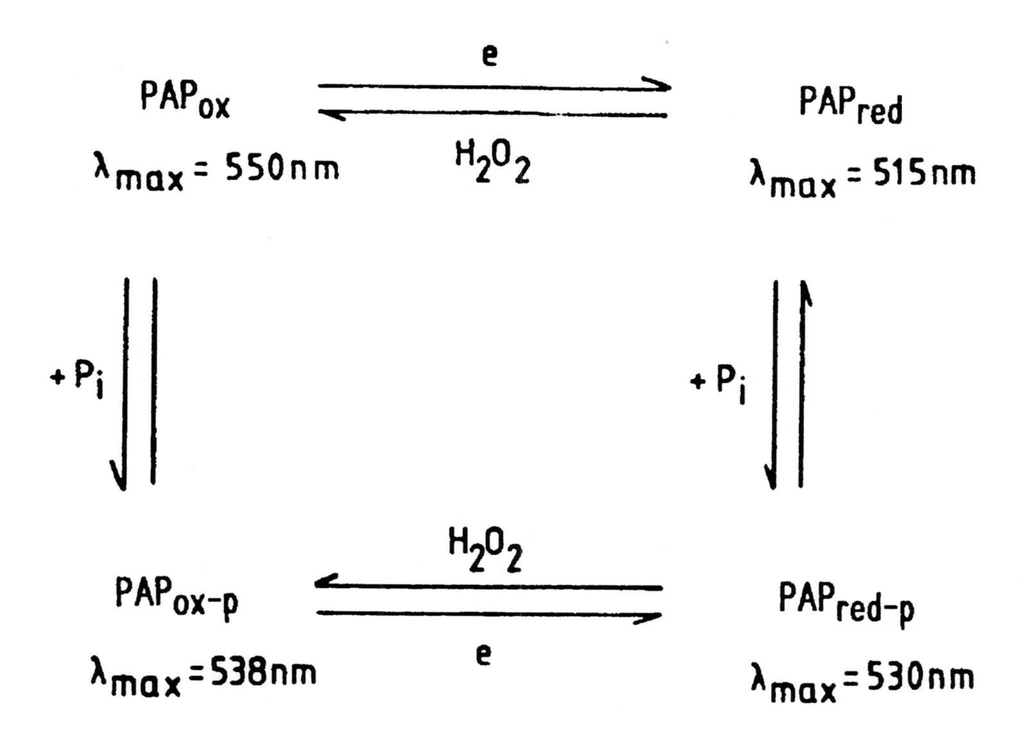

Fig. 4. PAP forms with UV/VIS absorption maxima.

Related enzymes are uteroferrin (Uf; (ref. 7)) with largely equivalent properties and ribonucleotid reductase in which the structure of the Fe_2–centre was confirmed by a X–ray analysis (ref. 44).

Concerning the enzymatically active form PAP_{red} two aspects were of special interest:

- Does a pH–variation which changes EPR and UV/VIS spectra (refs.41 and 45) also influence the spin coupling in the dinuclear iron centre, and
- is the coupling strength changed after treatment with the inhibitor phosphate ?

In the following the evaluation of exchange parameters for PAP and an interpretation are presented. The $\chi(T)$–data of samples of frozen solutions of PAP_{red} were measured using the same equipment as for hemocyanine. After thawing NaH_2PO_4–solution was added and the measurement repeated. Fits of a $\chi(T)$ equation from $\hat{H}=-2J\hat{S}_1\hat{S}_2$ to the experimental data result in (ref. 46)

	pH	J/cm^{-1}
PAP_{red}	3.9	-10(1)
PAP_{red}	4.9	-9(1)
PAP_{red}	5.6	-13(1)
$PAP_{red\text{-}P}$	3.9	-5(1)
$PAP_{red\text{-}P}$	5.6	-6(1)

The electron g–value was 1.77 (from EPR, (ref. 41)), an amount of 10-14 % of uncoupled paramagnetic Fe(III)–centres was indicated for a sufficient theoretical description of the $\chi(T)$–data.

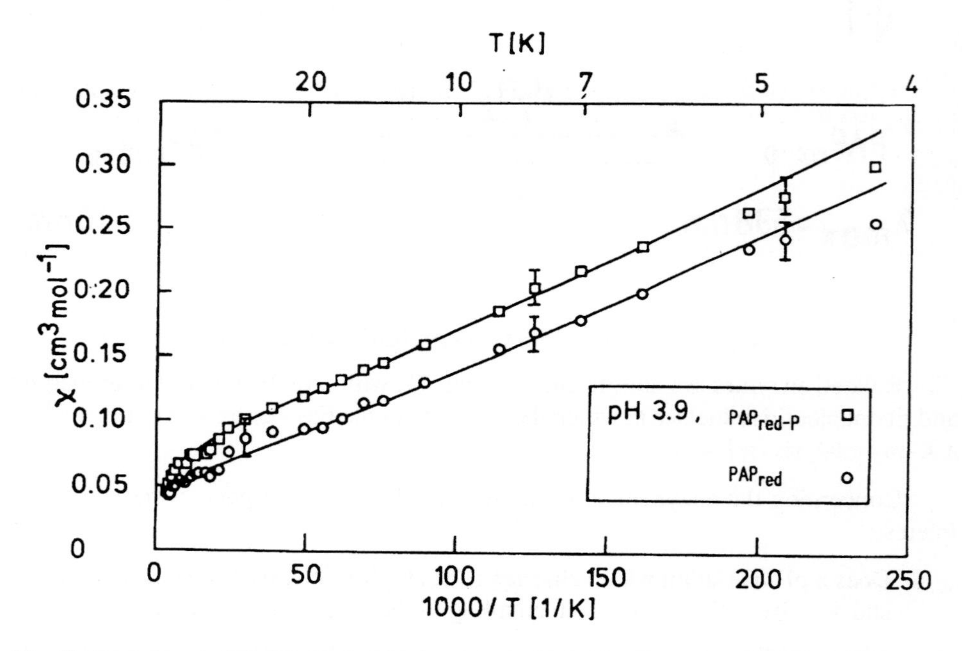

Fig. 5. Experimental (o,□) and calculated (——) temperature dependence of magnetic susceptibilities in PAP_{red} and $PAP_{red\text{-}P}$, pH=3.9

Fig. 5 shows experimental (o,□) and calculated (——) data for the system $PAP_{red} \rightarrow PAP_{red\text{-}P}$ at pH=3.9. The increased magnetic susceptibility is due to a weaker exchange coupling in the phosphate containing form.

The weak antiferromagentic exchange coupling for PAP_{red} is in accordance with previous reports ((ref. 41) and refs. therein). The treatment of PAP_{red} with phosphate leads to a reduction of J by the factor 2. In uteroferrin the spin coupling is reduced by the factor 3 at the transition $Uf_{red} \rightarrow Uf_{red\text{-}P}$ (ref. 39). No influence of pH–change on J could be determined. This may be explained by an equilibrium μ–OH/μ–OH_2 as in model complexes M-L-M (M=Mn(II), Fe(III), Ni(II), L=O, OH, OH_2) the exchange coupling is correlated to the bridging (ref. 47) as $J(\mu\text{–O}) >> J(\mu\text{–OH}) \approx J(\mu\text{–OH}_2)$. However, it should be mentioned that up to now an experimental evidence for a μ–oxo bridging is still missing (ref. 48).

Recently performed measurements (refs. 46,49) on PAP_{ox} and $PAP_{ox\text{-}X}$ ($X=PO_4^{3-}$ and related oxoanions) could not confirm the reported strong antiferromagnetic coupling in the oxidized species with $|J| \geq 150$ cm^{-1} (ref. 41). There, coupling constants one order of magnitude smaller were found (refs. 46 and 49).

4. Distance Dependence of the Exchange Coupling Constant

Some relations for distance dependences of exchange coupling parameters are known of which selected examples will be given here.

For clusters of aromatic free radicals Yamaguchi *et al.* (ref. 50) calculated the intermolecular effective through space exchange interaction using an ab initio spin projected UHF Møller-Plesset method. The authors show that the effective exchange interaction depends on the stacking mode. The distance dependence calculated was

$$J/cm^{-1} = Z \exp(p - qR_{12}) \tag{5}$$

with R_{12} as intermolecular distance and different parameter sets (Z/cm^{-1}, p, q/cm^{-1}) = = (-3.66, 20.7, -4.73) for the geminal and (2.32, 20.7,-4.75) for the para conformer of dimers of benzyl radicals.

The formula

$$2J/Gauss = 1.9 \cdot 10^{10}/Gauss \exp(-R_{12}/0.468 \text{ Å}) \tag{6}$$

was proposed by Kaptein (ref. 51) for calculating the exchange interaction in open–shell systems. At R_{12} = 10 Å a coupling constant J = 0.001 cm^{-1} is obtained.

The distance dependence of the exchange coupling constant in closed shell systems is of great importance for metallo-biomolecules and also for electron transfer reactions. A relationship was given by Coffman and Buettner (ref. 52) as a limit for the exchange coupling

$$2J/cm^{-1} = 1.35 \cdot 10^7 \exp(-1.8\, R_{12}/Å) \qquad (7)$$

The data presented are consistent with the data given by Hendrickson (ref. 53). For compounds at R_{12}=10Å a singlet–triplet splitting of about 0.4 cm^{-1} is obtained, for R_{12}=15Å about 0.0001 cm^{-1}.

As it was mentioned in the introduction $2J_{exp}$=-140 cm^{-1} was described for a terephthalato bridged dinuclear Cu(II) system (ref. 29). This means that under certain conditions stronger exchange coupling effects can occur in the range >10Å. This is due to the geometric and electronic structure and the bridging group. Theoretical calculations to this topic are in progress (ref. 54).

For electron transfer processes on a molecular level the exchange interaction can be assumed as a through space interaction or a superexchange one. The energy levels are influenced by dipolar interactions and Zeeman splitting, too.

Some aspects of long range electron transfer processes in the photosynthetic reaction centre will be discussed here since a superexchange mechanism is proposed to be responsible for the primary charge separation in the reaction centre of the purple photosynthetic bacteria (refs. 17, 55 and 56), but also in the photosystem I (ref. 18).

The photosynthetic process can be interpreted as follows: The reaction centre (RC) is excited by the energy of a photon absorbed by antenna systems of a multitude of pigment molecules, *e.g.* haem groups in cytochrome. The energy from the antenna system is transferred to the RC by excitation migration in a fast process.

The dimeric unit of bacteriochlorophyll ^1P in case of photosynthetic bacteria acts as a primary electron donor. From the ecxited $^1P^*$ there is a very fast electron transfer to the primary electron acceptor bacteriopheophytin (H) *via* the bacterio-chlorophyll (B) which acts as a transient ionic state, or in general by enhancing the electronic coupling between $^1P^*$ and H. The secondary electron acceptors are bound quinone molecules (Q) as menoquinone in *rhodobacter viridis* or ubiquinone in *rhodobacter sphaeroides.* The kinetics of these steps are not clear, as well as the mechanism of the electron transfer from Q^- to the various external acceptors *e.g.* the Fe/S centres.

The model for the primary part of the photosynthetic reaction process in a very rough approximation is given as

$$\underset{(a)}{^1[PBHQ]} \dashrightarrow \underset{(b)}{^1[P^*BHQ]} \rightleftharpoons \underset{(c)}{^1[P^+B^-HQ]} \rightleftharpoons \underset{(d)}{^1[P^+BH^-Q]} \rightleftharpoons \underset{(e)}{^1[P^+BHQ^-]} \qquad (8)$$

The reaction from (a) to (d) is a very fast process in the ps scale, from (d) to (e) about two orders of magnitude slower. Parallel to all of the singlet states three triplet sublevels caused by spin–spin interaction can occur. The interaction between the electron spins in the radical pairs (c)–(e) can be also due to dipolar spin–spin interaction or Zeeman interaction. The related exchange coupling energy leading to a singlet–triplet separation of $2J_{ij}$ in any of the transient states of eq. (8) is

$$2J_{ij} = E_T(^3RC) - E_S(^1RC) \tag{9}$$

with E_T, E_S as relative energies of the triplet and singlet states.

Michel–Beyerle *et al.* (ref. 17) suggested for the reaction (b) to (d) a one step electron transfer mechanism mediated through superexchange interaction *via* (c). Ogrodnik *et al.* (ref. 55) described the exchange constants using a second order pertubation theory. J_{ij} is proportional to the square of the electronic matrix coupling elements V_{ij} between the related triplet states divided by the sum of the related vertical energy differences ΔE,

$$J_{ij} \sim V_{(ij)}^2 / \Delta E \tag{10}$$

Using the models describing the matrix coupling elements V_{ij} the J_{ij} can be estimated.

The distance dependence of J_{ij} is related to the distance dependence of the electronic matrix coupling element *via*

$$V_{ij} = V_{ij}^0 \exp(-\beta(R_{12}-R_0)/2) \tag{11}$$

V_{ij}^0 is V_{ij} at $R_{12}=R_0$, R_0 is the close contact distance of the two radical sites, β is a measure for the decrease of the electronic coupling with increasing separation.

Due to the long distance — *e.g.* in purple bacteria about 17Å between P and H — the interaction can be only a superexchange one.

The absolute values of the exchange coupling constant for the photosynthetic reaction processes are very small, in the order of 10^{-3} cm^{-1} or even smaller. The experimental techniques to measure these data are indirect methods like EPR, Transient Electron Spin Polarization (ESP), fluorescence spectroscopy, time resolved spectroscopy *etc.*, which are used for processes with an acceptable time scale. It is evident that magnetic susceptibility measurements cannot be used for those extreme weakly coupled systems.

5. Mixed Valence and Resonance Delocalization

A mixed valence state in a dimeric homonuclear exchange coupled system can be realized in two different ways

- The oxidation states of both centres differ. This difference in the localized mixed valence complex is detectable by experimental techniques. An example is the dinuclear $Fe_A^{2+} - Fe_B^{3+}$. Here, the excess electron is localized on A.
- Each centre in the dimer carries an averaged charge due to two formal different oxidation states. Experiments reveal the same properties for each centre. This is the socalled resonance delocalization as *e.g.* in $Fe_A^{2.5+} - Fe_B^{2.5+}$.

For the localized mixed valence state the Heisenberg–Dirac–Van Vleck–model (see (refs. 25-27)) can be assumed as adequate. The relative energy states can be calculated from

$$E(S_T) = J\, S_T(S_T+1),\ S_T\text{: total spin} \tag{12}$$

A localized mixed valence state is realized in the reduced form of the purple acid phosphatase, *e.g.* in PAP_{red} as described in section 3.3. Evidences for a localized mixed valence state are given by Mössbauer data, magnetic susceptibility data as shown, and EPR results (ref. 41).

Some synthetic molecules with a Fe^{2+}–Fe^{3+}–unit may be assumed as models for this enzyme, an example is shown in Fig. 6 (refs. 57 and 58) as a $\mu(T)$ plot ($\mu/\mu_B = 2.828(\chi T)^{1/2}$).

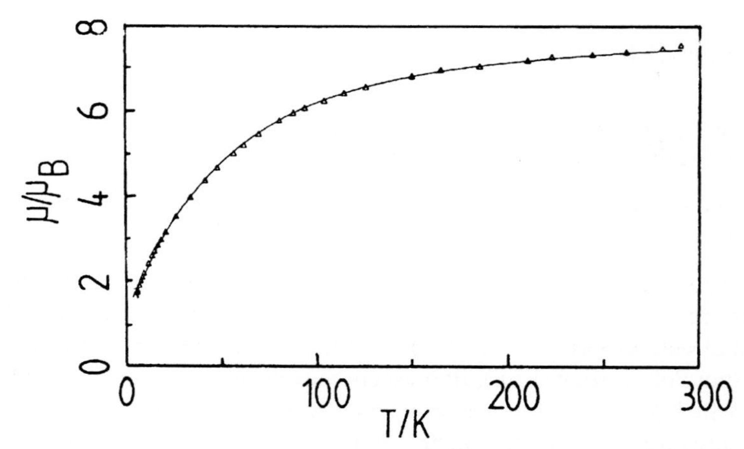

Fig. 6. Magnetic moment as function of temperature for [L'Fe(II)-(μ-OR)(μ-L)$_2$-Fe(III)], L = $O_2P(O\Phi)_2$, L'= macrocycle (ref. 57). The solid line is calculated with $J = -6.2(5)\ cm^{-1}$, $g = 2.06(4)$, $x_{PARA} = 0.003$, $N\alpha = 800\cdot10^{-6}\ cm^3mol^{-1}$.

In case of the resonance delocalized states each spin state splits into two different states due to the resonance delocalization characterized by a resonance parameter B.

$$E(S_T) = J\ S_T(S_T+1) \pm B(S_T+1/2) \tag{13}$$

The resonance delocalization first described as double exchange by Zener (ref. 59) and more detailed by Anderson and Hasegawa (ref. 60) was recently applied to exchange coupled molecules (refs. 61-64). For a $S_T = 1/2$ state the resonance splitting is smaller compared to that of a $S_T = 9/2$ state. Noodleman *et al.* (refs. 63 and 65) pointed out that a resonance delocalization of the remaining d–electron is easier when the spins are parallely (*i.e.* ferromagnetically) oriented. In fact, the resulting spin ground state is a competition between the Heisenberg exchange and the resonance delocalization. The Heisenberg exchange usually leads to an antiferromagnetic ground state whereas the resonance delocalization favours the parallel arrangement.

Molecules with resonance delocalization are very scarce. One example is the dimeric tris–hydroxo bridged iron complex $[L_2Fe_2(\mu\text{–}OH)_3](ClO_4)_2 \cdot CH_3OH \cdot 2H_2O$, L = N, N'N''–trimethyl–1,4,7–triazacyclononane (ref. 64). For this complex a delocalized valence state within the experimental time resolution limit was found. The temperature dependence of the magnetic moment is shown in Fig. 7, (ref. 66).

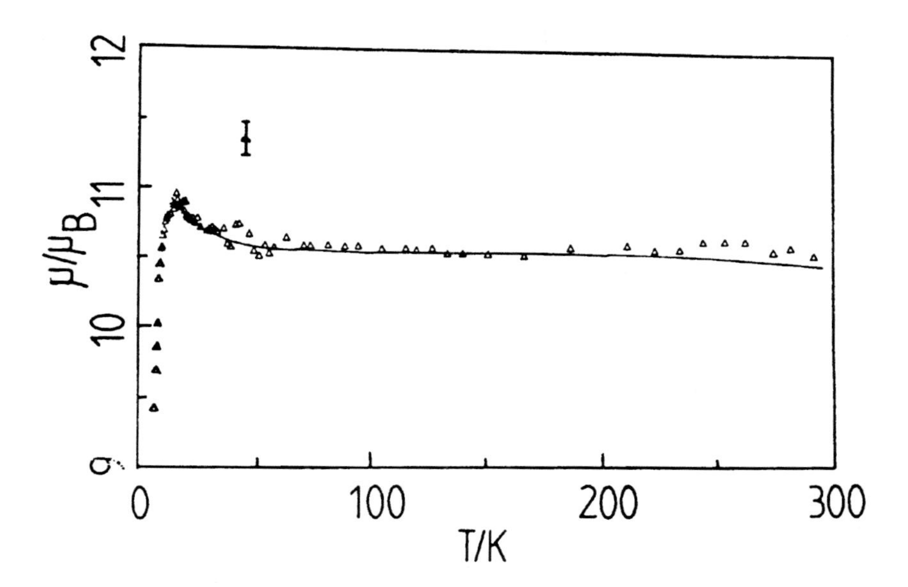

Fig. 7. Magnetic moment as function of temperature for $[L_2Fe_2(\mu\text{–}OH)_3]$ $(ClO_4)_2 \cdot CH_3OH \cdot 2H_2O$ (ref. 64). The solid line is calculated with $J=60\ cm^{-1}$, $g=2.12$, $\theta=0.86$ K (see text).

With decreasing temperature μ remains nearly constant at 10.5 μ_B and decreases rapidly below a maximum at T=10 K. A S_T=9/2 state is the spin ground state, the decrease of μ at T<10K may be attributed to it's zero–field splitting and intercluster interactions. Ding *et al.* (ref. 64) calculated $J \geq -132$ cm^{-1} (according to eq. (13)), B=1300 cm^{-1} with $J \geq -20/99$ B. Noodleman *et al.* (refs. 63, 65) reported for dimeric and trimeric ferredoxines B–values between 406-568 cm^{-1} using Xα– and LCAO–methods. For a HiPIP–protein B=598 cm^{-1} was found (ref. 67). The value of B=1300 cm^{-1} for the hydroxo–bridged complex is in the same order of magnitude but twice as large.

For resonance delocalized systems exchange parameters cannot be determined from susceptibility measurements alone due to the competition between J and B. This can be demonstrated by application of a susceptibility function deduced from the isotropic Heisenberg Hamilton (ref. 1). The data for the delocalized 4/2-5/2–spin system can be fitted with exchange parameters J=60 cm^{-1}, g=2.12, θ=0.86 K. The solid line in Fig. 7 represents this fit (T>10 K) with a fairly good agreement between experimental and calculated data, even using an inadequate model. This means an interpretation is possible only together with the results from UV/VIS–, EPR–, and Mössbauer–spectroscopy.

In the following some remarks on the groups of ferredoxines and HiPIP–proteins (refs. 10, 11 and 65) seem appropriate. There is a series of well characterized synthetic compounds and even investigated proteins.

Two oxidation states are known for the 2Fe–2S unit:

$$[Fe_2S_2(SR)_4]^{2-} + e^- \rightleftharpoons [Fe_2S_2(SR)_4]^{3-}$$

$$S_T = 0 \qquad\qquad S_T = 1/2$$

whereas the tetrameric cuban–type structures show three different oxidation states:

$$[Fe_4S_4(SR)_4]^{-} + e^- \rightleftharpoons [Fe_4S_4(SR)_4]^{2-} + e^- \rightleftharpoons [Fe_4S_4(SR)_4]^{3-}$$

$$S_T = 1/2 \qquad\qquad S_T = 0 \qquad\qquad S_T = 1/2,\ 3/2,\ 7/2$$

The first two core oxidation states are realized in the high potential iron–sulfur proteins, the later two belong to the group of ferredoxins. The high potential form with three Fe^{3+} and one Fe^{2+} show resonance delocalization in the Fe^{2+}–Fe^{3+} moiety with a S'=9/2 ground state leading to a total spin 1/2 for the tetrameric unit. In the reduced form nearly degenerated spin states of S_T=1/2, 3/2, or 7/2 occur.

6. Magneto–Structural Correlations from Experimental and Theoretical Data

The strength and type of the spin exchange interaction depends on different parameters. This was stressed in section 4 for the distance of the paramagnetic centres. But the distance is only one variable, well investigated are effects from bond angle variations on the size and sign of J (ref. 21).

With help of a SCF–CI method developed by de Loth *et al.* (ref. 68), and program extensions by Astheimer *et al.* (ref. 69) calculations for different dimeric Cu(II) systems were performed (refs. 23 and 68-71. Also for a Cu-VO system calculations were described (ref. 72)). The method starts with an open–shell SCF calculation to define singly occupied magnetic orbitals. The singlet–triplet splitting is obtained as sum of different contributions arising in the CI–treatment.

It should be mentioned that an exact agreement between theoretical and experimental data cannot be achieved due to questions for the basis set (ref. 69), for relevance of special CI–terms, the convergence of higher orders terms, the use of simplified model molecules caused by program limitations *etc.*

Here, the exchange interaction in two $Cu(II)_2O_2$–systems with alkoxo– and pyridine–N–oxide bridging (PNO; Fig. 8) are compared to demonstrate how sensitively the magnetic behaviour depends on structural and electronic features.

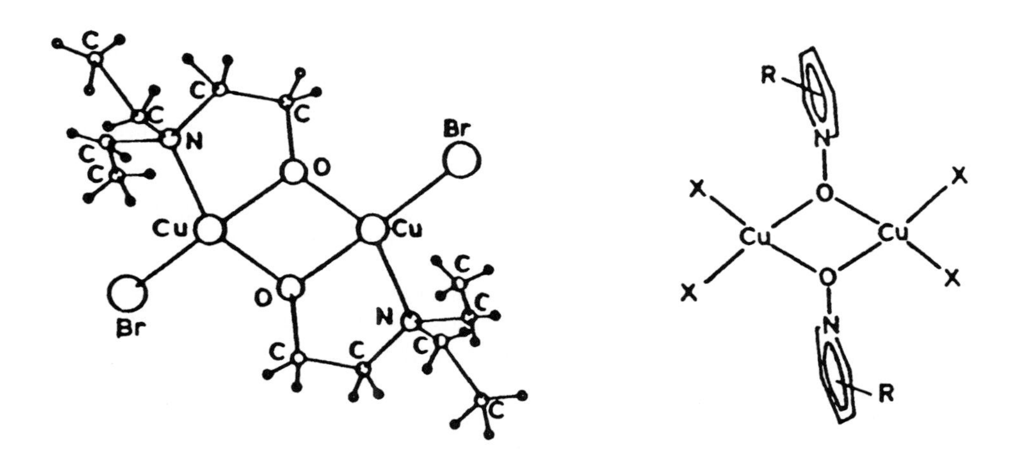

Fig. 8. $Cu(II)_2O_2$ model complexes for investigations of magneto–structural correlations (refs. 70, 71). *Left side*: Alkoxo bridged system; *right side:* Pyridine-N-oxide bridged system, X = Cl, Br, R = CH_3, C_6H_5.

TABLE 2.

Magneto–structural correlations in $Cu(II)_2O_2$–systems with alkoxo–bridging (AO, (ref. 70)) and pyridine–N–oxide–bridging (PNO, (ref. 71)).

AF means change towards antiferromagnetic coupling, F towards ferromagnetic coupling ($H = -2J\ S_1S_2$)

Increase of	Change in J		Towards
	AO	PNO	
angle Cu–O–Cu'	60-90 $cm^{-1}/°$	15 $cm^{-1}/°$	AF
distance Cu–O	40-70 $cm^{-1}/Å$	600 $cm^{-1}/Å$	F
torsion angle [L_2Cu/CuO_2]	15 $cm^{-1}/°$		F
Replacement of bridging ligands			
hydroxo→alkoxo	120 cm^{-1}		AF
Replacement of non–bridging ligands			
Br→Cl	130 cm^{-1}		F
Br→F	550 cm^{-1}		F

Table 2 presents the results from theoretical calculations (refs. 70, 71). The following points should be stressed:

- Structural properties like the bond angles and lengths in the superexchange pathway and local geometries of the spin centres influence J.
- Electronic effects on J arise from the substituents at the diamagnetic bridging ligands.
- In a second approximation influences of terminal non–bridging ligands are of importance. They may become relevant for electron transfer processes.

Another result from the SCF–CI-calculations are the electronic structures of the magnetic orbitals. As a typical example the orbitals of the bridging oxygen for the alkoxo– and PNO–compounds are compared: In the first ones the bridging oxygen orbital has sp_x character resulting in weak Cu–O bond length dependence and a strong angle dependence; in PNO–systems the s character dominates and therefore an expressed Cu–O bond length dependence and a minor Cu–O–Cu angle dependence are found in the calculations.

The results summarized in Table 2 are in good agreement with the experimental data. In experimental studies several parameters are changed simultaneously, therefore the calculations offer the chance to separate different influences.

The de Loth method in it's present form can be applied only on two S_i=1/2 centres with C_i or C_s symmetry. For spin systems S_i>1/2 or more than two centres other methods like those of Jansen and Block (ref. 73) or of Noodleman and Baerends (refs. 65, 74 and 75) are developed.

7. Conclusion

The exchange coupling — mainly as superexchange interaction — is a widespread property in multimetal proteins. At present time the magnetic behaviour of these biomolecules is investigated to characterize their active sites. Conclusions can be drawn from the comparison with inorganic model complexes. It is evident that redox potentials in the metallo proteins are influenced by the exchange interaction. Electron transfer processes seem to be facilitated by multimetal or multiradical centres. Also the kinetics of electron transfer processes can depend on different spin states. However, detailed information about the role of exchange coupling effects on the catalytic processes is missing. This lack of information gives rise to new investigations on this challenging field.

Acknowledgement

The authors thank Profs. B. Krebs, L. Noodleman, K. Wieghardt, H. Witzel, and Dipl.–Ing. P. Fleischhauer for stimulating discussions and cooperation. This work was supported by the Deutsche Forschungsgemeinschaft.

References

1 L. Que, Jr. (Ed.) *Metal Clusters in Proteins* ACS Symposium Series 372, American Chemical Society, Washington DC, 1988.

2 K.D. Karlin and J. Zubieta (Eds.) *Biological & Inorganic Copper Chemistry* Vol. I and II, Adenine Press Inc., New York, 1986.

3 E.I. Solomon and D.E. Wilcox in: R.D. Willett, D. Gatteschi, and O. Kahn (Eds.) *Magneto-Structural Correlations in Exchange Coupled Systems* NATO ASI Series, Vol. 140, D. Reidel, Dordrecht, 1985, p.463.

4 R.E. Stenkamp, L.C. Sieker and L.H. Jensen, J. Am. Chem. Soc. **106** (1984) 618.

5 J. Lorösch and W. Haase, Biochem. **25** (1986) 5850.

6 E.I. Stiefel, H. Thomann, H. Jin, R.E. Bare, T.V. Morgan, S.J.N. Burgmaer and C.L. Coyle in ref. (1), p.372.

7 L. Que, Jr. and R.C. Scarrow in ref. (1), p. 152.

8 B.G. Fox, K.K. Surerus, E. Münck and J.D. Lipscomb, J. Biol. Chem. **263** (1988) 10553.

9 P.A. Clark and D.E. Wilcox, Inorg. Chem. **28** (1989) 1326.

10 A.G. Sykes and J.D. Sinclair–Day in: A.V. Xavier (Ed.) *Frontiers in Bioinorganic Chemistry* VCH Verlagsgesellschaft, Weinheim, 1986, p. 326..

11 V. Papaefthymiou, M.M. Millar and E. Münck, Inorg. Chem. **25** (1986) 3010.

12 R.A. Scott, J.R. Schwartz and S.R. Cramer in ref (2), Vol. I, p. 41.

13 D.R. McMillin and H.R. Engeseth in ref (2), Vol. I, p. 1.

14 A. Messerschmidt, A. Rossi, R. Ladenstein, R. Huber, M. Bolognesi, G. Gatti, A. Marchesini, T. Petruzzelli and A. Finazzi-Agro, J. Mol. Biol. **206** (1989) 513.

15 K. Wieghardt, Angew. Chem. **101** (1989) 1179.

16 E.C. Theil in ref. (1), p. 179.

17 M.E. Michel–Beyerle, M. Bixon and J. Jortner, Chem. Phys. Lett. **151** (1988) 188.

18 D. Stehlik, C.H. Bock and J. Petersen, J. Phys. Chem. **93** (1989) 1612.

19 D.A. Hunter, A.J. Hoff and P.J. Hore, Chem. Phys. Lett. **134** (1987) 6.

20 R.A. Marcus, Chem. Phys. Lett. **146** (1988) 13.

21 R.D. Willett, D. Gatteschi and O. Kahn (Eds.) *Magneto–Structural Correlations in Exchange Coupled Systems* NATO ASI Series, Vol. 140, D. Reidel, Dordrecht, 1985.

22 O. Kahn, Angew. Chem. **97** (1985) 837.

23 S. Gehring and W. Haase, Mol. Cryst. Liq. Cryst. **176** (1989) 513.

24 J.F. Gibson, D.O. Hall, J.H.M. Thornley and F.R. Whatley, Proc. Nat. Acad. Sci. USA **56** (1966) 987.

25 W.E. Hatfield in: E.A. Boudreaux and L.N. Mulay (Eds.) *Theory and Applications of Molecular Paramagnetism* Wiley–Interscience, New York, 1976, p. 369.

26 R.L. Carlin *Magnetochemistry* Springer–Verlag, Berlin, Heidelberg, 1986.

27 A. Bencini and D. Gatteschi *EPR of Exchange Coupled Systems* Springer–Verlag, Berlin, Heidelberg, 1990.

28 H.A. Kramers, Physica **1** (1934) 182.

29 P. Chaudhuri, K. Oder, K. Wieghardt, S. Gehring, W. Haase, B. Nuber and J. Weiss, J. Am. Chem.Soc. **110** (1988) 3657.

30 U. Bossek, K. Wieghardt, B. Nuber and J. Weiss, Angew. Chem. **102** (1990) 1093.

31 K. Wieghardt, U. Bossek, B. Nuber, J. Weiss, S. Gehring and W. Haase, J. Chem. Soc., Chem. Commun. **1988**, 1145.

32 S. Drüeke, K. Wieghardt, B. Nuber, J. Weiss, P. Fleischhauer, S. Gehring and W. Haase, J. Am. Chem. Soc. **111** (1989) 8622.

33 P. Chaudhuri, M. Winter, K. Wieghardt, S. Gehring, W. Haase, B. Nuber and J. Weis, Inorg. Chem. **27** (1988) 1564.

34 S. Alvarez, M. Julve and M. Verdaguer, Inorg. Chem. **29** (1990) 4500.

35 F. Tinti, M. Verdaguer, O. Kahn and J.–M. Savariault, Inorg. Chem. **26** (1987) 2380.

36 L. Banci, A. Bencini and D. Gatteschi, Inorg. Chem. **22** (1983) 2681.

37 T.R. Felthouse, E.N. Duesler and D.N. Hendrickson, J. Am. Chem. Soc. **100** (1978) 618.

38 C.J. O'Connor, Progr. Inorg. Chem. **29** (1982) 203.

39 E.P. Day, S.S. David, J. Peterson, W.R. Dunham, J.J. Bonvoisin, R.H. Sands and L. Que, Jr., J. Biol. Chem. **263** (1988) 15561.

40 T.J. Dutton, T.F. Baumann and J.A. Larrabee, Inorg. Chem. **29** (1990) 2272.

41 B.A. Averill, J.C. Davis, S. Burman, T. Zirino, J. Sanders–Loehr, T.M. Loehr, J.T. Sage and P.G. Debrunner, J. Am. Chem. Soc. **109** (1987) 3760.

42 J.T. Sage, Y.–M. Xia, P.G. Debrunner, T.D. Keough, J. de Jersey and B. Zerner, J. Am. Chem. Soc. **111** (1989) 7239.

43 J. Lörsch, Thesis, TH Darmstadt, 1986.

44 P. Nordlund, B.–M. Sjöberg and H. Eklund, Nature **345** (1990) 593.

45 D. Münstermann, M. Dietrich, H. Suerbaum and H. Witzel, submitted for publication.

46 S. Gehring, P. Fleischhauer, W. Haase, M. Dietrich and H. Witzel, Biol. Chem. Hoppe–Seyler **371** (1990) 786.

47 A. Caneschi, F. Ferraro, D. Gatteschi, M.C. Melandri, P. Rey and R. Sessoli, Angew. Chem. **101** (1989) 1408.

48 J. Sanders–Loehr, W.D. Wheeler, A.K. Shiemke, B.A. Averill and T.M. Loehr, J. Am. Chem. Soc. **111** (1989) 8084.

49 P. Fleischhauer, S. Gehring, W. Haase, M. Dietrich and H.Witzel, in preparation.

50 K. Yamaguchi, H. Namimoto and T. Fueno, Mol. Cryst. Liq. Cryst. **176** (1989) 151.

51 F.J.J. de Kanter, J.A. den Hollander, A.H. Huizer and R. Kaptein, Mol. Phys. **34** (1977) 857.

52 R.E. Coffman and G.R. Buettner, J. Phys. Chem. **83** (1979) 2387.

53 D.N. Hendrickson in ref. (21), p. 523.

54 R. Bärthel, S. Gehring and W. Haase, unpublished.

55 A. Ogrodnik, N. Remy–Richter and M.E. Michel–Beyerle, Chem. Phys. Lett. **135** (1987) 576.

56 M.E. Michel–Beyerle (Ed.) *Antennas and Reaction Centers of Photosynthetic Bacteria – Structure, Interactions and Dynamics* Springer, Berlin, 1985.

57 K. Schepers, B. Bremer, B. Krebs, G. Henkel, E. Althaus, B. Mosel and W. Müller–Warmuth, Angew. Chem. **102** (1990) 582.

58 P. Fleischhauer, W. Haase, K. Schepers and B. Krebs, unpublished.

59 C. Zener, Phys. Rev. **82** (1951) 403.

60 P.W. Anderson and H. Hasegawa, Phys. Rev. **100** (1955) 675.

61 E. Münck, V. Papaefthymiou, K.K. Surerus and J.–J. Girerd in ref. (1), p. 302.

62 L. Noodleman and E.J. Baerends, J. Am. Chem. Soc. **106** (1984) 2316.

63 S.F. Sontum, L. Noodleman and D.A. Case in: D.R. Salahub and M.C. Zerner (Eds.) *The Challenge of d and f Electrons* ACS Symposium Series 394, American Chemical Society, Washington DC, 1989, p. 366.

64 X.–Q. Ding, E.L. Bominaar, E. Bill, H. Winkler, A.X. Trautwein, S. Drüeke, P. Chaudhuri and K. Wieghardt, J. Chem. Phys. **92** (1990) 178.

65 L. Noodleman, D. Case and E.J. Baerends in: *Theory and Application of Density Functional Approaches to Chemistry* in press.

66 P. Fleischhauer, S. Gehring and W. Haase, unpublished.
P. Fleischhauer, Diplom work, TH Darmstadt, 1989.

67 J. Jordanov, E.K.H. Roth, P.H. Fries and L. Noodleman, Inorg. Chem. **29** (1990) 4288.

68 P. de Loth, P. Cassoux, J.–P. Daudey and J.P. Malrieu, J. Am. Chem. Soc. **103** (1981) 4007.

69 P. de Loth, J.–P. Daudey, H. Astheimer, L. Walz and W. Haase, J. Chem. Phys. **82** (1985) 5048.

70 H. Astheimer and W. Haase, J. Chem. Phys. **85** (1986) 1427.

71 F. Nepveu, H. Astheimer and W. Haase, J. Chem. Soc., Faraday Trans. 2 **82** (1986) 551.

72 P. de Loth, P. Karafiloglou, J.–P. Daudey and O. Kahn, J. Am. Chem. Soc. **110** (1988) 5676.

73 E.L. Bominaar and R. Block, Phys. Rev. B, **33** (1986) 3672, and references therein.

74 L. Noodleman, D.A. Case and S.F. Stontum, J. Chimie Physique **86** (1989) 742.

75 L. Noodleman, Inorg. Chem., in press.

Electron and Proton Transfer in Chemistry and Biology, edited by A. Müller et al.
Studies in Physical and Theoretical Chemistry, Vol. 78

Multi-Electron Transfer Processes in Nitrogen Fixation and other Natural Systems

David J. Lowe

AFRC IPSR Nitrogen Fixation Laboratory
University of Sussex
Brighton BN1 9RQ (UK)

Summary

Multi–electron transfers in biological systems generally involve small inorganic molecules. A brief description is given, with references to enable the reader to access the literature on each system. Metal ions are involved in all these processes. Biological nitrogen fixation is treated in greater depth, with a discussion of the known information on the mechanism of nitrogenase. Reduction of dinitrogen by this enzyme probably proceeds by successive electron and proton transfers to substrate bound to a metal. In some other systems a concerted electron transfer probably follows a build–up of oxidizing or reducing equivalents on a metal cluster.

Introduction

In natural systems oxidation/reduction reactions generally involve either one or, at the most, two concerted electron transfers. However, the relatively small number of reactions requiring three or more concerted electron transfers, the multi-electron processes, are of profound metabolic importance since they generally occur at the point of interaction with small inorganic molecules from the environment. This chapter is primarily concerned with biological nitrogen fixation in which dinitrogen is converted to ammonia and which is the primary biological process whereby dinitrogen from the atmosphere enters the biosphere. The other principle component of the air, dioxygen, is also normally involved in multi–electron processes, whether in its reduction to water or the reverse of this, the photosynthetic 'water splitting' reaction. A number of other processes that only occur in bacteria belong to this class and include important steps in nitrogen and sulphur metabolism whereby these elements can be mobilized from the environment or lost to it. Only biological nitrogen fixation will be covered in depth here; many of the other topics are so vast that only a brief description and introduction to the literature can be given. Table 1 gives a summary of the reactions and enzymes discussed.

Table 1

Multi–electron transfer processes described in the text.

	Reaction	Protein(s)/Enzyme(s)
(1)	$O_2 + 4H^+ + 4e^- \rightarrow 2H_2O$	Cytochrome Oxidase Blue Copper Oxidases
(2)	$2H_2O \rightarrow O_2 + 4H^+ + 4e^-$	Water–Splitting Enzyme
(3)	$NH_2OH + H_2O \rightarrow NO_2^- + 5H^+ + 4e^-$	Hydroxylamine Dehydrogenase
(4)	$NO_2^- + 7H^+ + 6e^- \rightarrow NH_3 + 2H_2O$	Nitrite Reductase
(5)	$HSO_3^- + 5H^+ + 6e^- \rightarrow S^{2-} + 3H_2O$	Bisulphite Reductase
(6)	$N_2 + 8H^+ + 8e^- \rightarrow 2NH_3 + H_2$	Nitrogenase

The main problem faced by organisms carrying out these reactions is the coupling of the input single or double electron steps to the multi–electron process; this is invariably done with the help of metal ions bound to the proteins involved in the catalysis. The metals can provide binding sites for the substrates and also storage sites for reducing equivalents by changing their oxidation levels.

Oxygen Reduction and Water Oxidation

The reduction of dioxygen is an essential life process for all plants and animals and for aerobic bacteria. Dioxygen is used as a terminal electron acceptor for aerobic respiration in these organisms and is, in general, reduced by four electrons to water (reaction 1, Table 1). The photosynthetic water oxidation reaction is the reverse of this (reaction 2, Table 1).

These reactions are best performed as multi–electron processes because of the high reactivity, and hence toxicity, of potential intermediates such as superoxide ($O_2^{\cdot-}$), peroxide (O_2^{2-}) and hydroxyl radical ($OH^{\cdot}$). Indeed, aerobic organisms have efficient mechanisms for disposing of intermediates, for example superoxide dismutases (catalyzing $2O_2^- + 2H^+ \leftrightarrow O_2 + H_2O_2$) and catalases and peroxidases (catalyzing $2H_2O_2 \leftrightarrow 2H_2O + O_2$).

Cytochrome Oxidase

Reaction (1) is catalyzed by cytochrome oxidase in the mitochondrial respiration of eukaryotes (animals, plants and yeasts) and also in some prokaryotes (bacteria). This enzyme has been estimated to account for over 90% of the dioxygen consumption by living organisms (ref. 1). Mitochondrial cytochrome oxidase, in common with some of the bacterial ones, itself contains two irons in cytochromes a and a_3 as well as two (or possibly three) distinct types of copper. Cytochrome oxidase, as its name indicates, receives electrons from cytochrome c; this is a single electron process. Dioxygen binds and is reduced in a four–electron process at a binuclear centre comprising haem a_3 and one of the copper atoms and, although the exact mechanism remains obscure, it is clear that all intermediates normally remain bound until water is released, (ref. 2), and that both haem iron atoms and the two copper atoms are redox active.

Bacterial cytochrome oxidases are more diverse, and although some are very similar to that found in mitochondria, others contain different haems and some lack copper. Work in this area has been reviewed by Poole (ref. 3). The different oxidases are produced both by different species and in response to different environmental conditions. In cases where sufficient mechanistic details are available, the rule that no dioxygen reduction intermediates are released, always seems to apply.

"Blue" Copper Oxidases

This class of oxidases (ref. 4) are called "blue" because they contain a "blue" cupric ion with an intense visible absorption at about 600 mμ; however, this atom is not believed to be responsible for dioxygen binding and reduction. Again they catalyze reaction (1) as a single process. However many other copper containing oxidases (*e.g.* amine oxidases), as well as multi–component oxidases (*e.g.* xanthine oxidase) and flavoprotein oxidases (*e.g.* amino acid oxidases), only catalyze the two–electron reduction of dioxygen to peroxide or occasionally its one electron–reduction to superoxide (ref. 5).

The three main groups of the "blue" copper oxidases are the ceruloplasmins, from blood plasma and with indeterminate function, laccases, found in various plants and fungi and which oxidize phenols, and ascorbate oxidases from plants. All contain at least four copper atoms, an isolated "blue" copper and a group of three closely associated copper atoms (refs. 6 and 7), the geometry of which has recently been determined for ascorbate oxidase as a result of an X–ray crystal structure determination (ref. 8). The presumption is that electrons are transferred singly from the reducing substrate to the single "blue" copper, and then accumulate at three–copper clusters where dioxygen is bound and reduced to water without any intermediates being released.

Water Oxidation

When organisms use light energy for fixing carbon dioxide, to make the carbon skeletons of their constituent molecules in the process of photosynthesis, they require electrons and protons. Green plants, algae and cyanobacteria use a system for splitting water to provide these electrons and protons by reaction (2), with dioxygen as a side product. This reaction is the source of atmospheric dioxygen.

There is again the potential problem with reaction (2) that toxic intermediates can form from partial reactions. In response to this, the manganese–containing water–splitting enzyme has evolved to catalyse reaction (2) as a single concerted process (refs. 9

and 10). Electrons are transferred singly from a cluster of four manganese atoms *via* various intermediates to photosystem II and the four oxidizing equivalents build up on the manganese cluster by an alteration of the oxidation level of the individual metal ions. These electrons are then used together to oxidize two water molecules, bound to the cluster, to dioxygen (*eg*. ref. 11).

Nitrogen and Sulphur Metabolism:

Ammonia Oxidation

The initial reaction in the oxidation of ammonia to nitrite, which is the only known energy source for bacteria of a number of genera (ref. 12), is the two–electron reduction to hydroxylamine. The subsequent conversion of hydroxylamine to nitrite by hydroxylamine dehydrogenase is, however, a four–electron process (reaction 3, Table 1) with all intermediates remaining bound (ref. 13). Although the exact mechanism remains uncertain, it may proceed *via* bound NO^+:

$$E + NH_2OH + H_2O \rightarrow E - NO^+ + 3H^+ + 4e^- + H_2O \rightarrow E + NO_2^- + 5H^+ + 4e^-$$

The hydroxylamine dehydrogenase molecule has an $\alpha_3\beta_3$ subunit structure with each β subunit containing one c–type haem, and each α subunit containing one P–460 centre and six c–type haems. There are, therefore, plenty of redox centres available for storing electrons before they are passed on to cytochrome c_{554}, and then presumably on to the conventional electron transport chain.

Nitrite Reduction

Nitrite can be reduced by various bacteria under different physiological conditions. In denitrification, whereby fixed nitrogen is lost to the biosphere, NO, N_2O and N_2 can be formed by enzymes containing haem b, c and d, or copper ions. The route is not clear (ref.14), NO and N_2O may remain bound as intermediates and can be products, although more than one enzyme is probably involved (ref. 15). Note that these nitrite reductases also have weak cytochrome oxidase activity. In dissimilatory nitrite reduction the role of nitrite is to act as a terminal electron acceptor for anaerobic respiration. This is also true for the bacterial dissimilatory nitrite reductases that reduce nitrite further to ammonia without loss of NO, N_2O or N_2 to the atmosphere (reaction 4, Table 1). A number of proteins with this activity are known (refs. 14 and 16) and these contain six haem groups per molecule, or two haem groups per molecule, or two 4Fe–4S, two flavin adenine dinucleotides (FAD) and two sirohaems per molecule (sirohaem is an iron tetrahydroporphyrin). The mechanisms of these enzymes have not been elucidated, although the multiple metal centres presumably fulfill their usual role in storing electrons prior to their transfer to the substrate. The assimilatory nitrite reductase of eukaryotes, which is involved in nitrogen uptake, and which catalyzes the six–electron reduction of nitrite to ammonia appears to be very similar to the sirohaem containing dissimilatory enzyme mentioned above, but without the FAD; indeed, in prokaryotes, the distinction between the enzymes responsible for the assimilatory and dissimilatory processes is not always clear. The mechanism is probably very similar to that of the sirohaem containing sulfite oxido–reductases.

Sulphite Reduction and Oxidation

Sulphite reduction occurs in both dissimilatory and assimilatory pathways (ref. 17). In the dissimilatory process, which occurs in the sulphate–reducing bacteria, bisulphite is reduced in a six–electron process to sulphite (reaction 5, Table 1), and is used as a terminal electron acceptor. All of the four known bisulphite reductases contain sirohaem and iron–sulphur centres. Assimilatory sulphite oxido-reductases, which are widespread, also contain sirohaem and iron–sulphur centres as well as FMN, although an enzyme of this type lacking iron–sulphur has been reported (ref. 18).

A clue to the mechanism of these sirohaem containing enzymes has been given by structural studies (ref. 19). These showed that the sirohaem and Fe_4S_4 centres are very close to each other, possibly sharing a common ligand which would provide a route for the transfer of stored reducing equivalents to the haem bound substrate.

Dinitrogen Reduction

The stoichiometry of this process is a six–electron reduction of dinitrogen to ammonia, although it is generally believed that an obligatory concomitant reduction of two protons to dihydrogen also occurs, giving an overall stoichiometry represented by reaction (6), (Table 1). The enzyme responsible for catalyzing this reaction, nitrogenase, is only found in a relatively small number of bacterial species spread among genera from cyanobacteria, obligate aerobes, facultative anaerobes, strict anaerobes, symbiotic bacteria, photosynthetic bacteria and archaebacteria. All nitrogenases studied to date appear to have very similar physicochemical properties and mechanisms. The only free intermediate observed from reaction (6) is the early release of dihydrogen. A number of reviews have been published covering the structure and mechanism of nitrogenase (refs. 20, 21, 22, 23, 24, 25, 26 and 27).

Nitrogenase Proteins

All nitrogenases contain non–haem–iron and sulphur, and until the mid 1980's a consensus existed that all nitrogenases also contained molybdenum. However, a number of papers (refs. 28, 29, 30, 31, 32 and 33) demonstrated that molybdenum was not as essential component and that vanadium could be used instead; they also demonstrated the existence of a third nitrogenase containing neither molybdenum nor vanadium.

Each of the three types of nitrogenase comprises two proteins. The smaller one, of molecular mass about 65,000 D, is reduced by a flavodoxin or ferrodoxin and transfers electrons singly to the larger one, of molecular mass about 220,000 D, with concomitant hydrolysis of ATP to ADP and inorganic phosphate. The larger protein is believed to contain the site of dinitrogen binding and reduction. Most structural and mechanistic studies have used the molybdenum enzyme and the rest of this review refers mainly to this system, except where specifically stated. The limited work on the other two enzymes has revealed no significant mechanistic differences from molybdenum nitrogenase.

The larger protein of the molybdenum enzyme is an $\alpha_2\beta_2$ dimmer and contains 2Mo, ~32Fe and ~30 S^{2-} per molecule. A useful nomenclature for this protein is the initial letters of the genus and species from which it can be purified followed by the numeral 1; thus the larger protein purified from *Klebsiella pneumoniae* becomes Kp1. Alternative generic nomenclatures used are MoFe protein and dinitrogenase. The smaller protein is a γ_2 dimer containing a single [4Fe–4S] cluster, and is called by, analogy with the larger protein, Kp2 (from *Klebsiella pneumoniae*), Fe protein or dinitrogenase reductase.

A great advance in the study of nitrogenase has occurred recently with the release of preliminary X–ray crystallographic results on the structure of Av2, *Azotobacter vinelandii* (ref. 34), and of Cp1, *Clostridium pasteurianum* (ref. 35). Av2 has a two fold symmetry axis relating the two subunits, with the cluster bound between them at the surface of the molecule. The molybdenum atoms of the MoFe protein have for some time been known to be bound, together with iron and sulphide, in two clusters (FeMoco) with approximate stoichiometry $MoFe_{6-8}S_{4-10}$ that probably also contain homocitrate. The remaining iron and sulphide are found in four 'P' clusters per molecule which may be unusual [4Fe–4S] clusters. The X–ray data have shown that the two FeMoco clusters are separated by about 70Å and that the 'P' cluster atoms are in two groups, each about 19Å (centre to centre) from one of the FeMoco clusters and each containing about 8Fe atoms. At the present degree of resolution (5Å) the pairs of 'P' clusters within each group cannot be resolved, and it is possible that they are present as a single larger cluster.

An overall scheme (ref. 36) for the reactions involved in the transfer of electrons from an *in vitro* reducing substrate, dithionite, to the Fe protein and hence to the MoFe protein and dinitrogen has been derived for *Klebsiella pneumoniae* molybdenum nitrogenase (ref. 24). It involves a number of partial reactions, the individual rate constants for which have been independently measured, and has been useful in suggesting further experiments. It quantitatively predicts the results of steady-state and pre–steady–state kinetic experiments and is consistent with the results of all such experiments to date. The scheme has since been expanded to include the reduction of other oxidizing substrates as well as details of the hydrolysis of ATP. These reactions will be considered more fully in the following two sections.

A number of other schemes have been presented that address parts of the reaction sequence (refs. 37, 38 and 39) and these will be discussed at the relevant points.

The Fe–protein cycle

Scheme 1

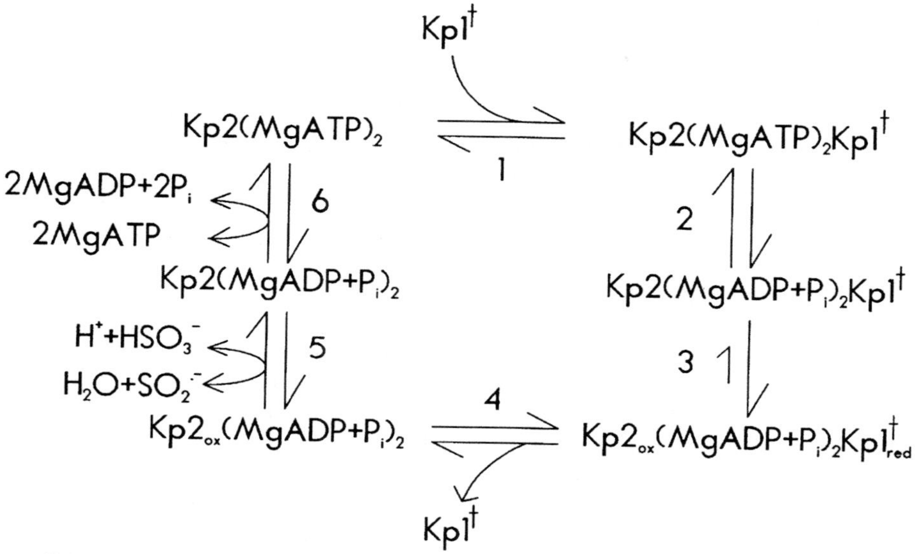

$Kp1^{\dagger}$

$Kp2(MgATP)_2 \rightleftharpoons^{1} Kp2(MgATP)_2Kp1^{\dagger}$

2: $Kp2(MgATP)_2Kp1^{\dagger} \rightleftharpoons Kp2(MgADP+P_i)_2Kp1^{\dagger}$

3: $Kp2(MgADP+P_i)_2Kp1^{\dagger} \rightleftharpoons Kp2_{ox}(MgADP+P_i)_2Kp1^{\dagger}_{red}$

4: $Kp2_{ox}(MgADP+P_i)_2 \rightleftharpoons Kp2_{ox}(MgADP+P_i)_2Kp1^{\dagger}_{red}$ (releasing $Kp1^{\dagger}$)

5: $Kp2_{ox}(MgADP+P_i)_2 \rightleftharpoons Kp2(MgADP+P_i)_2$ ($H_2O+SO_2^{\cdot-} \rightarrow H^{+}+HSO_3^{-}$)

6: $Kp2(MgADP+P_i)_2 \rightleftharpoons Kp2(MgATP)_2$ ($2MgATP \rightarrow 2MgADP+2P_i$)

Scheme 1 shows the Fe–protein cycle (refs. 40, 41 and 42). It involves an initial reversible association (reaction 1 of Scheme 1) of Kp2, with 2 molecules of MgATP bound, and half a Kp1 molecule ($Kp1^{+}$) containing one FeMoco (so that, for simplicity, the two halves are assumed to function independently). In reaction 2 (Scheme 1), the bound MgATP is hydrolyzed followed by the transfer of a single electron (reaction 3, Scheme 1) from Kp2 to $Kp1^{+}$ (ref. 43); these reactions appear to be reversible at 6°C but essentially irreversible at 23°C (ref. 42). The two proteins then undergo a reversible dissociation (equilibrium 4, Scheme 1), the dissociative step of which is the rate limiting step in enzyme activity when all components are at saturating concentrations. $Kp2_{ox}(MgADP + P_i)_2$ is then reduced by $SO_2^{\cdot-}$ (the dissociation product of dithionite, $S_2O_4^{2-} \leftrightarrow 2SO_2^{\cdot-}$), and finally MgATP replaces MgADP bound to Kp2. It is not yet clear at which point bound P_i is actually released and this may account for the apparent irreversibility of steps 2 and 3 of Scheme 1 during normal enzymic activity at 23°C (ref. 40). It should be noticed that the stoichiometry of 2ATPs hydrolyzed per electron transferred is the limiting value, and that under suboptimal conditions excess ATP is hydrolyzed (refs. 44, 45 and 43). In the quantitative model of Lowe and Thorneley (ref. 46), the individual rate constants for reactions 1 and 4 of Scheme 1 have been measured independently. However, reactions 2 and 3, and reactions 5 and 6 are treated as single steps with their rates measured when all components are at standard concentrations.

The net effect of one turn of the reactions of Scheme 1 is the transfer of a single electron from Kp2 to $Kp1^+$ with the concomitant hydrolysis of 2ATP to 2ADP + $2P_i$. These electrons accumulate on the MoFe protein (Kp1) until they can be used for the reduction of dinitrogen as described below.

The MoFe Protein Cycle

Scheme 2

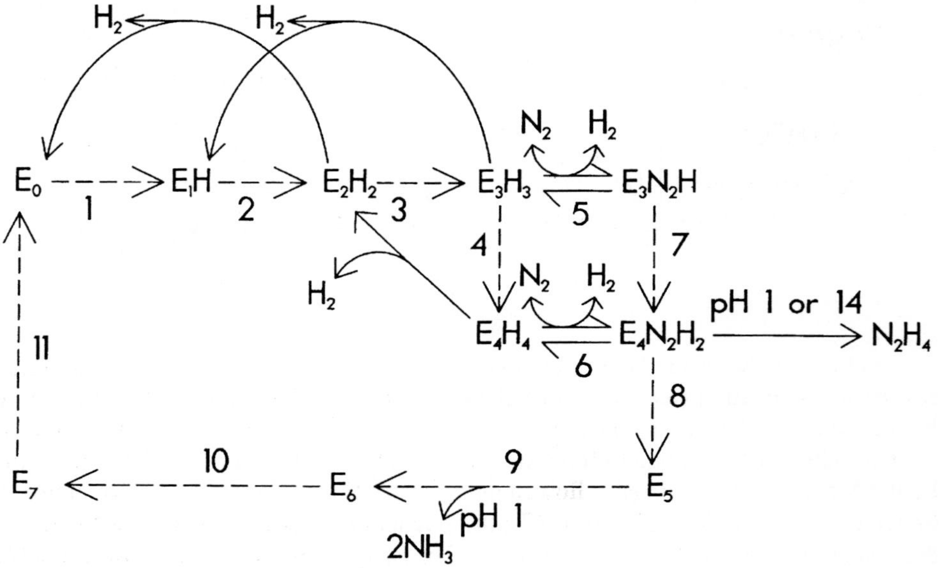

H_2 H_2

N_2 H_2

E_0 --1--> E_1H --2--> E_2H_2 --3--> E_3H_3 ⇌(5) E_3N_2H

H_2 4 7

N_2 H_2

11 E_4H_4 ⇌(6) $E_4N_2H_2$ —pH 1 or 14→ N_2H_4

8

E_7 <--10-- E_6 <--9-- E_5

pH 1

$2NH_3$

Scheme 2 shows the MoFe protein cycle of the model in (refs. 46, 47, 48 and 49). E represents a half molecule of the MoFe protein ($Kp1^+$) with a subscript giving the number of electrons that have been transferred to it (by one turn of the Fe protein cycle of Scheme 1 as represented by the broken arrows in Scheme 2); E_o is the protein as isolated in the presence of dithionite. Each turn of the Fe protein cycle is assumed to occur with the same rate constants, independent of the level of reduction of the MoFe protein. In one turn of Scheme 2, the eight single electron transfers from the Fe protein (reactions 1,2,3 and 4 or 7,8,9,10 and 11 of Scheme 2) provide the eight electrons required to complete the reaction of equation 4. Reducing equivalents can be 'lost' by the evolution of dihydrogen from E_2H_2, E_3H_3 or E_4H_4, and dinitrogen binds to E_3H_3 or E_4H_4 by displacing dihydrogen. There is no direct evidence concerning the nature of any of the enzyme bound intermediates as none has yet been isolated or observed spectroscopically. However, on quenching the functioning enzyme in acid, E_2H_2, E_3H_3 and E_4H_4 give dihydrogen, $E_4N_2H_2$ gives hydrazine, and E_6 and E_7 give two molecules of ammonia. We also know that dihydrogen cannot be released, even on acid quench, until not only two electrons have been transferred, but also until two rate–limiting dissociations have taken place. This led Lowe and Thorneley (ref. 46) to conclude that the second proton required for the formation of dihydrogen did not have access to the electron–storage site unless the MoFe protein was free in solution, *i.e.* not complexed by the Fe protein. They further concluded that many other kinetic observations could be explained if the active site for substrate reduction was only available for interactions with species in the solvent if the two proteins were not complexed.

The other reaction sequence schemes in the literature do not consider interactions with the Fe protein, and so by their nature are less complete and do not possess the quantitatively predictive ability of Schemes 1 and 2. Both Cleland (ref. 39) and Li and Burris (ref. 38) use dinitrogen displacement of dihydrogen (from a different reduction level of the MoFe protein than in Scheme 2) to explain the obligatory formation of dihydrogen; Cleland does not propose further intermediates in the formation of ammonia. The main scheme of Guth and Burris, (ref. 39), lacks any explanation of the stoichiometry of dihydrogen formation and again does not describe the dinitrogen reduction intermediates. The scheme of Burgess et al (ref. 37), which again does not address this dihydrogen formation, proposes an intermediate at the diimide levels, as in (ref. 38), suggesting a sequence of bound intermediates dinitrogen, diazene, hydrazine, ammonia:

$$E + N_2 \rightarrow E\cdots N{\equiv}N \rightarrow E\cdots NH{=}NH \rightarrow E\cdots NH_2{-}NH_2 \rightarrow E + 2NH_3$$

Neither the scheme of Burgess et al (ref. 37) nor that of Li and Burris (ref. 38) explains the competitive nature of dinitrogen reduction and $^2H^1H$ formation when dinitrogen reduction by nitrogenase is inhibited by $^2H_2$ in ^1H_2O.

The Substrate Reduction Site

There is a consensus that the dinitrogen binding and reduction site is at the FeMo cofactor of molybdenum nitrogenase, although there is no direct evidence for this. The most compelling reasons for accepting this consensus are first the uniqueness of the occurrence of FeMoco in nitrogenase, and secondly the observation by Hawkes *et al.* (ref. 50) that, when FeMoco extracted from protein isolated from *Klebsiella pneumoniae* with a NIF V^- phenotype is inserted into MoFe protein lacking FeMoco, the altered substrate specificity characteristic of the *nifV* mutant–derived species is also transferred to the reconstituted protein.

Although there has been kinetic evidence for some time that the two cofactor sites function independently (ref. 24), it is only the recent crystallographic results (ref. 35), showing that the two FeMocos in each molecule of the MoFe protein are 70Å apart, that definitively exclude the possibility of bound dinitrogen bridging the two molybdenum atoms. It is still possible, however, that it may bridge between molybdenum and iron within a single cofactor.

The importance of FeMoco has led a number of groups to expend great efforts in attempting to crystallise it so as to determine its structure using X–ray crystallography, so far without success. Our principal sources of information are chemical analysis, extended X–ray absorbance fine structure (EXAFS, refs. 51, 52 and 25) and electron nuclear double resonance (ENDOR, refs. 53, 54 and 55). These studies give us a picture of molybdenum and iron in a sulphur environment with about four Mo-S distances of 2.37Å, at least three Mo–Fe distances of about 2.69Å, about three Fe–S distances of 2.2Å, about two Fe–Fe distances at 2.2Å and a mean Fe–Fe interaction at 3.68Å. The irons are in at least five different environments and the molybdenum is possibly Mo(IV) in an unsymmetric octahedral environment with probably 3S and 3O or N ligands. At least one nitrogen ligand (presumably from the protein) to the cluster has been detected by electron spin echo spectroscopy (ref. 56) and ENDOR (ref. 53). There is evidence that the cofactor is associated with homocitrate (ref. 57) and although this has not been detected in isolated FeMoco, the incorporation of homocitrate analogues alters the substrate specificity of nitrogenase (ref. 58).

Not surprisingly, the vanadium nitrogenase from *Azotobacter chroococcum* has been shown to contain an extractable cofactor containing vanadium instead of molybdenum that can be inserted into the molybdenum protein polypeptides to yield an active protein (ref. 25). This cofactor appears similar to FeMoco in the limited studies to date (ref. 59). No cofactor has yet been extracted from the third nitrogenase.

The Nature of Intermediates

A discussion of the detailed mechanism of the multi–electron transfer process by which electrons and protons are used to reduce dinitrogen to ammonia turns on the nature of the intermediates in Scheme 2. At the time of writing, there have been no spectroscopic studies of any of these, except for the initial dithionite–reduced state E_0. Nor has there been evidence of binding of any reducible substrate or substrate analogue to E_0, which might give clues to the mechanism. Species E_0 itself shows an $S = \frac{3}{2}$ epr signal from FeMoco, but this is bleached after the transfer of the first electron from the Fe protein. We are left with observations of the results of quenching active enzyme in various ways, and the nature and time courses of appearance of the products after quenching. These can be correlated with studies of possible chemical analogues. It is generally assumed that dinitrogen binds and is reduced at a metal, possibly molybdenum in molybdenum nitrogenase (or vanadium in vanadium nitrogenase).

Intermediates in Scheme 2 that are written as E_2H_2, E_3H_3 and E_4H_4, which kinetic evidence suggests, give rise to dihydrogen on quenching in acid, are likely to contain metal hydrides or metal dihydrogen complexes. Such structures are well known in chemical systems (refs. 60 and 61) although there is no direct evidence of their occurrence during nitrogenase turnover. The subsequent step, the displacement of dihydrogen is crucial, and it is reasonable to speculate that this is required to prepare a site for dinitrogen binding in a protic medium. There are chemical precedents for such a reaction, (refs. 62, 63 and 64). The competitive inhibition of dinitrogen reduction by dihydrogen (ref. 48) is consistent with the reversibility of this displacement.

The only intermediate on which we have any significant information occurs at the E_4 state with N_2 bound. This yields hydrazine on quenching in acid or alkali (refs. 65 and 48), although no hydrazine is released under normal assay conditions (pH 7.4) so that it is not a free intermediate. As it would be expected from Scheme 2, its concentration during steady–state reduction of dinitrogen by nitrogenase is about one eighth of the total molybdenum concentration. The suggestion that the intermediate was a bound hydrazido (2–) group (=N–NH_2) was based on the reactivity of chemical analogues (ref. 66), with various dinitrogen and dinitrogen hydrides bound to a metal, since only the hydrazido (2–) complexes yielded hydrazine on both acid and base quench. This is consistent with the location of the intermediate in Scheme 2, where four electrons have been transferred to the MoFe protein, and two lost as dihydrogen, leaving two available for transfer to bound dinitrogen.

Scheme 3

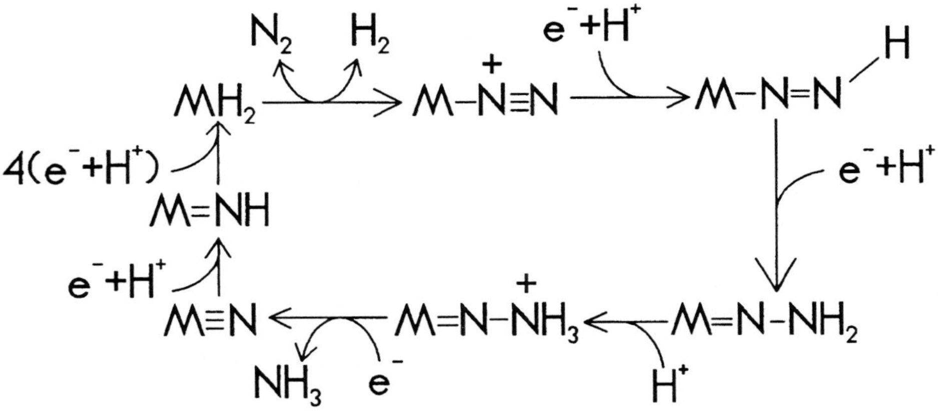

$$MH_2 \xrightarrow{N_2 \text{ in},\ H_2 \text{ out}} M{-}\overset{+}{N}{\equiv}N \xrightarrow{e^- + H^+} M{-}N{=}N{-}H$$

$$M{-}N{=}N{-}H \xrightarrow{e^- + H^+} M{=}N{-}NH_2 \xrightarrow{H^+} M{=}N{-}\overset{+}{N}H_3 \xrightarrow{e^-} M{\equiv}N + NH_3$$

$$M{\equiv}N \xrightarrow{e^- + H^+} M{=}NH \xrightarrow{4(e^- + H^+)} MH_2$$

Finally, after one or two more electrons have been transferred to the MoFe protein, a species occurs, that gives rise to two molecules of ammonia on quenching in acid. Again, by analogy with the chemical systems, it is less likely that intermediates at E_5 and E_6 are bound $—NH_2—NH_2$ or $—NH_2—N^+H_3$ since these might be expected to give hydrazine on quenching in acid. It is more likely that the intermediate present at E_5 is $=N—N^+H_3$, and that at E_6 is $\equiv N + NH_3$. It is not yet possible to determine where ammonia is released at pH 7.4 during normal nitrogenase turnover.

The reaction sequence above strongly suggests that the overall mechanism of dinitrogen reduction by nitrogenase is similar to that suggested by Chat and Richards (ref. 67). This was that end–on bound dinitrogen is successively reduced three times by one electron, each followed by one protonation at the β–nitrogen, to give ammonia plus a nitride. A further three single electrons and three protonations would then give a second ammonia molecule and regenerate the binding site. It has been shown that such a series of steps can indeed form a catalytic cycle in an electrochemical cell (refs. 68 and 69) using gaseous dinitrogen binding to a $W(Ph_2PCH_2CH_2PPh_2)_2$ core as a starting material. Chemical analogues have also been synthesized and characterized for all the potential intermediates shown in Scheme 3 (ref. 70).

Other Possible Substrates

As well as catalyzing reaction 6, and the reduction of protons to dihydrogen, nitrogenase also catalyzes the reduction of a number of other small multiply–bonded molecules (Table 2). Cyanide and acetylene reduction have recently been fitted into schemes similar to Scheme 2 (refs. 71 and 72). A common feature of such reduction is that they all could occur after binding of the substrate, subsequent transfers of electrons followed by protonation, and then release of products. Chemical models are known for many these reductions (ref. 73). In the case of substrates giving products after different numbers of reductions, release of intermediates before the complete reduction could explain the results of the smaller number of electron reductions.

How are Electrons Transferred to Reducible Substrates?

The simple answer to this questions is that we do not know. It seems likely that electrons that have come singly from the Fe protein are also transferred to bound substrate singly, and that this substrate is also protonated. maintaining the overall charge neutrality. The enzyme then retains bound intermediates until the product is formed and released. There would appear to be little requirement for the storage of multiple reducing equivalents on metal clusters in the enzyme itself.

What, then, is the role of the non–FeMoco iron in the 'P' clusters? It may be that they serve to store reducing equivalents transiently, although there is little evidence for their undergoing redox reactions while nitrogenase is functioning. Smith *et al.* (ref. 74)

Table 2

Some reduction reactions catalyzed by nitrogenase products after reduction by the given number of electrons.

Substrate	2	4	6	8	Ref.
$2H^+$	H_2				
D_2 (with N_2)	2HD				77
N_2O	$N_2 + H_2O$				78
C_2H_2	C_2H_4				72
C_2H_4	C_2H_6				79
Cyclopropane	$\frac{1}{3}$ cyclopropane $\frac{2}{3}$ propene				80
HCN	?	CH_4+CH_3NH_2	CH_4+NH_3		71
CH_3NC			CH_4+CH_3NH_2		82
N_3^-	N_2+NH_3		N_2H_4+NH_3		83
NO_2^-			NH_3+H_2O		84
NH_2-C≡N			CH_3NH_2+NH_3	CH_4+$2NH_3$	81
$N_2 2NH_3$+H_2					

observed that an epr signal, probably associated with 'P'–clusters, induced by oxidation with ferricyanide decayed at a rate close to the rate of dissociation of the Fe and MoFe proteins. It was proposed that this was an intramolecular electron transfer process from 'P' clusters to FeMoco, and that the coincidence of rates could mean that electron transfer triggered (or was triggered by) the protein dissociation. This still remains an open question. Some of the epr signals observed by Lowe *et al.* (ref. 75) may also be associated with redox changes in the 'P' clusters.

Summary and Forward Look

From the above, it is clear that much is still to be learnt about the detailed mechanisms of multi–electron transfers in biological systems. Although metals appear to be involved in all cases, there are apparently qualitative differences between the likely transfer of single electrons to bound substrate and intermediates in nitrogenase, and the build–up of reducing equivalents by oxidation level changes of manganese ions in the water–splitting complex associated with photosystem II. The involvement of metals, however, considerably extends the potential of spectroscopic probes, such as epr (and the related techniques such as ENDOR and electron spin echo), circular dichroism, magnetic circular dichroism, magnetic susceptibility and Mössbauer spectroscopy, in investigating these processes.

As far as nitrogenase is concerned, the (hopefully) imminent publication of a crystal structure for a MoFe protein should answer many questions. It should also pose many more, in terms of interactions of the Fe–protein and substrates with neighbouring amino acid residue side chains and the electronic structures of the 'P' and MoFe clusters under various conditions. The potential of the various physical techniques in conjunction with the relatively new method of site directed mutagenesis (ref. 76) has also yet be realized.

References

1 M. Wikström, K Krab and M. Saraste *Cytochrome Oxidase. A Synthesis* Academic Press, London, New York, 1981.

2 K. Krab and M. Wikström, Biochim. Biophys. Acta **895** (1987) 25.

3 R.K. Poole, in: C. Anthony (Ed.) *Bacterial Energy Transduction* Academic Press, London, New York, 1988, pp. 231.

4 R. Malkin and B.G. Malmström, Advan. Enzymol. **33** (1970) 177.

5 P.F. Knowles, J.F. Gibson, F.M. Pick and R.C. Bray, Biochem. J. **111** (1969) 53.

6 L. Calabrese, N. Carbonara and G. Musci, J. Biol. Chem. **264** (1989) 6183.

7 L. Morpurgo, Life Chem. Rep. **5** (1-4) (1987) 277.

8 A. Messerschmidt, A. Rossi, R. Ludenstein, R. Huber, M. Bolognesi, G. Gatti, A. Marchesini, R. Petruzzelli and A. Finazzi-Agro, J. Mol. Biol. **206** (1989) 513.

9 J.G. Metz, H. Pakrasi, C.J. Arntzen and M. Seibert, in: J. Biggins (Ed.) *Progress in Photosynthesis* Vol. 4, Martinus Nijhoff, The Hague, 1987, pp. 679-682.

10 J. Barber, Trends Biochem. Sci. (1987) 321.

11 M. Sivaraja, J.S. Philo, J. Lang and G.C. Dismukes, J. Am. Chem. Soc. **111** (1989) 3221.

12 P.M. Wood, in: J.A. Cole and S.J. Ferguson (Eds.) *The Nitrogen and Sulphur Cycles* Cambridge, New York & Cambridge, 1988, pp. 219-244.

13 P.M. Wood, in: C. Anthony (Ed.) *Bacterial Energy Transduction* Academic Press, London, New York, 1988, pp. 183-230.

14 A.H. Stouthamer, in: *Biology of Anaerobic Microorganisms* Wiley, New York & London, 1988, pp. 245-303.

15 G.J. Carr and S.J. Ferguson, Biochem. J. **269** (1990) 423.

16 J.A. Cole, in: J.A. Cole and S.J. Ferguson (Eds.) *The Nitrogen and Sulphur Cycles* Cambridge University Press, New York & Cambridge, 1988, pp. 281-330.

17 H.D. Peck and T. Lissolo, in: J.A. Cole and S.J. Ferguson (Eds.), *The Nitrogen and Sulphur Cycles* Cambridge, New York & Cambridge, 1988, pp. 99-132.

18 L. Kang, Ph.D. Dissertation, University of Georgia, 1986.

19 D.E. McKee, J.S. Richardson and L.M. Siegel, J. Biol. Chem. **261** (1986) 10277.

20 B.K. Burgess, in: T.G. Spiro (Ed.) *Metal Ions in Biology Series* Vol. 7: *Molybdenum Enzymes* Wiley-Interscience, New York & London, 1985, pp. 161-219.

21 D.J. Lowe, R.N.F. Thorneley and B.E. Smith, in: P. Harrison (Ed.) *Metalloproteins* Vol. 1, Macmillan, London, 1985, pp. 207-249.

22 W.H. Orme–Johnson, Annu. Rev. Biophys. Chem. **14** (1985) 419.

23 P.J.Stephens, in: T.G. Spiro (Ed.) *Metal Ions in Biology Series* Vol. 7: *Molybdenum Enzymes* Willey–Interscience, New York & London, 1985, pp. 117-159.

24 R.N.F. Thorneley and D.J. Lowe *Metal Ions in Biology Series* Vol. 7: *Molybdenum Enzymes* in: T.G. Spiro (Ed.), Wiley–Interscience, New York & London, 1985, pp. 117-159.

25 B.E. Smith, M. Buck, R.R. Eady, D.J. Lowe, R.N.F. Thorneley, G.A. Ashby, J. Deistung, M. Eldridge, K. Fisher, C. Gormal, I. Ionnidis, H. Kent, J. Arber, A. Flood, C.D. Garner, S. Hasnain and R.W. Miller, in: H. Boethe, F. de Bruijn and W.E. Newton (Eds.), *Nitrogen Fixation: Hundred Years After* Gustav Fischer, New York & Stuttgart, 1988, pp. 91-100.

26 R.R. Eady, R.L. Robson and B.E. Smith, in: J.A. Cole and S.J. Ferguson (Eds.), *The Nitrogen and Sulphur Cycles* Cambridge University Press, New York & Cambridge, 1988, pp. 363-382.

27 A.R. Glenn and M.J. Dilworth (Eds.) *Biology and Biochemistry of Nitrogen Fixation* Elsevier, Amsterdam, 1991.

28 P.E. Bishop, M.E. Hawkins and R.R. Eady, Biochem. J. **238** (1986) 437.

29 P.E. Bishop, R. Premakumar, D.R. Dean, M.R. Jacobson, J.R. Chisnell, T.M. Rizzo and J. Kopzynski, Science **232** (1986) 92.

30 R.L. Robson, Arch. Microbiol. **146** (1986) 74.

31 R.L. Robson, R.R. Eady, T.H. Richardson, R.W. Miller, M. Hawkins and J.R. Postgate, Nature **322** (1986) 388.

32 B.J. Hales, E.E. Case, J.E. Morningstar, M.F. Ozeda and L. Mauterer, Biochemistry **25** (1986) 7251.

33 B.J. Hales, D.J. Langosch and E.E. Case, J. Biol. Chem. **261** (1986) 15301.

34 M.M. Georgiadis, P. Chakrabarti and D.C. Rees, in: P.M. Gresshoff, G. Stacey, L.E. Roth and W.E. Newton (Eds.) *Nitrogen Fixation: Achievements and Objectives* Chapman & Hall, New York, 1990, pp. 111-116.

35 J.T. Bolin and A.E. Ronco, in: P.M. Gresshoff, G. Stacey, L.E. Roth and W.E. Newton (Eds.) *Nitrogen fixation: Achievements and Objectives* Chapman & Hall, New York, 1990, pp. 117-124.

36 R.N.F. Thorneley and D.J. Lowe, Israel J. Botany **31** (1981) 61.

37 B.K. Burgess, S. Wherland, W.E. Newton and E.I. Steifel, Biochemistry **20** (1981) 5140.

38 J. Li and R.H. Burris, Biochemistry **22** (1983) 4472.

39 J. Guth and R.H. Burris, Biochemistry **22** (1983) 5111.

40 R.N.F. Thorneley and D.J. Lowe, Biochem. J. **215** (1983) 393.

41 G.A. Ashby and R.N.F. Thorneley, Biochem. J. **246** (1987) 455.

42 R.N.F. Thorneley, G.A. Ashby, J.V. Howarth, N. Millar and H. Gutfreund, Biochem. J. **264** (1989) 657.

43 J. Cordewener, A. Asbroek, H. Wassink, R.R. Eady, H. Haaker and C. Veeger, Eur. J. Biochem. **162** (1987) 265.

44 D.Y. Jeng, J.A. Morris and L.E. Mortenson, J. Biol. Chem. **245** (1970) 2809.

45 S. Imam and R.R. Eady, FEBS Lett. **110** (1980) 35.

46 D.J. Lowe and R.N.F. Thorneley, Biochem. J. **224** (1984) 877.

47 D.J. Lowe and R.N.F. Thorneley, Biochem. J. **224** (1984) 895.

48 R.N.F. Thorneley and D.J. Lowe, Biochem. J. **224** (1984) 887.

49 R.N.F. Thorneley and D.J. Lowe, Biochem. J. **224** (1984) 903.

50 T.R. Hawkes, P.A. McLean and B.E. Smith, Biochem. J. **217** (1984) 317.

51 S.D. Conradson, B.K. Burgess, W.E. Newton, L.E. Mortenson and K.O. Hodgson, J. Am. Chem. Soc. **109** (1987) 7507.

52 A.M. Flank, M. Weininger, L.E. Mortenson and S.P. Cramer, J. Am. Chem. Soc. **108** (1986) 1049.

53 A.E. True, P. McLean, M.J. Nelson, W.H. Orme–Johnson and B.M. Hoffman, J. Am. Chem. Soc. **112** (1990) 651.

54 A.E. True, M.J. Nelson, R.A. Venters, W.H. Orme–Johnson and B.M. Hoffman, J. Am. Chem. Soc. **110** (1988) 1935.

55 B.M. Hoffman, J.E. Roberts and W.H. Orme-Johnson, J. Am. Chem. Soc. **104** (1982) 860.

56 H. Thomann, T.V. Morgan, H. Jin, S.J.N. Burgmayer, R.E. Bare and E.I. Steifel, J. Am. Chem. Soc. **109** (1987) 7913.

57 T.R. Hoover, J. Imperial, P.W. Ludden and V.K. Shah, Biochemistry **28** (1989) 2768.

58 T.R. Hoover, J. Imperial, J. Liang, P.W. Ludden and V.K. Shah, Biochemistry **27** (1988) 3647.

59 S. Hasnain, R.R. Eady, B.R. Dobson, J.M. Arber, M. Nomura, T. Matsushita, C.D. Garner and B.E. Smith, Biochem. J. **258** (1989) 733.

60 R.G. Teller and R. Bau, Structure & Bonding **44** (1981) 1.

61 R.A. Henderson, Transition Met. Chem. **13** (1988) 474.

62 A. Frigo, G. Puosi and A. Turco, Gazz. Chim. Ital. **101** (1971) 637.

63 R.H. Morris, J.F. Sawyer, M. Shiralian and J.D. Zubkowski, J. Am. Chem. Soc. **107** (1985) 5581.

64 R. Ellis, R.A. Henderson, A. Hills and D.L. Hughes, J. Organomet. Chem. **333** (1987) C6.

65 R.N.F. Thorneley, R.R. Eady and D.J. Lowe, Nature **272** (1978) 557.

66 S.N. Anderson, M.E. Fakley and R.L. Richards, J. Chem. Soc. Dalton Trans., (1981) 1973.

67 J. Chatt and R.L. Richards, J. Organomet. Chem. **239** (1982) 65.

68 C.J. Pickett and J. Talarmin, Nature **217** (1985) 652.

69 C.J. Pickett, K.S. Ryder and J. Talarmin, J. Chem. Soc. Dalton Trans., (1986) 1453.

70 R.L. Richards, Chem. Brit., (1987) 133.

71 D.J. Lowe, K. Fisher, R.N.F. Thorneley, S.A. Vaughn and B.K. Burgess, Biochemistry **28** (1989) 8460.

72 D.J. Lowe, K. Fisher and R.N.F. Thorneley, Biochem. J. **272** (1990) 621.

73 A.J.L. Pombiero and R.L. Richards, Coord. Chem. Rev. **104** (1990) 13.

74 B.E. Smith, D.J. Lowe, G-X. Chen, M.J. O'Donnell and T.R. Hawkes, in: A. Mueller and W.E. Newton (Eds.), *Nitrogen Fixation: Chemistry–Biochemistry–Genetics Interface* Plenum, New York & London, 1983, pp. 23-62.

75 D.J. Lowe, R.R. Eady and R.N.F. Thorneley, Biochem. J. **173** (1978) 277.

76 B.E. Smith, in: P.M. Gresshoff, G. Stacey, L.E. Roth and W.E. Newton (Eds.), *Nitrogen Fixation: Achievements and Objectives* Chapman & Hall, New York, 1990, pp. 3-13.

77 B.B. Jensen and R.H. Burris, Biochemistry **24** (1985) 1141.

78 B.B. Jensen and R.H. Burris, Biochemistry **25** (1986) 1083.

79 G.A. Ashby, M.J. Dilworth and R.N.F. Thorneley, Biochem. J. **247** (1987) 547.

80 G.J. Leigh, C.E. McKenna, M. Bravo, J.P. Gemoets and B.E. Smith, Biochem. J. **258** (1989) 487.

81 R.W. Miller and R.R. Eady, Biochim. Biophys. Acta **952** (1988) 679.

82 J.F. Rubinson, J.L. Corbin and B.K. Burgess, Biochemistry **22** (1983) 6260.

83 J.F. Rubinson, B.K. Burgess, J.L. Corbin and M.J. Dilworth, Biochemistry **24** (1985) 273.

84 S.A. Vaughn and B.K. Burgess, Biochemistry **28** (1989) 419.

Electron and Proton Transfer in Chemistry and Biology, edited by A. Müller et al.
Studies in Physical and Theoretical Chemistry, Vol. 78

Electron Transfer in Anaerobic Microorganisms

A. Kröger, C. Körtner, F. Lauterbach, F. Droß, M. Bokranz,
O. Klimmek, T. Krafft and M. Gutmann

Institut für Mikrobiologie
J.W.Goethe–Universität
Frankfurt am Main (Germany)

Summary

Like many other bacteria *Wolinella succinogenes* performs electron transport coupled phosphorylation with either fumarate or polysulphide as terminal electron transport acceptors under anaerobic conditions. The mechanism of electron transport was studied with the aim of understanding the coupling between electron transport and phosphorylation. For this purpose the enzymes making up the electron transport chains were isolated and characterized. Furthermore, the corresponding genes were cloned and sequenced. This communication reports on the structure and function of the enzymes.

Introduction

Wolinella succinogenes grows at the expense of either reaction (a), (b) or (c) (refs. 1 and 2). The reactions

$$H_2 + \text{Fumarate} \rightarrow \text{Succinate} \quad (a)$$

$$HCO_2^- + \text{Fumarate} + H^+ \rightarrow CO_2 + \text{Succinate} \quad (b)$$

$$HCO_2^- + S^0 \rightarrow CO_2 + HS^- \quad (c)$$

are catalyzed by membrane–integrated electron transport chains (refs. 3 and 4) and are coupled to ATP synthesis from ADP and inorganic phosphate (refs. 5, 6, 7 and 8). The electron transport chains catalyzing reactions (a) – (c) have been investigated with the aim of elucidating the mechanisms of electron transfer and of coupled phosphorylation.

Abbreviations:
MK, menaquinone
MKH_2, menahydroquinone
Fe/S, iron–sulphur center
FAD_b, covalently bound FAD
Mr, relative molecular weight
MW, molecular mass
SDS–PAGE, gel electrophoresis in the presence of dodecylsulphate
[S], polysulphide
DMN, 2,3-dimethyl-1,4-naphthoquinone

Electron Transport Chains

The enzymes constituting the electron transport chains were isolated and characterized. Incorporation of the isolated enzymes into liposomes led to preparations catalyzing reactions (a) – (c) with turnover numbers commensurate to those of the enzymes in the bacterial membrane (refs. 8, 9 and 4). Thus the composition of the individual electron transport chains was elucidated.

The chain catalyzing reaction (a) is made up of hydrogenase, fumarate reductase and MK which serves as a redox mediator between the two enzymes (Fig. 1), (ref. 8).

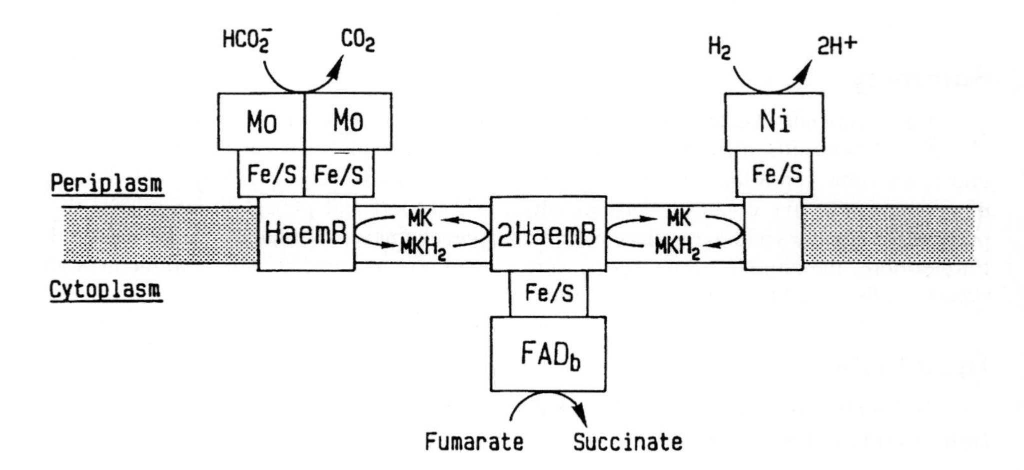

Fig. 1. Components of the electron transport chains catalyzing reaction (a) and (b). Formate dehydrogenase (ref. 11) and hydrogenase (ref. 5) are oriented towards the periplasmic side and fumarate reductase (ref. 11) towards the cytoplasmic side of the bacterial membrane.

Similary, reaction (b) is catalyzed by an electron transport chain consisting of formate dehydrogenase, fumarate reductase and MK (ref. 9). Formate dehydrogenase catalyzes the reduction of MK by formate and fumarate reductase the oxidation of MKH_2 by fumarate.The fumarate reductase involved in the catalysis of reaction (a) is identical with that present in the chain catalyzing reaction (b).

Elemental sulphur reacts with sulphide to give polysulphide (reaction d) which is used by the bacteria instead of sulphur as the acceptor of formate oxidation (reaction c) (ref. 10).

$$nS + HS^- \rightarrow S_{n+1}^{2-} + H^+ \qquad (d)$$

The electron transport chain consists of formate dehydrogenase and polysulphide reductase (Fig. 2), (ref. 4). A quinone is not involved. The formate dehydrogenase is identical with that involved in the catalysis of reaction (b).

The properties of the four membrane–bound enzymes are summarized in Table 1.

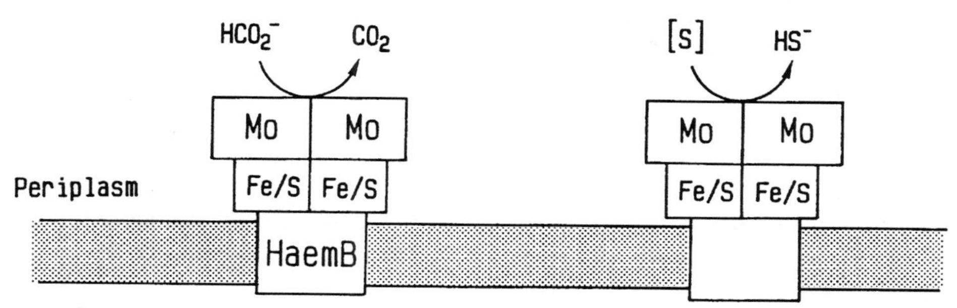

Fig. 2. Components of the electron transport chain catalyzing reaction (c). Both formate dehydrogenase and polysulphide ([S]) reductase are oriented towards the periplasmic side of the bacterial membrane.

Table 1

Subunit composition and prosthetic groups of four membraneous respiratory enzymes of *W. succinogenes.* The genes *frdA,B,C* of fumarate reductase, *fdhA,B,C* of formate dehydrogenase, *hydA,B* of hydrogenase and *srdA* of polysulphide reductase correspond to subunits of the respective enzyme.

Subunit	MW from the gene	MR by SDS–PAGE	Amino acid composition	Prosthetic groups	Location
FrdA	72 769	79 000	hydrophilic	FAD 3Fe-4S 4Fe-4S	cytoplasmic
FrdB	27 152	31 000	hydrophilic	2Fe-2S	cytoplasmic
FrdC	29 719	25 000	lipophilic	2 haemB	membrane
FdhA	101 140	110 000	hydrophilic	Mo–cofactor	periplasmic
FdhB	21 761	20 000	hydrophilic	Fe/S	periplasmic
FdhC	31 781	25 000	lipophilic	haemB	membrane
FdhD	31 694	missing	hydrophilic	?	?
HydB	63 997	60 000	hydrophilic	Ni	periplasmic
HydA	31 897	30 000	hydrophilic	Fe/S	periplasmic
HydC	18 765	missing	lipophilic	?	membrane
SrdA	81 263	85 000	hydrophilic	Mo-cofactor	periplasmic
SrdB	20 942	missing	hydrophilic	Fe/S	perilasmic
SrdC	?	missing	lipophilic	?	membrane

The genes of the individual subunits have been cloned and sequenced. Their identity was verified by comparison with the N–terminal amino acid sequences of the corresponding subunits in most cases. The genes encoding the subunits of each enzyme were found to form transcriptional units on the genome of *W.succinogenes.* The genes of the fumarate reductase (*frdA,B,C*), hydrogenase (*hydA,B,C*) and of polysulphide reductase (*srdA,B,C*) occur only once, while there is a second set of *fdh* genes on the genome which was detected by hybridization with DNA–probes derived from the first one.

The enzymes appear to be built according to a common principle.The substrate site is localized on the bigger hydrophilic subunit,while the lipophilic one anchors the enzyme in the membrane. The smaller hydrophilic subunit which contains one or more iron-sulphur centers,serves as a redox mediator between the two other subunits (Fig.1).

Fumarate Reductase

The fumarate reductase is oriented towards the cytoplasmic side of the membrane (ref. 11). It is made up of one copy each of the three different subunits (refs. 12, 13, 14, 15 and 16). In electronmigraphs of negatively stained liposomes containing fumarate reductase, the enzyme appears as a cylinder of 7 nm lenght and 5 nm diameter which is linked to the liposomal membrane.The dimensions are consistent with the sum of the molecular weights of FrdA and FrdB (ref. 17).

FrdA carries the substrate site and FAD which covalently linked to His-43 in the polypeptide chain. The enzyme contains two heme B groups with distinctly different redox potentials. These are bound to the lipophilic FrdC which reacts with the menaquinone present in the bacterial membrane. Seperation of the diheme cytochrome *b* yields an enzyme catalyzing fumarate reduction with artificial redox donors, but does not react with quinones. On coprecipitation of the cytochrome–deficient enzyme with the isolated cytochrome *b*, quinone reactivity is restored. The enzyme carries three different iron–sulphur centers ([2Fe-2S], [3Fe-4S] and [4Fe-4S] (refs. 16, 18 and 19). The binuclear center is bound to FrdB, while the location of the two other centers is not clear. It is possible that the ligands are contributed both by FrdA and FrdB. FrdB serves as a redoxmediator between FrdA and FrdC (ref.13).

Formate Dehydrogenase

When incorporated in liposomes, formate dehydrogenase appears as a sphere of 7.5 nm diameter which is linked to the membrane by means of a stalk. The size of the sphere is consistent with the molecular weight of the dimer of FdhA.In the bacterial membrane,the formate site of the enzyme is oriented towards the periplasmic side (Fig. 1), (ref. 11). The isolated enzyme is made up of two molecules each of FdhA and FdhB and one molecule of FdhC (refs. 20 and 21). FdhD is missing from the preparation, and its function is not known. The enzyme contains two atoms of molybdenum which are probably bound to the subunit carrying the formate site (FdhA),

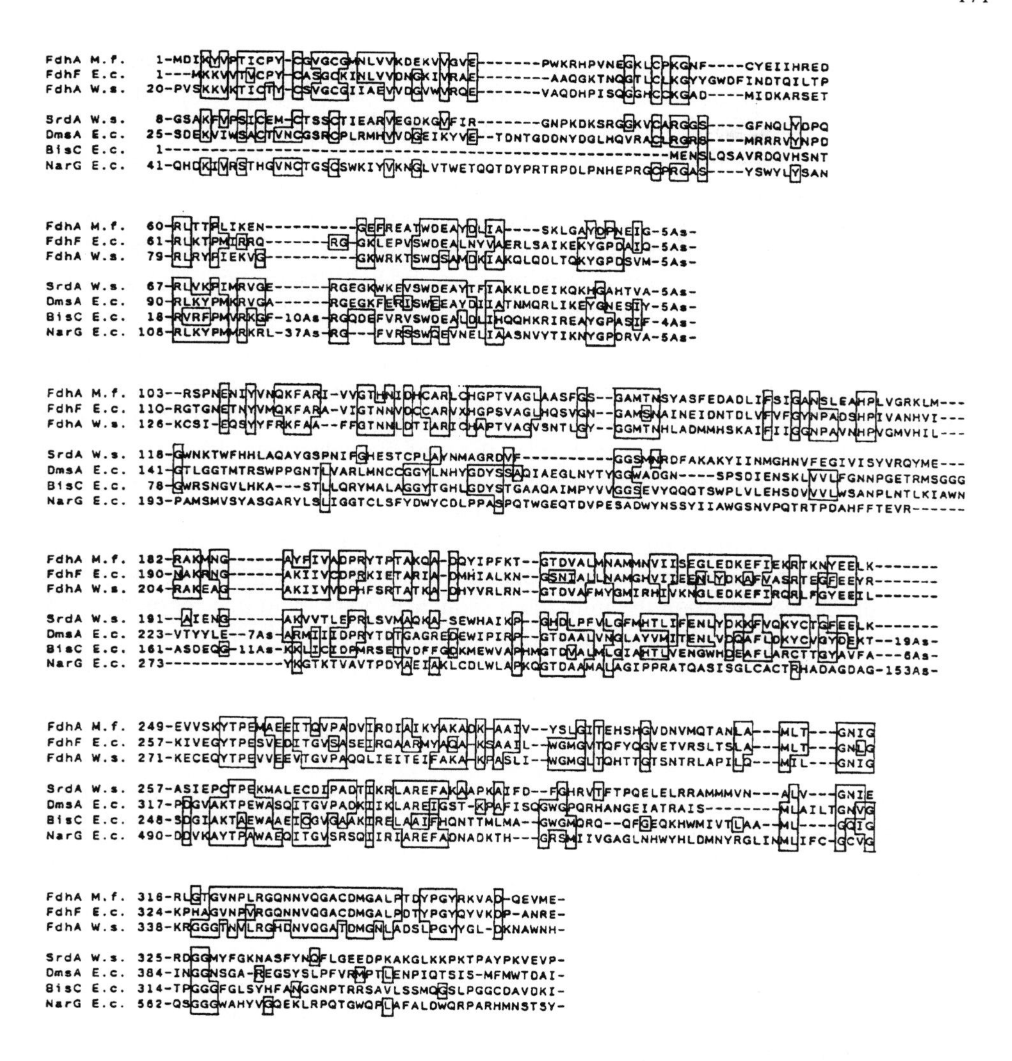

Fig. 3. Alignment of the N–terminal halfs of the amino acid sequences of FdhA (ref. 21) and SrdA (ref. 24) of *W.succinogenes* to those of five other molybdo–enzymes. Fdh, formate dehydrogenase. Srd, polysulphide reductase. Dms, dimethylsulphoxide reductase (ref. 31). Bis, biotin sulphoxide reductase (ref. 32). Nar, Nitrate reductase (ref. 33). M.f., *Methanobacterium formicicum* (ref. 34). E.c., *Escherichia coli*. W.s., *Wolinella succinogenes*. FdhF (ref. 35).

and one haem B group with a redox potential below -200mV (as compared to the standard hydrogen electrode). The heam B is linked to FdhC which represents a cytochrome *b*. FdhC forms the membrane anchor of the enzyme and react with the bacterial menaquinone, the redox acceptor of the enzyme (ref. 22). The isolated formate dehydrogenase contains free iron and sulphide, the amounts of which would be consistent with the presence of at least four tri- or tetranuclear iron–sulphur centers. Most of these centers are presumably ligated by FdhB. Twelve cysteine residues of FdhB are conserved among homologous enzymes and have the proper spacing typical of structurally known ferredoxins (ref. 21).

In Fig. 3 the N–terminal half of the amino acid sequence of FdhA as deduced from the corresponding gene, is aligned to those of six other enzymes, which are known to contain molybdenum. The sequences of the three formate dehydrogenases are more similar (40-47% identical amino acids) than those of the reductases (16-28% identical amino acids). The opposite holds for the phylogenetic relation of the corresponding bacteria. While three of the reductases are enzymes of *E.coli*, the formate dehydrogenases belong to distantly related bacteria. This suggests that the homologies signify functional rather than phylogenetic relationships. This view is supported by the finding that the homologies between the individual sequences vary from one segment (I-VI) to the other. Segment I which is missing from BisC, contains a conserved cysteine cluster typical of ferredoxins carrying 4Fe-4S iron–sulphur centers. Segments II and V may be involved in the binding of molybdenum,since their homologies concern all the enzymes. The molybdenum is probably bound to a pterin dinucleotide similar as in other molybdo-enzymes (ref. 23). Segment III appears to serve a specific function in the formate dehydrogenases and another one in the two sulphoxide reductases. Segment IV is conserved among all the enzymes with the exception of nitrate reductase (NarG), while the sequence of segment VI is conserved mainly among the formate dehydrogenases and may constitute part of their substrate sites.

Polysulphide Reductase

The polysulphide reductase (formerly sulphur reductase) of *W. succinogenes* is the first enzyme of this type that has been isolated and the amino acid sequence of which has been determined (refs. 4, 10 and 24). The isolated enzyme has been reported to consist merely of SrdA, while the corresponding gene is localized in an operon comprising two additonal genes (SrdB and C). The molybdenum present in the isolated enzyme is probably bound to a pterin dinucleotide (ref. 23). Considering the similarities of the amino acid sequence with those of other molybdoenzymes (Fig. 3), it is likely that the molybdenum cofactor is linked to SrdA.

SrdB contains 16 cysteine residues located in three cystein clusters that are conserved among the corresponding subunits of the molybdoenzymes listed in Fig. 3. This suggests that SrdB represents an iron–sulphur protein which is involved in electron transfer. Although SrdB has not been detected in the isolated enzyme, the preparation contains free iron and sulphide equivalent to the presence of three (4Fe-4S) iron–sulphur centers, while only one such center is predicted to be present in SrdA (Fig.3).

One of the cysteine clusters of SrdB is situated in a stretch of mainly hydrophobic amino acids which should be integrated in the membrane, while the corresponding stretches in the homologous enzymes (see Fig. 3) are hydrophilic. It may be speculated that this cluster ligates an iron–sulphur center that reacts as direct acceptor of the electrons provided by formate dehydrogenase (Fig. 2). This would explain the electron transport from formate dehydrogenase to polysulphide reductase in the absence of a quinone. In contrast, the chains catalyzing the reduction of fumarate, nitrate or dimethylsulphoxide by formate do contain a quinone (ref. 25).

Hydrogenase

The hydrogenase of *W. succinogenes* reacts both with viologens and dimethylnaphtoquinone (DMN) as acceptor (refs. 26 and 8). The reactivity towards DMN is lost during purification, while that with the viologens is retained. The final preparation consists of HydA and HydB (Table 1). The loss of quinone reactivity may be caused by the seperation a hydrophobic subunit, the gene (*hydC*) of which is present in the *hyd* operon (ref. 27). HydC possibly anchors the enzyme in the membrane and carries the quinone–reactive site. The hydrogen site and the nickel are located on HydB. This is concluded from the observation that HydB retains nickel together with part of the activity of viologen reduction on separation from the enzyme. The nickel is ligated by one or two cysteine residues (ref. 28). Cys-546 of HydB is presumbly this ligand. It is located in a stretch of amino acids that is highly conserved among nine hydrogenases, and is replaced by seleno–cysteine in HydB of *Desulfovibrio baculatus*. In the latter enzyme, the selenium is reported to be a ligand to nickel (refs. 29 and 30). HydA contains eight cysteine residues that are conserved in the corresponding subunits of seven other hydrogenases. These cysteines could ligate the [3Fe-xS] center which is present in the enzyme of *W. succinogenes* (ref. 28) and probably one or two more centers.

Acknowledgements

This work was supported by grants from the Deutsche Forschungsgemeinschaft and from the Fonds der Chemischen Industrie to A.K.

References

1 M. Bronder, H. Mell, E. Stupperich and A. Kröger, Arch. Microbiol. **131** (1982) 216.

2 J.M. Macy, I. Schröder, R.K. Thauer and A. Kröger, Arch. Microbiol. **144** (1986) 147.

3 A. Kröger and A. Innerhofer, Eur. J. Biochem. **69** (1976) 487.

4 I. Schröder, A. Kröger and J.M. Macy, Arch. Microbiol. **149** (1988) 572.

5 A. Kröger and E. Winkler, Arch. Microbiol. **129** (1981) 100.

6 H. Mell, C. Wellnitz and A. Kröger, Biochim. Biophys. Acta **852** (1986) 212.

7 M. Bokranz, E. Mörschel and A. Kröger, Biochim. Biophys. Acta **810** (1985) 332.

8 M. Graf, M. Bokranz, R. Böcher, P. Friedl and A. Kröger, FEBS Lett. **184**(1985)100.

9 G. Unden and A. Kröger, Methods in Enzymology **126** (1986) 387.

10 O. Klimmek, A. Kröger, R. Steudel and G. Holdt, (Arch. Microbiol.), in press.

11 A. Kröger, E. Dorrer and E. Winkler, Biochim. Biophys. Acta **589** (1980) 118

12 G. Unden, H. Hackenberg and A. Kröger, Biochim. Biophys. Acta **591** (1980) 275.

13 G. Unden, A. Kröger, Eur. J. Biochem. **120** (1981)577.

14 F. Lauterbach, C. Körtner, D. Tripier and G.Unden, Eur. J. Biochem. **166** (1987) 447.

15 C. Körtner, F. Lauterbach, D. Tripier, G. Unden and A.Kröger, Molec. Microbiol. **4** (1990) 855.

16 F. Lauterbach, C. Körtner, S.P.J. Albracht, G. Unden and A. Kröger, Arch. Microbiol. **154** (1990) 386.

17 G. Unden, E. Mörschel, M. Bokranz and A. Kröger, Biochim. Biophys. Acta **725** (1983) 41.

18 S.P.J. Albracht, G. Unden and A. Kröger, Biochim. Biophys. Acta **661** (1981) 295.

19 G. Unden, S.P.J. Albracht and A. Kröger, Biochim. Biophys. Acta **767** (1984) 460.

20 A. Kröger, E. Winkler, A. Innerhofer, H. Hackenberg and H. Schagger, Eur. J. Biochem. **94** (1979) 465.

21 M. Bokranz, M. Gutmann, C. Körtner, E. Kojro, F. Fahrenholz, A. Kröger and F. Lauterbach, (Arch. Microbiol.), in press.

22 G. Unden and A. Kröger, Biochim. Biophys. Acta **725** (1983),325.

23 J.L. Johnson, N.R. Bastian and K.V. Rajagoplan, Proc. Natl. Acad. Sci. USA **87** (1990) 3190.

24 T. Krafft, M. Bokranz, E. Kojro, F. Fahrenholz and A.Kröger, in preparation.

25 U. Wissenbach, A. Kröger and G. Unden, Arch. Microbiol. **154** (1990) 60.

26 G. Unden, R. Böcher, J. Knecht and A. Kröger, FEBS Lett. **145** (1982) 230.

27 F. Droß, E. Kojro, F. Fahrenholz and A. Kröger, (Molec. Microbiol.), in press.

28 S.P.J. Albracht, A. Kröger, J.W. van der Zwaan, G. Unden, R. Böcher, H. Mell, and R.D. Fontijn, Biochim. Biophys. Acta **874** (1986) 116.

29 S.H. He, M. Teixeira, J. LeGall, D.S. Patil, I. Moura, J.J.G. Moura, D.V. DerVartanian, B.H. Huynh and H.D. Peck, J. Biol. Chem. **264** (1989) 2678.

30 M.K. Eidsness, R.A. Scott, B.C. Prickril, D.V. DerVartanian, J. LeGall, I. Moura, J.J.G. Moura and H.D. Peck, Proc. Natl. Acad. Sci. USA **86** (1989) 147.

31 P.T. Bilous, S.T. Cole, W.F. Anderson and J.H. Weiner, Molec. Microbiol. **2** (1988) 785.

32 D.E. Pierson and A. Campbell, J. Bacteriol. **172** (1990) 2194.

33 F. Blasco, C. Jobbi, G. Giordano, M. Chippaux and V.Bonnefoy, Mol. Gen. Genet. **218** (1989) 249.

34 A.P. Shuber, E.C. Orr, M.A. Recny, P.F. Schendel, H.D. May, N.L. Schauer and J.G. Ferry, J. Biol. Chem. **261** (1986) 12942.

35 F. Zinoni, A. Birkman, T.C. Stadtman and A. Böck, Proc. Natl. Acad. Sci. USA **83** (1986) 4650.

Electron and Proton Transfer in Chemistry and Biology, edited by A. Müller et al.
Studies in Physical and Theoretical Chemistry, Vol. 78
1992 Elsevier Science Publishers B.V.

The Importance of Inhibitors as an Analytic Tool for the Study of the Quinol Oxidation Centre and the Quinol Oxidase Reaction

G. von Jagow and U. Brandt

Univ. Klinikum
Frankfurt ZBC-Therapeutische Biochemie
Theodor-Stern-Kai 7, Haus 25 B
6000 Frankfurt am Main 70
(Germany)

Dedicated to Dr. Peter Mitchell on the Occasion of the 25th Anniversary of the Glynn Research Institute

Summary

The short review gives an overview of the recent developments in the use of natural and synthetic inhibitors for enzymological studies of the structure/function relationship of mitochondrial cytochrome *c* reductase (bc_1 complex). A model of the gross structure of the dimeric bc_1 complex with a tentative arrangement of its eleven subunits per monomer is given. The model is discussed in terms of the characteristic requirement of 80 molecules of phospholipid per dimer for the catalytic activity of this enzyme.

Source, structure and site of action of the most important inhibitors, which block individual reaction steps of the protonmotive Q cycle are described. A revised classification of these compounds is given: Group I inhibitors, which block centre P are subdivided into the E–β–methoxyacrylates (Ia), the chromones (Ib) and the hydroxynaphthoquinones (Ic). Group II inhibitors, which block centre N are subdivided into inhibitors of the antimycin type (IIa) and the quinolin–N–oxides (IIb).

The last part of the article focuses on the outer ubiquinol oxidation centre (Centre P), which establishes a bifurcation of the electron flow involving two different redox components of the bc_1 complex. Based on enzymological studies using isolated bc_1 complex, more detailed insight into the Q cycle mechanism is derived: A "catalytic switch", synonymus with a conformational transition, is postulated which is triggered by reduction of the "Rieske" iron–sulfur cluster. Blockage of the trigger could explain the noncompetitive mechanism of the E–β–methoxyacrylate inhibitors. Moreover, such a mechanism would prevent the second electron of ubiquinol to pass onto the iron–sulfur cluster. In reality such an unfavourable side–reaction, which would be equivalent to a 'slip' of the protonmotive action of cytochrome *c* reductase, has never been observed, even under physiological and experimental conditions favouring such a side reaction such as antimycin inhibition of the electron flow.

Introduction

About ten years ago we started with the first inhibitor of the quinol oxidation centre in our hands. We did not expect this natural compound, called oudemansin A, to be of greater help in understanding the branched electron transfer chain of the ubiquitous bc complexes (ref. 1). However, during the last few years the rising number of such inhibitors has played a substantial role in the understanding of the catalytic cycle of these complexes (refs. 2-4). At the same time remarkable contributions have been made by protein chemical and molecular genetic studies. Thus, for instance, cloning and sequencing of the structural genes and characterisation of inhibitor resistant mutants and revertants have provided a great amount of useful data (refs. 5-7). Combining the experimental results gathered in both functional and structural studies, one now possesses the most detailed picture available so far for any multiprotein complex of an electron transfer route.

A short discussion of the gross protein structure of the bc complexes as far as it concerns the specific role of the so called 'supernumerary' subunits and the phospholipid should be given prior to the compilation of the molecular structure and function of these novel inhibitors.

Architecture of the Dimeric bc_1 Complex

Ultracentrifugal studies of the bovine and *Neurospora crassa* complexes and later electron microscopic analysis of two–dimensional crystals of the *Neurospora crassa* complex in more detail have shown that the bc complexes are present in the form of dimers formed by two identical multisubunit monomers (ref. 8). The monomers are arranged around a two–fold axis of symmetry, each monomer of the bovine enzyme consisting of eleven subunits (ref. 9).

In each monomeric half, at one side the front, at the other side the back of the protein mass appearing (Fig. 1), ten subunits surround an essential functional nucleus, namely cytochrome *b*. Two of the surrounding subunits, the Rieske iron sulfur protein and cytochrome c_1 form, as will be discussed later, a connecting electron transfer chain towards cytochrome *c*. The question then arises as to the requirement for such a complicated arrangement, especially what is the function of these numerous subunits? Why does such a high number of apparently 'supernumerary' subunits, lacking any redox centres, shield the catalytic core, whereas this is not the case in the bacterial complexes consisting of merely the three redox subunits? This question remains at present mainly unanswered.

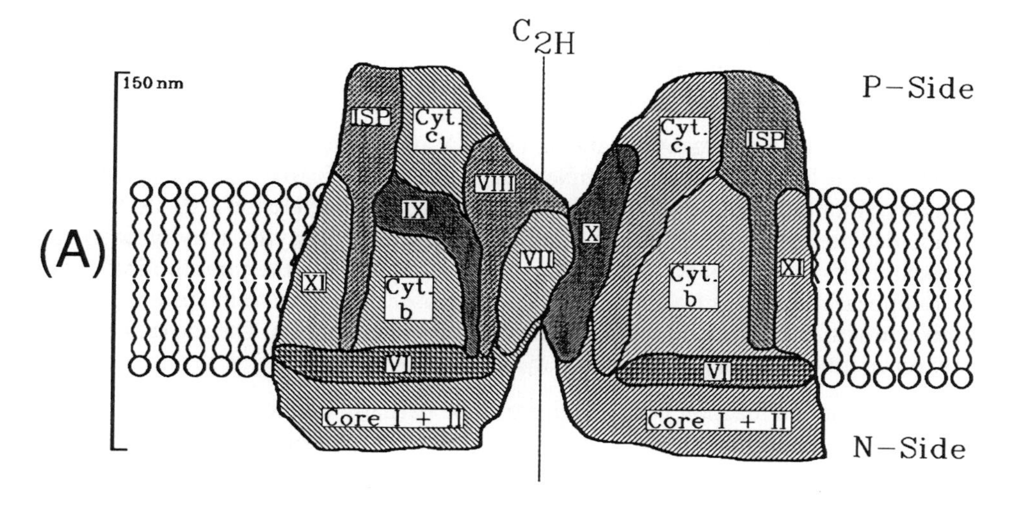

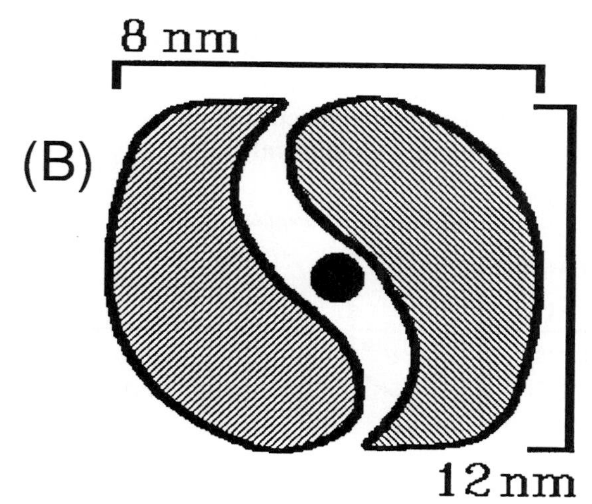

Fig.1. Cartoon of the dimeric bc_1 complex. Front (A) and top (B) view of the bc_1 complex dimer. The gross structure was adopted from the low resolution structure of the *Neurospora crassa* enzyme. The subunits are arranged according to data from protease digestion and immunological studies (refs. 26 and 27). The left side of A shows the "front" of one monomer, the right side the "back" of the other monomer. The two monomers are arranged around an axis of twofold symmetry indicated by a dashed line in A and a dot in B.

Phospholipid Requirement for Catalytic Activity

For its conformational stability the isolated dimer requires the presence of about 20 molecules of cardiolipin (ref. 10). In addition as a minimal condition for full catalytic activity about 80 molecules of PE and/or PC have to be bound. If one assumes 4Å for the surface width of one phospholipid molecule and 320Å for the circumference of the dimer, it would be presumable that the complex is surrounded by a phospholipid bilayer annulus at its equator — so to speak, by a phospholipid halo formed by 80 molecules of phospholipid, the cardiolipin molecules being more specifically bound and not part of the annulus. The function of the annulus could be to guarantee full lateral and transversal mobility of the quinone molecules.

Apparently, a detergent micelle of Triton X–100 which intimately surrounds the completely delipidated dimeric complex cannot substitute for the phospholipid bilayer annulus, since this complex is enzymatically completely inactive (ref. 10). On the other hand, the detergent micelle seems not to impair full catalytic activity when the phospholipid annulus is present. Under these circumstances the phospholipid molecules appear to be — at least kinetically — in closer proximity to the protein surface and therewith to the reaction centres than the detergent molecules. The activation would rely on an inhomogenity of the mixed micelle.

The quinone reaction pockets appear to be partially buried in and connected by the bulk phospholipid. Maybe some molecules of cardiolipin are more tightly and specifically arranged around the entrances of the quinone pockets.

Site of Action of the Inhibitors on the Functioning of the Dimeric bc_1 Complex

In order to simplify matters the dimer will be treated functionally as non–cooperative, *i.e.* the two monomeric halves as functioning independently.

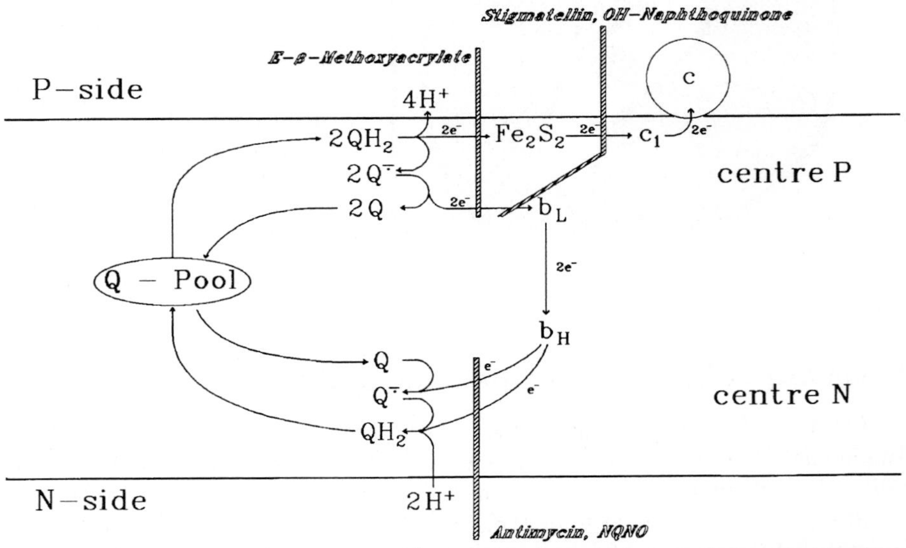

Fig. 2. Q–cycle scheme of the monomer. The scheme shows the general electron transfer and protonmotive reactions of the bc_1 complex. Abbreviations: Q, ubiquinone; QH_2, Ubihydroquinone; $Q^{\cdot-}$, Ubisemiquinone; heme b_L, low–potential heme b (b_{566} of bovine heart); heme b_H, high–potential heme b (b_{562} of bovine heart); FeS, Fe_2S_2 iron–sulfur cluster.

We are aware that only during the last month recent data have revealed some new hints of functional cooperativity (refs. 11 and 12). However, the dimeric bovine complex prepared in Triton X–100 has given no indication for functional cooperativity so far. But still the functional dimer is a controversial issue.

The three bars drawn into a Q–cycle reaction scheme (refs. 2 and 3), compile the present state of knowledge about the site of action of the different classes of inhibitors (Fig. 2). Independent sites of inhibitor action at the inner quinone reduction centre — called centre N according to the negative charge of the membrane — and at the outer quinol oxidation centre — called centre P according to the positive charge of the membrane — have been revealed (ref. 1). We will concentrate on the relations at centre P. The scheme of electron flow given in Fig. 2 should serve as a guide through the following text. The details of electron and proton flow will be pointed out by a more elaborated flow scheme at the end of the article.

First the molecular structure of the inhibitor molecules together with their origin will be given, secondly functional studies will be described allowing some more insight into the Q cycle mechanism and the possible role of conformational changes.

Molecular Structure of the Inhibitors of the bc_1 Complexes

The first natural compound which came into our hands was found by T. Anke. It is produced in the basidiomycete *Oudemansiella mucida*, a fungus growing on the bark of rotten beech trees. The poison oudemansin A was extracted from submerse cultures of *Oudemansiella mucida* (ref. 13). The molecule possesses two predominant structural segments: E–β–methoxy–acrylic acid — abbreviated as MOA — in the form of the methyl ester; and secondly a phenyl ring. These two almost planar parts are connected by an alkenyl chain. They form an angle of about 72° (ref. 13). This V–shaped structure may be a prerequisit for a good fit in the respective quinone reaction pocket. Oudemansin A is the weakest of the known MOA–inhibitors. It appears that the MOA part binds to the MOA receptor site, whereas the phenyl ring projects more freely into pocket P.

The second antibiotic came again from T. Anke (ref. 14). He detected it in the basidiomycete *Strobilurus tenacellus*, a fungus growing on fir cones. The poison which is produced in this fungus is strobilurin A, which is also known under the incorrect name Mucidin (ref. 15). The structure of strobilurin A is close to that of oudemansin A: again the molecule contains the MOA system and a phenyl ring system, *i.e.*, the V–shaped structure is again present, but the bridging alkenyl chain is more rigid than in oudemansin A due to the presence of two conjugated double bonds. This may be the reason for extended binding and higher efficacy of strobilurin A when compared to oudemansin A.

Table 1

<u>Structures of the inhibitors of the bc_1 complex.</u> The structural formulas are given for representatives of every class of bc_1 complex inhibitors. Group I are inhibitors of the ubihydroquinone oxidating Q_P centre, Group II are inhibitors of the ubiquinone reducing Q_N centre.

Group I

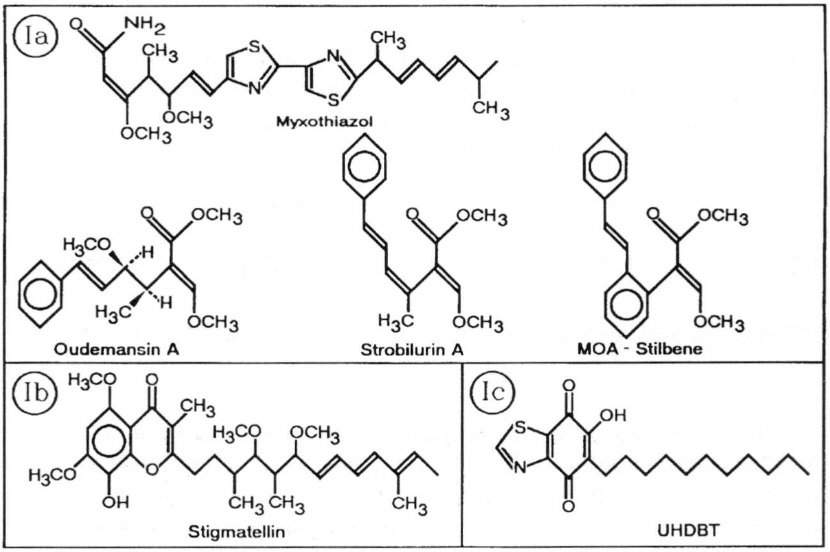

(Ia) Myxothiazol; Oudemansin A; Strobilurin A; MOA - Stilbene

(Ib) Stigmatellin

(Ic) UHDBT

Group II

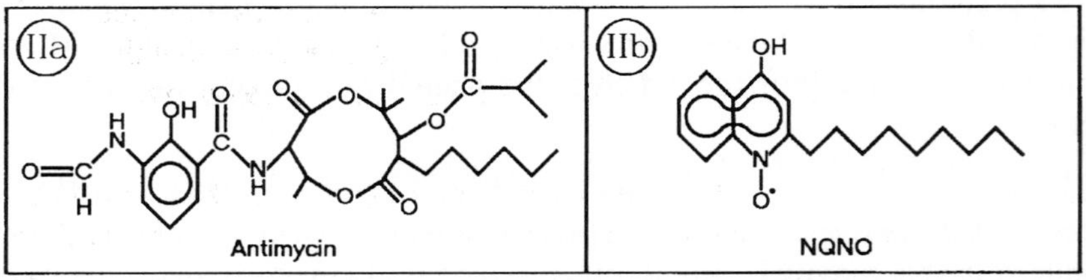

(IIa) Antimycin

(IIb) NQNO

From the working group of Hans Reichenbach we obtained another centre P inhibitor. He isolated from the small fruit bodies of the gliding bacterium *Myxococcus fulvus* a more exotic compound, known as myxothiazol (refs. 16 and 17). Although coming from completely different source, the molecule also possesses the MOA system, but now in the form of an amide. The second half of the molecule, the side chain, is quite different. As a new variation of the theme it includes a bisthiazole ring system, which makes myxothiazol bind considerably more tightly than the two previous simpler MOA inhibitors by undergoing some additional interactions with the protein matrix of the reaction pocket. It is remarkable that such a big molecule can fit into the reaction pocket. In our view the alkenyl side chain connected to the bisthiazole ring should project into the bulk phospholipid.

The last natural compound we present here, has been extracted from the fruit bodies of the gliding bacterium *Stigmatella aurantiaca.* Stigmatellin has a structure quite different from the former molecules (ref. 18). It contains a substituted chromone ring, a spacious alkenyl side chain substituted with methoxy and methyl groups. It should already be mentioned here, that this antibiotic — as one would expect — is bound to a receptor site different to that of the MOA inhibitor receptor site. As already indicated in Fig. 2 stigmatellin inhibitors act in a mode different to that of the MOA inhibitors. However, as will be demonstrated later, both kinds of inhibitors bind in one and the same quinone reaction pocket, namely in pocket P (cf. Fig. 2).

Interestingly, a variety of synthetical hydroxynaphthoquinone analogues, which also inhibit electron flow, bind either to or very close to the Stigmatellin receptor site. One example is undecyl–hydroxy–dioxo–benzothiazol. The hydroxynaphthoquinones are less effective than the natural inhibitors. They were known before the inhibitors from biological sources, (ref. 19) and have been studied in detail by B. Trumpower, (ref. 20) and T. Crofts, (ref. 21).

It is beyond the scope of this article to describe comprehensively the multitude of natural antibiotics and the mass of synthetic molecules which are now at our disposal, but the collection of antibiotics herein described might already have given some impression of the variability, complexity and the respective sizes of the different molecules.

We have delineated the library of inhibitors into 5 main classes. Three classes belong to group I, that is they act at centre P. Class Ia are inhibitors which contain E-β-methoxyacrylate, *i.e.* these are MOA inhibitors. Stigmatellin is a representative of class Ib. The hydroxynaphthoquinones belong to class Ic. Inhibitors of centre N (group II) are divided in class IIa, *i.e.* Antimycin, and class IIb, *i.e.* quinolin–N–oxides.

Binding and Action of the Inhibitors at Centre P

For a better understanding of the specific inhibitor action, a few general remarks should be made about the similarities and dissimilarities of the quinone reaction centres. On first sight a bipartite classification into quinone oxidation and quinone reduction centres seems to be sufficient. The oxidation centres are assumed to serve simply as electron entries standing at the upstream end of an electron transfer chain and transferring electrons from ubihydroquinone to just one electron acceptor of the following redox chain. The reduction centres are supposed to serve as electron exits standing at the downstream end of an electron transfer chain and transferring electrons on quinone by just one terminal electron donating redox centre. However, as known so far the rule is broken by two exceptions : i) by the Q_A centres of photosynthetic electron transfer chains and ii) by all kinds of quinone oxidation centres of the bc/b_6f complexes. The Q_A centres possess the peculiar feature that they hold the quinone molecule in the pocket as a prosthetic group serving as intrinsic electron carrier, whereas all other quinone centres bind and release the quinone molecules. The quinone oxidation centres of bc/b_6f complexes do not transfer the electrons to just one electron acceptor, but alternatively to two different electron acceptors in a branched electron transfer chain, which is a prerequisite for the protonmotive action of this device. This point will be discussed below. Moreover they are formed by an agglomeration of at least two proteins.

Function of the Inhibitors

Two effective spectroscopic techniques will be described by which binding of the inhibitors has been studied: one technique is based on the use of the bathochromic absorbance shift spectra of the heme *b* centres as a quantitative signal (ref. 22), the other technique uses fluorescing inhibitor molecules, the fluorescence of which becoming quenched when they are bound in the pocket (ref. 23).

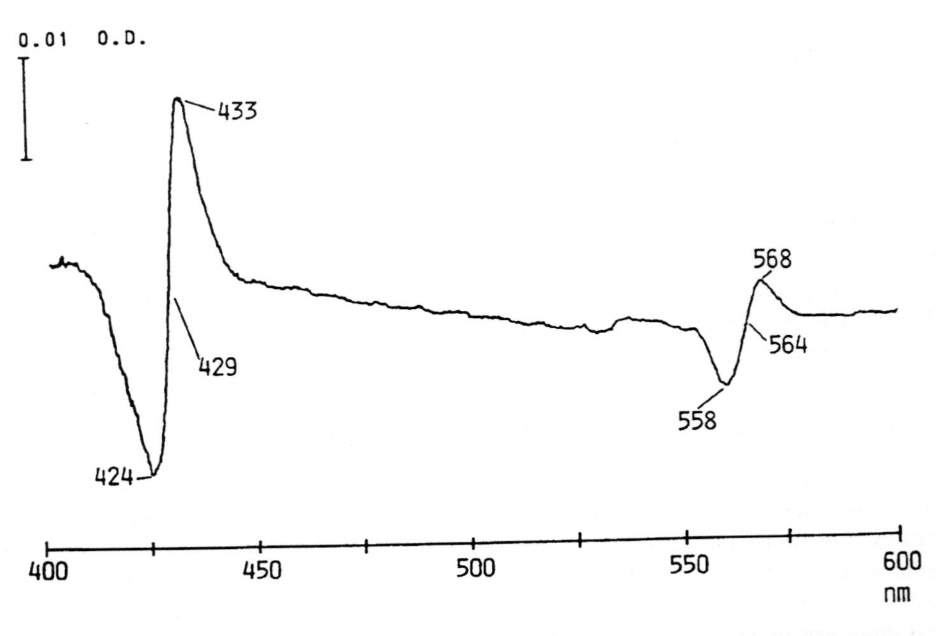

Fig. 3. Red–shift spectrum of oudemansin A. The spectrum was recorded as the difference between dithionite reduced bc_1 complex without and with oudemansin A present. Concentrations: bc_1 complex, 3μM; Oudemansin A, 30μM. Buffer: 100mM K^+/Mops, 0.05% Triton X-100, 100mM NaCl, 2mM NaN_3, pH 7.2.

Binding of MOA inhibitors to reduced bc1 complex produces a red shift of the absorbance maximum due to a distortion of the ligand field of mainly heme *b*–566. The absorbance maximum is shifted from 566 to 568 nm. A more or less symmetric shift spectrum is obtained which resembles a first derivative spectrum (Fig. 3). As long as binding proceeds, the red shift increases and as soon as the binding site is saturated, the red shift is maximal. All questions of the binding of an inhibitor and the

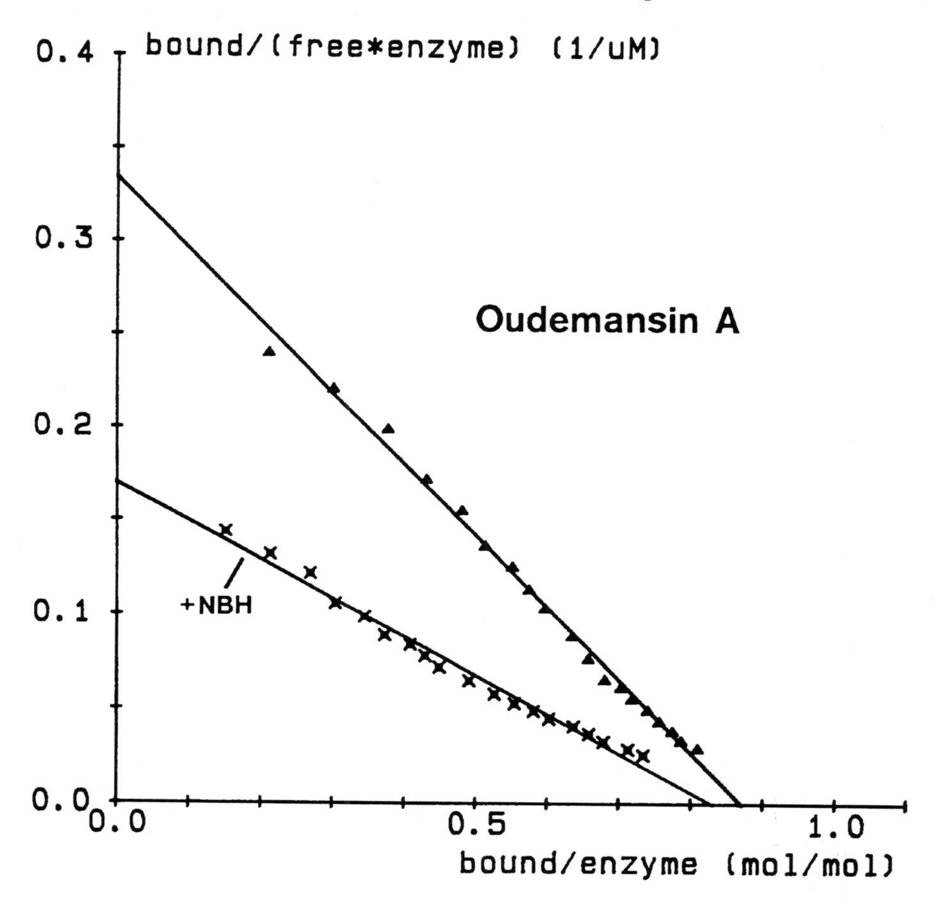

Fig. 4. Scatchard plots of oudemansin A binding. The binding was monitored as absorption difference at 568-558nm. Other conditions as in Fig. 3. (▲) no addition, apparent K_d 2.6μM; (x) 0.8mM nonylubihydroquinone (NBH) added, apparent K_d 4.6μM.

displacement of the inhibitors with each other can be conveniently tested by this method. The disadvantage of the method is the restriction to fully reduced state of the enzyme and some lack of sensitivity when compared to the fluorescence quench method.

Fig. 4 demonstrates Scatchard plots of data obtained by red–shift titrations of oudemansin A in the presence and absence of ubihydroquinone. For a single binding site the number of receptor sites amounted to one per centre P for all inhibitors tested, the binding affinity increased from oudemansin A, over strobilurin A to MOA–stilbene. The plots indicate a non–competitive mode of inhibition. The inhibitor was not displaced by increasing concentrations of ubihydroquinone. Even the highest quinone concentrations only lowered the binding affinity to a constant and definite value as demonstrated in Fig. 5:

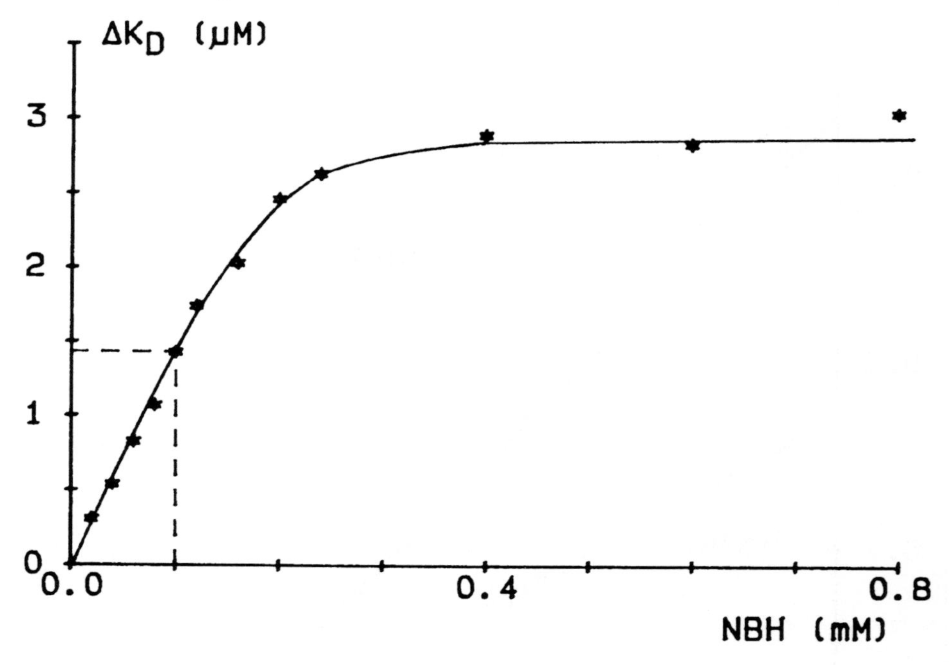

Fig. 5. K_d shift by increasing concentrations of nonlyubihydroquinone. Replot of dissociation constants determined as described in Fig. 4 in the presence increasing concentrations of nonylubihydroquinone. The differences between the K_d determined without quinone (2.6μM) and in the presence of 0.02-0.8mM nonylubihydroquinone are plotted.

A replot of the K_d values shows that with increasing concentrations of quinone the K_d for oudemansin A was raised only by a factor of two after saturation of the quinone binding site. The same was valid for all other MOA–inhibitors (ref. 22).

Since the result of a non–competitive inhibition mode is surprising, especially in view of the fact that the inhibitor molecules are large and that in the Q_B centre of *Rhodobacter capsulatus* — as H. Michel and coworkers have shown directly by X–ray crystallographic studies, (ref. 24) — the quinone is displaced, we rechecked the binding by means of classical enzyme kinetics.

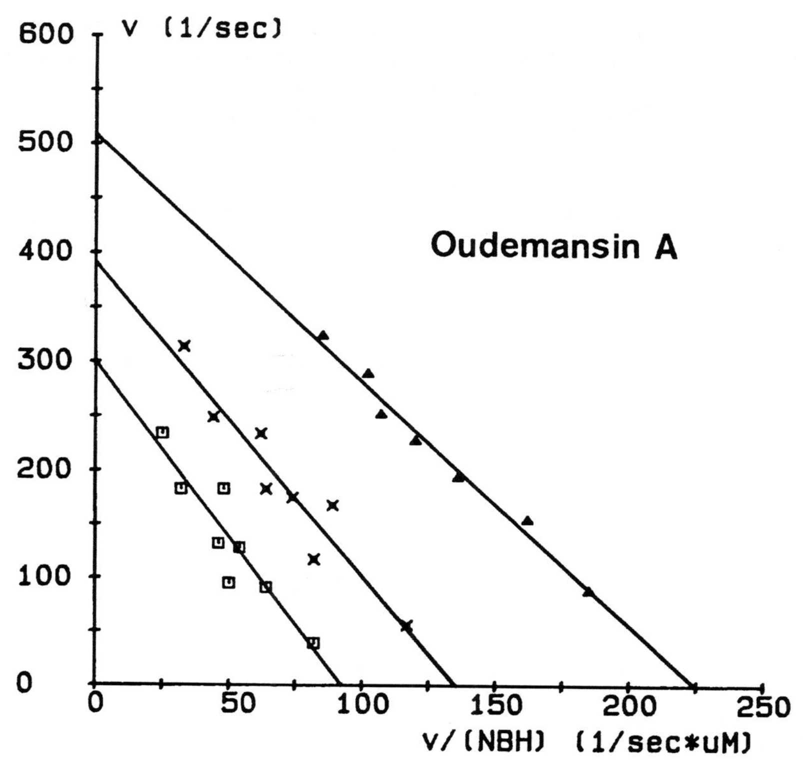

Fig. 6. Eadie–Hofstee plot of inhibition kinetics with oudemansin A. The kinetics of bc_1 complex reconstituted into proteoliposomes were measured at 25°C with nonylubihydroquinone as substrate. Buffer: 50mM potassium phosphate, 200mM Sucrose, 2mM KCN, 0.2mM EDTA, 1 M carbonylcyanide p–trifluoromethoxyphenylhydrazone, pH 7.2. The final concentration of bc1complex was approx. 1nM. (▲) no inhibitor, K_m=2.2μM, V_{max}=510s^{-1}; (x) 170nM oudemanisn A; (□) 510nM oudemansin A. Assuming non–competitive inhibition apparent oudemansin A dissociation constants of 340±60nM for the free enzyme and 630±160nM for the enzyme substrate complex were calculated from experiments using 5 different inhibitor concentrations.

As demonstrated in Fig. 6 by Eadie–Hofstee plots in the presence of increasing amounts of oudemansin A, V_{Max} was changed by the inhibitor, *i.e.* the inhibition is non–competitive. This again was valid for oudemansin A, strobilurin A and for the synthetic inhibitor MOA–stilbene (ref. 22).

The question whether it is possible to bind two inhibitors simultaneously, belonging to the same or to different classes was studied by the fluorescence quench method, (ref. 23) as described in the following. Using this method it turned out that the inhibitors displace each other in all cases, even when they were bound to different domains of centre P.

At present we possess only one fluorescing inhibitor for centre P. That is MOA–stilbene. It has an excitation maximum at 308 nm and an emission maximum at 380 nm. The fluorescence is reliable, since it is not significantly quenched or interfered by protein or by heme absorbance and/or fluorescence.

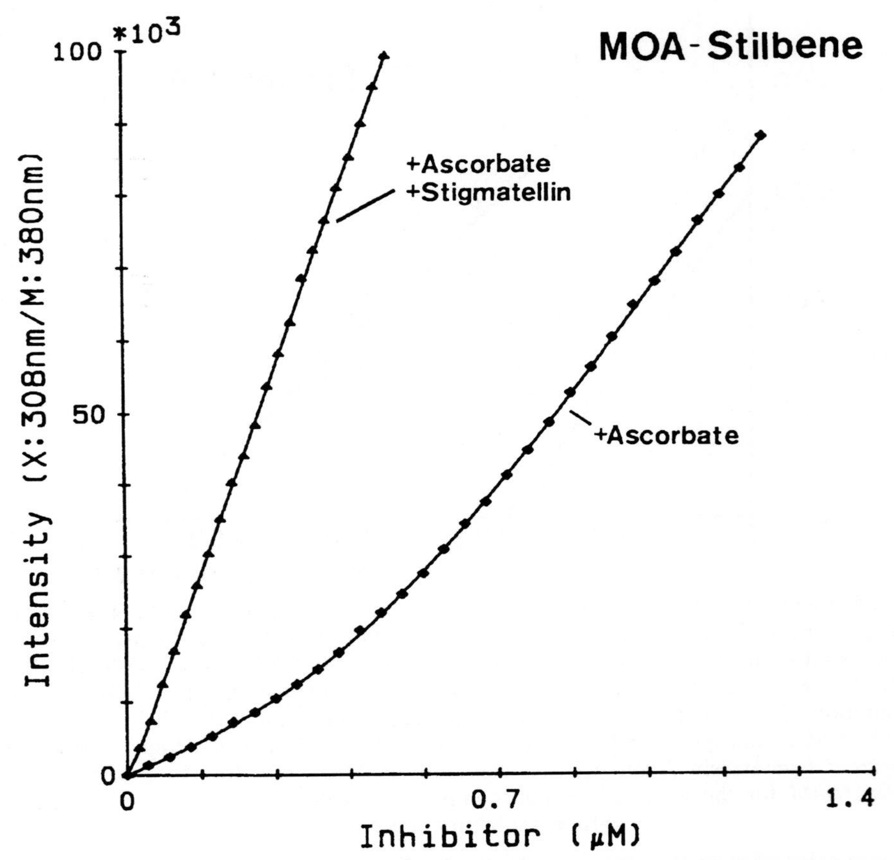

Fig. 7. Fluorescence titration of MOA–stilbene and competition with stigmatellin. When stigmatellin was added before titration no quench due to binding of MOA–stilbene to the complex was observed. The bc_1 complex concentration was 0.5M. 1mM sodium–ascorbate was added before the titration to reduce the high–potential chain. Other experimental conditions as in Fig. 3. (◆) no additions; (▲) 5μM stigmatellin added.

The analysis is based on a quench of fluorescence of the inhibitor which dominates as long as it is bound. When the receptor site is saturated, full fluorescence of the free inhibitor molecule ensues. The titration curves were analysed online by a numerical fit. Based on a standard binding equation, the number of binding sites, the binding affinity and the specific fluorescence of the bound and free inhibitor were used as parameters for matching the data points by a theoretical curve (ref. 23).

Binding of MOA–stilbene was abolished by stigmatellin as indicated by the abolition of the fluorescence quench when stigmatellin was bound prior to MOA-stilbene (Fig.7). It should be mentioned that as expected all MOA inhibitors were competitive to each other. But it should be recalled, that above when the structures of the group I inhibitors were introduced, it was anticipated that MOA–stilbene and stigmatellin do not bind to the same receptor site, but nevertheless in the same pocket. How has this been elucidated? How could we find out that we were dealing with two different binding sites?

Given just by the Q–cycle mechanism (see Fig.2) at least cytochrome b and the iron sulfur protein contribute to centre P. Using a method of protein chemistry developed earlier (ref. 25), we were able to dissociate exclusively the iron sulfur protein from the isolated mammalian complex. Quite logically, after dissociation an incomplete centre P was received. All other parts of the multisubunit complex remained intact, thus for instance centre N stayed fully active.

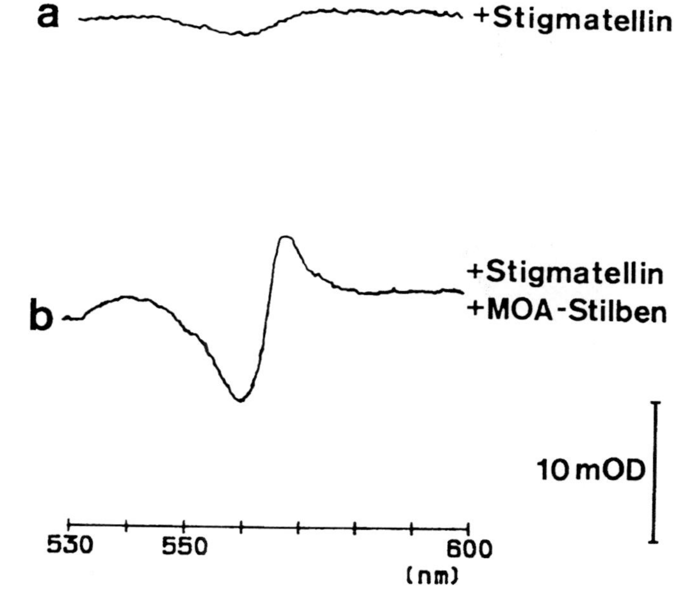

Fig. 8. Red–shift spectra of the iron–sulfur protein depleted bc_1 complex. No stigmatellin red–shift The spectrum was obtained (a) and formation of a normal red–shift of MOA–stilbene was not prevented by stigmatellin (b). The iron–sulfur depleted complex was prepared as described in (ref. 22). Experimental conditions as described in Fig. 2.

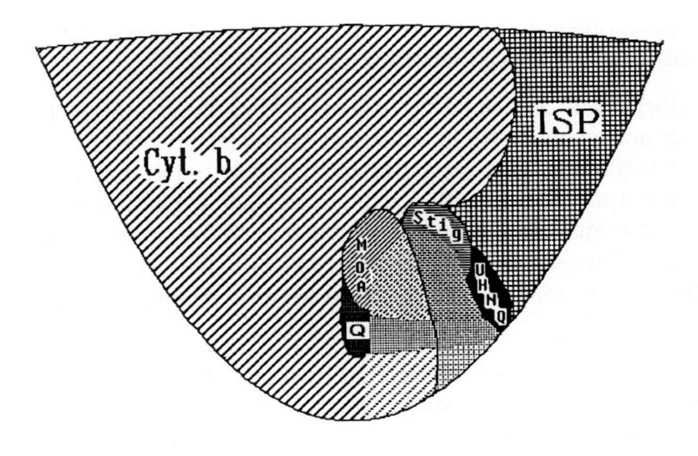

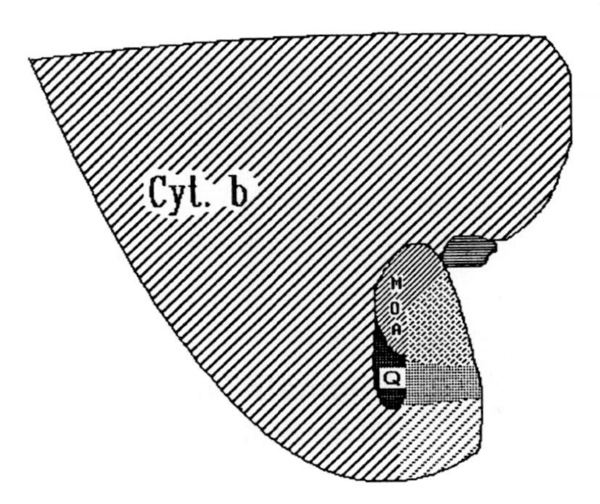

Fig. 9. Cartoon of the functional organization of the reaction pocket of centre P. The cartoon tries to describe the functional organization of centre P, as derived from protein–chemical and inhibitor studies. Different patterns grossly indicate the binding sites for the three classes of inhibitors and ubihydroquinone. No experimental evidence exists to decide whether the binding site of the substrate is on cytochrome *b* or the iron–sulfur protein or is formed by domains of both subunits.

The incomplete pocket revealed no red shift by stigmatellin (Fig. 8), but when MOA–stilbene was subsequently added, the red shift of MOA–stilbene occurred, indicating not only that this inhibitor was bound, but also indirectly, the lack of stigmatellin binding, because — as shown before — binding of stigmatellin would prevent MOA–stilbene binding, since stigmatellin binds much more tightly than MOA–stilbene. The incomplete centre P not only showed the same red–shift of the MOA inhibitors, but by quantitative analysis also the same number of binding sites and the same binding affinity as obtained for the complete pocket (ref. 22).

The cartoon of Fig. 9 summarizes the result: after removal of the iron sulfur protein, the MOA inhibitors are still bound, whereas stigmatellin and the hydroxynaphthoquinones are not. Unfortunately at present we are still awaiting for a method with which we could check whether quinone is still bound to the incomplete reaction pocket or not.

The experiments tell us that centre P is a spacious pocket encompassing several binding sites, so that the incomplete pocket still possesses the full number of side chain ligands for the MOA–inhibitors but no longer for stigmatellin.

In our view, the non–competitive typ of inhibition and the binding of different large inhibitor molecules to a centre formed by two proteins gives the picture of a quite spacious pocket capable of the bifurcated oxidation of ubihydroquinone. The pocket must be so spacious as to bind the substrate and an inhibitor molecule simultaneously (Fig. 9). As a consequence blockage of electron flow by these inhibitors has to be explained by an indirect conformational mechanism. If this assumption holds true one has to postulate an additional interaction. If the inhibitors of centre P all interfer with a conformational process relevant for the redox reaction of centre P one would expect that a redox change of the redox centres would influence the binding affinity of the inhibitors, but in a different way for the different classes of inhibitors. This has been proven to be the case (Table2, (ref. 23)):

Table 2

K_d changes of centre P inhibitors on reduction of the high–potential chain. Apparent K_d values for MOA–stilbene (upper panel) were determined by direct fit of fluorescence titrations as described in the text. Apparent K_d values for UHNQ (3n–undecyl–2–hydroxy–1,4–naphthoquinone) were indirectly determined from a competition analysis between MOA–stilbene and UHNQ in fluorescence quench titrations. The UHNQ concentration was varied between 0.1 and 10μM. Experimental conditions as in Fig. 7.

MOA–stilbene

reducing agent	K_d	n
	nM	
none	19 ± 7	50
Na-ascorbate	51 ± 12	21
Na-dithionite	49 ± 23	13

UHNQ

reducing agent	K_d	n
	μM	
none	4.7 ± 1.0	7
Na-ascorbate	1.2 ± 0.2	8

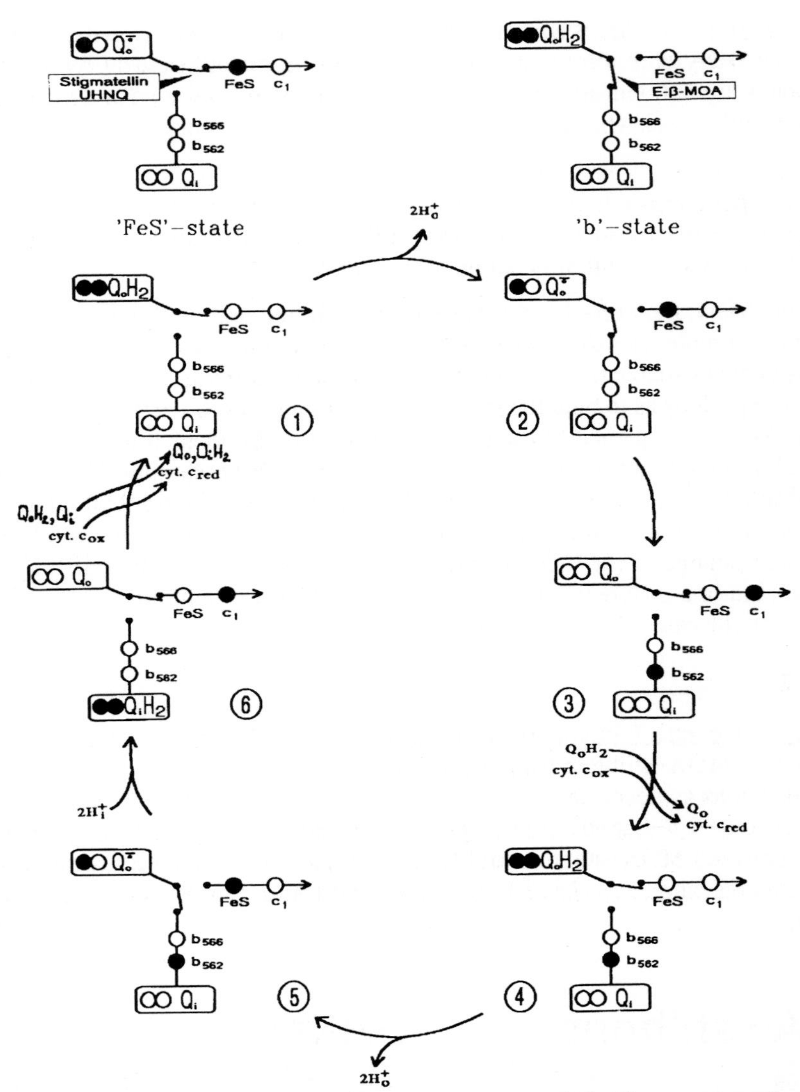

Fig. 10. Q–cycle scheme including the role of a catalytic switch. The representation alludes to an electrical switching scheme. The quinone centres are shown to be occupied either by fully reduced quinone – indicated by two full dots, by half reduced quinone – indicated by one full dot, or by oxidised quinone – signified by two empty dots. Similarly a reduced redox centre is marked by a full dot, an oxidised one by an empty dot, *etc.* A few intermediary states of the cycle have been omitted, as for instance those where the quinone centres are empty. Abbreviations used as in Fig. 2. The indices 'i' and 'o' for 'in' and 'out' correspond to the matrix (N–side) and cytosolic side (P–side) of the inner mitochondrial membrane, respectively. See text for further details.

The K_d value of MOA–stilbene was increased by a factor higher than two on reduction of the iron–sulfur centre. On the contrary, the K_d value of undecylhydroxy-naphthoquinone was decreased by a factor of four when the iron–sulfur centre was reduced.

From all our inhibitor studies we confer that the "specific next mode" of electron transfer is realised by a conformational transition of the quinone oxidation pocket, occuring twice during one catalytic cycle. The "conformational switch" (Fig. 10) must ensure that electrons at low energy level flow along the high potential pathway, whereas electrons at high energy level flow along the low potential pathway.

Centre P is alternatively linked to the high or to the low potential pathway. On the contrary, centre N is firmly linked to the low potential chain.

How must the pacing of the switch be visualised? The first electron — after deprotonation of quinol — arrives at the iron–sulfur centre. This event is believed to trigger the switch from "iron–sulfur position" to "b position". In other words the change of the iron-sulfur centre from oxidised to reduced state is associated with a conformational change of centre P. This makes sure that the second electron, that of the highly negative $Q^{\cdot -}/Q$ redox couple, does not arrive at the high potential chain, but at centre b_{566}.

In the meantime the electron has migrated from the iron–sulfur centre to centre c_1; hence the switch is triggered back to the "iron–sulfur position" by the oxidised iron–sulfur centre or maybe by the reoxidation of one of the heme *b*'s. Quinone is released from centre P, a second quinol collides with the centre.

Now the second half of the catalytic cycle starts: again after deprotonation of quinol the first electron flows along the high potential pathway, after conformational transition the second electron arrives at centre b_{566}. In the balance during a whole catalytic cycle flow of two electrons *via* the cytochrome *b* pathway (here the single steps are omitted) under uptake of $2H^+$ from matrix space leads to reduction of one quinone molecule in centre N, in other words the recycling leads to regain of one ubihydroquinone molecule. The molecule may encounter either centre P directly or the quinone pool. Thus the machinery is ready for the next cycle.

Altogether, oxidation of one quinol molecule leads to reduction of two molecules of cytochrome *c*, release of $4H^+$ to the cytosolic space and uptake of $2H^+$ from the matrix space.

At this point one could ask: "Is there really any need to incorporate an 'electrical' switch into the Q–cycle and what is the underlying idea of this implication?"

Our answer may be the following: not only experimental data forced us to imply the switch, but also theoretical considerations. Amongst other things, it had to be explained why all centre N inhibitors completely block electron flow, independent of the respiratory chain status. In other words, why does the highly negative semiquinone not donate its electron to the high potential chain. To us a switch seems to be a straightforward solution of the problem.

We know that the MOA–inhibitors prevent electron transfer to the high as well as to the low potential chain. Therefore we assume that the MOA–inhibitors freeze the "b–position" of the switch or a position very similar to the "b–position".

On the other hand it is claimed that stigmatellin as well as the hydroxynaphthoquinone inhibitors still allow reduction of the iron sulfur centre, but not of heme b, therefore we assume that those inhibitors freeze the "iron sulfur position" of the switch and simultaneously lower the redox potential of the iron sulfur protein making the iron sulfur centre eventually an electron trap.

During the normal catalytic cycle the switch channels the electrons in the proper way by timing the bifurcated electron flow at centre P.

Outlook

The aim of this article is to give some insight into the regime of the bc complexes and to describe an extended model of the Q–cycle, which has been created in such an ingenious way by P. Mitchell in 1975, (ref. 2), one should not forget, lacking most of the experimental data available today and therefore mainly on theoretical basis.

We hope that our extended model of the Q cycle might provoke some controversy, but also guide experimental studies. We have the feeling that no other part of the respiratory chain has been analysed in such detail by means of inhibitors. At this state of the art, cooperation with molecular geneticists who have created point mutations inducing inhibitor–resistant mutants and revertants might be most useful. The available techniques and systems challenge our immediate progress, in particular, we would next be very interested to learn more about the function of the non–redox ('supernumerary') subunits.

Acknowledgements

This work was supported by grant No. Ja284/7-4 of the Deutsche Forschungsgemeinschaft to GvJ. Additional financial support from the Fonds der Chemischen Industrie is gratefully acknowledged.

We are indebted to Dr. Sewell for carefully revising the manuscript.

References

1 W.F. Becker, G. Von Jagow, T. Anke and W. Steglich, FEBS Lett. **132** (1981) 329.

2 P. Mitchell, FEBS Lett. **59** (1975) 137.

3 P. Mitchell, J. Theor. Biol. **62** (1976) 327.

4 G. Von Jagow and T.A. Link, Methods Enzymol. **126** (1986) 253.

5 J.-P. Di Rago, J.-Y. Coppe and A.-M. Colson, J. Biol. Chem. **264** (1989) 14543.

6 F. Daldal, M.K. Tokito, E. Davidson and M. Faham, EMBO J. **8** (1989) 3951.

7 N. Howell, J. Mol. Evol. **29** (1989) 157.

8 K. Leonard, P. Wingfield, T. Arad and H. Weiss, J. Mol. Biol. **149** (1981) 259.

9 H. Schägger, T.A. Link, W.D. Engel and G. Von Jagow, Methods Enzymol. **126** (1986) 224.

10 H. Schägger, Th. Hagen, B. Roth, U. Brandt, T.A. Link and G. Von Jagow, Eur. J. Biochem. **190** (1990) 123.

11 C.A.M. Marres and S. de Vries, EBEC-Short. Rep. **6** (1990) 5.

12 G. Bechmann and H. Weiss, EBEC-Short. Rep. **6** (1990) 55.

13 T. Anke, H.J. Hecht, G. Schramm and W. Steglich, J. Antibiot. **32** (1979) 1112.

14 T. Anke, F. Oberwinkler, W. Steglich and G. Schramm, J. Antibiot. **30** (1977) 806.

15 G. Von Jagow, G.W. Gribble and B.L. Trumpower, Biochemistry **25** (1986) 775.

16 G. Thierbach and H. Reichenbach, Biochim. Biophys. Acta **638** (1981) 282.

17 W. Trowitzsch, G. Höfle and W.S. Sheldrick, Tetrahedron Lett. **22** (1981) 3829.

18 G. Höfle, B. Kunze, C. Zorzin and H. Reichenbach, Liebigs Ann. Chem. (1984) 1883.

19 E.G. Ball, C.B. Anfinsen and O. Cooper, J. Biol. Chem. **168** (1949) 257.

20 B.L. Trumpower and J.G. Haggerty, J. Bioenerg. Biomembr. **12** (1980) 151.

21 J.R. Bowyer, P.L. Dutton, R.C. Prince and A.R. Crofts, Biochim. Biophys. Acta **592** (1980) 445.

22 U. Brandt, H. Schägger and G. Von Jagow, Eur. J. Biochem. **173** (1988) 499.

23 U. Brandt and G. Von Jagow, Eur. J. Biochem. **195** (1991) 163.

24 I. Sinning, H. Michel, P. Mathis and A.W. Rutherford, Biochemistry **28** (1989) 5544.

25 W.D. Engel, C. Michalski and G. Von Jagow, Eur. J. Biochem. **132** (1983) 395.

26 T.A. Link *Zur Struktur des mitochondrialen bc1–Komplexes* Ph.D. Thesis, Ludwig–Maximilians–Universität, München, 1988,

27 D. Gonzlez–Halphen, M.A. Lindorfer and R.A. Capaldi, Biochemistry **27** (1988) 7021.

Electron and Proton Transfer in Chemistry and Biology, edited by A. Müller et al.
Studies in Physical and Theoretical Chemistry, Vol. 78
1992 Elsevier Science Publishers B.V.

Electron and Proton Transfer through the Mitochondrial Respiratory Chain

Thomas A. Link

Universitätsklinikum Frankfurt
ZBC, Therapeutische Biochemie,
Theodor–Stern–Kai 7, Haus 25B,
D-6000 Frankfurt/M. 70 (Germany)

Dedicated to Dr. P. Mitchell on the occasion of the Silver Jubilee of Glynn Research

Summary

The mitochondrial respiratory chain catalyses the final reaction steps during the oxidation of substrates within the metabolism of cells. It transfers reducing equivalents coming mainly from the citric acid cycle and from β–oxidation of fatty acids to molecular oxygen. The energy released in this redox reaction is converted into an electrochemical membrane potential through the translocation of protons across the inner mitochondrial membrane. The series of coupled hydrogen, electron, and proton translocations is the main source of energy in the eucaryotic cell. Similar reactions occur in procaryotic cells (bacteria); therefore, the chemiosmotic mechanism proposed by P. Mitchell is one of the fundamental principles of life. The general mechanisms and structures underlying these reactions have been unraveled to a great extent during the last twenty years. This review will describe our present picture of the different types of mechanisms (redox loops, redox cycles, proton pumps) as well as of the electron transfer components involved and their protein environment.

Introduction

The mitochondrial respiratory chain is functionally equivalent to the electron transfer chains of aerobic bacteria like *Rhodobacter* and *Paracoccus denitrificans.* Their function is to funnel reducing equivalents of substrates deriving from a diversity of metabolic pathways into a single terminal oxidation reaction and to utilize the energy released during this reaction for the formation of an electrochemical proton gradient. The electrochemical proton gradient (protonmotive force) thus formed may then be utilized by other enzyme systems for production of ATP (F_0F_1 ATPase), secondary active transport (*e.g.*, export of ATP and import of ADP and phosphate in mitochondria, uptake of substrates and formation of ion gradients in mitochondria and bacteria), motion (flagellar motion in bacteria, muscle contraction in specialised animal cells), *etc.* (ref. 1).

ATP production driven by an electrochemical membrane potential generated by an electron transfer chain must be regarded as one of the essential reactions of living systems. This is indicated, *e.g.*, by the universal nature of the ATPases which convert the membrane potential (proton or sodium motive force) into chemical energy, ATP, and by the fact that for a number of bacteria electron transfer is the only way of driving ATP synthesis since these organisms do not possess any substrate chain phosphorylation system.

The electron transfer complexes of the mitochondrial and of the bacterial electron transfer chains contain similar redox centers and perform the same reactions. However, the bacterial complexes have a simpler protein composition and molecular biological studies of these complexes are easier in bacteria since bacteria usually code for all protein subunits of each electron transfer complex within one operon. Eucaryotic complexes contain a larger number of subunits, some of which are mitochondrially encoded, and the other subunits are encoded on different chromosomes. In photosynthetic bacteria like *Rhodobacter*, light driven electron transfer from the photosystem (reaction center) through the electron transfer chain has allowed the measurement of fast electron transfer kinetics. Therefore, some of our present knowledge of the mitochondrial respiratory chain has been deduced from the study of bacterial electron transfer chains and from comparative studies of the mitochondrial, bacterial, and chloroplast systems (ref. 2).

General Concepts and Mechanisms

The general reaction catalyzed by the mitochondrial respiratory chain is the oxidation of a hydrogen donor, *e.g.*, NADH, and the reduction of dioxygen to water (see Fig. 1). This reaction involves the net transfer of two hydrogen atoms, *i.e.*, two electrons and two protons, per oxygen atom reduced. In catalyzing what is essentially the oxyhydrogen reaction, the respiratory chain performs two main functions: (i) control of the reaction, and (ii) utilization of the energy released in the redox reaction.

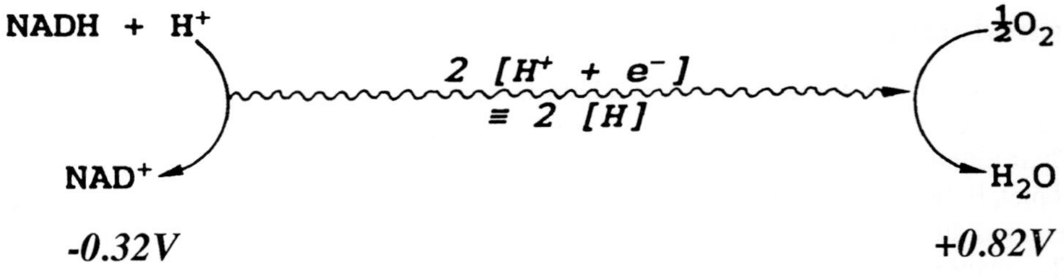

NADH + H^+ → NAD^+

$2\,[H^+ + e^-] \equiv 2\,[H]$

$\frac{1}{2}O_2$ → H_2O

-0.32V *+0.82V*

Fig. 1. General reaction catalyzed by the mitochondrial respiratory chain. The numbers indicate the redox potentials of NADH and oxygen, respectively.

(i) Control.

In order to achieve control, nature has developed an *electron transfer chain* by splitting the reaction into three steps and by introducing two additional electron carriers with *pool function*, *i.e.*, ubiquinone and cytochrome *c*. The electron transfer components are organized with respect to the inner mitochondrial membrane: ubiquinone and oxygen react within the membrane, NADH and cytochrome *c* on opposing sides of the

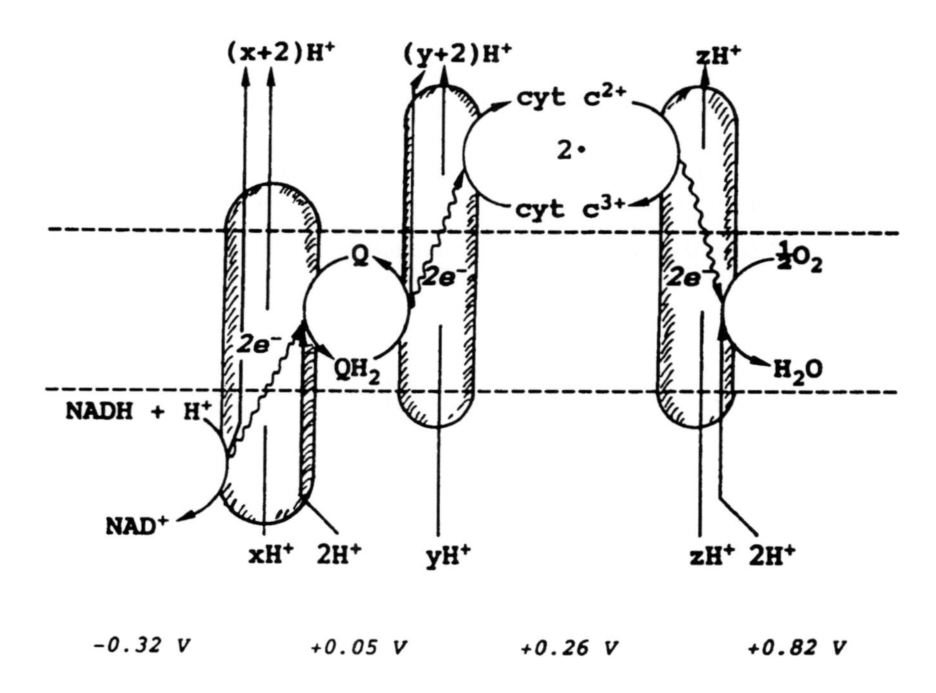

Fig. 2. Scheme of the mitochondrial respiratory chain. The dashed lines indicate the inner mitochondrial membrane; the matrix (inside) is on the bottom, the intermembrane space (outside) on top. The ellipsoids represent electron transfer complexes I, III and IV which catalyze electron transfer between the mobile electron carriers and concomitant proton translocation. x, y and z are the number of protons translocated by complexes I, III and IV, respectively, in addition to the 2 redox loops (see text). The redox potentials of ubiquinone (Q) and cytochrome *c* (cyt c) are shown. '2·' indicates that two molecules of cytochrome *c* are required for transfer of two electrons.

membrane: NADH on the inside (matrix side) and cytochrome *c* on the outside (intermembrane space between inner and outer membrane).

Electron transfer between these components is mediated by three *electron transfer complexes* (Fig. 2):

- Complex I (NADH:ubiquinone oxidoreductase)
- Complex III (ubihydroquinone:cytochrome *c* oxidoreductase, also called *bc1* complex, according to its cytochrome components)
- Complex IV (cytochrome *c*:O_2 oxidoreductase or cytochrome oxidase, the *Atmungsferment* discovered by Otto Warburg).

These electron transfer complexes are all *multiprotein complexes* consisting of from 9 up to more than 30 subunits each, depending on the nature of the complex and of the organism (refs. 3-4). The complexes are embedded within the membrane and have domains exposed to both aqueous phases. As membrane proteins, their activity depends on the phospholipid environment (ref. 5).

Complex II (succinate:ubiquinone oxidoreductase) serves as an additional input source by transferring 2 hydrogens at a higher redox potential level from succinate to ubiquinone.

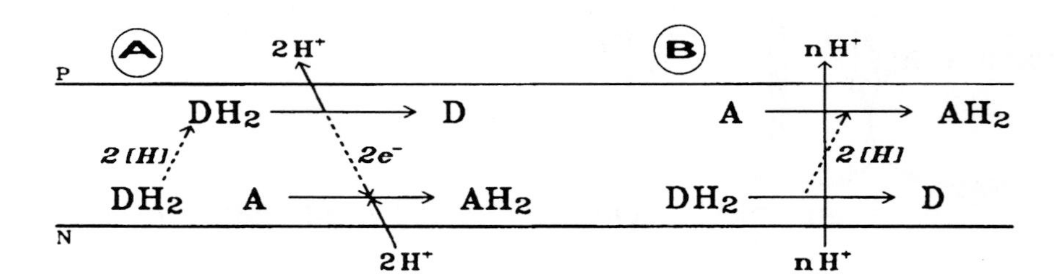

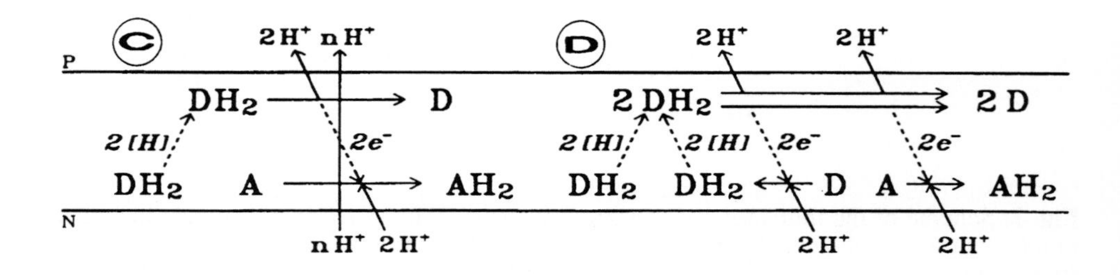

Fig. 3. Mechanisms of proton translocation. (A) Redox loop. (**B**) Redox pump. (**C**) Redox loop + pump. (**D**) Redox cycle. **P** and **N** indicate the outside (positive) and inside (negative) of the inner mitochondrial membrane. **DH_2**: hydrogen donor; **A**: hydrogen acceptor . In **B**, the direction of the hydrogen transfer (not of proton translocation!) is arbitrary.

(ii) Energy conservation.

The redox energy released in the reaction is conserved by transformation into an electrochemical gradient of protons across the inner mitochondrial membrane, *i.e.*, the *protonmotive* force (see Fig. 2), (ref. 6). The *electroneutral* transfer of hydrogens from the respective electron donor to the respective electron acceptor is linked to a *charge translocation* across the membrane, creating a *membrane potential* (positive outside) as well as a chemical *concentration gradient* of protons (higher proton concentration outside).

Three different types of mechanisms have been conceived for this *redox–reaction coupled charge translocation*: (a) redox loops (see Fig. 3A), (b) proton pumps (see Figs. 3B and 3C), and (c) redox cycles (see Fig. 3D), (ref. 7).

(a) Redox loops (Fig. 3A)

The concept of the *redox loop* involves a separation of the pathways of electrons and protons and the organization of the redox transfer components across the membrane so that they are in electrical contact with different sides of the membrane (ref. 8). On oxidation of the electron donor, $\mathbf{DH_2}$, two protons are released to the P–side of the membrane (outside, positively charged). Subsequently, two electrons are transferred to the acceptor reduction site and the reduction of the acceptor, **A**, is coupled to proton uptake from the N–side (inside, negatively charged). The reduced acceptor, $\mathbf{AH_2}$, may then serve as donor in the next redox loop. The loop, therefore, consists of two transmembrane transfers: two *hydrogens* are translocated from the N– to the P–side and 2 *electrons* are transferred back. By this mechanism, a constant $H^+/2e^-$ ratio of 2 (two protons translocated per two electrons transferred from DH_2 to A) is achieved.

In the mitochondrial respiratory chain, the nature of the electron transfer components and their organization within the inner mitochondrial membrane suggests that two redox loops exist, accounting for the translocation of 4 $H^+/2e^-$ transferred from NADH to oxygen: one redox loop between NADH and ubiquinone, and the second between ubiquinone and oxygen. However, since the total number of protons translocated in the respiratory chain is 10-12 $H^+/2e^-$, this mechanism is not sufficient to explain the observed $H^+/2e^-$ ratio (ref. 9). Therefore, the two other types of mechanisms are supposed to operate in addition to the redox loop mechanism.

(b) Proton pumps (Figs. 3B and 3C)

The *proton pump* mechanism involves conformational transitions of the protein coupled to the redox reaction in such a way that a protonatable group of the protein is exposed to different sides of the membrane during the reaction cycle. This mechanism which has been demonstrated, *e.g.*, for bacteriorhodopsin (ref. 10) has for a long time been considered to be the most likely mechanism of proton translocation of complex III due to the *redox Bohr effect* displayed by cytochrome *b*, *i.e.*, the redox–dependent shift of the pK of a protonatable group which is a prerequisite for this type of mechanism. According to this model (see Fig. 4), proton translocation may be achieved by a concerted series of reactions involving reduction/oxidation (1,3), protonation/deprotonation at different Pk values (4,6), and conformational changes (2,5), (ref. 11).

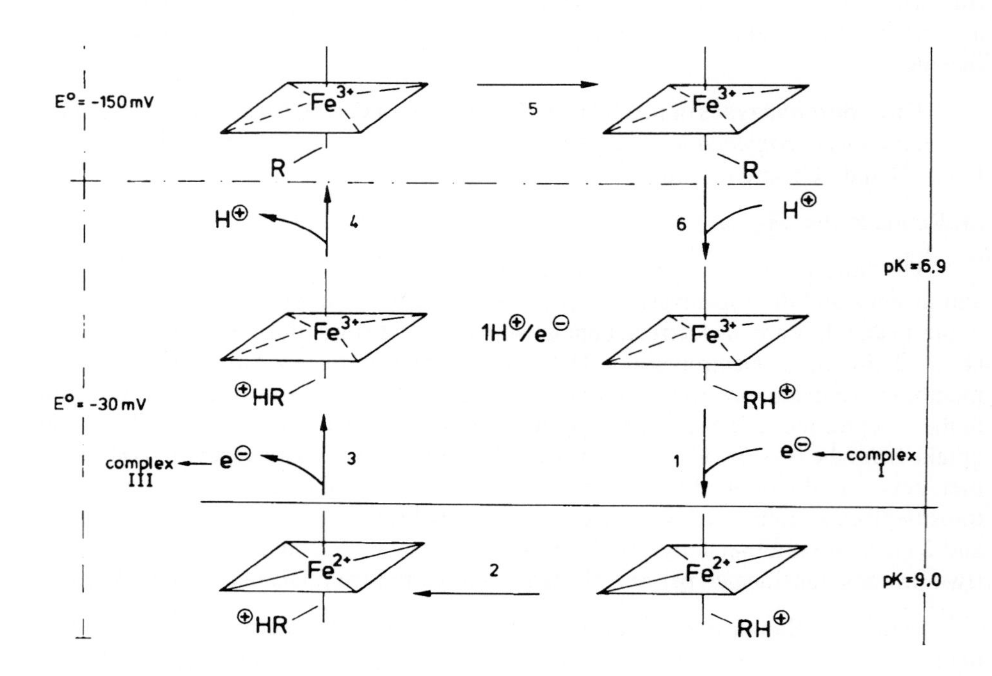

Fig. 4. Hypothetic model of cytochrome *b* as a proton pump from (ref. 11).

The proton pump mechanism may operate as sole mechanism of proton translocation (Fig. 3B) or in combination with the redox loop mechanism (Fig. 3C).

(c) Redox cycles (Fig. 3D)

The alternative mechanism to the proton pump is the redox cycle in which the efficiency of proton translocation is increased by cycling part of the electrons back to the electron donor. This mechanism is now accepted by most workers in the field to be operating in the cytochrome bc_1–Fe/S region of the respiratory chain, in the form of the *Q* (quinone) *cycle* proposed by P. Mitchell in 1975 (ref. 12).

Like the redox loop mechanism, the redox cycle requires a transmembrane arrangement of the oxidation and reduction sites. During oxidation of the electron donor, $\mathbf{DH_2}$, one electron goes to the electron acceptor, **A**, the other electron to an oxidized donor molecule serving as additional electron acceptor, and two protons are released to the P–side. A second donor molecule, $\mathbf{DH_2}$, is then oxidized and again two protons are released to the P–side. Thereby, the acceptor and the half–reduced donor are fully reduced, both with the uptake of two protons each from the N–side. Since two molecules of $\mathbf{DH_2}$ are oxidized and one **D** is re–reduced, only one $\mathbf{DH_2}$ is effectively oxidized during the full cycle. Alltogether, four protons are transported per donor molecule being net oxidized, thus resulting in a $H^+/2e^-$ ratio of 4.

The mechanisms are presented in an idealised form as no electrogenic reactions during the overall reaction ($DH_2+A \rightarrow D+AH_2$) are considered. However, in the mitochondrial respiratory chain the situation is much more complex since an alternation between hydrogen donors/acceptors (NADH, flavins, quinone, and oxygen) and electron carriers (iron sulfur clusters, cytochromes) occur. Therefore, the reactions of the individual electron transfer complexes do not always result in an electroneutral scalar reaction. As an example, this shall be briefly discussed for the "Q cycle" of complex III. On transfer of two electrons from hydroquinone to cytochrome *c*, two "scalar" protons are released to the outside. The cycling of electrons *via* the *b* bus results in two additional "vectorial" protons being pumped. Therefore, complex III transfers two electrons with the uptake of two protons and the release of four protons. The additional two protons are taken up during the reduction of oxygen to water by complex IV which constitutes the second half of the redox loop reaction.

Proton pumping (see Fig. 3B) is generally coupled to the *electrogenic* transfer of electrons rather than to the *electroneutral* transfer of hydrogens, *i.e.*, to the reaction $D^-+A^+ \rightarrow D+A$. Different types of pumping mechanisms, *electrophoretic* (parallel) and *electrogenic* (antiparallel), have therefore been developed. These aspects have been discussed in great detail by Mitchell (ref. 13).

Electron Transfer Components and Reactions

The electron transfer components of the mitochondrial respiratory chain (see Fig. 5) can be divided into:

- Diffusible carriers with pool function
  - within the membrane: ubiquinone
  - in the aqueous phase: cytochrome *c*
- Prosthetic groups of the electron transfer complexes: flavins; iron sulfur centers; cytochromes *a*, a_3 , *b*, and c_1; copper atoms.

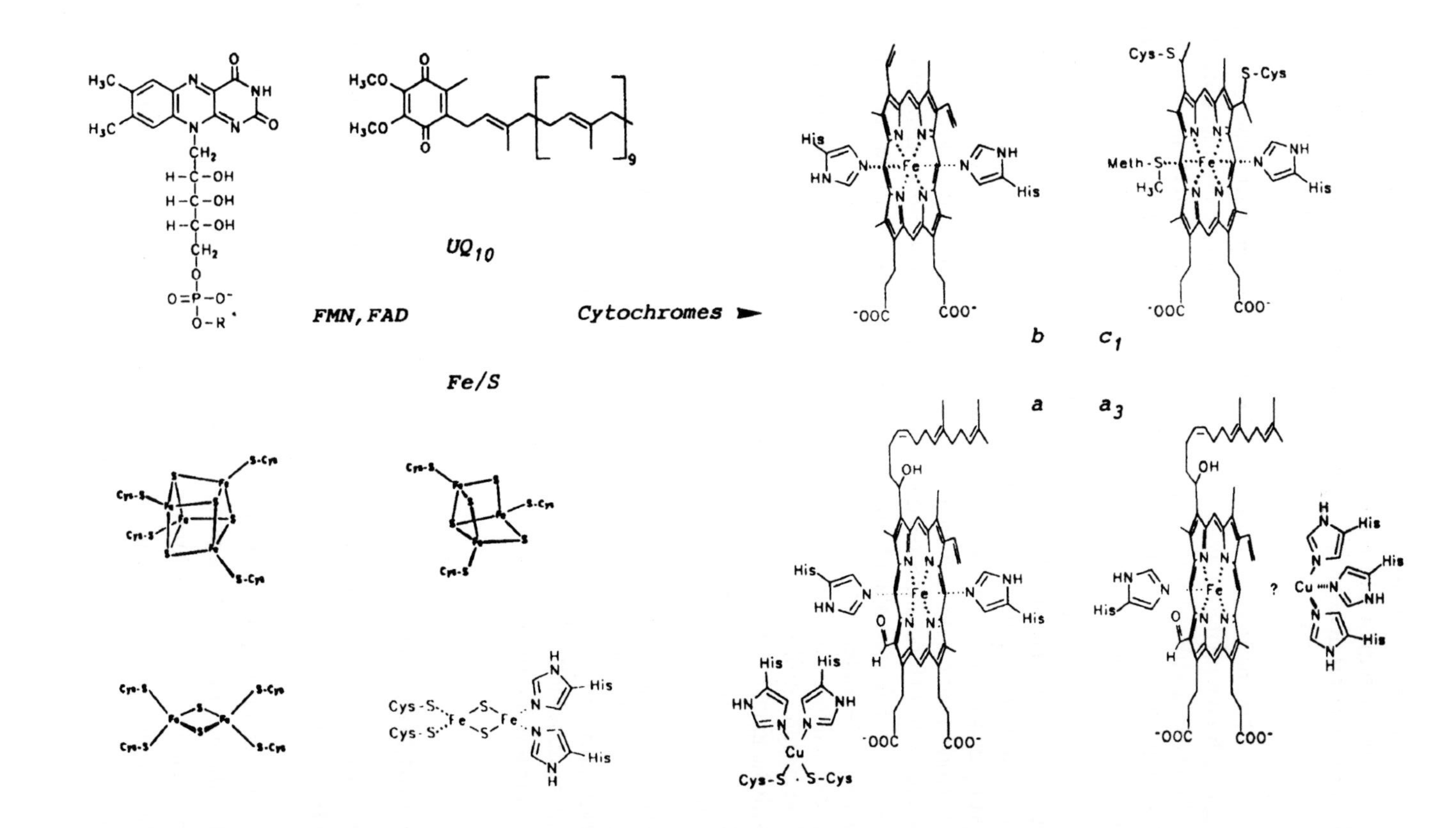

Fig. 5. Electron transfer components of the mitochondrial respiratory chain. FMN,FAD: flavins; UQ_{10}: ubiquinone–10; Fe/S: iron sulfur clusters.

TABLE 1

Electron transfer components of the mitochondrial respiratory chain.

Complex	Redox center	Redox potential (mV)	Reference
I	FMN	-320	(3)
	1 Fe_2S_2	-370	(14,15)
	1 Fe_2S_2	-245	
	1 Fe_4S_4	-140 .. -20	
	2 Fe_4S_4	-245	
	x Fe_2S_2	?	
II	FAD	-50	(36)
	1 Fe_2S_2	0	(16)
	1 Fe_3S_4 or Fe_4S_4	+65	
	1 Fe_4S_4	-270	
	cytochrome *b*	-200	
	ubiquinone	+45	(26)
III	cytochrome b_l	-50	(11)
	cytochrome b_h	+50	
	1 Fe_2S_2	+280	(37)
	cytochrome c_1	+250	(26)
	cytochrome *c*	+260	
IV	cytochrome *a*	+220	(38)
	cytochrome a_3	+380	
	Cu_A	+240	
	Cu_B	+340	

They can also be classified according to the species translocated into:
- Hydrogen carriers: flavins; ubiquinone
- Electron carriers: iron sulfur centers; cytochromes; copper atoms.

The electron transfer components are arranged in the electron transfer chain basicly according to their redox potential with the most negative components on the donor side (complex I) and the most positive components participating in oxygen reduction (complex IV). Table 1 gives a list of the components and of their redox potentials.

Flavins (FMN and FAD in complexes I and II, respectively) serve as entry sides for two hydrogens each coming from NADH and succinate, respectively. In addition, complexes I and II contain several iron sulfur clusters (Fe_2S_2, Fe_3S_4 and Fe_4S_4) as low potential electron carriers (refs. 14-16). Complex II does also contain a cytochrome *b* with, so far, unknown function.

The mobile electron carrier between complex I, II, and III is ubiquinone which is present in approximately 10–fold excess over complex III. Ubiquinone is highly hydrophobic due to its isoprenoid side chain comprising 6 (yeast) to 10 (mammals) isoprenoid units which correspond to 30-50 carbon atoms.

Complexes I and II each donate two electrons to ubiquinone. Reduction of ubiquinone is accompanied by proton uptake from the (inner) N–side. In complex I, electron transfer to quinone can be regarded as the second (acceptor) part of a redox loop when the protons coming from NADH are released *via* FMN to the (outer) P–side.

Complex III contains two electron transfer chains, one potential chain along the P–side of the membrane formed by a "Rieske" iron sulfur protein with a high redox potential (+280mV in beef heart mitochondria) and by cytochrome c_1 (+250mV), and a low potential chain formed by two heme *b* centers (heme b_l, -50 mV, and heme b_h, +50 mV) contained in a single apocytochrome *b* polypeptide (ref. 17). The two heme *b* centers of the low potential chain are arranged across the membrane to give a transmembrane electron transfer pathway essential for the *Q cycle* mechanism (see below) while the high potential chain transfers electrons along the P–side of the membrane to cytochrome *c* (ref. 18).

The "Rieske" Fe_2S_2 cluster differs from the plant ferredoxin type Fe_2S_2 clusters not only by its redox potential, but also by its optical and EPR spectra. Spectroscopy of "Rieske" type Fe_2S_2 clusters from bacterial dioxygenases has established that the iron sulfur cluster is not ligated by four cysteine residues (like in the plant ferredoxins) but rather by two cysteine and two histidine residues (ref. 19).

Cytochrome *b* type hemes are iron protoporphyrin IX systems bound only *via* two histidines as fifth and sixth ligands of the central iron atom. In contrast, *c*–type hemes are covalently attached to the apoprotein through two thioether bonds resulting from the addition of cysteine side chains to the two vinyl groups of the porphyrin ring. Heme ligands of cytochromes *c* and c_1 are one histidine and one methionine.

Since all six ligand positions of the iron atom are occupied in the *b*– and *c*– type cytochromes of the mitochondrial respiratory chain, these cytochromes transport one electron by reducing the central iron atom from the Fe^{3+} to the Fe^{2+} redox state and do not react with oxygen, cyanide, or carbon monoxide in their native state. Cytochrome *a* of cytochrome *c* oxidase (complex IV) also has a bis–histidine ligation (ref. 20) and is therefore also an electron carrier, accepting electrons from cytochrome *c* and donating them to a *binuclear center* formed by cytochrome a_3 and one of the copper atoms, Cu_B. Like heme *a*, the other copper atom, Cu_A acts as an internal electron carrier within cytochrome oxidase.

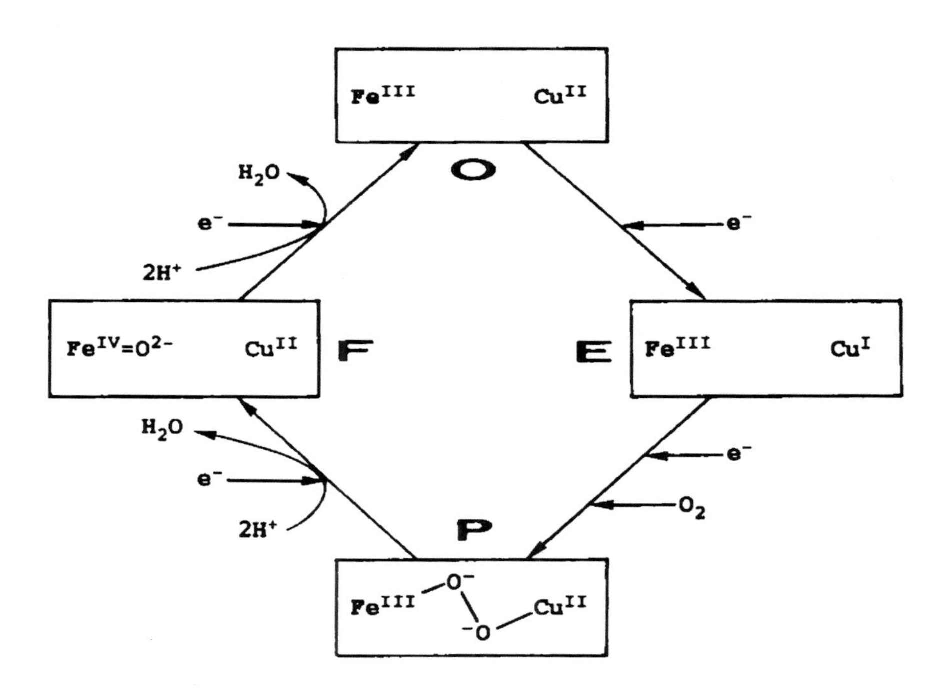

Fig. 6. Dioxygen reduction by the P/F cycle of cytochrome *c* oxidase, occurring at the binuclear center. Fe=heme a_3, Cu=Cu_B.

Heme a_3 has only one permanent ligand (histidine) with the sixth ligand position being open. In this direction, Cu_B is located at a distance of 3.8Å (ref. 21). Both heme a_3 and Cu_B can take up one electron each as well as an additional ligand. Between the two redox centers, oxygen is bound and reduced to two water molecules in four successive steps (see Fig. 6), (ref. 22). During each step, one electron is donated into the binuclear center by cytochrome *a* and/or Cu_A. Starting from the oxidized enzyme (**O**), addition of the first electron gives the electronated state (**E**) recently described by Moody *et al.*, (ref. 23). Addition of the second electron and binding of oxygen results in the transfer of two electrons from heme a_3 and Cu_B to oxygen, so that oxygen is bound in the form of the bridged peroxy (**P**) compound. Transfer of the third electron results in reduction of one oxygen atom to water with the uptake of two protons, the remaining oxygen being bound in the ferryl (**F**) state [$Fe(IV)=O^{2-}$]. Upon transfer of the fourth electron, the second oxygen atom is reduced and reacts with two protons to water, while the binuclear center returns to the oxidized (**O**) state.

Since two protons are taken up from the N–side (ref. 24), this reaction constitutes one half reaction of a proton motive redox loop. In addition, cytochrome oxidase pumps two additional protons per two electrons transferred to oxygen; the mechanism of this additional proton pumping is not known so far.

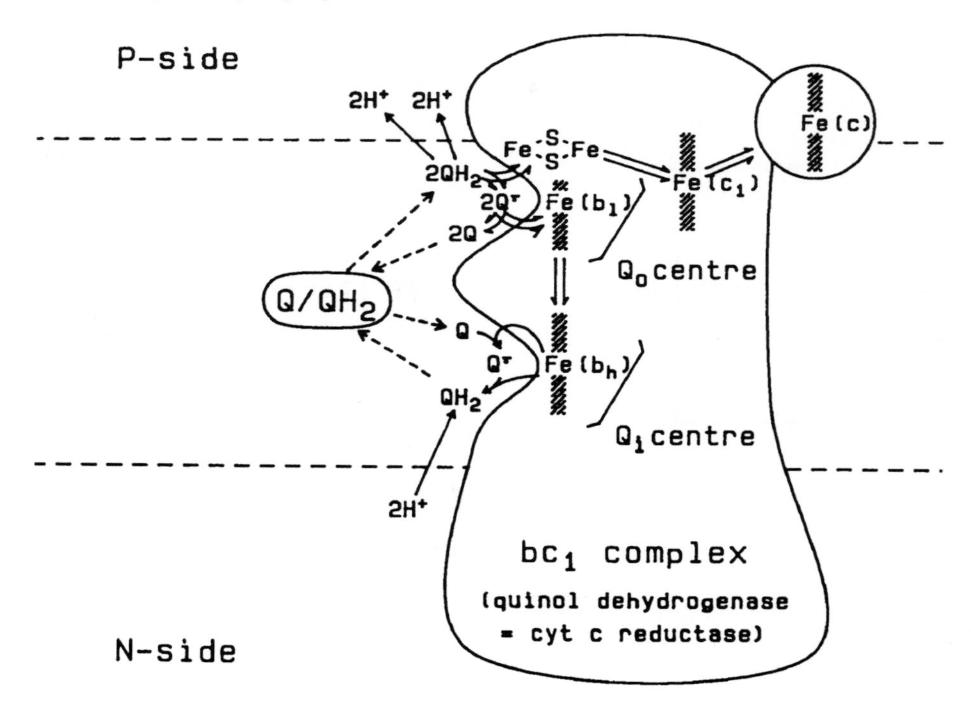

Fig. 7. "Q cycle" mechanism of the bc_1 complex.

The proton translocation of the bc_1 complex (complex III) occurs by a redox cycle mechanism, the so called *Q* (quinone) *cycle* proposed by P. Mitchell in 1976 (see Fig. 7), (ref. 12). The bc_1 complex contains two reaction centers for ubiquinone, one in contact with the outer (P–) side where ubihydroquinone oxidation takes place, *i.e.*, the $Q_o(Q_P)$ center, the other in contact with the inner (N–) side where quinone re–reduction takes place, *i.e.*, the Q_i (Q_N=Q_r) center (ref. 25). The Q_o center is close to heme b_l, the Q_i center close to heme b_h. On oxidation of ubihydroquinone at the Q_o center, the first electron is transferred into the "high potential chain", *i.e.*, onto the iron sulfur cluster and from there *via* cytochrome c_1 to cytochrome *c*. The second electron goes to heme b_l and from there across the membrane *via* heme b_h to a molecule of ubiquinone in the Q_i center forming a stabilized ubisemiquinone. Both protons are released from the Q_o center to the P–side. A second molecule of ubihydroquinone is oxidized at the Q_o center, the first electron again going to high potential chain and the second electron *via* cytochrome *b* to the Q_i center where it fully reduces ubisemiquinone to ubihydroquinone with the concomitant uptake of two protons from the N–side. Summing up, one ubihydroquinone is net oxidized during one full turnover (two oxidized and one reduced), two molecules of cytochrome *c* are reduced, four protons are released to the P–side and two protons are taken up from the N–side.

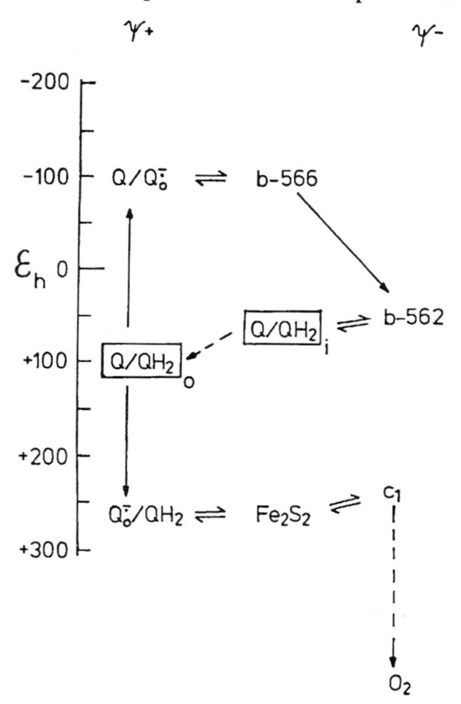

Fig. 8. Energetics of the "Q cycle" mechanism of the bc_1 complex. Indices o and i indicate reactions occurring at the outer (P) and inner (N) quinone reaction center, respectively. b–566≡b_l; b–562≡b_h.

This Q cycle mechanism serves several functions:

- Providing a mechanism for proton translocation not requiring a proton pumping device within the protein but rather using the protein as a framework for an organized sequence of chemical reactions;
- Doubling of the efficiency by cycling half of the electrons back to the electron donor (the $H^+/2e^-$ ratio is 4, compared to 2 for a redox loop);
- Switching from the *two hydrogen* (two electron plus two proton) carrier ubiquinone to the *one electron* carrier cytochrome *c*.

The endergonic reduction of ubiquinone by electrons coming from ubihydroquinone is driven by the exergonic transfer of electrons into the high potential chain. This involves a separation of the potentials of the redox couples $QH_2/Q^{-\cdot}$ and $Q^{-\cdot}/Q$ at the Q_o center (Fig. 8). In contrast, at the Q_i center both electrons are donated to the quinone by the same donor (heme b_h) at the same redox potential; $Q^{-\cdot}$ is therefore tightly bound and stabilized at the Q_i center and the redox potentials of the $QH_2/Q^{-\cdot}$ and $Q^{-\cdot}/Q$ couples are similar (ref. 26). In establishing this mechanism, specific inhibitors of either the Q_o or the Q_i center have been extremely valuable (ref. 27).

Structural Basis of Electron and Proton Transfer

The dependence of the functional properties of the electron transfer components and of the mechanistic details of the reaction on the structure of the protein complexes will be discussed for the bc_1 complex since this complex is best understood at present. The mitochondrial complex consists of 10 (fungi) or 11 (mammals) subunits, only three of which carry the redox centers (ref. 4). The subunits not containing redox centers are not present in the homologous bacterial bc_1 complexes. Some of these subunits contribute to the formation of the reaction or substrate (ubiquinone) binding sites; others seem to be essential for assembly of the complex from its protein precursors or for maintaining the structural integrity of the highly organized complex. Their function is not completely understood at present although genetic experiments (gene disruption) in yeast have shown that most of them are essential for function and/or assembly of the complex (ref. 28). The further discussion will focus on the three subunits containing the redox centers; however, the reader should bear in mind that these subunits do not exist (and are unstable) as isolated proteins but rather within the context of the multiprotein complex.

The three subunits carrying the four redox centers are conserved through all species:

- cytochrome *b*, a hydrophobic protein of approximately 380 amino acids (in mitochondria), containing both the b_l and b_h heme center;
- cytochrome c_1, containing covalently bound heme c_1;

- the "Rieske" iron sulfur protein containing a high potential Fe_2S_2 cluster with unusual ligation.

Cytochrome *b* contains nine hydrophobic helical domains (I–IX), eight of which form transmembrane α–helices (refs. 17, 29). The arrangement of cytochrome *b* has recently been established by gene fusion experiments and by sequencing of cytochrome *b* revertants. The two *b* hemes are bound by four histidine residues which are present as two pairs each 13 residues apart in helices II and V with an iron–iron distance of about 20Å. Conserved arginine residues at the ends of the helices may assist in binding of the hemes by interaction with their propionic acid side chains. The two hemes form the transmembrane electron path essential for the *Q cycle* mechanism.

Although being chemically identical species, *i.e.*, bis–histidine coordinated protoporphyrins IX, the two hemes differ in their redox potentials, optical and EPR spectra (Table 2).

TABLE 2

Properties of beef heart cytochrome *b.*

(G. von Jagow and S. Albracht, unpublished; cf. ref. 31).

		native	**+SDS**
absorbance maximum	b_l	566 nm	560 nm
	b_h	562 nm	560 nm
EPR (g_z)	b_l	3.8	2.9
	b_h	3.4	2.9
Redox potential	b_l	-50 mV	-240 mV
	b_h	+50 mV	-140 mV

These properties all depend on the *"steric strain"* exerted by the protein environment (ref. 30); on addition of small concentrations of the denaturing detergent, SDS, the strain is released and the spectral properties are changed (G. von Jagow, unpublished; Table 2), (ref. 31). The most direct explanation for these changes of the EPR–spectra is a change of the dihedral angle of the histidines coordinating the hemes (Fig. 9); the observed g values > 3.5 indicate a dihedral angle of almost 90° corresponding to a highly strained state, while a low angle corresponds to an almost parallel orientation of the histidines (ref. 32). A similar transition can be observed in chloroplast cytochrome b_{559} (ref. 33). The model for the "catalytic core" of cytochrome *b*, *i.e.*, helices I–V with the two *b* hemes bound in between is that of a "4–α–helical bundle"

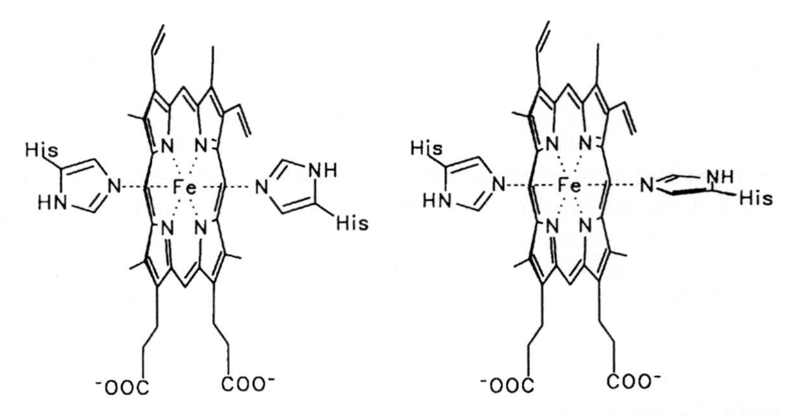

Fig. 9. Possible orientations of the histidine ligands of the *b* hemes: left: parallel, relaxed, corresponding to g_z=2.9; right: 90°, strained, corresponding to g_z > 3.5 (ref. 32).

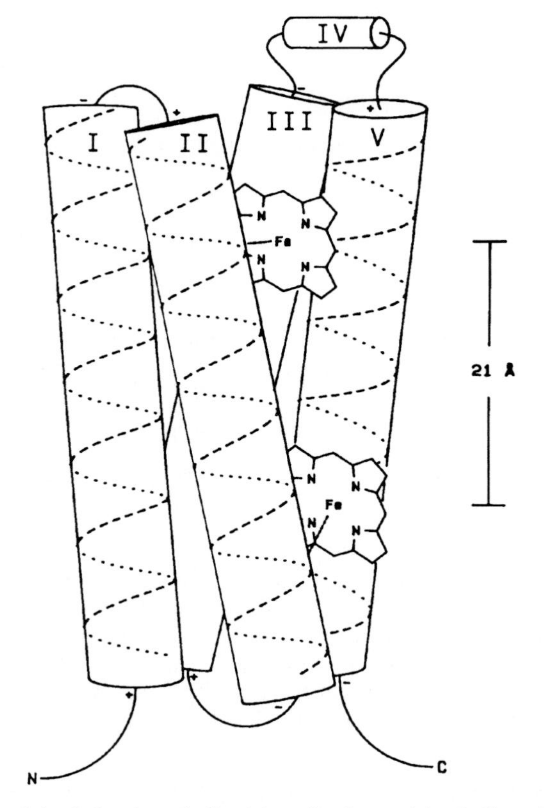

Fig. 10. Proposed model of the 4–α–helical bundle formed by helices I, II, III and V of cytochrome *b*. The model has been modified from ref. 31 by omission of helix IV from the membrane. The net dipole charges of the helices are indicated by + and -, respectively.

(Fig. 10) containing four transmembrane helices in a twisted bundle stabilized by electrostatic interactions between the helices and by interaction with other protein subunits (ref. 31). Through the steric strain exerted by twisted helical bundle the protein modifies and controls the properties of the redox centers.

Cytochrome c_1 transfers one electron to the water soluble electron carrier cytochrome *c* and, therefore, contains the heme center bound in a hydrophilic protein domain. This domain is bound to the membrane and to the rest of the complex by a hydrophobic membrane anchor at the C–terminus of the protein (Fig. 11). Cytochrome c_1 has no detectable homology to cytochrome *c* except the heme binding pentapeptide

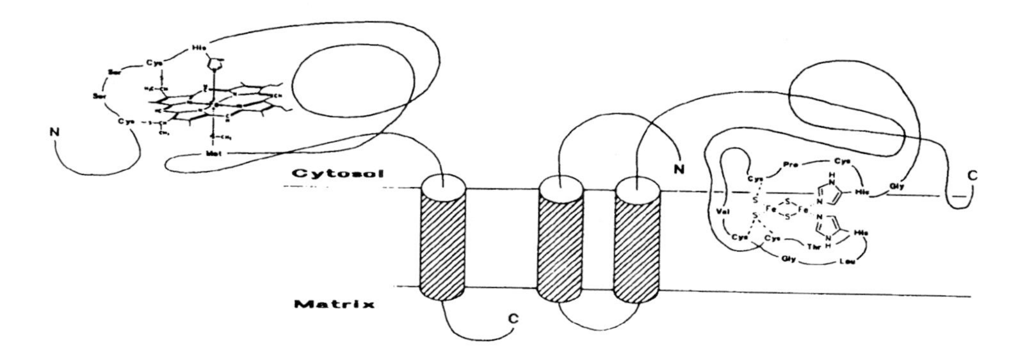

Fig. 11. Predicted folding patterns of cytochrome c_1 and the "Rieske" iron sulfur protein in the inner mitochondrial membrane. The cylinders represent transmembrane α–helical segments. Since the ligands of the Fe_2S_2 cluster are not known, three possible cysteine ligands are indicated by dashed lines, two of which bind the cluster.

Cys–X–X–Cys–His; the two cysteine residues form thioether linkages with the porphyrin side chains, while histidine is the fifth ligand to the iron atom. The sixth ligand position is occupied by a methionine from a distant part of the protein.

The "Rieske" iron sulfur protein forms part of the Q_o reaction center together with cytochrome *b* and transfers one electron from hydroquinone to the hydrophilic cytochrome c_1 head. A comparison of the known sequences from mitochondria, bacteria, and chloroplasts reveals two strongly conserved stretches near the C–terminus of the protein (ref. 34). These two stretches contain the ligands of the iron sulfur cluster which is bound between a hydrophobic and an amphipathic part of the protein.

The conserved residues include four cysteine residues which were initially thought to be the ligands of the iron sulfur cluster, like in the Fe_2S_2 clusters of the plant ferredoxins. However, spectroscopic measurements showed that the cluster was bound by one or two non–cysteine ligands (ref. 35). By ^{15}N ENDOR of ^{15}N–histidine enriched phthalate dioxygenase of *Pseudomonas cepacia* containing a "Rieske" type Fe_2S_2 cluster with identical spectroscopical properties, Gurbiel *et al.* have recently demonstrated that two histidine residues bind to one iron (ref. 19). The two conserved stretches contain besides the four conserved cysteines two conserved histidines; two of the cysteines and the two histidines are, therefore, the ligands of the iron sulfur cluster (Fig. 12).

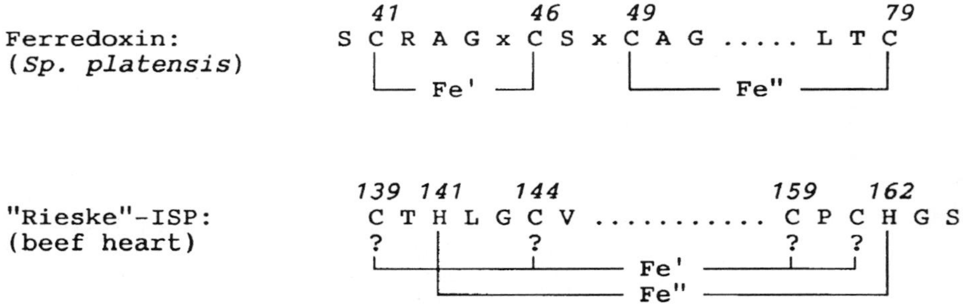

```
                              41          46     49                    79
Ferredoxin:               S C R A G x C S x C A G . . . . . L T C
(Sp. platensis)             |__  Fe'  __|     |______  Fe"  ______|

                          139 141     144                  159    162
"Rieske"-ISP:             C T H L G C V . . . . . . . . . . C P C H G S
(beef heart)              ?   |       ?                     ?   ?  |
                          |___|_______|______ Fe' _________|___|  |
                              |______________ Fe" ________________|
```

Fig. 12. Comparison of the Fe_2S_2 cluster binding sequences of the ferredoxins (top), and of the "Rieske" iron sulfur protein (bottom). In the "Rieske" protein, the two histidines bind to one iron while two of the four cysteines which are conserved within all species bind to the other iron atom.

The different ligation of the ferredoxin and the "Rieske" type Fe_2S_2 clusters corresponds to the observed differences, *i.e.*, of the redox potentials (-420 and +280 mV, respectively), EPR g–values, *etc.*

Except for the partially hydrophobic environment of the iron sulfur cluster, the major part of the iron sulfur protein is hydrophilic. In the N–terminal part, the protein contains the membrane anchor: a hydrophobic domain of approximately 40 residues is predicted to form two α–helices with a hairpin structure (Fig. 11).

The protein structure of the *bc₁* complex thus controls the properties of the redox centers as well as the topology of the electron transfer reactions; the low potential electron pathway across the membrane is formed by a single hydrophobic membrane protein comprising two heme *b* centers in a transmembrane arrangement while the high potential pathway along the P–side of the membrane is formed by two hydrophilic proteins bound to the complex by hydrophobic membrane anchors.

Summary and Outlook

The multiprotein complexes of the mitochondrial respiratory chain catalyze the transfer of reducing equivalents from substrates (NADH, succinate) to molecular oxygen; this reaction is coupled to the electrogenic translocation of protons across the inner mitochondrial membrane. Different mechanistic models for the energy conservation reactions have been developed: ligand conduction (mobile carrier, redox loops, and redox cycles) and primary proton pumping (mobile barrier). Both types of mechanisms seem to operate within the respiratory chain. The best understood enzyme is the bc_1 complex (complex III), where a ligand conduction mechanism ("Q cycle") is well established.

The complexity of the reactions (hydrogen, electron, and proton transfers with different donors and acceptors) is paralleled by the complexity of the enzyme systems. All enzyme complexes are multiprotein complexes and each contains several electron transfer components. The properties of these redox centers are controlled by the (largely hydrophobic) protein environment; they are arranged within the enzyme complexes to form pathways for electrons and protons, from specific acceptor to specific donor sites, either across the membrane or along its surface.

A large body of knowledge has been obtained by classical enzymology, applying protein chemistry, spectroscopical and kinetic methods. Recently, modern molecular biological methods have offered new approaches. At present, these methods are combined and enzymological studies of genetically engineered systems will provide further insight in the near future. However, a true molecular understanding will require high–resolution structural information, which is not available so far since crystallization experiments have not yet yielded satisfactory results. Therefore, crystal structures of respiratory chain complexes will have similar impact on their understanding as the structure of the bacterial reaction center had on the understanding of photosynthesis.

Acknowledgements

I thank Prof. Dr. G. von Jagow for his consistent support during the last years and for critically reading the manuscript.

References

1 S. Soboll, T.A. Link and G. von Jagow, in: D. Häussinger (Ed.) *pH Homeostasis – Mechanisms and Control* Academic Press, 1988, p. 97.

2 D.B. Knaff, Trends Biochem. Sci. **15** (1990) 289.

3 Y. Hatefi, C.I. Ragan and Y.M. Galante, in: A.N. Martonosi (Ed.) *The Enzymes of Biological Membranes* Plenum Press, New York/London, 2nd Edit., Vol. 4, 1985, p. 1.

4 H. Schägger, T.A. Link, W.D. Engel and G. von Jagow, Methods Enzymol. **126** (1986) 224.

5 H. Schägger, Th. Hagen, B. Roth, U. Brandt, T.A. Link and G. von Jagow, Eur. J. Biochem. **190** (1990) 123.

6 P. Mitchell, Nature **191** (1961) 144.

7 P. Mitchell, J. Theor. Biol. **62** (1976) 327.

8 P. Mitchell, Biol. Rev. **41** (1966) 445.

9 C.D. Stoner, J. Biol. Chem. **262** (1987) 10445.

10 M. Engelhard, K. Gerwert, B. Hess, W. Kreutz and F. Siebert, Biochemistry **24** (1985) 400.

11 G. von Jagow, W.D. Engel, H. Schägger, W. Machleidt and I. Machleidt, in: F. Palmieri, E. Quagliariello, N. Siliprandi and E.C. Slater (Eds.) *Vectorial Reactions in Electron and Ion Transport in Mitochondria and Bacteria* Elsevier, Amsterdam, 1981, p. 149.

12 P. Mitchell FEBS Lett. **59** (1975) 137.

13 P. Mitchell, in: C.H. Kim, H. Tedeschi, J.J. Diwan and J.C. Salerno (Eds.) *Advances in Membrane Biochemistry and Bioenergetics* Plenum Press, New York/London, 1988, p. 25.

14 H. Beinert and S.P.J. Albracht, Biochim. Biophys. Acta **683** (1982) 245.

15 T. Ohnishi and J.C. Salerno, in: T.G. Spiro (Ed.) *Iron-sulfur Proteins* Vol. 4, John Wiley & Sons, New York, 1982, p. 285.

16 T. Ohnishi, Current Top. Bioenerg. **15** (1987) 37.

17 W.R. Widger, W.A. Cramer, R.G. Herrmann and A. Trebst, Proc. Natl. Acad. Sci. USA **81** (1984) 674.

18 T. Ohnishi, H. Schägger, S.W. Meinhardt, R. LoBrutto, T.A. Link and G. von Jagow, J. Biol. Chem. **264** (1989) 735.

19 R.J. Gurbiel, C.J. Batie, M. Sivaraja, A.E. True, J.A. Fee, B.M. Hoffmann and D.P. Ballou, J. Biol. Chem. **28** (1989) 4861.

20 C.T. Martin, C.P. Scholes and S.I. Chan, J. Biol. Chem. **260** (1985) 2857.

21 L. Powers, B. Chance, Y.–C. Ching and P. Angiolillo, Biophys. J. **34** (1981) 465.

22 M.K.F. Wikström, Proc.Natl. Acad. Sci. USA **78** (1981) 4051.

23 A.J. Moody, U. Brandt and P.R. Rich, in: IUB-IUPAB Bioenergetics Group (Ed.) *EBEC Reports* Elsevier, Amsterdam, Vol. 6, 1990, p. 22.

24 M.K.F. Wikström, FEBS Lett. **231** (1988) 247.

25 G. von Jagow, T.A. Link and T. Ohnishi, J. Bioenerg. Biomembr. **18** (1986) 157.

26 P.R. Rich, Biochim. Biophys. Acta **768** (1984) 53.

27 G. von Jagow and T.A. Link, Methods Enzymol. **126** (1986) 253.

28 S. de Vries and C.A.M. Marres, Biochim. Biophys. Acta **895** (1987) 205.

29 A.R. Crofts, H. Robinson, K. Andrews, S. van Doren and E.A. Berry, in: S. Papa, B. Chance and L. Ernster (Eds.) *Cytochrome Systems: Molecular Biology and Bioenergetics* Plenum Press, New York/London, 1987, p. 617.

30 K.R. Carter, A.–l. T'sai and G. Palmer, FEBS Lett. **132** (1981) 243.

31 T.A. Link, H. Schägger and G. von Jagow, FEBS Lett. **204** (1986) p. 9.

32 G. Palmer, Biochem. Soc. Trans. **13** (1985) 548.

33 G.T. Babcock, W.R. Widger, W.A. Cramer, W.A. Oertling and J.G. Metz, Biochemistry **24** (1985) 3638.

34 H. Schägger, U. Borchart, W. Machleidt, T.A. Link and G. von Jagow, FEBS Lett. **219** (1987) 161.

35 J. Telser, B.M. Hoffmann, R. LoBrutto, T. Ohnishi, A.–l. T'sai, D. Simpkin and G. Palmer, FEBS Lett. **214** (1987) 117.

36 A. Tzagoloff *Mitochondria* Plenum Press, New York/London, 1982.

37 G. von Jagow and T. Ohnishi, FEBS Lett. **185** (1985) 311.

38 M.K.F. Wikström and M. Saraste, in: L. Ernster (Ed.) *Bioenergetics* Elsevier, Amsterdam, 1984, p. 49.

Electron and Proton Transfer in Chemistry and Biology, edited by A. Müller et al.
Studies in Physical and Theoretical Chemistry, Vol. 78

Coupled Proton and Electron Transfer Pathways in the Acceptor Quinone Complex of Reaction Centers from Rhodobacter Sphaeroides

Eiji Takahashi[1], Pétér Maróti[3] and Colin A. Wraight[1, 2]

[1]Department of Plant Biology and
[2]Department of Physiology and Biophysics
University of Illinois
Urbana, IL 61801 (U.S.A.)
and
[3]Department of Biophysics
Joszef Attila University
Szeged (Hungary)

Summary

The events leading to the reduction of the primary and secondary acceptor quinones, Q_A and Q_B , in reaction centers from the photosynthetic bacterium, *Rhodobacter spheroides*, are described. When Q_A and Q_B are reduced to the semiquinone state, several ionizable groups in the protein undergo pK shifts resulting in the uptake of protons over wide pH range. The kinetics of proton uptake in response to Q_A- are rapid but they are not diffusion limited. It is speculated that rate limiting processes occur in the protein, possibly involving conformational changes that control the accessibility of buried residues that account for most of the proton binding stoichiometry. Electron transfer from Q_A- to Q_B is accompanied by readjustments of the charge compensating polarization of the protein and redistribution of the protonation state. The electron transfer and the associated proton uptake are biphasic, with the rates and relative amplitudes of fast and slow phases dependent on pH. This is interpreted in terms of a fast electron transfer equilibrium, dependent on the initial protonation state of the protein, followed by a slower phase of net proton uptake that facilitates a larger electron transfer equilibrium. When A_B- is reduced further to the quinol, the quinone headgroup, *per se*, must take protons — ultimately from the solution, through the protein matrix of the L and H subunits. Specific residues that might be crucial to the delivery of protons to Q_B were identified by inspection of the X-ray structure, at atomic resolution. Glutamate and aspartate at positions 212 and 213 of the L-subunit are two carboxylic acid residues close (5-6Å) to Q_B. These residues have been altered by site-directed mutagenesis to the non-ionizable amides, glutamine and asparagine, singly and together in a double mutant. The characterization of reaction centers from these mutants is described and it is concluded that the first proton delivered to Q_B, just before or after the second electron, comes from Asp^{L213}, while the second proton comes from Glu^{L212}, possibly *via* Asp^{L213}. The two residues interact electrostatically with each other and with the anionic states of Q_B, and Asp^- is suggested to contribute significantly to the anomalous pK of Glu^{L212}, which has a pK of about 9.5. Mutation of Asp^{L213} to aspargine causes an especially dramatic block in the functioning of the acceptor quinones causing a steady state inhibition of at least 10^4 fold. This mutational lesion is partially alleviated at lo pH and it is suggested that the protein matrix can be penetrated by aqueous protons to some extent. Glu^{L212} and Asp^{L213} are also good candidates for major contribution to the protonation accompanying the formation of the anionic semiquinones and are suitably buried to account for the anomalous kinetics of proton uptake.

Introduction

The reaction center complex (RC) of the purple photosynthetic bacterium *Rhodobacter sphaeroides* is composed of three polypeptide subunits L, M and H. The L and M subunits, which bind all the cofactors involved in photochemical charge separation, are encoded by *puf* L and *puf* M genes within the *puf* operon (ref. 1). Light activation of the RC initiates charge separation between a dimer of bacteriochlorophyll (special pair, P) and a bacteriopheophytin with a very high (close to 1) quantum yield. The charge–separated state is stabilized by reduction of a tightly bound primary quinone (Q_A) to semiquinone. Further stabilization occurs if the electron can be transferred to the secondary quinone. Q_B, which is loosely bound (ref. 2). Protonation state changes are also associated with the electron transfer events, especially the reduction of the quinones. Under normal circumstances P^+ is rereduced by secondary donors (c–type cytochromes *in vivo*) and a second, flash–induced turnover of the RC results in the double reduction of Q_B to quinol with the uptake of two H^+–ions (refs. 3-6). In a series of flashes, the two quinones function as a two–electron gate, exhibiting binary oscillations in the formation and disappearance of semiquinone and in proton binding and release of reducing equivalents (see Fig. 1):

$$\text{odd flashes:}\quad Q_AQ_B \xrightarrow{h\nu} Q_A^-Q_B \underset{(H^+)}{\overset{K_2}{\longleftrightarrow}} Q_AQ_B^-\,(H^+)$$

$$\text{even flashes:}\quad Q_AQ_B^-(H^+) \xrightarrow{h\nu} Q_A^-Q_B^-(H^+) \underset{(H^+)}{\overset{K_3}{\longleftrightarrow}} Q_AQ_BH_2 \overset{Q\ \ QH_2}{\longleftrightarrow} Q_AQ_B$$

Release of the fully protonated quinol and rebinding of a quinone from the membrane pool completes the turnover of activities at the reducing side of the RC. It is noteworthy that in many species the two quinones are chemically identical (ubiquinones in *Rb. sphaeroides*), but their physicochemical properties are very different, due to their interactions with the protein environment.

In isolated RCs from *Rb. sphaeroides*, there is no endogenous electron donor to rereduce P^+ and, unless an exogenous donor is added, the charge separation state recombines after each flash in 0.1 - 1 s, depending on the availability of functional Q_B. Recombination occurs exclusively *via* the $P^+Q_A{}^-$ state (refs. 2, 3 and 7):

$$PQ_AQ_B \underset{k_{QA}}{\overset{hv}{\rightleftarrows}} P^+Q_A^-Q_B \overset{K_e}{\longleftrightarrow} P^+Q_AQ_B^-$$

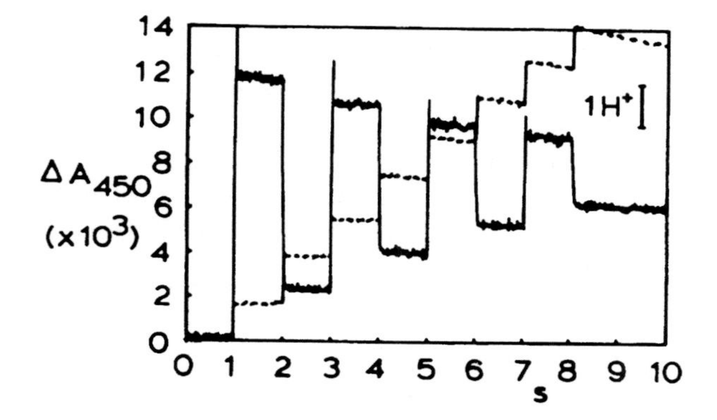

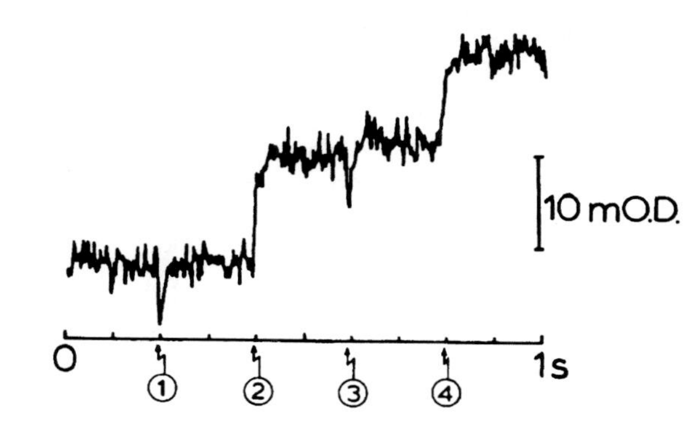

Fig. 1. Oscillatory characteristics of the two electron gate in reaction centers. Top: simultaneous measurement of semiquinone formation and disappearance at 450nm (——) and H^+ binding at 586nm, with chlorphenol red as indicator (----); pH 6.4. Bottom: quinol formation and release, measured at 288 nm; pH 7.4.

Thus, in the absence of electron donors to rereduce P^+, the observed rate of charge recombination with functional Q_B, is given by (refs. 3 and 7):

$$k_{QB} = k_{QA}(1 + K_e)^{-1} \tag{1}$$

and the recombination kinetics are a convenient assay of the Q_A-$Q_B \leftrightarrow Q_AQ_B$-equilibrium.

Stoichiometries of Proton Binding

As indicated in the scheme above, the one electron states (Q_A- and Q_B-) are accompanied by H^+–binding, but the protonation targets are protonatable residues of the protein rather than the semiquinones themselves, and both Q_A- and Q_B- exhibit clearly anionic absorption spectra in the visible region (refs. 3 and 8). Flash–induced proton–binding by the RC can be interpreted, crudely, by assuming a higher pK value for a protonatable group in the reduced than in the oxidized states of the quinones (ref. 3). However, recent proton–binding stoichiometry measurements showed that this simple picture needs to be extended: a minimum of four separate acid–base groups are involved in the proton uptake after a flash, and the exact stoichiometry depends on the protonation and redox states of the species P, as well as Q_A and Q_B (refs. 4-6). Thus, if no donor is present for P+ reduction, significant amounts of H^+–binding ($>0.3H^+/P^+$) to the semiquinone states of the RC were observed only in the alkaline pH range ($8<pH<11$) (refs. 4 and 5). This is shown in Fig. 2 for the states P^+Q_B- and PQ_B-.

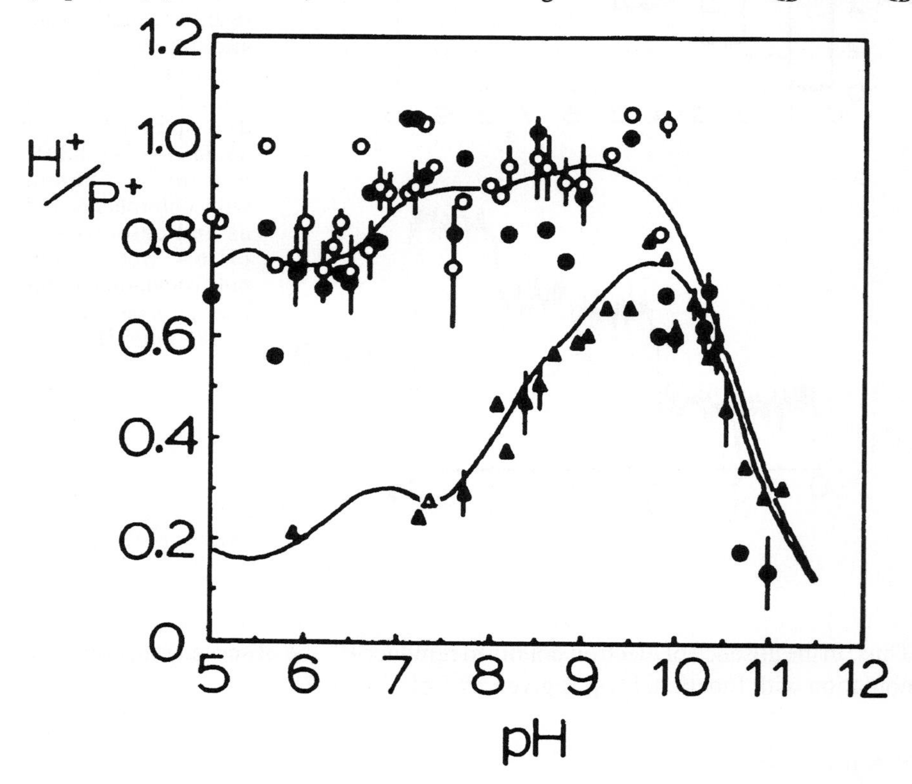

Fig. 2. Proton binding stoichiometries for reaction centers with functional Q_B activity. Circles (o,●): with external donors (various, but mostly ferrocene or cytochrome c), corresponding to binding to the PQ_B- state.

The pK values are sensitive to which quinone is charged (Q_A or Q_B) and whether the Q_B pocket is occupied — by quinone or by competitive inhibitors (ref. 5).

The concept of multiple protonation states considered in our work was elaborated by McPherson *et al.* (ref. 6) in a similar study, using their knowledge of the X–ray structure. They estimated the electrostatic interaction between the semiquinone charges and all available titratable amino acid residues, using Coulomb's law with a distance–dependent dielectric constant, and concluded that the number of protonatable groups contributing to the H^+ uptake in RC could be much larger than four, but with correspondingly smaller pK shifts.

Kinetics of Proton Binding

The P^+Q_A- State

The kinetics of H^+ binding by RCs are generally fast and can only be followed readily by spectroscopic methods using pH indicators. This makes the detection somewhat indirect as the signal arises from deprotonation of the indicator, which may be several steps removed from the original protonation of the RC. Assuming the rate of proton binding to be determined by H^+ or OH^- diffusion yields calculated bimolecular rate constants that are pH dependent and have improbably high values ($>10^{13}$ $M^{-1}s^{-1}$ at pH 10) (Fig. 3).

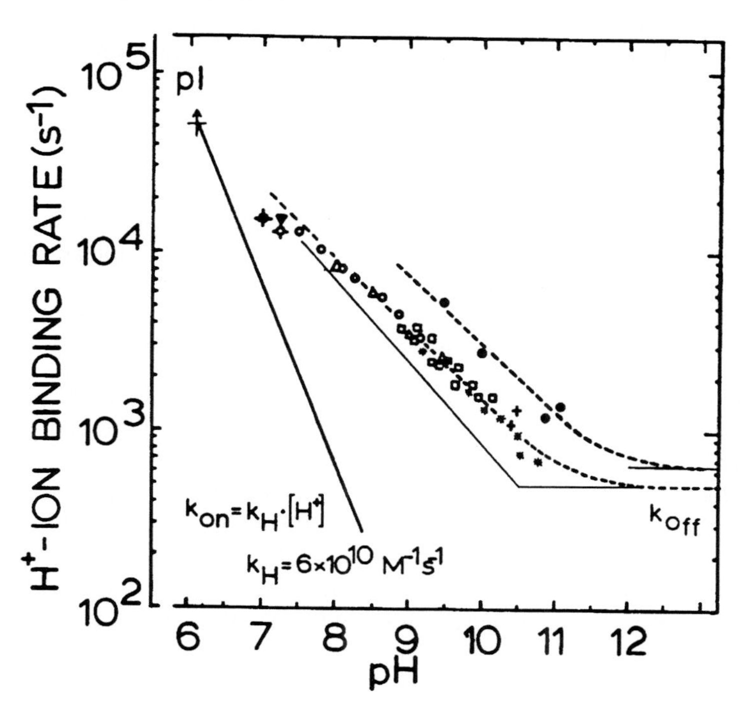

Fig. 3. pH–dependence of observed H^+ binding rates. ● -conductimetric method (ref. 4), low salt (10 mM buffer). Other symbols -○, Δ, ⊕, +, ✳, ▢, spectrophotometric assay, with various pH indicators, 100 mM NaCl. The heavy solid line indicates the expected behaviour for a diffusion controlled protonation reaction, with bimolecular rate constant, $k_H = 6 \cdot 10^{10}$ $M^{-1}s^{-1}$. It is drawn through the isoelectric point (pI) for the RC.

However, from the original work of Eigen and coworkers (ref. 9) and, more recently, the extensive studies by Gutman and colleagues (ref. 10), we know that the pathways of proton transfer and detection are diverse and the rates are determined by diffusion controlled reactions between various proton–carrier species, such as indicators and buffers. The kinetics are rarely limited by diffusion of H^+ ions (hydronium, *etc.*) and are generally complete in no more than a few microseconds. This is significantly faster than we observe for H^+ binding by RCs. In the case of proton binding to the P^+Q_A- state, no electron transfer event can be rate limiting (all steps are complete in one nanosecond), and we suppose that the limiting process is a collisional reaction between indicators bound (partially immobilized?) at the protein surface and proton acceptor groups that may only be accessible through conformational equilibria of the protein:

$$RC \xrightarrow[]{K} RC' \xrightarrow[]{k\cdot[HIn]} RC'H^+$$

The observed rate is then given by:

$$k^{obs} = k\ [HIn\}\ K\ [1 + K]^{-1} \qquad (2)$$

For small values of K, the free energy of the conformational equilibrium will contribute significantly to the apparent activation energy of the proton transfer. The measured temperature dependence of the overall reaction is large, equivalent to $E_a \geq 40$ kJ/mol, which is consistent with a non-diffusion controlled process. Similarly, the viscosity dependence is small.

The P^+Q_B- State

In the presence of Q_B, the coupling of protonation events to electron transfer from Q_A- to Q_B makes the kinetics of proton binding potentially more complex. Indeed, between pH 7 and 9.5, the kinetics of both electron transfer and proton uptake are generally biphasic (Fig. 4), each phase exhibiting pH dependent rates. The pH dependence of the electron transfer rates are shown in Fig. 5. Similar behavior is seen for the first and second flash, and we will discuss the behavior in terms of the first flash only.

The electron transfer equilibrium, Q_A-$Q_B \leftrightarrow Q_AQ_B$-, is well known to be pH dependent, being most favorable at low pH ($K > 10$) and becoming unfavorable at high pH ($K < 1$). The net rate of transfer exhibits a similar pH dependence. This is generally interpreted as implicating a group which, on ionization, presents an electrostatic inhibition to the electron transfer and which must be reprotonated in order to stabilize the electron on Q_B- (see below). The pH dependent regions of the fast phases of electron transfer and the associated H^+ binding, yield calculated rate constants greater than 10^{12} M^{-1} s^{-1} —a similar anomaly to that encountered for H^+ binding to the P^+Q_A- state. The electron transfer kinetics demonstrate that the arrival of the proton at some functional location within the RC occurs on the same time scale as pH indicators report the uptake from the external medium. The definite pH dependence of the electron transfer from Q_A- to Q_B, compared to the weak pH dependence of the rate of H^+ uptake to the P^+Q_A- state, implies the establishment of a pH dependent equilibrium that controls the electron transfer:

$$Q_A^- Q_B \underset{H^+}{\overset{K_{HA}}{\longleftrightarrow}} Q_A^- Q_B(H^+) \overset{K_e'}{\longleftrightarrow} Q_A Q_B^-(H^+)$$

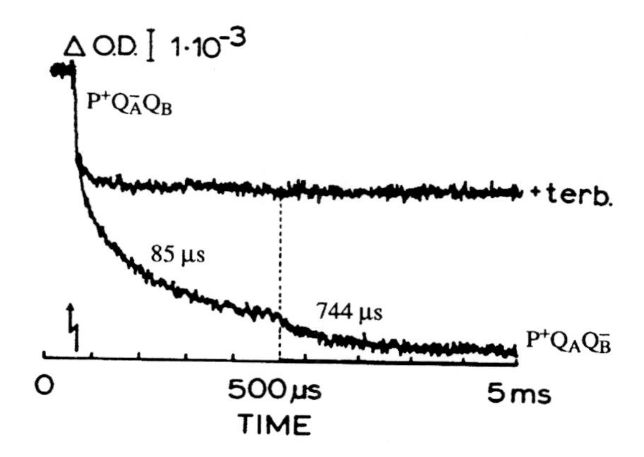

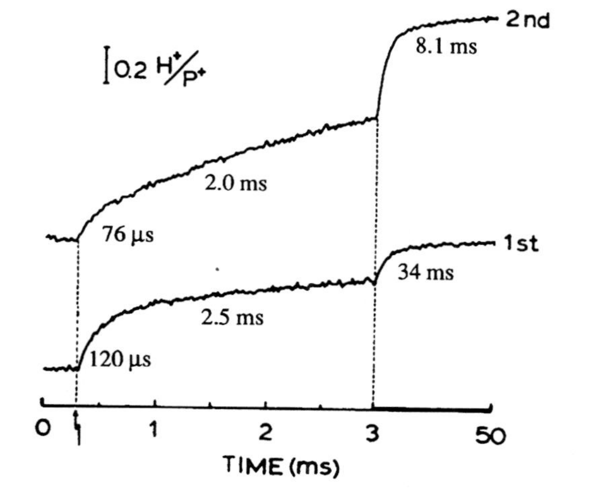

Fig. 4. Kinetics of electron and proton transfer in reaction centers with functional Q_B. Top: Q_A-Q_B $\rightarrow$ Q_AQ_B- electron transfer kinetics, measured at 397 nm; pH 8.0. (Terbutryn, present for the upper trace, is a potent inhibitor of Q_B activity). Note split time base. Bottom: proton binding kinetics after two consecutive flashes, with donor; pH 8.45, cresol red as indicator. Note split time base. The third, very slow phase of H^+ binding is associated with the rereduction of P^+ by exogenous donor (ferrocene).

If the first equilibrium is rapid, the observed rate of electron transfer is given by:

$$k_e^{obs} = ke' \{[H^+]/K_{HA}\}/\{1 + [H^+]K_{HA}\} \sim k_e' [H^+]/K_{HA} \qquad (3)$$

(when pH > pKA)

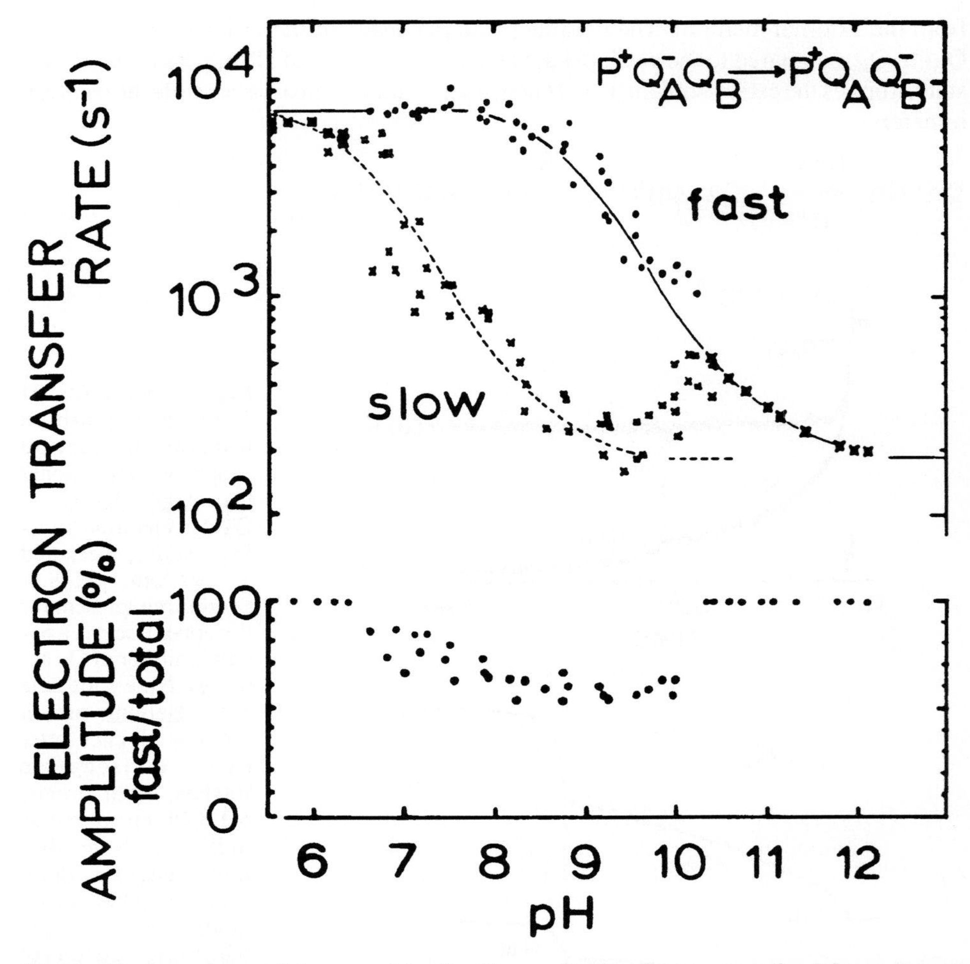

Fig. 5. pH–dependence of the rate of the Q_A-Q_B $\rightarrow$ Q_AQ_B- electron transfer. Measurements as in Fig. 4, Top. Two component (exponential) analyses were performed by a modified Marquardt algorithm. The two components were designated fast (●) or slow (x) and plotted accordingly. At low and high pH extremes, biexponential analysis did not give reproducible components and a single component analysis was used (x).

This source of pH–dependence is quite distinct from that expected for rate limiting proton diffusion. At high pH ($\geq$ 9), the rate of proton binding, as evidenced in the P^+Q_A- state, is rapid enough to allow a pH dependent equilibrium to be established before the electron transfer. At lower pH, the fast phase of electron transfer is not pH dependent and the observed proton binding in the P^+Q_B- state, appears to be rate limited in the same way as the P^+Q_A- state.

The slow phases of electron transfer and proton binding exhibit similar general behavior to the fast phases, but the onset of pH dependence occurs at significantly lower pH (between 6 and 7). The existence of two phases of electron transfer may imply either two populations of the RC reacting in parallel, or two states of the RC operating in series during the electron transfer, and connected by a slow reaction. However, due to the known relationship between the Q_A-$Q_B \leftrightarrow Q_AQ_B$- equilibrium and the recombination rate (Eq. 1), the first case would give rise to biphasic charge recombination kinetics above pH 7, contrary to observation.

In the second case, we suppose that in any given state of the RC, some portion of the Q_A-$Q_B \leftrightarrow Q_AQ_B$- electron transfer occurs rapidly, attaining an equilibrium position dependent on the protonation state, and that the equilibrium is generally favored in the more fully protonated states. Above pH 7, the instantaneous value of the equilibrium constant decreases from about 12, under our conditions, to 3. Thus, rapid electron transfer occurs to about 50% reduction of Q_B, followed by a slower transfer as readjustment of the protonation state pulls the equilibrium over, *i.e.*, $pK_{Q_B^-} > pK_{Q_A^-}$ (ref. 7):

$$
\begin{array}{ccccc}
 & & K_e^{\circ} & & \\
 & Q_A^- Q_B & \longleftrightarrow & Q_A Q_B^- & \\
pK_{Q_A^-} & H^+ \updownarrow & & \updownarrow H^+ & pK_{Q_B^-} \\
 & Q_A^- Q_B(H^+) & \longleftrightarrow & Q_A Q_B^- (H^+) & \\
 & & K_e' & &
\end{array}
$$

At even higher pH (> 9.2), further ionization decreases the instantaneous equilibrium constant for electron transfer to < 1. A small proportion of electron transfer occurs rapidly and subsequent readjustments of the protonation state again pull the equilibrium over. This should yield a steadily decreasing amplitude for the fast phase of the electron transfer, which is roughly as observed between pH 7 and 10. However, the fast phase amplitude does not appear to decrease enough. This is partly due to the lack of any fixed identity of the phases provided by the biexponential analysis, *i.e.*, the nature of the fast phase may change at high pH. In fact, the relative amplitude of fast phase should increase again at high pH: above $pK_{Q_B^-}$ the net amplitude of electron transfer will be small (equivalent only to $K_e' < 1$), but the kinetics will be all fast. Prior to this limit the amplitude of the slow phase declines and the fast phase gets slower, so that the kinetic analysis cannot reliably distinguish between them. This is clearly seen in the range of pH 9-10 where the analysis implies that the slow phase actually accelerates as the pH is raised.

Studies on Site Directed Mutants of the Q_B Pocket

We may now consider possible structural components responsible for the observed proton binding behavior of the acceptor quinones. This question has been addressed by us and by Okamura and Feher and coworkers, through mutagenesis of specific residues in the quinone binding domains of the protein. One of the more obvious differences in the environments of Q_A and Q_B, evident from the three dimensional structure of the RC (refs. 11 and 12), is the presence of two ionizable residues, glutamate (Glu^{L212}) and aspartate (Asp^{L213}), in close proximity to Q_B (Fig. 6).

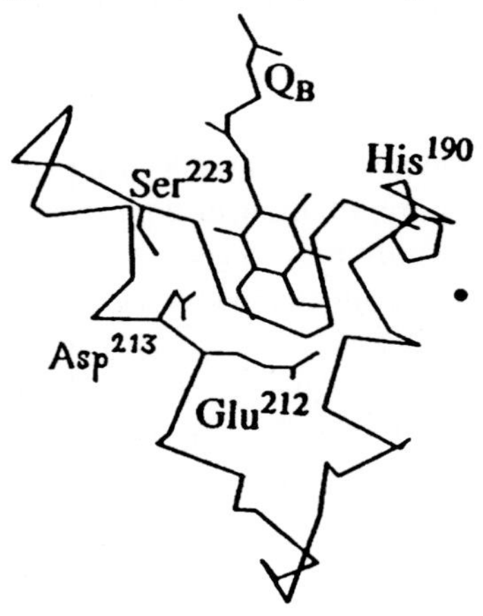

Fig 6. Sketch of the Q_B binding loop between the D and E helices of the L–subunit. Only the first two isoprene units of the ubiquinone side chain are shown. To decrease the complexity of the figure, α–carbons are connected directly by line segments, and only a few relevant side chains are shown.

Mutagenesis of Glu^{L212} to the non–ionizable residue glutamione (Gln) in RCs of *Rb. sphaeroides*, by Paddock *et al.* (ref. 13), essentially eliminated the pH dependence of the Q_A-$Q_B \leftrightarrow Q_AQ_B$- equilibrium and electron transfer rate, above pH 7. This result identifies Glu^{L212} as the residue responsible for inhibiting the electron transfer when ionized. It implies that Glu^{L212} has an unusually high pK (approx. 9.6) due, in part, to its somewhat hydrophobic environment. This mutation also resulted in a much lower steady state rate of cytochrome c oxidation when compared to the wild type (Wt) RC. The decrease by a factor of 25 was attributed to a reduced rate of proton transfer to Q_B^{2-} and the resulting inhibition of release of quinol (QH_2). However, the RCs were capable of undergoing a total of three rapid turnovers with the rapid uptake of approximately one proton. Thus, the RC probably becomes "blocked" in the state Q_A-Q_BH^-, as follows:

$$\text{1st flash:}\quad Q_AQ_B \xrightarrow{h\nu} Q_A^-Q_B \underset{(H^+)}{\overset{K_{e1}}{\longleftrightarrow}} Q_AQ_B^-\,(H^+)$$

$$\text{2nd flash:}\quad Q_AQ_B^-\,(H^+) \xrightarrow{h\nu} Q_A^-Q_B^-\,(H^+) \overset{K_{e2}}{\longleftrightarrow} Q_AQ_BH^-$$

$$\text{3rd flash:}\quad Q_AQ_BH^- \xrightarrow{h\nu} Q_A^-Q_BH^-$$

The failure of the quinol to unbind is consistent with the expectations of tight binding by anionic species in the Q_B pocket. The anionic semiquinone of Q_B also binds tightly, contributing significantly to its functionally important redox properties (refs. 2 and 14).

The internal transfer of a proton to Q_B is indicated in the scheme above, because of the likely unfavorability of electron transfer to generate the Q_B^{2-} state. Since this occurs readily in the L212EQ mutant, even in the absence of Glu^{L212}, we have recently examined the role the other acidic residue, Asp^{L213}, in the *Rb. sphaeroides* RC function. Both residues, Glu^{L212} and Asp^{L213}, were altered, separately and together, by site directed mutagenesis, to the non-protonatable residues Gln and Asn, yielding mutants L212EQ, L213DN and L212EQ/L213DN.

In isolated RCs from L213DN and L212EQ/L213DN, in the absence of added ubiquinone or any exogenous donors, the decay of P^+ after a flash was fast, with $t_{1/2}$ ~ 60-80 ms, corresponding to charge recombination of the P^+Q_A- state. When ubiquinone was added the kinetics became very slow, with $t_{1/2} > 10$ s at pH 7.0, indicative of reconstitution of Q_B activity with altered properties. The slow phase titrated in with a half–saturation concentration of < 2μM for Q–10, and in thoroughly dark–adapted samples the maximum amplitude of the slow phase reached 85-90%. Apart from the very slow decay, these parameters are similar to the WT (ref. 15). In the L212EQ mutant, the rate of recombination was quite fast, even with quinone added, as reported previously (ref. 13).

The pH dependences of the P^+ recovery rates ($P^+Q_AQ_B$- → PQ_AQ_B charge recombination) for Wt and mutant RCs are shown in Fig. 7.

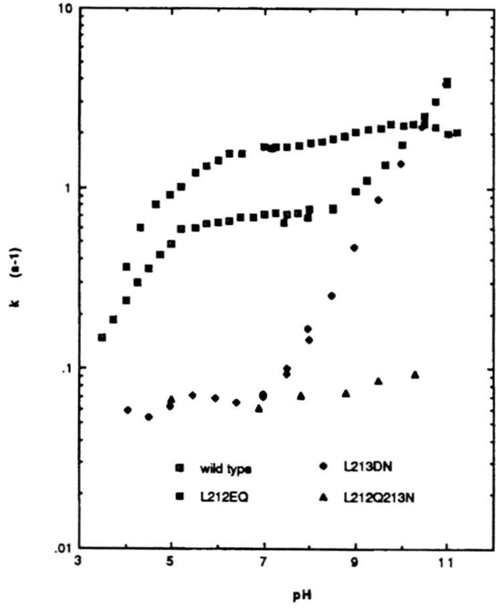

Fig. 7. pH dependence of the $P^+Q_AQ_B$- charge recombination kinetics in isolated reaction centers from wild type and various mutants of *Rb. sphaeroides*.

In the range accessible to study (pH 4-11), Wt RCs exhibited two pH–dependent regions — below pH 5 and above pH 9. In contrast, the charge recombination rate for L213DN was pH independent below pH 7.0 and was continuously pH dependent from pH 7.0 to 10.5. It may then become pH– independent again above pH 10.5. The onset of pH dependence for Wt RCs at about pH 9 has been attributed to the ionization of Glu^{L212} with an apparent pK value of pH ~9.5, and the resulting electrostatic interaction with Q_B/Q_B- (ref. 13). Below pH 7.0 the L213DN charge recombination rate was essentially pH independent, in contrast to the Wt RC, which displayed a second region of pH dependence below pH 5.0. Thus, Asp^{L213} could be responsible for the pH–dependent in the Wt RC, with an apparent pK value of about 4.5.

It is generally accepted for the Wt that the decay of $P^+Q_AQ_B$- proceeds by charge recombination *via* P^+Q_A-Q_B, as described above (refs. 3 and 7). However, in the L213DN and L212EQ/L213DN mutants, the observed decay is sufficiently slow that the direct charge recombination between P^+ and Q_B- cannot be ruled out, *a priori*, as contributing to the decay. Thus, the tendency towards pH independence may be misleading. However, taking the data at face–value, the region of pH–dependence of the charge recombination rate in L213DN RCs corresponds with the high pH region of the Wt, but with the pK shifted down from 9.5 to 7.5, or lower. Thus, at least 2 pH units of the pK shift of Glu^{L212} in the Wt may result from interaction with Asp^{L213}. The behavior of the L212EQ/L213DN mutant is consistent with this. It exhibits slow, almost pH independent, recombination kinetics over a wide pH range, as expected for the absence of any ionizable residues in the Q_B pocket.

The recombination rates for L213DN and L212EQ/L213DN are substantially slower than the Wt over most of the measured pH range, and especially above pH 5 where the rates differ by more than an order of magnitude. This indicates a significant additional stabilization of the electron on Q_B-. Assuming that the redox midpoint potential (E_m) of Q_A/Q_A- is unchanged in the mutant, application of the relationship of Equation 1 yields an E_m for Q_B/Q_B- that is at least 80 mV higher than in the WT. If the slow decay in these mutants occurs by direct recombination between P^+ and Q_B-, then this represents a lower limit for the Em of Q_B/Q_B-. It is likely that the average reduction potential (the E_m for Q/QH_2) is unchanged, in which case we expect the Em for the second electron (Q_B-/Q_BH_2) to be equivalently lowered relative to the Wt. This could contribute to the failure of transfer of the second electron in these mutants.

The pH dependence of the Q_A-$Q_B \rightarrow Q_AQ_B$- electron transfer rate (k_e) for Wt and mutant RCs is shown in Fig. 8.

As with the charge recombination rate, the pH–dependent region of the electron transfer is evident at lower pH in L213DN, indicating again the influence of this mutation on the pK of ionizable residues (*e.g.* Glu^{L212}) near Q_B. A surprising aspect of the forward electron transfer rate, in light of the charge recombination behavior, is that it is slower than the Wt by about an order of magnitude for L213DN. Since the $P^+Q_AQ_B$- charge separation state of the L213DN RC is considerably stabilized relative to the Wt, a rapid transfer of an electron from Q_A- to Q_B might be expected. The fact that it is not fast, together with the large equilibrium constant for the Q_A-Q_B $\rightarrow$ Q_AQ_B- electron transfer (K_e), implies that the reverse rate (k_{-e}) is greatly slowed in this mutant.

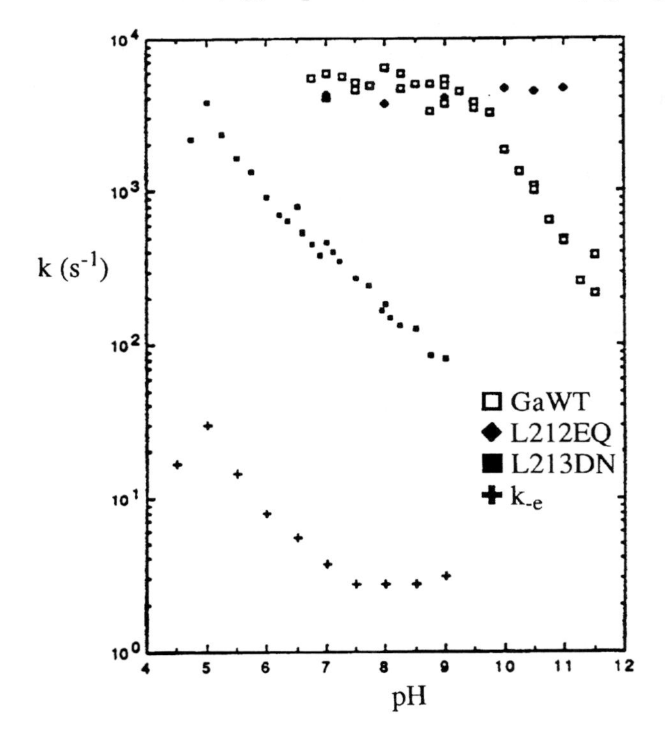

Fig. 8. pH dependence of the Q_A-Q_B $\rightarrow$ Q_AQ_B- electron transfer kinetics in isolated reaction centers from wild type (GaWT) and two mutants (L212EQ and L213DN) of *Rb. sphaeroides*. Kinetics measured as in Fig. 4, top; mutant kinetics were analysed as a single component.

In contrast to L213DN, L212EQ shows essentially Wt rates of electron transfer, except at high pH where the mutant remains fast while the Wt decelerates. This is entirely consistent with inhibition of the Wt electron transfer equilibrium by ionized Glu^{L212} above its pK, and requiring a rate limiting reprotonation to permit significant electron transfer at high pH.

When semiquinone behavior was examined with multiple saturating flashes in the presence of exogenous donor (ferrocene), the typical oscillatory behavior seen for the Wt RC (refs. 3, 7, 8) was not observed with L213DN RCs at pH values above 6.5 (Fig. 9) or in L212EQ/L213DN above pH 7.5.

After the first flash, a stable semiquinone absorbance signal was observed, but on all subsequent flashes the additional absorbance change decayed slowly to the first flash level. The spectrum of the decaying signal, produced by the second flash, showed it to arise from an anionic semiquinone species. The decay was very slow at pH $\geq$ 8 with $t_{1/2}$ > 1 s, indicating a drastically inhibited rate for the second electron transfer to

Q_B when compared to the Wt RC [$t_{1/2}$ ~ 200 μs (refs. 3 and 16)]. At pH values below 7, the rate of transfer of the second electron increased and distinct oscillations of the stable semiquinone signal were observed. The slow decay of the semiquinone signal after the second and subsequent flashes at pH ≥ 7.0 is due, at least in part, to direct oxidation of Q_A- by exogenous oxidants, including oxidized ferricenium and dissolved oxygen. Thus, the steady state inhibition factor for these mutants may be substantially greater than 10^4, compared to 25 for L212EQ.

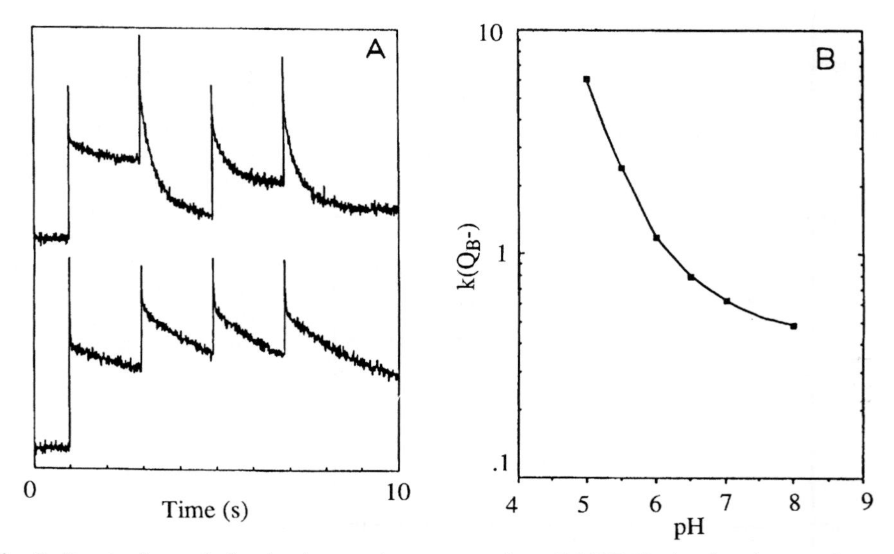

Fig. 9. Semiquinone behavior in reaction centers from L213DN. **A**: absorbance changes at 450 nm in a series of flashes, with ferrocene as donor. Top: pH 6.0. Bottom: pH 8.0. **B**: pH dependence of the kinetics of decay of Q_A-Q_B- after the second flash.

The remarkably potent inhibition of the second electron transfer in L213DN and L212EQ/L213DN was also apparent in an impaired ability of these RCs to oxidize cytochrome c in a series of flashes. At pH > 6.5, normal cytochrome oxidation was observed only on the first two flashes, and very little after subsequent flashes (not shown).

Flash–induced proton–binding by L213DN RCs is fully consistent with the observed semiquinone behavior. At pH > 6.5, small rapid uptake of H^+ occurs after the first flash, but very little after subsequent flashes (Fig. 10, Top).

At pH < 6.5, where oscillations in the semiquinone signal are observed, the stoichiometry of H^+ binding is very similar to that of the Wt — approximately 0.6 H^+ on the first flash and 1.4 H^+ on the second (refs. 4-6), with low amplitude oscillations apparent in a series of flashes (Fig. 10, Bottom). The major component of the kinetics of H^+ binding on the second flash was very similar in rate to the Q_A-Q_B- → $Q_AQ_B^{2-}$ electron transfer ($t_{1/2}$ ~ 100 ms at pH 6.0). It is noteworthy that at pH 8.0, where the

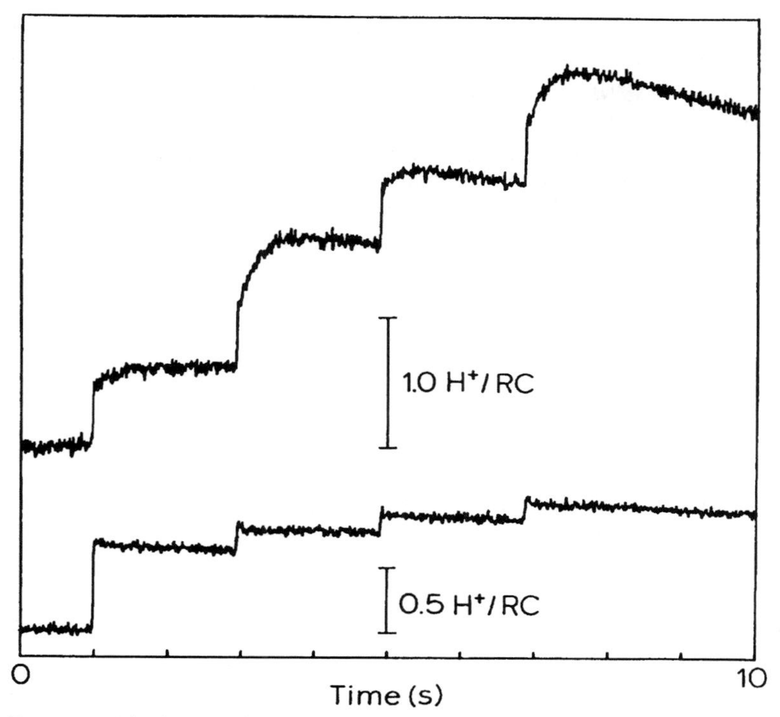

Fig. 10. Proton uptake by reaction centers from L213DN. Top: pH 6.0; chlorphenol red as indicator. Bottom: pH 8.0; cresol red as indicator.

Q_A-Q_B- → $Q_AQ_B{}^{2-}$ electron transfer in this mutant is blocked, the H^+ binding on the second and subsequent flashes is very small. Allowing for reoxidation of Q_A- between flashes, the stoichiometry of H^+ binding is less than 0.1 H^+/RC in the Q_AQ_B- → Q_A-Q_B- transition.

Pathways of Proton Transfer to the Secondary Quinone

The observed behavior of L213DN and L212EQ/L213DN RCs leads us to suggest that the primary lesion is a severe obstruction of the normal uptake of the first proton associated with, or necessary for, the double reduction of Q_B after the second flash, leading to a failure of the transfer of the second electron to Q_B. The process is almost completely blocked at pH ≥ 7.0, but is facilitated at lower pH. Protonation of Q_B is likely to involve a number of protonatable residues which cooperate to transfer protons from the aqueous phase to the Q_B site, and some residues which may be involved in such a scheme have been indicated by Allen *et al.* (ref. 12). The side chain of Asp^{L213} is located close to a Q_B carbonyl group (~ 5 Å) and to the hydroxyl group of Ser^{L223}, which is thought to hydrogen bond to the nearby quinone carbonyl (2.7Å) (ref. 11). Either of these groups could act as the direct proton–donor to Q_B. Even if Ser^{L223} serves as the primary proton donor to the quinone, Asp^{L213} may play a crucial role in the reprotonation of Ser^{L223} to permit net proton transfer to the quinone. From the pH–dependence of the $Q_A^-Q_B \leftrightarrow Q_AQ_B^-$ equilibrium (K_e) in Wt RCs, we suggest that ionization of Asp^{L213} may determine the behavior below pH 6.0, in which case this residue would be fully ionized in the $Q_AQ_B^-$ state above pH 6.0. In order to participate in proton donation on the second transfer, therefore, it would have to be rapidly reprotonated, perhaps by internal transfer, in the $Q_A^-Q_B^-$ or $Q_AQ_B^{2-}$ states. The exact order of events cannot be established from any available data, and two possibilities must be considered (ref. 3):

$$Q_A^-Q_B^- \xleftrightarrow{K_{H1} \ll 1} Q_A^-Q_BH \xleftrightarrow{K_{e2} \gg 1} Q_AQ_BH^-$$

$$Q_A^-Q_B^- \xleftrightarrow{K_{e2} \ll 1} Q_AQ_B^{2-} \xleftrightarrow{K_{H1} \gg 1} Q_AQ_BH^-$$

where K_{H1} and K_{e2} are equilibrium constants for the transfer of the first proton (from Asp^{L213}) and the second electron, respectively. In either case, an unfavorable equilibrium is followed by a very favorable one, thereby pulling the net electron transfer over. Both situations have been widely encountered in the electrochemistry of quinones in solution (ref. 17). The first case (A) is attractive because the electron transfer to the neutral semiquinone is easily imagined to proceed rapidly. The second case (B) is attractive because the dianion, Q_B^{2-}, is expected to have a very high pK (> 13) probably capable of obtaining a proton from Ser^{L223} with the help of the protonated Asp^{L213} to reprotonate the serine hydroxyl.

Turnover in the acceptor quinone complex is completed by the transfer of a second H^+–ion to form Q_BH_2, which is then released from the Q_B site as quinol and replaced by an oxidized quinone (refs. 2, 3, 15 and 18). From the behavior of the $Glu^{L212} \rightarrow$ Gln mutant, L212EQ, as also reported by Paddock *et al.* (ref. 13), it is likely that Glu^{L212}, located ~ 6 Å from the other carbonyl group of Q_B, is involved in donation of the second proton, *i.e.*, mutation to a non–ionizable residue permits three turnovers of the RC leading to accumulation of the state $PQ_A^-Q_BH^-$. This is strongly supported by the double mutant, L212EQ/L213DN, which behaves very similarly to the single mutant, L213DN, supporting the more primary function of Asp^{L213}.

This overall description implies two slightly different pathways for donation of the first and second protons in the formation of Q_BH_2 in RCs of *Rb. sphaeroides*, such as has been described by Allen *et al.* (ref. 12). However, in *Rps. viridis* (ref. 19) and *R. rubrum* (ref. 20), the residue L213 (or equivalent) is naturally asparagine. This is a perplexing fact, especially in view of the very dramatic effects of mutation of L213 in *Rb. sphaeroides* — a decrease in proton and electron transfer rates of 10^4 fold. Presumably other compensating differences allow the participation of other residues in this function and it is noteworthy that a relatively nearby residue (~ 8Å) is Asn in

Rb. sphaeroides (M44), but Asp in *Rps. viridis* (M 43) and *R. rubrum* (M 43). In this context it is relevant that both L213DN and L212EQ/L213DN become functional at moderate to low pH, with rather similar activities. Apparently both lesions can be slowly circumvented under such conditions. This may involve the activation of a distinct alternate pathway or it may simply reflect penetration of the RC structure by water. The fact that the double mutant actually functions better, *i.e.*, at higher pH, than L213DN, may favor the notion of simple water penetration as any structural perturbations might be expected to be greater in the double mutant.

However, structural perturbations accompanying these mutations must, generally, be quite small as the limiting value of the electron transfer equilibrium constant at low pH is remarkably similar for the Wt and all three mutants. On the other hand, the value for L212EQ above pH 6 is considerably smaller (faster recombination rate) than the plateau value for Wt, between pH 6 and 9. This suggests that the protonated Glu^{L212} in the Wt partially alleviates the inhibitory effect of the negative charge of Asp^{L213} in a way that glutamine cannot. This may reflect the greater H–bonding potential of glutamic acid.

Acknowledgements

This work was supported by grants from the National Science Foundation (DMB 86-01744 and DMB 89-04991) and from the United States Department of Agriculture Competitive Grants Office (AG 86-1-CRCR-2149 and AG 89-37262-4462) and the McKnight Foundation.

References

1 J.C. Williams, L.A. Steiner and G. Feher, Proteins **1** (1986) 312.

2 A.R. Crofts and C.A. Wraight, Biochim. Biophys. Acta **726** (1983) 149.

3 C.A. Wraight, Biochim. Biophys. Acta **548** (1979) 309.

4 P. Maroti and C.A. Wraight, Biochim. Biophys. Acta **934** (1988) 314.

5 P. Maroti and C.A. Wraight, Biochim. Biophys. Acta **934** (1988) 329.

6 P.H. McPherson, M.Y. Okamura and G. Feher, Biochim. Biophys. Acta **934** (1988) 348.

7 D. Kleinfeld, M.Y. Okamura and G. Feher, Biochim. Biophys. Acta **766** (1984) 126.

8 A. Vermeglio, Biochim. Biophys. Acta **459** (1977) 516.

9 M. Eigen, W. Kruse and G. Maasse, Prog. React. Kinet. **2** (1964) 286.

10 M. Gutman and E. Nachliel, Biochim. Biophys. Acta **1015** (1990) 391.

11 T.O. Yeates, H. Komiya, A. Chirino, D.C. Rees, J.P. Allen and G. Feher, Proc. Natl. Acad Sci. USA **85** (1988) 7993.

12 J.P. Allen, G. Feher, T.O. Yeates, H. Komiya and D.C. Rees, Proc. Natl. Acad. Sci. USA **85** (1988) 8487.

13 M.L. Paddock, S.H. Rongey, G. Feher and M.Y. Okamura, Proc. Natl. Acad. Sci. USA **86** (1989) 6602.

14 C.A. Wraight in: B.L. Trumpower (Ed.) *Function of Quinones in Energy Conserving Systems* Academic Press, 1982, pp. 182-192.

15 J.C. McComb, R.R. Stein and C.A. Wraight, Biochim. Biophys. Acta **1015** (1990) 156.

16 A. Vermeglio and R.K. Clayton, Biochim. Biophys. Acta **461** (1977) 159.

17 *The Chemistry of Quinoid Compounds* S. Patai (Ed.), Wiley Interscience, New York, 1974.

18 P.H. McPherson, M.Y. Okamura and G. Feher, Biochim. Biophys. Acta **1016** (1990) 289.

19 H. Michel, K.A. Weyer, H. Gruenberg, I. Dunger, D. Oesterhelt and F. Lottspeich, EMBO J. **5** (1986) 1149.

20 G. Belanger, J. Berard, P. Corriveau and G. Gingras, J. Biol. Chem. **263** (1988) 7632.

Electron and Proton Transfer in Chemistry and Biology, edited by A. Müller et al.
Studies in Physical and Theoretical Chemistry, Vol. 78

The Water Oxidizing Enzyme – An Alternative Model

Elfriede K. Pistorius

Lehrstuhl Zellphysiologie, Fakultät für Biologie
Universität Bielefeld
4800 Bielefeld 1 (Germany)

Summary

The photosystem II complex can catalyze a photochemical charge separation as all the other photosystems can, and in addition it can catalyze the unique photodriven water oxidation reaction. The process of charge separation seems to be fairly well understood, while the process of water oxidation has remained uncertain in several aspects. In the first part of this paper a brief review on the present knowledge of photosystem II will be given, and in the second part an alternative model to the more generally accepted model of the water oxidizing enzyme will be presented. This alternative model which is based on our results with cyanobacteria, suggests that the water oxidizing enzyme has evolved from a substrate dehydrogenase type enzyme and that it is a separate protein (distinct from D1 and D2) in photosystem II.

Introduction

Photosynthetic organisms can be divided into two major classes: 1. the anoxygenic organisms which consist of a wide range of photosynthetic bacteria and which use organic acids or reduced sulfur compounds as electron donor, and 2. the oxygenic organisms which consist of cyanobacteria, prochlorophyta, green algae and higher plants and which developed the ability to utilize water as electron donor (refs. 1, 2, 3 and 4). In oxygenic photosynthesis light energy is converted into chemical energy within two specialized chlorophyll–protein complexes called photosystem I (PS I) and photosystem II (PS II). PS I generates a weak oxidant and a strong reductant, while PS II generates a strong oxidant and a weak reductant. The latter complex — sometimes referred to as water plastoquinone oxidoreductase (ref. 5) — is that part of oxygenic photosynthesis which catalyzes the photoinduced transfer of electrons from water to plastoquinone and which evolves O_2 as a by–product. The minimal scheme to describe the PS II reaction sequence is given in Fig. 1.

PS II has probably been subjected to more investigations than any other photosystem, since PS II can catalyze a photochemical charge separation as all the other photosystems (PS I and the photosystems of purple and green bacteria), and in addition it can catalyze the unique photodriven water oxidizing reaction. It is also the site of action of a large number of commercial herbicides, and it is the site of damage when plants are exposed to high light intensities (the so called photoinhibition). Several

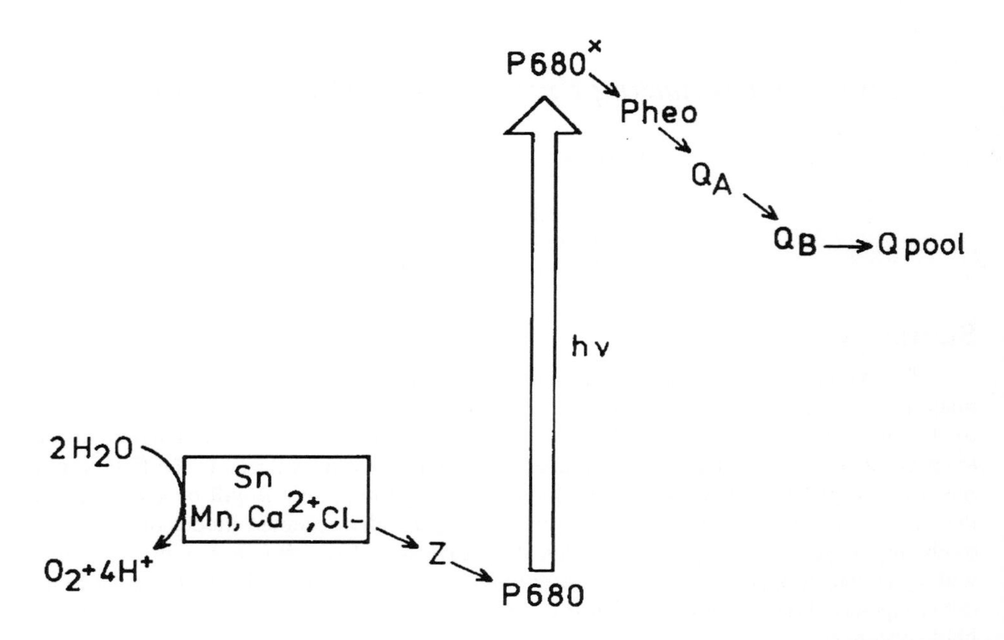

Fig. 1. Minimal scheme of the PS II reaction sequence. The symbols are: Sn, water oxidizing enzyme; Z: electron donor to P680 (tyrosine radical on D1); P680, reaction center chlorophyll a; Pheo, pheophytin; Q_A and Q_B, first and second bound plastoquinone acceptor; Q_{pool}, plastoquinone pool.

general reviews on the current knowledge of PS II have appeared during the last years (refs. 6, 7, 8, 9 and 10) and an extended in–depth review on PS II has just recently been written by Hansson and Wydrzynski, (ref. 11). That review also cites a number of reviews on special issues of PS II. For this reason I shall only give a brief summary of the structural and functional aspects of PS II and shall mainly discuss some aspects concerning the still unanswered question of the identity of the water oxidizing enzyme. Because of the vast literature about PS II, reviews rather than the original papers will be mainly cited in the first part of this paper. In the second part of this paper a model of the water oxidizing enzyme based on our results with cyanobacteria will be presented.

A. Structural and Functional Aspects of PS II.

In recent years rapid progress in the understanding of PS II has been achieved — mainly for two reasons: There have been great improvements in the techniques for isolating functional PS II complexes and a structural and functional analogy has been established between PS II and the reaction center of purple bacteria (refs. 6, 7, 8, 9, 10 and 11). The successful preparation of distinct detergent–lipid–protein complexes capable of light–dependent water oxidation has shown that the PS II complex is composed of a fairly large number of integral and peripheral membrane proteins (ref. 12). In higher plants the number of polypeptides in PS II seems to exceed 22 (ref. 13). Associated with these proteins are several essential organic and inorganic cofactors. From such complexes a minimal complex which can still catalyze water oxidation, was isolated, and such a complex apparently consists of only 8 polypeptides (refs. 11 and 14). These are the two chlorophyll a binding peptides of 44-51 and 40-44 kDa (CP 47 and CP 43) which are assumed to be the inner antenna of PS II (although it cannot be totally excluded that they might have additional functions), two 30-34 kDa peptides (D1 and D2) which are the reaction center peptides, two apoproteins of cytochrome b559 (9 and 4.3 kDa), a 4.8 kDa peptide of unknown function, and the extrinsic manganese stabilizing peptide (MSP) of 27-36 kDa (Table 1).

Table 1.

Genes and proteins of minimal O_2 evolving PS II complexes.

Genes	Product and apparent MW (kDa)		Function/Cofactors
psb A	D1:	30-34	Reaction core subunit: P680, pheophytin, Q_B, Z
psb B	CP 47:	44-51	Proximal antenna: chlorophyll a, β–carotine
psb C	CP 43:	40-44	Proximal antenna: chlorophyll a, β–carotine
psb D	D2:	30-34	Reaction core subunit: P680, pheophytin, Q_A, D
psb E		9.0	Cytochrome b559–binding
psb F		4.3	peptides
psb I		4.8	?
psb O	MSP:	27-36	Extrinsic manganese stabilizing peptide

These polypeptides have been shown to be present in all PS II complexes so far investigated from plants and cyanobacteria. The variations among organisms and the changes during evolution are mainly associated with the light–harvesting pigment–

protein complexes (*e.g.* being phycobili proteins in cyanobacteria and chlorophyll a/b–protein complexes in plants) and with some small extrinsic or intrinsic peptides which are thought to have either a structural, stabilizing or regulatory role (*e.g.* the extrinsic peptides of 16 and 23 kDa which are present in plants, have so far not been detected in cyanobacteria) (refs. 12, 4, 15 and 16). Associated with the minimal O_2 evolving PS II complex are the following cofactors: the reaction center chlorophyll P680, pheophytin, carotenoids, two plastquionones (Q_A and Q_B), a non–heme iron (component Q_{400}), cytochrome b559, and Mn, Ca^{2+} and Cl^-. The components Z and D on the donor side of PS II — giving rise to the ESR signals $II_{very\ fast}$ and II_{slow} — have recently been assigned to tyrosine radicals on the D1 and D2 peptides, respectively (refs. 17 and 18).

Investigations about the function of the various polypeptides in PS II have greatly advanced when it became obvious that a structural and functional analogy exists between the reaction center of purple bacteria and PS II (refs. 19, 20 and 21). Knowledge about the structure of photosynthetic reaction centers is based in large on the crystallization and determination of the three dimensional structure of the reaction center from the purple bacteria *Rhodopseudomonas viridis* (ref. 19) and *Rhodobacter sphaeroides* (ref. 22) where 4 bacteria chlorophylls, 2 bacteria pheophytins with 1 carotenoid, 1 non–heme iron and 2 quinones comprise the components of the reaction center. The cofactors which are located on the L and M subunits of the reaction center complex, are arranged in two branches with a twofold symmetry which runs from the chlorophyll dimer to the iron atom.

In the absence of detailed structural models for the reaction center of PS II, analogies to the bacterial reaction center have been proposed — mainly based on amino acid sequence homologies between the D1 and D2 polypeptides of PS II and the L and M subunits of the reaction center of purple bacteria (refs. 20, 21 and 23). Experimental support for the similarity between the bacterial reaction center and PS II has been provided by Nanba and Satoh, (ref. 24), who isolated a complex which only consisted of D1 and D2 and the two cytochrome b559 binding peptides (and a 4.8 kDa peptide - shown later to be present (ref. 25)) and which retained the ability to catalyze charge separation. The complex isolated by Nanba and Satoh, (ref. 24), contained 4 to 5 chlorophylls, 2 pheophytins, 1 to 2 cytochrome b559 along with carotenoids and nonheme iron. Based on these results, it seems that the minimal unit which can catalyze the electron transfer process leading to charge separation, is quite similar in both centers.

However, despite these similarities there are also substantial differences between the reaction center of PS II and purple bacteria (refs. 26 and 27). The pigment binding properties of the two types of reaction centers seem to be different, since the reaction center complex of PS II binds more chlorophylls (probably up to 12 chlorophylls) (ref. 28) than the reaction center of purple bacteria (4 bacteriachlorophylls). The isolated reaction center of *Rhodobacter sphaeroides* does not contain bound cytochrome, while *Rhodopseudomonas viridis* contains a polypeptide which binds 4 hemes. This cytochrome functions as electron donor to the reaction center chlorophyll. On the other hand, the isolated reaction center complex of PS II contains a heterodimer which binds cytochrome b559 and which has in contrast to the cytochrome binding peptide in *R. viridis* transmembrane segments. The role of cytochrome b559 in PS II is unclear, but it seems certain that it does not function as main electron donor to P680. All purple bacteria reaction centers contain the subunit H which has one transmembrane segment. This subunit binds no chromophores or prosthetic groups but is probably important in optimizing the comformation of the L and M heterodimer and in helping to stabilize the quinones on the acceptor side. No equivalent to the bacterial H subunit has yet been identified in PS II, although a 9-10 kDa or 22 kDa protein might play a role in stabilizing Q_A and Q_B in PS II. Possibly as a consequence of the loss of the H–like subunit, the plastoquinone molecules Q_A and Q_B are also lost during isolation of the reaction center complex of PS II. Another significant difference between the two reaction centers is that the D1 subunit of PS II is continously turning over, while the corresponding L subunit of the reaction center of purple bacteria does not (refs. 26 and 29).

However, the most significant difference is associated with the donor side of the two reaction centers (refs. 26 and 27). In purple bacteria, the reaction center bacteria-chlorophyll becomes reduced by cytochrome c, completing the electron transfer cycle. In some purple bacteria it is possible that electrons may be obtained from reduced sulfur compounds *via* a branched pathway. On the other hand, the redox chemistry in PS II is quite different from that in purple bacteria. PS II creates a more powerful oxidizing potential of about 1.0 Volt compared to 0.4 Volt in bacteria. The generation of a powerful oxidant in PS II is necessary in order to oxidize water and to produce molecular oxygen. It is quite amazing to realize that in spite of an immensely increased knowledge about PS II during the last years, the identity of the water oxidizing enzyme has remained uncertain.

The water oxidizing enzyme is an unique enzyme containing a polynuclear Mn cluster (consisting of 2 or 4 Mn) and Ca^{2+} and Cl^- as cofactors (refs. 7 and 11). The enzyme can exist in 5 oxidation states (so called S–states - S_0 to S_4) and catalyzes the 4–step oxidation of 2 water molecules to molecular O_2 (refs. 30 and 31). The active site of the enzyme is connected by the Z/Z^+ couple to the reaction center chlorophyll P680. In general, there are two views on how water oxidation occurs, either as a concerted reaction on the S_3 to (S_4) to S_0 transition (refs. 32 and 33) or in a stepwise reaction which involves a bound water oxidation intermediate (ref. 34). It is generally assumed that the Mn ions are the place where two water molecules are bound to form the dioxygen bond. Although it is quite clear that Mn is the central constituent of the water oxidizing enzyme (ref. 35), the oxidizing equivalents accumulated during the reaction do not need to be restricted to Mn ions exclusively but could be located on additional redox active ligands, including water oxidation intermediates, redox groups within the protein matrix (ref. 36) or an additional not yet identified organic prosthetic group (see later). The majority of the groups working in this area is in favor of 4 Mn being involved in water oxidation, but the Mn ions seem to be inequivalent, since they are released and reincorporated differently in the PS II complex. There is substantial information in the literature on the structure of the Mn cluster. However, none of it provides a complete picture, and the interpretations are somewhat contradictory. Possible models for the Mn cluster of the water oxidizing enzyme have *e.g.* been suggested by Brudvig and Crabtree (ref. 37) or George et al. (ref. 38).

Besides Mn, Cl^- is a well established cofactor of the water oxidizing enzyme, and the third inorganic ion which is required for efficient functioning of the water oxidizing enzyme is Ca^{2+} (refs. 39, 40, 41, 42). In some cyanobacteria, such as *Synechocystis PCC6714*, Na^+ is required instead of Ca^{2+} (ref. 43). A number of experiments have shown that in the absence of Ca^{2+} (ref. 44) or Cl^- (*e.g.* ref. 45) the last step of the S-cycle transition is blocked. In the absence of Cl^- possibly also earlier steps of oxidant accumulation might be blocked (ref. 11). The cofactor action of Ca^{2+} and Cl^- in the mechanism of the water oxidizing enzyme has remained uncertain. For Ca^{2+}, a conformational role has been suggested, while for Cl^- some groups suggest a function as bridging ligand in the Mn cluster (ref. 46).

At the present time it is unknown, where the cofactors Mn, Ca^{2+} and Cl^- bind or interact with the water oxidizing system. For a while it was thought that the extrinsic peptides of 33, 23 and 16 kDa were the binding sites of the inorganic cofactors of water oxidation. However, it has been clearly shown that all three subunits are not obligatory for water oxidation and can be removed from the PS II complex without loss of water oxidation. The 33 kDa (MSP) has a role in stabilizing Mn at the water oxidation side, and the 23 and 16 kDa peptides modulate the binding of Ca^{2+} and Cl^- (refs. 27 and 47).

Since the extrinsic polypeptides are not involved in the binding of Mn, most groups believe that the reaction center proteins D1 and D2 have become the best candidates as the Mn binding proteins (ref. 10). There is some experimental evidence in favor of this hypothesis. A D1 mutant of *Scenedesmus* which has an unprocessed D1 peptide, has a reduced Mn binding capacity (refs. 48, 49, 50). Moreover, photodestruction which leads to loss of D1 also leads to loss of Mn (ref. 51). However, the reduced binding of Mn or the loss of Mn could also be an indirect consequence caused by conformational changes in the other PS II peptides or by changes in the Ca^{2+} and/or Cl^-–binding due to improper insertion or the loss of D1, and this then indirectly might effect the binding of Mn. A simple model based on the above results was suggested by Barber (ref. 20) and later also by Dismukes(ref. 52). This model locates the Mn at the interface of D1 and D2, and it would imply that the peptides (D1 and D2) which originally in evolution were probably only able to catalyze the photochemical charge separation (based on the structural similarity of D1 and D2 to the L and M subunits in the reaction center of purple bacteria) have acquired additional inorganic cofactors (Mn, Ca^{2+} and Cl^-) which enable these peptides now to catalyze water oxidation in addition to the process of photochemical charge separation. This model would imply that a separate water oxidizing enzyme does not exist. Although the model suggested by Barber (ref. 20) and Dismukes (ref. 52) seems to be favored by the majority of groups working in this area, it still remains somewhat speculative. Therefore, the question arises whether experiments exist which might point to an alternative model. Based on our results with the cyanobacterium *Anacystis nidulans* we would like to suggest such an alternative model. For better understanding, I will explain our hypothesis at the beginning before giving the experimental evidence.

B. An Alternative Model of the Water Oxidizing Enzyme.

Our model predicts that the water oxidizing enzyme is a separate polypeptide in the PS II complex (distinct from D1 and D2), and that this enzyme has evolved from a substrate dehydrogenase type enzyme which originally could oxidize basic L–amino acids (such as L–arginine) and which could mediate electron flow from L–arginine to the plastoquinone pool in the thylakoid membrane of cyanobacteria (Fig. 2).

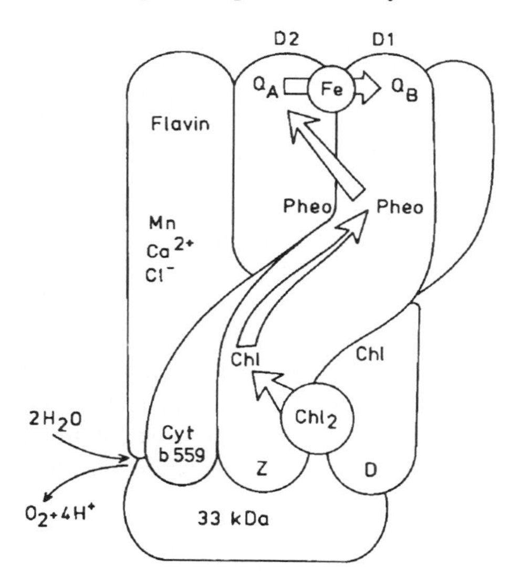

Fig. 2. Hypothetical model of the PS II complex. Alternative model of the PS II complex suggesting that the water oxidizing enzyme is a separate enzyme (distinct from D1 and D2) and evolved from a substrate dehydrogenase type enzyme (ref. 53).

Since in cyanobacteria the photosynthetic electron transport chain and the respiratory electron transport chain are located together in the thylakoid membrane and partially use the same electron carriers (plastoquinone and the cytochrome b/f complex) the electrons from the organic substrate could subsequently be transferred either to PS I or to the cytochrome oxidase (refs. 54, 55 and 56). Amino acid dehydrogenases have been shown to be able to couple to respiratory chains, as *e.g.* in *Escherichia coli*, where under certain conditions a D–amino acid dehydrogenase or a proline oxidase mediates electron flow from the corresponding amino acid to the ubiquinone pool of the respiratory chain (ref. 57).

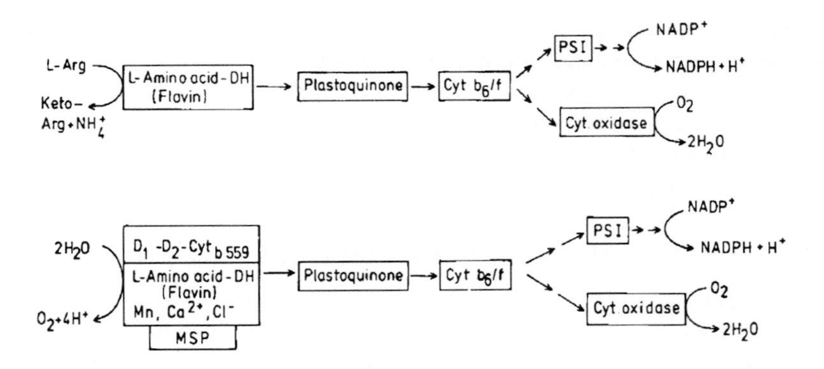

Fig. 3. Schematic presentation showing the suggested conversion of an L–arginine oxidizing enzyme to the water oxidizing enzyme.

As shown in Fig. 3, our model further predicts that such a flavoenzyme with L–amino acid dehydrogenase (oxidase)–activity was modulated during evolution with additional inorganic cofactors (Mn, Ca^{2+} and Cl^-) and additional peptides (MSP and later in evolution also with the 23 and 16 kDa peptides) in such a way that the flavin in the enzyme could interact with the Mn to form a charge transfer complex. As a consequence, flavin would become a better oxidant as has been shown to be the case in experiments with charge transfer complexes consisting of flavin model compounds and transient metals (ref. 58). Our model further suggests that such a modified flavoenzyme came in tight contact with a reaction center complex which originally was probably only able to catalyze a cyclic electron flow. The combination of such a modified substrate dehydrogenase type enzyme with a reaction center complex then eventually became the present–day water plastoquinone oxidoreductase. Thus, in our model (Fig. 2 and 3) a separate Mn enzyme which evolved from a substrate dehydrogenase, exists, and the reaction center complex (D1/D2/cytochrome b559) can only catalyze the photochemical charge separation.

Several reasons might exist why in cyanobacteria, such as *Anacystis nidulans*, an L–amino acid dehydrogenase with high specificity for L–arginine and other basic L-amino acids was originally present. Later in evolution after accumulation of O_2 in the atmosphere, this enzyme (at least in *A. nidulans*) could also interact with O_2 to become an L–amino acid oxidase, and this is the activity which we mainly measure in *A. nidulans*. Cyanobacteria have an incomplete citric acid cycle (the α–ketoglutarate dehydrogenase seems to be missing) (ref. 59). Therefore, generation of NADH for the respiratory chain *via* this cycle is limited and amino acids, such as L–arginine, might have had an essential role as electron donors. In addition to the above suggested role of L–arginine as electron donor for the electron transport chain, this L–amino acid is known to have special functions in cyanobacteria. It has been shown that most cyanobacteria contain a storage peptide consisting of L–arginine and L–aspartic acid (ref. 60), and CO_2 fixation in cyanobacteria partially occurs *via* formation of L–citrulline (and then arginine) from carbamoyl–phosphate and L–ornithine (ref. 61). For reasons presently unknown, we think that such an L–arginine metabolizing enzyme might have been especially suited for modification with additional cofactors and peptides to eventually become the water oxidizing enzyme (Fig. 3). Therefore, if our hypothesis is correct, it can be assumed that during evolution the mechanisms for suppressing this original L–arginine metabolizing activity of the protein, greatly improved, but that in some of the earlier oxygenic organisms (such as cyanobacteria) this activity might still be detectable.

The main evidence for our model as shown in Figs. 2 and 3 is based on experiments with the cyanobacterium *A. nidulans*. This cyanobacterium contains an L-amino acid oxidase which after cell breakage is partly present in the soluble fraction and partly tightly associated with the thylakoid membrane. We have isolated and partially characterized the enzyme present in the soluble fraction of the French press extract (refs. 53, 62, 63 and 64). The enzyme has a molecular weight of 98 kDa and consists of two subunits of equal molecular weight of 49-50 kDa. It catalyzes the regular oxidative deamination reaction with O_2 as electron acceptor and has a high specificity for basic L–amino acids (L-Arg > L-Lys > L-Orn > L-His). The enyzme has an unusual high turn–over number of 70 000 (mole L–arginine oxidized per mole enzyme/min), and the L–amino acid oxidase activity of this protein is strongly inhibited by cations (M^{3+}> M^{2+} > M^{+}) and less strongly by anions (refs. 63 and 65). The isolated enyzme contains FAD as prosthetic group, but besides authentic oxidized FAD a flavin derivative of unknown structure is present in variable amounts in different enzyme preparations (ref. 64). It seems that the flavin in the enzyme can rapidly undergo modification reactions, and those modification reactions are irreversible (at least those ones which occur during the purification) and lead to an inactivation of the L–amino acid oxidase activity. The modified flavin in the enzyme has some similarities with $10,10^a$–ring opened flavin derivatives which have been suggested by Mager and Addink (ref. 66) to be possible intermediates in flavin catalyzed oxygenase reactions.

Based on activity measurements and based on experiments with the antiserum raised against the purified L–amino acid oxidase, we could show that the L–amino acid oxidase protein is present in *Anacystis* thylakoid membranes as well as in purified PS II complexes (refs. 67, 68 and 69). Moreover, an organic prosthetic group which has a similarity to the modified flavin present in the soluble enzyme, could be extracted from purified PS II complexes (ref. 64). Our initial reason to believe that a connection might exist between the water oxidizing enzyme and the L–amino acid oxidase activity in the thylakoid membranes of *A.nidulans* was based on the observation that $CaCl_2$ has an antagonistic effect on the two activities: $CaCl_2$ stimulates the photosynthetic water oxidation but inhibits the L–amino acid oxidase activity (refs. 53, 65, 67 and 70). Those results seemed to indicate that the L–amino acid oxidase has to become suppressed before water oxidation can occur. Moreover, some inhibitors of water oxidation, such as o–phenanthroline, chlorpromazine and hexanitrodiphenylamines, also inhibited the L–amino acid oxidase activity (ref. 71).

Those initial results had shown that the L–amino acid oxidase protein was present in PS II complexes and was possibly functional in water oxidation. Therefore, we further investigated whether the L–amino acid oxidase protein was identical to one of the known PS II polypeptides (see Table 1) or whether it was an additional peptide which was not yet shown to be a constituent of PS II. Those results have shown that the antiserum raised against the purified L–amino acid oxidase protein clearly recognizes a 36 kDa peptide in highly active, purified PS II complexes of *A. nidulans* (ref. 72). In O_2 evolving PS II complexes, three polypeptides in the 30 kDa region have so far been identified (D1, D2 and MSP). With the corresponding antisera we could show that the 36 kDa peptide is neither D1 nor the MSP. More recent results with an antiserum raised against D2 (kind gift from Prof. Trebst –unpublished results) have shown that the 36 kDa peptide is neither D2. Under our conditions which were optimized to obtain the best separation of the 36 kDa peptide from the other peptides, the D2 peptide is exclusively located in the upper aggregation bands of the SDS polyacrylamide gel. The difference in molecular weight of the isolated L–amino acid oxidase protein (50 kDa) and the peptide of 36 kDa recognized by the antiserum in PS II complexes, could possibly be best explained by suggesting that the membrane bound form is a processed or otherwise modified form of the soluble protein. It is also possible that the peptide in PS II complexes is associated with lipids, and this might lead to an alteration of the molecular weight as has been shown for other membrane proteins (ref. 73).

Since those results indicated that PS II complexes from *Anacystis nidulans* contain an additional peptide in the 30 kDa region (besides D1, D2 and MSP), it was the question whether any information in the literature exists about such an additional peptide. Riethman and Sherman (ref. 74) have shown that in PS II preparations from *A. nidulans R2* (PCC 7942) a 36 kDa peptide of unknown function is present. This peptide which is not D1, D2 or the MSP, is one component of a chlorophyll protein complex consisting of at least 3 peptides (36, 34 and 12 kDa). This complex increases when the cells are grown under iron deficiency. Moreover, Mei *et al.*, (ref. 75) have shown that a complex which was isolated from spinach PS II complexes after cross–linking the MSP to the complex, consisted of D1, D2, the MSP, and an additional 34 kDa polypeptide of unknown function — so far not shown to be present in PS II.

From our results with *A. nidulans* (PCC 6301) it seems justified to conclude that an additional peptide (besides the known ones) is present in PS II, and the papers cited above seem to support this conclusion. Therefore, it was the question whether we could find such a peptide (and possibly also such an L–amino acid oxidase activity in other cyanobacteria). Those investigations have shown, that the antiserum raised against the L–amino acid oxidase protein crossreacted with crude extracts of a number of cyanobacteria, such as *A. nidulans R2* (PCC 7942), *Synechocystis PCC 6803*, *Anabaena variabilis* (ATCC 29413) and the thermophilic cyanobacterium *Synechococcus sp.*. However, an L–amino acid oxidase activity with high specificity for basic L–amino acids, has so far only been detected in the closely related cyanobacterium *A. nidulans R2* (unpublished results). This could mean that the mechanisms for suppressing the L–amino acid oxidase activity in favor of the water oxidizing activity might be better developed in the other cyanobacteria.

To investigate this aspect further, we have chosen the thermophilic cyanobacterium *Synechococcus sp.* which has been shown to have very stable PS II complexes (refs. 76 and 77). This cyanobacterium has also an L–arginine metabolizing enzyme which is immunologically related to the L–amino acid oxidase from *A. nidulans* (refs. 78 and 79). As in *A. nidulans* the enyzme is also partly present in the soluble fraction and partly present in thylakoid membranes and in purified PS II complexes. However, a number of properties of this enzyme are quite different from the enzyme in *A. nidulans*. The enzyme catalyzes the conversion of L–arginine to ornithine and NH_4^+ and requires no O_2 or other added electron acceptor for this reaction. Moreover, the turn–over number of 85 for this reaction is very low in constrast to the high turn-over number of 70 000 for the *Anacystis* enzyme. Superficially, the activity of the enzyme from *Synechococcus sp.* could be classified as a dihydrolase activity, but a number of results indicate that the enzyme might have a rather complex reaction mechanism which we do not yet understand. Since the two enzymes are immunologically related, it could mean that the initial reaction in both cases is the oxidation of L–arginine to imino–arginine but that the subsequent reactions are different –depending on whether the enzyme can or can not interact with O_2 in the atmosphere. In the *Anacystis* enzyme the reduced flavin can interact with O_2 to form H_2O_2, and the over–all reaction is a regular amino acid oxidase reaction. In contrast, the reduced form of the *Synechococcus sp.* enzyme (the prosthetic group of this enzyme is unknown) can not interact with O_2, and as a consequence of a complex internal reaction the guanidino group is split off and hydrolyzed to NH_4^+. Subsequently the formed imino–ornithine becomes reduced by the reduced form of the enzyme to give ornithine as product. This aspect requires further work.

In conclusion, it can be said, that an L–arginine metabolizing enzyme is associated with PS II preparations from the mesophilic cyanobacterium *A. nidulans* and the thermophilic cyanobacterium *Synechococcus sp.*. Both enzymes metabolize L–arginine, they are immunologically related, but otherwise they show substantial differences in properties. If our hypothesis is correct, then these activities should represent different activation stages and/or different evolutionary stages of this enzyme (from an originally L–arginine metabolizing activity eventually leading to the water oxidizing activity). These differences of the two described enzymes might also indicate that during evolution various organisms have developed slightly different strategies for supressing the L–arginine metabolizing activity and for optimizing the water oxidizing activity of this protein. If we suggest that an enzyme which originally metabolized L–arginine, became the water oxidizing enzyme after proper modifications, then the binding site(s) for L–arginine on the enzyme must have been suited for binding two water molecules. Possibly it could be imaged that the two water molecules could bind at the binding sites for the α–amino group and the basic R–group of L–arginine —related to the fact that water and ammonia are isoelectronic.

To prove our hypothesis that the water oxidizing enzyme has evolved from a substrate dehydrogenase type enzyme will require to clearly establish that an additional polypeptide in the 30 kDa region (besides D1, D2 and MSP) is present in PS II complexes from cyanobacteria as well as higher plants. Moreover it is necessary to identify the corresponding gene. By performing the inactivation experiments of the gene and the corresponding transformation experiments in cyanobacteria, we hope eventually to answer the question whether our hypothesis is or is not correct. However, based on our results we think that it is justified to suggest such an alternative model of the elusive water oxidizing enzyme of PS II.

Acknowledgements.

The financial support of the Deutsche Forschungsgemeinschaft is gratefully acknowledged. I thank A.E. Gau., M. Kuhlmann, M. Ruff and G. Wälzlein for critical reading of the manuscript and valuable discussions.

References

1 P.L. Dutton *Energy Transduction in Anoxygenic Photosynthesis* In: L.A. Staehelin and C.J. Arntzen (Eds) *Photosynthesis III, Encyclopedia of Plant Physiology* New Series, Vol. 19, pp. 197-237, SpringerVerlag Berlin, Heidelberg, New York, Tokyo, 1986.

2 D.R. Ort *Energy Transduction in Oxygenic Photosynthesis: An Overview of Structure and Mechanism* In: L.A. Staehelin and C.J. Arntzen (Eds.) *Photosynthesis III, Encyclopedia of Plant Physiology* New Series, Vol. 19, pp. 143-196, SpringerVerlagBerlin, Heidelberg, New York, Tokyo, 1986.

3 L.–E. Andréasson and T. Vänngård, Ann. Rev. Plant Physiol. Plant Mol. Biol. **39** (1988) 379.

4 D.A. Bryant *The Cyanobacterial Photosynthetic Apparatus: Comparison to Those of Higher Plants and Photosynthetic Bacteria* In: T. Platt, and W.K.W. Li (Eds.), Canadian Bulletin of Fisheries and Aquatic Sciences **214** (1987) 423., Dept. of Fisheries and Oceans Ottawa, Canada.

5 K. Satoh, T. Ohno and S. Katoh, FEBS Lett. **180** (1985) 326.

6 H.J. Van Gorkom, Photosynth. Res. **6** (1985) 97.

7 J.P. Dekker and H.J. Van Gorkom, J.Bioenergetics Biomembranes **19** (1987) 125.

8 G.T. Babcock *The Photosynthetic Oxygen–Evolving Process* In: J. Amesz (Ed.) *Photosynthesis, New Comprehensive Biochemistry* Vol. 15, pp. 125-158. Elsevier, Amsterdam, New York, Oxford, 1987.

9 P.H. Homann, Plant Physiol. **88** (1988) 1.

10 A.W. Rutherford, Trends Biochem. Sci. **14** (1989) 227.

11 Ö. Hansson and T. Wydrzynski, Photosynth.Res. **23** (1990) 131.

12 B. Andersson and H.E. Åkerlund *Proteins of the Oxygen Evolving Complex* In: J. Barber (Ed.) *Topics in Photosynthesis* Vol. 8, *The Light Reactions* pp. 379-420, Elsevier Amsterdam,1987.

13 J. Masojidek, M. Droppa and G. Horvath, Eur. J. Biochem. **169** (1987) 283.

14 W.F.J. Vermaas *Photosystem II Function as Probed by Mutagenesis* In: S.E. Stevens Jr., and D.A. Bryant (Eds.) *Light–Energy Transduction in Photosynthesis: Higher Plant and Bacterial Models* pp. 197-214. American Society of Plant Physiologists, Rockville, Maryland, USA, 1988.

15 A.C. Stewart, U. Ljungberg, H.E. Åkerlund and B. Andersson, Biochim. Biophys. Acta **808** (1985) 353.

16 S. Specht, M. Kuhlmann and E.K. Pistorius, Photosynth. Res. **24** (1990) 15.

17 B.A. Barry and G.T. Babcock, Chemica Scripta **28A** (1988) 117.

18 R.J. Debus, B.A. Barry, I. Sithole, G.T. Babcock and L. McIntosh, Biochemistry **27** (1988) 9071.

19 J. Deisenhofer, O. Epp, K. Miki, R. Huber and H. Michel, Nature (London) **318** (1985) 618.

20 J. Barber, Trends Biochem. Sci. **12** (1987) 321.

21 H. Michel and J. Deisenhofer, Biochemistry **27** (1988) 1.

22 J.P. Allen, G. Feher, T.O. Yeates, H. Komiya and D.C. Rees *Structure of the Reaction Center from* Rhodobacter sphaeroides *R-26 and 2.4.1.* In: J.Breton and A.Vermeglio (Eds.) *The Photosynthetic Bacterial Reaction Center* pp. 5-11, Plenum Press New York, 1988.

23 A. Trebst, Z. Naturforsch. **41c** (1986) 240.

24 O. Nanba and K. Satoh, Proc. Natl. Acad. Sci. USA **84** (1987) 109.

25 M. Ikeuchi and M. Inoue, FEBS Lett. **241** (1988) 99.

26 J. Barber *Similarities and Differences between the Photosystem Two and Purple Bacterial Reaction Centers* In: S.E. Stevens Jr. and D.A. Bryant (Eds.) *Light–Energy Transduction in Photosynthesis: Higher Plant and Bacterial Models* pp.178-196, American Society of Plant Physiologists Rockville, Maryland, USA,1988.

27 A.W. Rutherford *Photosystem II, The Oxygen Evolving Photosystem* In: S.E. Stevens Jr. and D.A. Bryant (Eds.) *Light–Energy Transduction in Photosynthesis: Higher Plant and Bacterial Models* pp. 163-177, American Society of Plant Physiologists Rockville, Maryland, USA,1988.

28 J.P. Dekker, N.R. Bowlby and C.F. Yocum, FEBS Lett. **254** (1989) 150.

29 A.K. Mattoo,J.B. Marder and M. Edelman, Cell **56** (1989) 241.

30 P. Joliot, G. Barbieri and R. Chabaud, Photochem.Photobiol. **10** (1969) 309.

31 B. Kok, B. Forbush and M. McGloin, Photochem. Photobiol. **11** (1970) 457.

32 R.Radmer and O.Ollinger, FEBS Lett. **195** (1986) 285.

33 K.P. Bader, P. Thibault and G.H. Schmid, Biochim. Biophys. Acta **893** (1987) 564.

34 G. Renger and Govindjee, Photosynth. Res. **6** (1985) 33.

35 K. Sauer, R.D. Guiles, A.E. McDermott and J.L. Cole, Chemica Scripta **28A** (1988) 87.

36 A. Boussac, J.–L. Zimmermann, A.W. Rutherford and J. Lavergne, Nature **347** (1990) 303.

37 G.W. Brudvig and R.H. Crabtree, Proc. Natl. Acad. Sci. USA **83** (1986) 4586.

38 G.N. George, R.C. Prince and S.P. Cramer, Science **243** (1989) 789.

39 C.F. Yocum *Electron Transfer on the Oxidizing Side of Photosystem II: Components and Mechanism* In: L.A. Staehelin and C.J. Arntzen (Eds.) *Photosynthesis III, Encyclopedia of Plant Physiology* New Series, Vol. 19, pp. 437-446. Springer Verlag Berlin, Heidelberg, New York, Tokyo, 1986.

40 P.H. Homann, J. Bioenergetics Biomembranes **19** (1987) 105.

41 J.J. Brand and D.W. Becker, Meth. Enzymology **167** (1988) 280.

42 A.–F. Miller and G.W. Brudvig, Biochemistry **28** (1989) 8181.

43 J. Zhao and J.J. Brand, Arch. Biochem.Biophys. **264** (1988) 657.

44 A. Boussac and A.W. Rutherford, Chemica Scripta **28A** (1988) 123.

45 E.K. Pistorius and G.H. Schmid, Biochim.Biophys. Acta **890** (1987) 352.

46 C. Critchley, Biochim. Biophys. Acta **811** (1985) 3346.

47 N. Murata and N. Miyao, Trends Biochem. Sci. **10** (1985) 122.

48 N.I. Bishop *Evidence for Multiple Functions of the Intrinsic, 32-34 kDa Chloroplast Membrane Polypeptide of* Scenedesmus *in Photosystem II Reactions* In: W. Wiessner, D.G. Robinson and R.C. Starr (Eds.) *Algal Development (Molecular and Cellular Aspects)* pp. 150-155, Springer Verlag Berlin, Heidelberg, 1987.

49 M.A. Taylor, J.C.L. Packer and J.R. Bowyer, FEBS Lett. **237** (1988) 229.

50 C. Preston and M. Seibert, Photosynth. Res., **22** (1989) 101.

51 I. Virgin, S. Styring and B. Andersson, FEBS Lett. **233** (1988) 408.

52 G.C. Dismukes, Chimica Scripta **28A** (1988) 99.

53 E.K. Pistorius, R. Kertsch and S. Faby, Z. Naturforsch. **44c** (1989) 370.

54 M. Hirano, K. Satoh and S. Katoh, Photosynth. Res. **1** (1980) 149.

55 G.A. Peschek *Respiratory Electron Transport* In: P. Fay and C. Van Baalen (Eds.) *The Cyanobacteria* pp. 119-161, Elsevier, Amsterdam, New York, Oxford, 1987.

56 S. Scherer, H. Almon and P. Böger, Photosynth. Res. **15** (1988) 95.

57 Y. Anraku and R.B. Gennis, Trends Biochem. Sci. **12** (1987) 262.

58 S. Shinkai, H. Nakao, N. Honda and O. Manabe, J. Chem. Soc. Perkin Trans. **1** (1986) 1825.

59 A.J. Smith, J. London and R.Y. Stanier, J. Bact. **94** (1967) 972.

60 R.D. Simon *Inclusion Bodies in the Cyanobacteria: Cyanophycin, Polyphosphate, Polyhedral Bodies* In: P. Fay and C. Van Baalen (Eds.) *The Cyanobacteria* pp. 199-225, Elsevier, Amsterdam, New York, Oxford, 1987.

61 F.R. Tabita *Carbon Dioxide Fixation and Its Regulation in Cyanobacteria* in: P. Fay and C. Van Baalen (Eds.) *The Cyanobacteria* pp. 95-117, Elsevier Amsterdam, New York, Oxford, 1987.

62 E.K. Pistorius, K. Jetschmann, H. Voss and B. Vennesland, Biochim, Biophys. Acta **585** (1979) 630.

63 E.K. Pistorius and H. Voss, Biochim. Biophys. Acta **611** (1980) 227.

64 G. Wälzlein, A.E. Gau and E.K. Pistorius, Z. Naturforsch. **43c** (1988) 545.

65 E.K. Pistorius, Z. Naturforsch. **40c** (1985) 806.

66 H.I.X. Mager and R. Addink, Tetrahedron **39** (1983) 3359.

67 E.K. Pistorius and H. Voss, Eur. J. Biochem. **126** (1982) 203.

68 E.K. Pistorius, Eur. J. Biochem. **135** (1983) 217.

69 E.K. Pistorius and A.E. Gau, Biochim. Biophys. Acta **849** (1986) 203.

70 E.K. Pistorius and G.H. Schmid, FEBS Lett. **171** (1984) 173.

71 E.K. Pistorius and A.E. Gau, FEBS Lett. **206** (1986) 243.

72 A.E. Gau, G. Wälzlein, S. Gärtner, M. Kuhlmann, S. Specht and E.K. Pistorius, Z. Naturforsch. **44c** (1989) 971.

73 R.B. Gennis *Biomembranes. Molecular Structure and Function: Structural Principles of Membrane Proteins* (Chapter 3), pp. 85-137, Springer–Verlag New York, Berlin, Heidelberg, London, Paris, Tokyo, 1989.

74 H.C. Riethman and L.A. Sherman, Biochim. Biophys. Acta **935** (1988) 141.

75 R. Mei, J.P. Green, R.T. Sayre and W.D. Frash, Biochemistry **28** (1989) 5560.

76 T. Ohno, K. Satoh and S. Katoh, Biochim. Biophys. Acta **852** (1986) 1.

77 J.P. Dekker, E.J. Boekema, H.T. Witt and M. Rögner, Biochim. Biophys. Acta **936** (1988) 307.

78 R. Meyer and E.K. Pistorius, Biochim. Biophys. Acta **893** (1987) 426.

79 R. Meyer and E.K. Pistorius, Biochim. Biophys. Acta **975** (1989) 80.

Electron and Proton Transfer in Chemistry and Biology, edited by A. Müller et al.
Studies in Physical and Theoretical Chemistry, Vol. 78

Proton Pumps, Proton Flow and Proton ATP Synthases in Photosynthesis of Green Plants

W. Junge, G. Althoff, P. Jahns, S. Engelbrecht, H. Lill
and G. Schönknecht

Biophysik, Universität Osnabrück
D-4500 Osnabrück (Germany)

Summary

Photosynthesis by bacteria and green plants is the basis for life on earth. In oxygenic photosynthesis by green plants three fourth of the useful work derived from light is delivered as redox couples and one fourth as electrochemical potential difference of the proton across the thylakoid membrane. The latter drives ATP synthesis. The article is a survey of the generation and the use of the protonmotive force. It describes the electrochemical relaxation in thylakoids, both along and across the coupling membrane. And it reviews the properties of proton flow through the channel portion, CFO, of the ATP synthase and through the complete enzyme, CFOCF1. The detailed kinetic analysis of these processes is mainly owed to flash spectrophotometry with intrinsic and added molecular probes, and with patch clamp techniques.

Introduction

Light driven proton pumps and proton translocating ATP synthases probably stood at the beginning of photoautotrophic life. Oxygenic photosynthesis, its most advanced form, delivers the greater portion of the useful work derived from light as redox potential, but still one quarter is transduced *via* protons to ATP. The evolution of photosynthesis from its prokaryotic origin to green plants may have proceeded in the steps that are schematically illustrated as "seven inventions" in Fig.1. A primitive proton pump, based on a retinal protein, (ref. 1), is found in one particular archaebacterium (*Halobacterium halobium*). Its plasma membrane also incorporates an ATP synthase. This belongs to the superfamily of F– and V–ATPases, (ref. 2), that is ubiquitous in higher plants and animals. While the ATP synthase has been conserved through further evolution, another type of a proton pump has been invented that is based on the interplay between light driven, and chlorophyll-mediated vectorial electron transfer, followed by protonation/deprotonation and vectorial hydrogen transfer (electron/hydrogen loops (ref. 3)). In purple bacteria it is composed of two membrane proteins: the photochemical reaction centre with bacteriochlorophyll as primary electron donor and bound quinone (Q) as acceptor and cytochrome–b_6.

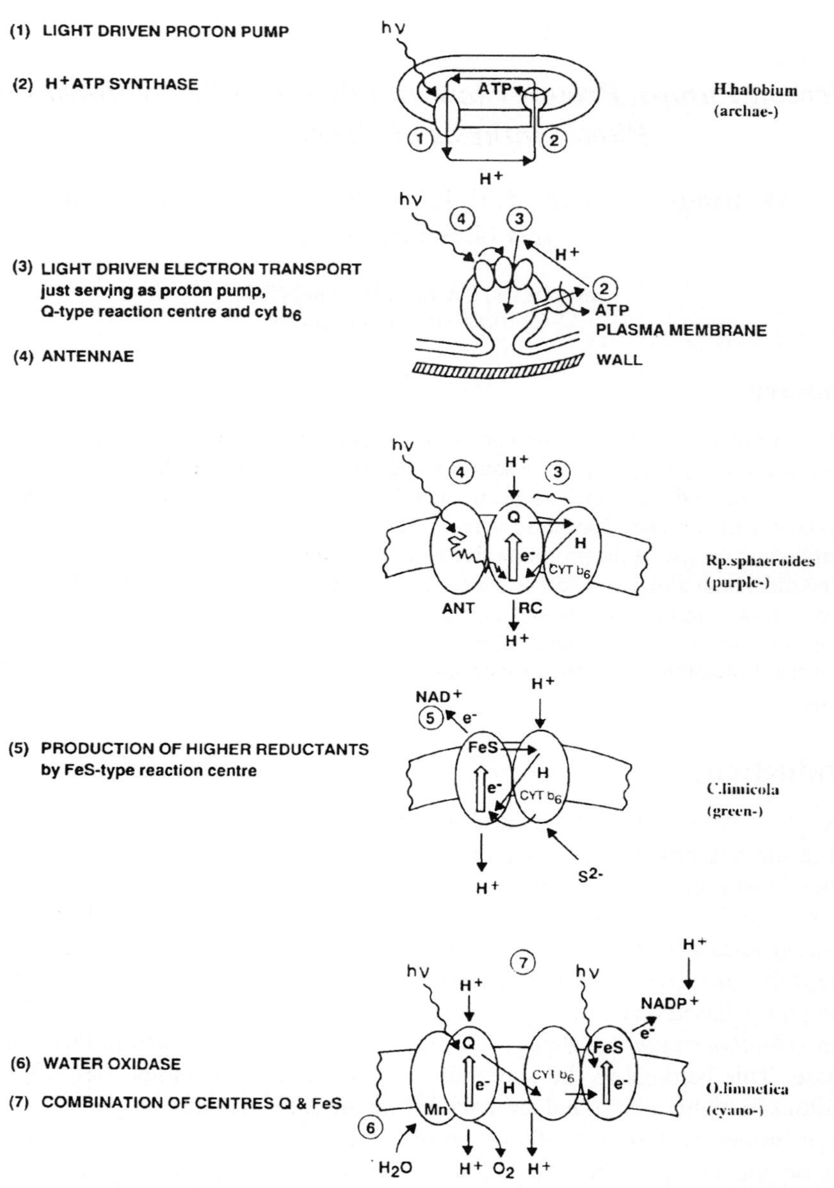

Fig. 1. Elements of photosynthesis as present in recent archaebacteria, purple bacteria, green bacteria and cyanobacteria (likewise in chloroplasts of green plants). From top to bottom the scheme scans a conceivable path of evolution towards oxygenic photosynthesis.

Electron transfer cycles, not futile, however, owing to net proton pumping. To improve the absorption cross section of the photochemically reactive bacteriochlorophyll each reaction centre has been supplemented by pigment–proteins serving as antennae. Another type of a reaction centre has been invented with iron–sulfur clusters as secondary electron acceptors. These centres are found in recent green sulphur bacteria. In addition to contribute to proton pumping, they drive electron transport from substances with more positive to those with more negative redox potential (*e.g.* from H_2S to NAD). The final steps might have been the modification of a Q–type reaction centre (as from purple bacteria) to an oxidizing power of about plus 1 volt and the addition of a manganese catalyst to oxidize water. The combination of the modified Q–type centre with a FeS–type centre (as from green bacteria) finally promotes electron transfer from ubiquitous water to NADP. This form of photosynthesis is present in cyanobacteria, algae and green plants. The ATP synthase which utilizes the electrochemical potential difference of the proton has been conserved throughout.

This article describes three aspects of the protonic reactions of photosynthesis in green plants in some detail: 1.) the electric and the protonic relaxation along and across the coupling membrane, 2.) the protolytic reactions of water oxidizing photosystem II, and 3.) proton flow through the ATP synthase. It starts with a short description of measuring techniques and of the complicated membrane structure of thylakoid membranes in green plants.

The System and the Techniques

Chloroplasts of higher plants carry an inner lamellar system, thylakoids, that is folded into densely packed membranes (grana, stacked thylakoids) and interconnecting membranes (stroma membranes). The photosynthetic apparatus consists out of four large protein complexes: photosystem II oxidizes water, photosystem I reduces NADP, the cytochrome b_6,f–complex mediates electron transfer between the latter two, and CF_OCF_1 synthesizes ATP at the expense of the transmembrane electrochemical potential difference of the proton. Fig. 2 illustrates this structure. The protein complexes are seggregated according to the folding pattern. Apposed membrane portions contain photosystem II, exposed membrane portions contain photosystem I and the ATP synthase (ref. 4). The upper inset shows an electron micrograph of a stacked domain. Each of the stacked disk–like structures (diameter about 0.5 μm) contains about 100 photosystem II molecules. The inner lumen and the outer partitions between stacked thylakoid membranes are very narrow, only about 5 nm wide slabs. In the middle of Fig. 2 the electron transport is sketched. In reaction centres the transmembrane transfer of electrons is very rapid (some 100 ps). The sites of proton uptake from the outer phase (the chloroplast stroma) and proton release into the inner phase (thylakoid lumen) are indicated. The lower portion of Fig. 2 shows a simplistic equivalent circuit for cyclic proton flow between pumps and ATP synthases.

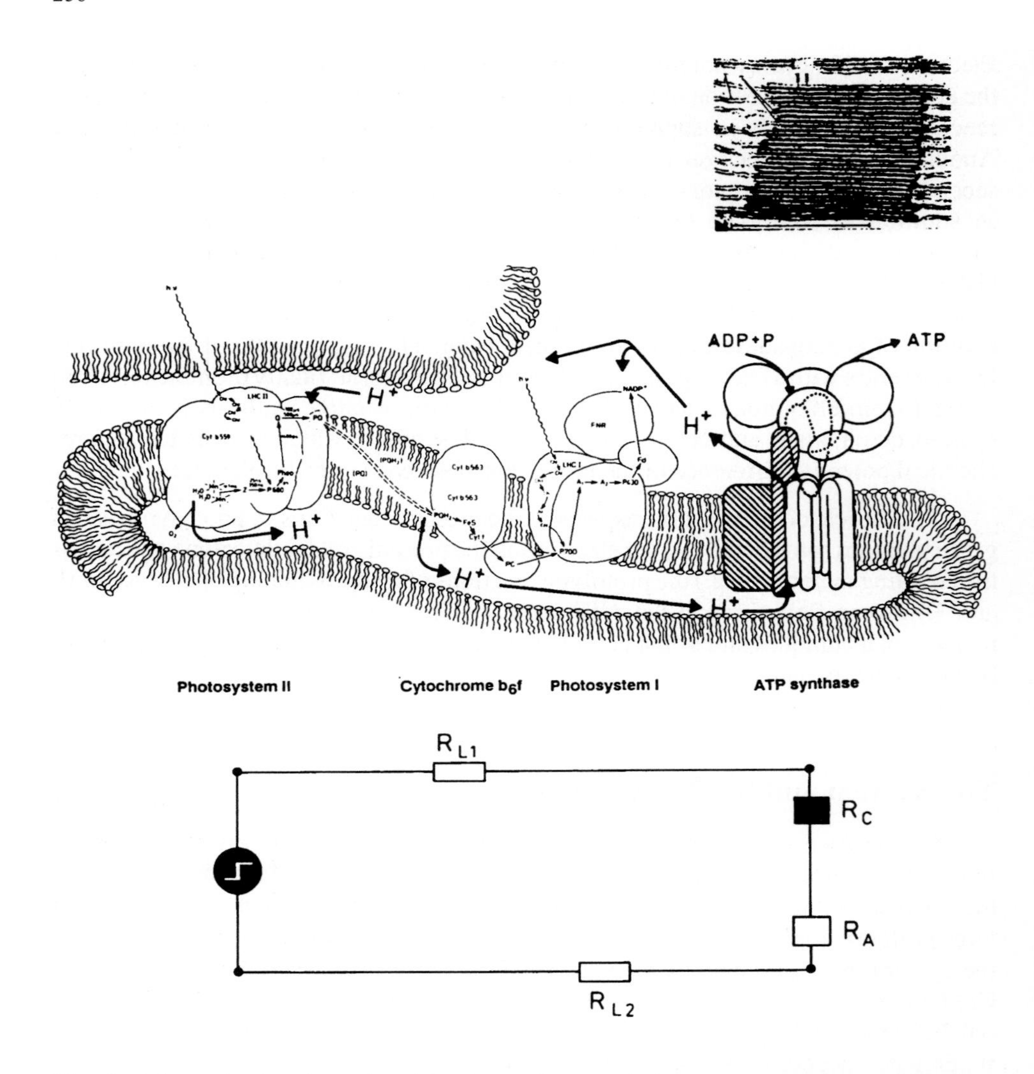

Fig. 2. Schematic drawing of thylakoid membranes: their arrangement (upper insert) — protein complexes, electron transport and proton flow (middle) — and an equivalent circuit for cyclic proton flow between pumps and ATP synthases (bottom). For details, see text.

Photoinduced protolytic reactions have been studied at high time resolution by laser–flash spectrophotometry. Thylakoids are a favourable object for such studies. Photosystems are to be stimulated for a single turnover by short laser flash (<10 ns). Transients of the transmembrane voltage can be measured at ns time resolution by electrochromism of intrinsic pigments (refs. 5 and 6). Hydrophilic pH–indicating dyes are practically selective for pH–transients in the large external bulk phase (ref. 7) while membrane adsorbed indicators (mainly neutral red) react to pH transients at both surfaces of the membrane. In the presence of appropriate non–permeating buffers they can be selective for pH transients in the very narrow (only 5 nm wide) lumen (refs. 8, 9, 10, 11, 81 and see below). While the response time of electrochromism so far has been instrument limited and in the ns time range, pH indicating dyes respond in the time range of 100 μs (around neutral pH). With these three observables, namely pH(lumen), pH(bulk medium) and transmembrane voltage, the spectrophotometric techniques have allowed "complete tracking of proton flow" (refs. 12 and 13).

Passive Electric and Protonic Relaxation at the Coupling Membrane

The **passive leak conductance** of the thylakoid membrane has been determined under flashing light. Single–turnover flashes of light generate a transmembrane voltage of below 50 mV and a transmembrane pH–difference of 0.05 units (around neutral pH, (ref. 9)). As the stoichiometry of net proton pumping over transmembrane electron transfer under flashing light is 1:1 (equ/equ) this ratio reflects the fact that the proton buffering capacity of the thylakoid lumen per unit area is about 12 times greater than the area–specific electrical capacitance (expressed in the same dimensions, see also (ref. 12)). In the absence of ATP synthase activity the transmembrane electric potential difference is relaxed with an exponential decay time of about 250 ms (20 mM 1:1 electrolyte). With the specific electric capacitance of $C = 1\ \mu F/cm^2$ (ref. 14) and applying the capacitor equation ($s = C/G$) this implies a specific leak conductance of $G = 4\ \mu S/cm^2$. The characterization of specific ion channels that might account for the rapid electric relaxation of thylakoid membranes has just begun. A voltage-gated anion channel that has been detected by patch-clamping of thylakoid membranes from giant chloroplasts of *Peperomia metallica* (ref. 15), and cation channels have been detected by fusion of thylakoids into planar bilayer membranes (ref. 82). Whether these channels are relevant under physiological conditions remains to be established. In particular, cation channels have been shown to be generated from components of the ATP synthase, CFOCF1, when thylakoid membranes or reconstituted CFOCF1 or even the proteolipid of CFO, alone, were incorporated into planar bilayers or giant liposomes (*eg*. ref. 64). The light induced transmembrane pH–difference is relaxed much more slowly, in about 12 s (ref. 11). The fast electric relaxation is carried by other ions than the proton. This is why a steadily growing pH–difference (of up to more than 3 units) is built up under steady illumination. The slower relaxation of the pH–difference is an approximately electroneutral process which is dominated by the lumenal buffering capacity. According to the above figure for the former, the specific leak conductance for protons is about four times smaller than the overall conductance, $1\ \mu S/cm^2$.

The **size of the electrically coupled unit** has been determined by asking how many (dimer–) molecules of the channel forming antibiotic gramicidin are minimally required to more rapidly discharge all membranes (ref. 16). Fig. 3 shows transient electrochromism with the concentration of added gramicidin as parameter.

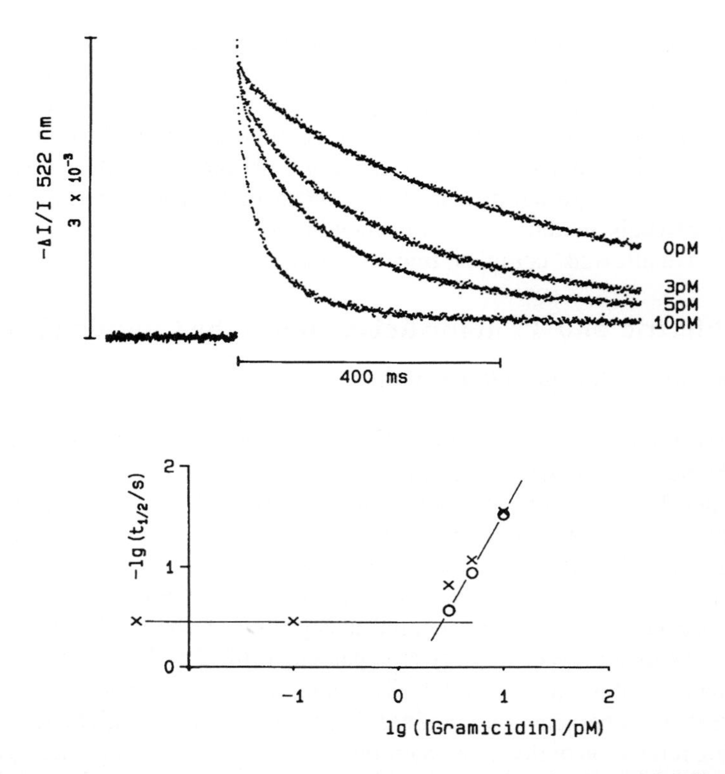

Fig. 3. Acceleration of the decay of the flash light induced electric potential difference across EDTA–washed thylakoid membranes as function of added gramicidin. TOP: Electrochromic absorption changes at 520 nm, BOTTOM: double–log plot of the decay rate. Average concentration of chlorophyll, 20 μM. (for details see ref. 16).

At the bottom, the logarithm of the inverse decay half-time is plotted over the logarithm of the gramicidin concentration. There is a sharp onset of the acceleration at about 3 pM followed by a continuous increase. Then the rate of decay rises with the second power (n=1.8) of the gramicidin concentration in agreement with the notion that the dimer acts as a channel (see (ref. 17)). At higher concentrations, when dimerization is complete, the dependence on the gramicidin concentration is linear. The quadratic dependence implies that most of the gramicidin molecules are still in their monomeric form. In comparison with chlorophyll concentration of 20 μM in these experiments the onset of acceleration at 3 pM gramicidin monomers, *i.e.* much less than 1.5 pM dimers, implies that the electrically connected membrane area contains much more than $2 \cdot 10^7$ chlorophyll molecules, conservatively estimating more than $5 \cdot 10^7$ (ref. 16). With the accepted specific area per chlorophyll of 2 nm^2 (ref. 18) the electrically connected area is at least 100 μm^2 large, almost as large as the estimated total area of thylakoid membranes in a chloroplast. Broadly speaking, the electric field is relaxed laterally in less than a few milliseconds over the whole membrane area, and thus the aqueous phases separated by the thylakoid membrane are both electrically equipotential.

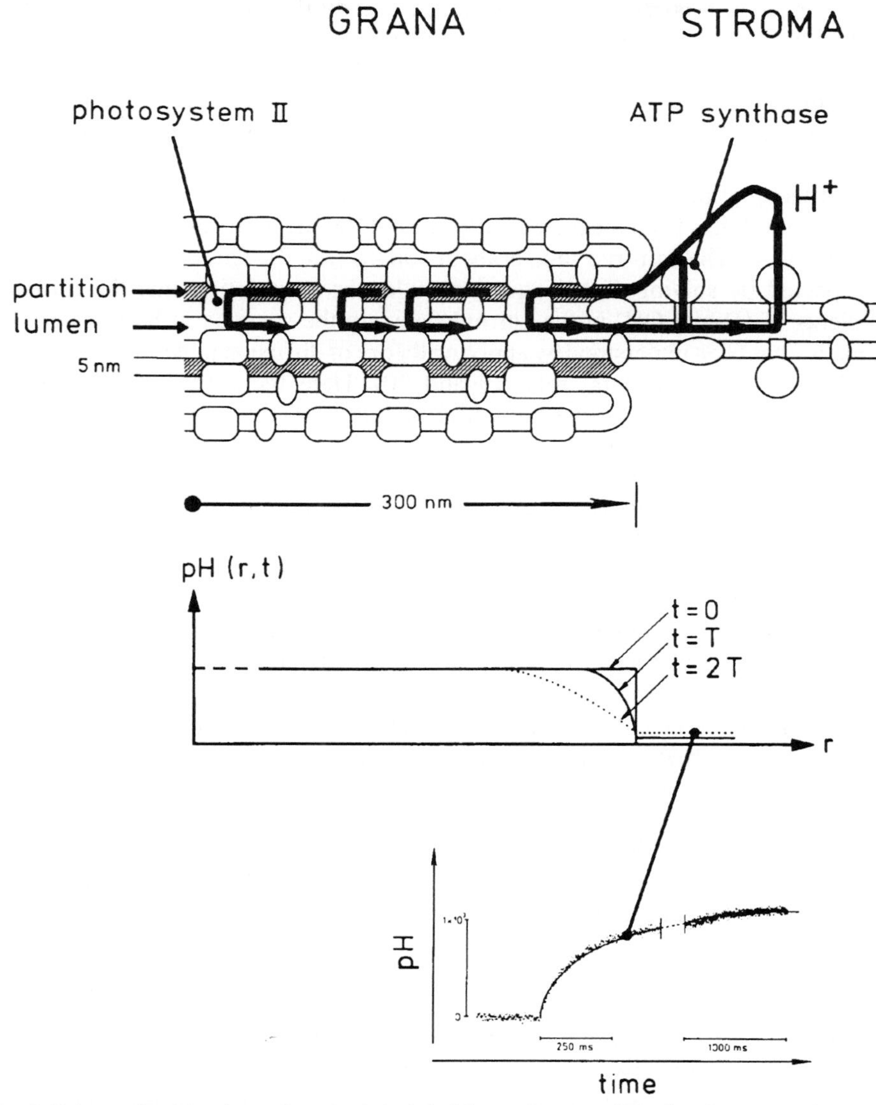

Fig. 4. Schematic side view of stacked thylakoid membranes with closely apposed grana and interconnecting stroma lamellae (ref. 79). The narrow outer space between thylakoids in a granum, called partition, is hatched. TOP: The fat arrows illustrate cyclic proton flow. MIDDLE: pH–profile in partitions after a single–turnover flash and its relaxation by diffusion into the medium. BOTTOM: Original data and theoretical fit of the pH–transient as reported by hydrophilic pH–indicating dye in the suspending medium (for details, see refs. 7 and11).

The **lateral relaxation of pH–differences** is expected to occur much more slowly than the electric one in particular around neutral pH where the sum of the concentrations of both contributors to diffusive relaxation, H^+ and OH^-, is minimum. The lateral relaxation has been studied in the narrow aqueous slabs separating the outer sides of stacked thylakoid membranes. Fig. 4 illustrates the layout of the experiments, original data and theoretical fit. Excitation with a short flash of light causes proton uptake by any photosystem II reaction center and from the narrow domain (partition) between stacked thylakoid membranes. The disk–like alkalinization jump is then relaxed by lateral diffusion of protons, hydroxyl anions and mobile buffers. This causes pH–rise in the medium. It has been measured with hydrophilic pH–indicators (see trace at the bottom of Fig.4). The half–rise time is about 100 ms (ref. 7). This is 105–times longer than expected by solving Fick's equation under the assumption that the diffusion coefficients of H^+ and OH^- in the narrow slabs are the same as in bulk water. The discrepancy has been understood by applying theory of diffusion in domains with fixed (ref. 11) and mobile buffers (refs. 19 and 20). Here the diffusive relaxation is determined by an effective diffusion coefficient. It is the weighted sum of the diffusion coeffients of proton, hydroxyl and mobile buffers. The weight factors are proportional to the ratio of the concentration of a given diffusing species over the effective concentration of all buffers (fixed and mobile) in this domain (ref. 20). Reasonable estimates for the total buffering capacity have explained the above factor of 105. Another implication is that the diffusion coefficients for proton and hydroxyl cation are rather smaller than larger than those in bulk water. For this protonic coupling membrane the above results have eliminated the possibility of enhanced surface diffusion of protons that has been postulated in the literature (ref. 21). Based on the foregoing the ohmic losses of protonmotive driving force between pumps and ATP synthases have been calculated (ref. 11). Under steady illumination, with the ATP synthase running at high speed and assuming the upper limit for the diffusion coefficients (the same as in bulk water) the ohmic losses are in the order of one tenth of the transmembrane pH–difference between a particular pump in the middle of a disk and an ATP synthase at the fringes. This loss is small but not negligible. Accordingly, the "drainage bassin" of an ATP synthase for protons is smaller than the very large interconnected membrane area with its rapid electric relaxation.

Protolytic Reactions of the Water Oxidase

Water oxidation is driven by Photosystem II. The primary photochemical reactions are carried by two pigmented polypeptides that resemble the reaction centres of purple bacteria regarding the redox cofactors, their position in the membrane, kinetic behaviour and even the sequence of the two major polypeptides (ref. 22). The most detailed cristal structure has been worked out for one purple bacterium, *R. viridis* (ref. 23). Water oxidation has not been observed in a unit as small as this reaction centre, but only in aggregates containing at least further five membrane proteins (see Fig. 5).

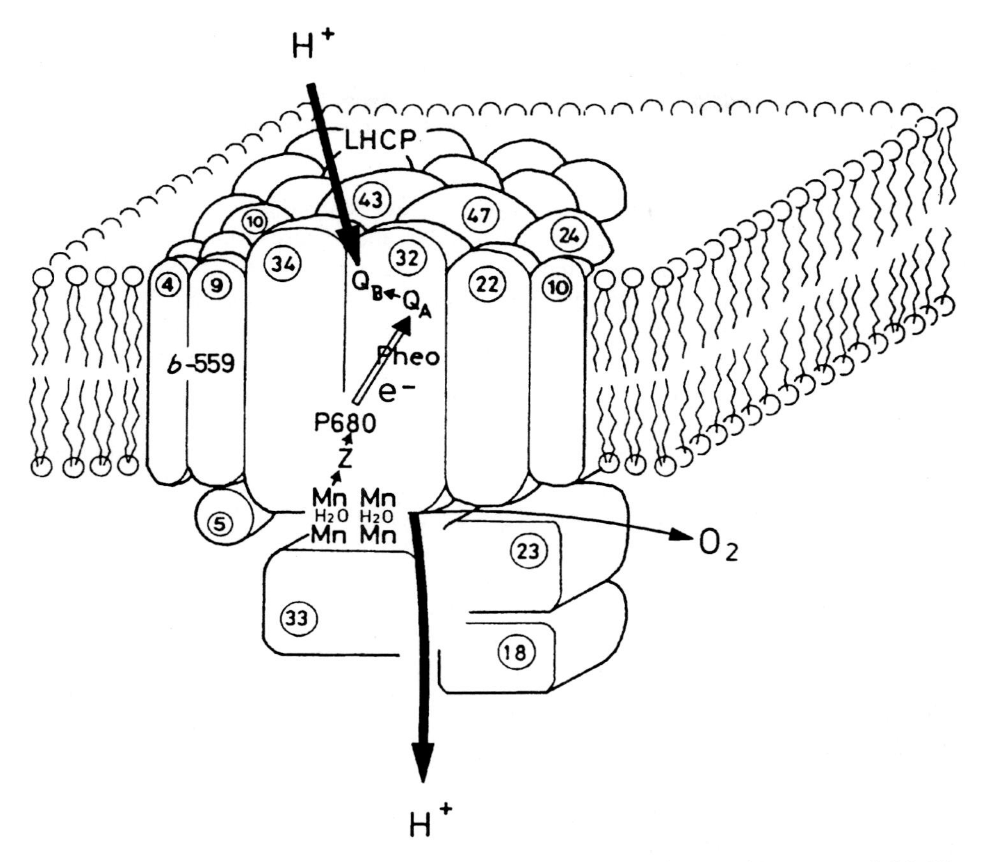

Fig. 5. Schematic drawing of Photosystem II adapted from Miyao and Murata, ref. 80. The apparent molecular masses of the proteins are indicated. The open arrows represent electron transport and the fat arrows proton uptake at the outer side of the thylakoid membrane and proton release into the lumen.

In thylakoids from green plants more than 15 intrinsic and extrinsic polypeptides are associated with the centre (ref. 24). This is illustrated in Fig.5.

The primary photochemical reaction in Photosystem II is the transfer of one electron from special chlorophyll a (P 680) at the lumen side of the thylakoid membrane to a pair of bound quinones (Q_A and Q_B) at the outer side (for the analogous reaction in reaction centres of purple bacteria, see Dutton, and Wraight this volume). The reduction of the quinones to quinole is two stepped. The doubly reduced Q_B leaves its binding site as QH_2 after association with two protons. Proton uptake shows no oscillations as function of the first, second, third, .. electron transfer step (ref. 7) although no protonation is spectroscopically appararent on the singly reduced (semi-)quinone. At this stage proton uptake likely reflects the electrostatic response of the protein environment to negative charge on Q rather than the pairing of electrons and protons at the cofactor. Only after a second reduction step is protonation direct and the protonated quinole leaves its binding pocket. This situation is similar to the one encountered in reaction centres from photosynthetic bacteria (see Wraight, see Dutton this volume).

The oxidation of water to yield dioxygen is a four stepped process driven by four quanta of light. It consumes two molecules of water, and releases four protons into the thylakoid lumen. After prolonged darkness the catalytic centres are synchronized mainly in state S_1 (out of the increasingly oxidized states S_0, S_1, S_2, S_3, $S_4 \rightarrow S_0$...). Excitation with a series of light flashes steps the centers forward and around the cycle. Oxygen is liberated during the transition $S_3 \Rightarrow S_4 \rightarrow S_0$ (ref. 25). Protons are liberated not only during this transition but, with a stoichiometric pattern to be discussed, during several transitions. The following observations have suggested that the pattern of proton release reflects the electrostatic response of the protein environment rather than the pattern of electron abstraction from water, which is supposed to be bound to the manganese centre. 1.) During one particular transition ($S_2 \Rightarrow S_3$) proton release (200 μs, ref. 26) precedes electron abstraction from the catalytic centre (350 μs ref. 27). 2.) The pattern of proton release as function of flash number but not the pattern of oxygen release varies depending on whether measurements are carried out with thylakoids or with photosystem II core preparations (ref. 28). 3.) It also varies in dependence of controlled trypsination of oxygen evolving membrane fragments (ref. 29). Thus it is probable that **protons are transiently liberated from amino–acid side groups** which are only reprotonated after the reaction of the catalytic centre with water.

For quite a while several laboratories have agreed that the **stoichiometric pattern of protons released over electrons abstracted from the water oxidase** is 1:0:1:2 when progressing from the transition $S_0 \Rightarrow S_1$ to $S_3 \Rightarrow S_4 S_0$. (refs. 30, 31, 32, 33 and 26). Recently, Lavergne and Rappaport, (ref. 34) have challenged this by reinterpreting the absorption changes of neutral red (when applied to stacked membranes). Their argument is based on the above cited slow relaxation of the alkalinization jump in the partition region (ref. 7). This region is not accessible to added non-permeant buffers, which are used to quench pH–transients in the suspending medium. Then the response of neutral red is composite from dye molecules located at the lumen side and at partitions. The superimposition is mended in totally destacked membranes, where the outer side of the thylakoid membrane is readily accessible to added non–permeant buffers. Here a non–integer, 1:0.5:1:1.5 (ref. 35), and pH–dependent pattern of proton release (Lavergne and Rappaport, unpublished) has been observed.

The older, but now questionable pattern of 1:0:1:2 (as found in thylakoids from green plants) has agreed very well with the pattern of charge storage (as observed in oxygen evolving reaction centres from a cyanobacterium). The latter has been detected by local electrochromism (ref. 36) and it has also been inferred from the effect of stored positive charges on the reduction kinetics of P680 (ref. 37). The new non–integer pattern of proton release calls for a reinvestigation of proton release and the internal electrostatic balance of the catalytic centre.

How to pin down the so far ill–defined proteins and amino–acid side chains that are responsible for the storing and channelling of protons out of the catalytic centre of the water oxidase? A possible clue is **the short–circuit of the proton–pumping activity of photosystem II** which has been observed after covalent modification of two polypeptides with N,N'–dicyclohexylcarbodiimide (ref. 38). The modification has induced the following effects: Proton release by water oxidation into the lumen and proton uptake from the outer phase upon reduction of bound quinone are diminished to the same extent. This parallelism holds for all redox transitions of the water oxidase. The seemingly non–released protons account for the lack of proton uptake from the outer phase (ref. 39). A rapid electrogenic back reaction discharges one portion of the transmembrane voltage (ref. 38). The oxygen evolving capacity is unaffected. These effects have been interpreted as an intra–protein short–circuit for protons that is not accompanied by proton–leakiness of the membrane. This effect is associated with the covalent modification of chlorophyll a/b containing antennae (LHC–) proteins. According to accepted structural models the modified residue(s) are located at the lumen side of a transmembrane helix (ref. 40). It seems that some light harvesting antennae proteins serve the additional function to channel protons from the catalytic centre of the water oxidase into the lumen. This function is probably not essential for water oxidation, but it may be essential for proton pumping by photosystem II. In this respect it will be interesting to search for proteins with such a function in cyanobacteria that lack LHC–proteins.

Proton Flow through the ATP Synthase and its Channel Portion, CF_O

The proton translocating ATP synthase is built from two distinct portions, the intrinsic membrane protein CF_O is supposed to act as access channel for protons, and the peripheral CF_1 contains the nucleotide binding sites. ATP synthases of the F_OF_1–type are ubiquitous in bioenergetics, they are not only found in photosynthetic organisms but also in mitochondria and non–photosynthetic bacteria (re. 41). A remarkable feature of these enzymes is a three–fold pseudosymmetry of the larger subunits of the headpiece (3α and 3β) which are arranged as a hexagonal chair (ref. 42). This aggregate interacts with nucleotides and phosphate. Another remarkable feature is the abundance of a very short and very hydrophobic subunit, a proteolipid, in the channel portion, which is probably present in more than 9 copies (refs. 43 and 44). The mechanism of coupling between proton flow and ATP synthesis is not known. The direct interference of protons with the catalytic events at the nucleotide binding sites is now considered as improbable. A more indirect (mechanical) coupling between proton flow through the transmembrane domain of CF_0CF_1 and the ejection of bound ATP is considered as the better working hypothesis. There is evidence for the spontaneous formation of bound ATP from bound ADP and bound phosphate. In this respect the role of the smaller subunits at the interface between the channel portion, CF_0, and the catalytic portion, CF_1, is of particular interest. We have focussed on two particular aspects, proton flow, both, through the intact enzyme and through the exposed channel portion, and the role of one particular subunit, δ.

Proton flow through the active enzyme is documented in Fig.6.

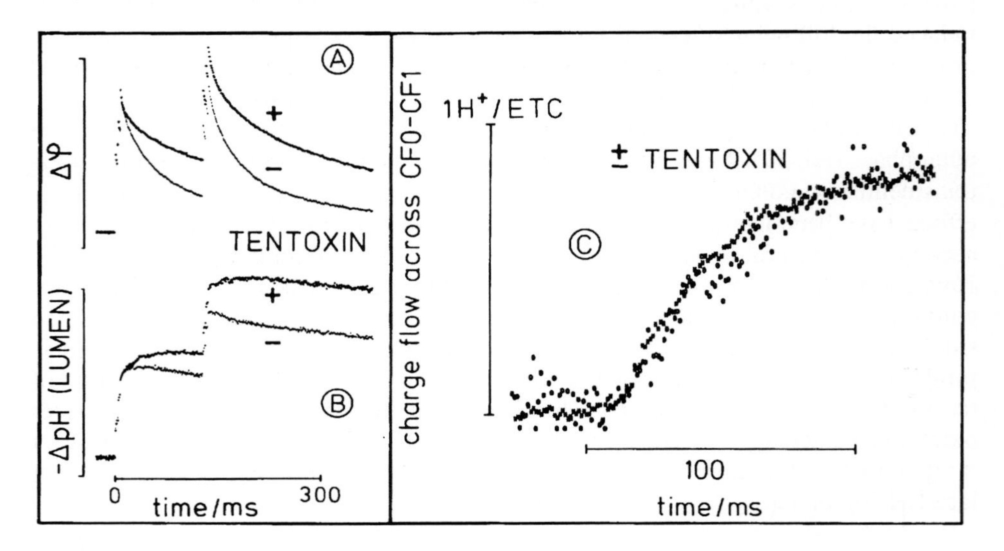

Fig. 6. Transients of the transmembrane voltage (A) and of the lumenal pH (B) in the presence of 20 μM ADP and 60 μM Pi and with and without tentoxin (specific blocker of ATP synthase). Reproduced from ref. 79. Suspension of thylakoids was excited with two groups of three short flashes each. The transmembrane voltage was recorded by electrochromisms and the pH transient by neutral red. Part C gives a superimposition of two data sets, namely, the number of protons entering CF_1CF_0 from the lumen and the number of charges crossing CF_1CF_0 in the time interval between the first and the second group of three short flashes. For details, see ref. 19.

Fig. 6 shows transients of the transmembrane voltage (A) and of the lumenal pH (B) as observed under excitation of thylakoids with two groups of three closely spaced light flashes. In the presence of a specific poison of the ATP synthase, tentoxin, both observables are relaxed slowly. The transmembrane voltage (A) decays more rapidly than the acidification of the thylakoid lumen. This reflects an electric leak conductance that is carried by ions other than the proton. In the absence of tentoxin and with the substrates (ADP and P) present, ATP synthesis has been turned on. This causes an accelerated decay of the transmembrane voltage and of the lumenal acidification. The difference between the two traces in Fig.6A indicates the extra transmembrane charge flow that accompanies ATP synthesis. The difference between the two traces in Fig.6B indicates extra proton uptake from the lumen. These two differences have been calibrated according to objective criteria and, for a short time segment, plotted on top of each other in Fig.6C. That they coincided within noise limits has shown that extra charges crossing the thylakoid membrane *via* CF_0CF_1 are protons taken up from the lumen (ref. 45). Another feature is evident from inspection of Fig.6. The accelerated charge flow stops at a certain driving force. This is due to the gating of the enzyme activity by the transmembrane voltage (ref. 45) or more generally by the transmembrane protonmotive force (ref. 46).

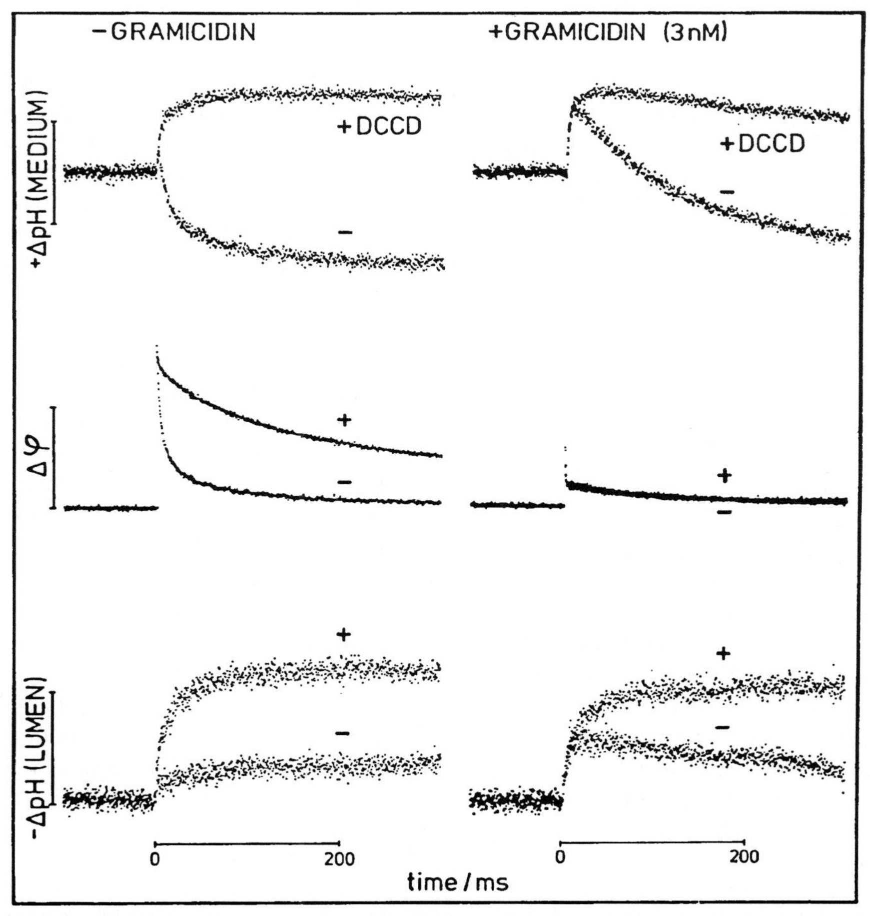

Fig. 7. Complete tracking of proton flow through CFO (reproduced from ref. 79, based on data from ref. 12). TOP: Transients of the pH in the medium, induced by a single short flash of light. MIDDLE: Transients of the transmembrane voltage. BOTTOM: pH–transients in the lumen. Thylakoids have been treated with EDTA to remove some CF_1. DCCD has been added, when indicated, to block proton flow through CF_O. LEFT: without and RIGHT: with added gramicidin.

Proton flow through the channel portion of the enzyme has been underestimated for quite a while. The turnover number for ATP synthesis of CF_0CF_1 can range up to 400 s^{-1} (ref. 47). With the accepted stoichiometry of 3 protons per ATP this implies a turnover number for protons of 1200 s^{-1}. Compared to this the turnover numbers of reconstituted F_O–channels from various organisms fell short by two orders of magnitude (refs. 48, 49 and 50). This has been incommensurate with the proposed role of F_O as low–impedance access to the coupling site in the enzyme (ref. 51). The shortcoming might have been caused by insufficient time resolution of the pH–electrode (as used in the cited studies) or by the survival during reconstitution of only a few active channels.

In an alternative attempt we have reinvestigated the protonic conduction of CF_O by flash spectrophotometry with thylakoids. The catalytic portion, CF_O, has been removed from thylakoid membranes by treatment with EDTA. The number of CF_1–-molecules which have been removed from the membrane has been determined by electro–immunmodiffusion. Thylakoids have been excited by short light flashes and the relaxation of the transmembrane voltage and of the pH–transients in the lumen and in the suspending medium have been monitored (refs. 12, 52 and 53). One set of experiments is documented in Fig. 7. It shows transients of the pH in the lumen (top), transients of the transmembrane voltage (middle) and transients of the pH in the medium (bottom). The signals have been recorded in pairs, with and without a blocking agent, here DCCD. The traces in the left column show the result in the absence of added gramicidin. Broadly speaking the differences between each pair of traces indicates: proton uptake from the lumen (top), charge flow across the membrane (middle) and proton release into the lumen (bottom), all of them mediated by CF_O as evident from the sensitivity to DCCD (and to different, more specific blocking agents). These curves have been quantitatively evaluated under considering the specific electric capacitance of thylakoid membranes and the number of CF_1 molecules removed (ref. 12). Under the assumption that any CF_1 which has been removed, leaves a conducting CF_O behind, the **protonic conductance per exposed CF_O is 10 fS, in the average**. The average refers to two domains, to the ensemble of all those CF_0, which have lost their CF_1–counterpart, and to the time (counting open and close intervals of the channels). At 100 mV driving force the conductance is equivalent of a turnover of about 6000 protons per second. This figure is fully compatible with the proposed role of CF_0 as proton channel of the enzyme.

Is this channel proton–specific? Considering that the relaxation of the transmembrane voltage has been measured by electrochromism and by two different pH–indicating dyes the answer is yes. The right column in Fig.7 shows what happens if the conductance for another ion becomes dominating. Gramicidin has been added. The electric potential difference has been collapsed by K^+–current very rapidly (Fig.7, middle, right). The relaxation of the transmembrane pH–difference has been slowed down. This is because the specific buffering capacity is larger than the specific electric capacity of the membrane (see above). This observation, in turn, has proven the proton specificity in the absence of gramicidin, where the pH–transients decayed as rapidly as the electric potential difference. The proton specificity has been scrutinized further and it persisted even at pH 8 and with 300 mM NaCl or KCl in the medium (ref. 53). This seeming selectivity over more than seven orders of magnitude is unparalleled for other known proton channels (*e.g.* gramicidin, selectivity for protons over other cations less than 100). It is the more astounding, as a homologous enzyme to CF_0CF_1 has been described for *Propionigenium modestum*, that can act both as a sodium or as a proton translocating ATPase (ref. 54).

The above described average proton conductance per exposed CF_0 is a lower limit. It has become clear that some exposed CF_0 are not conducting (ref. 55). We have attempted to determine the conductance of the *subset of conducting* CF_0–channels. The rationale has been to use vesicles as small as to contain only 0, 1, 2,.. active channels. Smaller vesicles have been obtained by EDTA– treatment of thylakoids. Their size has been determined by titration with gramicidin in the concentration range of 30 pM (ref. 52). For the experiments with gramicidin the conductance induced by exposure of CF_0 has been sealed with DCCD. These vesicles contain only about $5 \cdot 10^5$ chlorophyll molecules. In the presence of low concentrations of gramicidin the decay of the transmembrane voltage has been biphasic, which is understood by Poisson's distribution. The decay curves of the electrochromic absorption changes that result from an ensemble of 10^{11} vesicles have been fitted with the following equation:

$$U_{app}(t) = U_o exp(-n) exp\{exp[n(-tG/C)\}$$

wherein n denotes the average number of channels per vesicle, G is the time averaged conductance of one channel and C the vesicle capacitance. The equation describes the behaviour of ohmic channels without voltage–dependent gating which are distributed over vesicles according to Poisson's distribution. For gramicidin the analysis has reproduced literature figures of the open conductance of this antibiotic under given concentration of monovalent cations (ref. 52). The same type of analysis has also been applied to chromatophores of purple bacteria to determine their size. It is quivalent to 770 molecules of bacteriochlorophyll (ref. 56).

To the same vesicles, but in the absence of gramicidin and of the blocking agent to CF_O, a similar analysis has been applied to elucidate the conductance of CF_0. It has yielded a figure of 1 pS. This is equivalent to the turnover of $6 \cdot 10^5$ protons per second (refs. 52 and 53). This conductance is hundredfold greater than the one that has been determined (as average conductance) under the assumption that every *exposed* CF_0 is actually a *conducting* CF_0 (see above). The result means that only one out of hundred exposed CF_0 is conducting.

The conductance has been investigated as function of pH, pD (isotopic substitution), concentration of other cations, addition of glycerol to perturb the water structure, and under variation of temperature (ref. 53). It is proton specific even at pH8 (10^{-8}M!)

and against a background of 300 mM NaCl or KCl, or 30 mM $MgCl_2$ in the medium. The conductance is independent of pH in the range between 5.6 and 8. In the same pH–range there is a constant hydrogen/deuterium–isotope effect of 1.7. Addition of glycerol decreases the conductance and it abolishes the isotope effect. This suggests that the isotope effect may be caused by events in the channel, and that the channel operates close to limitation by events in the water phase. The Arrhenius activation energy of proton conduction by CF_0 is 42 kJ/mol, intermediate between the respective figures of a channel–type (30 kJ/mol, gramicidin) and a carrier–type antibiotic (65 kJ/mol, valinimycin) (see ref. 53 for details).

A proton conductance of 1 pS around neutral pH exceeds by orders of magnitude the calculated convergence conductance to a pore mouth of reasonable diameter (say 1 nm) (ref. 57). It exceeds calculated rates of net proton transfer mediated by short hydrogen bonded chains (ref. 58) and measured rates of proton transfer through other proton specific channels (refs. 17 and 59). Higher rates of net proton transport are not a problem at acid nor at more alkaline pH, where water hydrolysis is the dominant proton donor and OH^- the acceptor, but they are around neutral pH. At the moment one can only speculate about the origin of the extremely high conductance: Short selectivity filter and large channel mouth have been discussed in the context of K^+–maxi channels (ref. 60). The drag force in the coulomb cage (ref. 57) and mobile buffers (refs. 61 and 62) are further factors that could rise the diffusion limitation for protons to the channel mouth.

Patch clamp experiments with reconstituted CF_0CF_1 (ref. 63) have revealed gated conductance in the order of some pS that probaly is protonic, as inferred from measurements in asymmetric bath solutions. The opening of channels has required relatively high voltage (150 mV) and the opening probability has been reduced by addition of ADP. The most probable interpretation is an over–voltage valve reaction of the intact enzyme, CF_0CF_1, where CF_1 is reversibly displaced from CF_0 to expose the conducting channel. The magnitude of the conductance has been determined in a far more direct way by patch clamp than by flash spectrophotometry. On the other hand, the evidence for the proton specificity is perfect by flash spectrophotometry and less so by patch clamp, as the key experiment, determination of the reversal potentials under asymmetric pH, is not to be performed, because the channels are closed at low voltages.

Patch clamp experiments with CF_0, with EF_O (from *E.coli*) or with the proteolipid subunit alone have revealed cation channels but no proton specificity. These channels have been attributed to disintegrated or deranged channels which have lost their proton specificity (ref. 64).

In conclusion, CF_O is a competent proton channel with a turnover number that greatly exceeds the one of the coupled enzyme. Thus it can act as a low impedance access channel to the coupling site. Its proton conductance is probably much higher than compatible with current understanding of conduction into, through and out of hydrogen bonded chains, or through diffusion limited channels. It is certain, however, that CF_O is endowed with an extremely specific selectivity filter for protons, unparalleled by any other proton channel. This filter is part of CF_0 (and not only present in the integral enzyme).

Attempts have been made to identify those **subunits of CF_0CF_1 that throttle** the very high rate of **proton conduction** through the channel portion to the lower rate of the coupled enzyme. Three subunits usually purified with CF_1 (namely γ, δ and ε) are known to regulate proton flow through the enzyme (refs. 65, 66 and 67). We have focussed on the role of subunit δ. After hints that δ may remain back on CF_O after removal of CF_1, where it keeps CF_O non–conducting (ref. 68) we have found that isolated δ (ref. 69), when added back to CF_1–depleted thylakoids, blocks proton flow (ref. 70) and thereby restores photophosphorylation (ref. 71). This "stopcock" action has to be relieved in the operating enzyme. It is probable that δ then acts either as a valve, admitting access of protons from the channel further on to the coupling site, or as part of the conformational transducer between protons and ATP. It is compatible with this role that δ, that is not necessary for the binding of CF_1 to CF_O, does bind not only to CF_O but also to CF_1 with one high affinity and one or two low affinity sites (ref. 72).

F_OF_1–ATP synthases from different organisms are very similar in their quarternary structure and the primary structures differ. The δ subunits from *E.coli* and spinach chloroplasts, for example, are similar only to 36%, conservative replacements included (ref. 73). Nevertheless, attempts to construct functional hybrids between the *E.coli* EF_1(-δ) and chloroplast δ and *vice versa* have been successful (ref. 74). The hybrid constructs plugged proton conduction through the respective F_O–channel. This is highly suggestive of a mechanical mode of coupling. On the same line, a genetically constructed hybrid, with the largest subunit of chloroplast CF_O as substitute for the largest subunit of *E.coli* EF_O, has yielded a functional EF_0EF_1‘ (ref. 75). Another remarkable example, the one acid residue on the abundant proteolipid subunit of F_O–channels, that is most critical for proton conduction, has been genetically displaced from one leg of this small hairpin–shaped, transmembrane protein to the other one, but at same depth in the membrane, according to hydropathy plots. Proton conduction has been recovered, that is absent without this acid residue in either position (ref. 76).

The molecular mechanism by which F_0F_1–type ATP synthases couple proton flow to ATP synthesis is still unknown. Its elucidation is a formidable task. Even if a structural model at atomic resolution was available for this very large protein (MW 550 kD), the lack of intrinsic reporter groups for partial reactions (as the pigments are in photosynthetic reaction centres) would hamper rapid progress. It has been proposed that the translocated protons do directly enter into the catalytic site (ref. 77), however, without any evidence. It is also conceivable, that the coupling mode is conformational or "mechanical". The approximate threefold symmetry of the catalytic headpiece of the enzyme has appealed to the common wisdom that structural symmetry implies functional symmetry and "rotational catalysis" has been proposed (ref. 78). This is far from being established. Further progress requires a wide attack by genetical, protein chemical, structural and kinetic techniques.

Acknowledgements

This work has been supported by the Deutsche Forschungsgemeinschaft (SFB 171, TP A2 and B3).

References

1 D. Oesterhelt and J. Tittor, Trends Biochem.Sci. **14** (1989) 57.

2 N. Nelson and L. Taiz, Trends Biochem.Sci. **14** (1989) 113.

3 P. Mitchell, Nature (London) **191** (1961) 144.

4 B. Andersson and J.M. Anderson, Biochim. Biophys. Acta **593** (1980) 427.

5 W. Junge and H.T. Witt, Z. Naturforsch. **23b** (1968) 244.

6 H.T. Witt, Biochim. Biophys. Acta **505** (1979) 355.

7 A. Polle, and W. Junge, Biochim Biophys. Acta **848** (1986) 257.

8 W. Ausländer and W. Junge, FEBS Lett. **59** (1975) 310.

9 W. Junge, A.G. McGeer, W. Ausländer and T. Runge, Biochim. Biophys. Acta **546** (1979) 121.

10 YQ. Hong and W. Junge, Biochim. Biophys. Acta **722** (1983) 197.

11 W. Junge and A. Polle, Biochim. Biophys. Acta **848** (1986) 265.

12 G. Schönknecht, W. Junge, H. Lill and S. Engelbrecht, FEBS Lett. **203** (1986) 289.

13 W. Junge, Proc. Natl. Acad. Sci. USA **48** (1987) 7084.

14 D.L. Farkas, R. Korenstein and S. Malkin, Biophys. J. **45** (1984) 363.

15 G. Schönknecht, R. Hedrich, W. Junge and K. Raschke, Nature **336** (1988) 589.

16 G. Schönknecht, G. Althoff and W. Junge, FEBS Lett. 277 (1990) 65.

17 W.R. Veatch, R. Mathies, M. Eisenberg and L. Stryer, J. Mol. Biol. **99** (1975) 75.

18 B. Stolz and D. Walz, Mol. Cell. Biol. **7** (1988) 83.

19 W. Junge and S. McLaughlin, Biochim Biophys. Acta **890** (1987) 1.

20 A. Polle and W.Junge, Biophys. J. **56** (1989) 27.

21 M. Prats, J. Teissie and J.S. Tocanne, Nature **322** (1986) 756.

22 A. Trebst, Z. Naturforsch. **C42** (1987) 742

23 J. Deisenhofer and H. Michel, EMBO J., **8** (1989) 2149.

24 N. Murata and M. Miyao, Trends Biochem. Sci. **10** (1985) 122.

25 B. Kok, B. Forbush and M. McGloin, Photochem. Photobiol. **11** (1970) 741.

26 V. Förster and W. Junge, Photochem. Photobiol. **41** (1985) 183.

27 J.P. Dekker, J.J. Plijter, L. Ouwekand and H.J. van Gorkom, Biochim Biohys. Acta **767** (1984) 1.

28 K. Lübbers and W. Junge, in: M. Baltscheffsky (Ed.) *Current Research in Photosynthesis* Kluwer Academic Publishers, Dordrecht, Vol.1, pp. 877-880, 1990.

29 U. Wacker, E. Haag and G. Renger, in: *Current Research in Photosynthesis* (M. Baltscheffsky, Ed.) Kluwer Academic Publishers, Dordrecht, Vol.1, pp. 869-872, 1990.

30 C.F. Fowler, Biochim Biophys. Acta **462** (1977) 414.

31 S. Saphon and A.R. Crofts, Z. Naturforsch. **32c** (1977) 617.

32 B.R. Velthuys, FEBS Lett. **115** (1980) 167

33 B. Wille and J. Lavergne, Photobiochim. Photobiophys. **4** (1982) 131.

34 J. Lavergne and S. Rappaport, in: *Current Research in Photosynthesis* (M. Baltscheffsky, Ed.) Kluwer Academic Publishers, Dordrecht, Vol.1, pp. 873-876, 1990.

35 P. Jahns, J. Lavergne, S.Rappaport and W. Junge, Biochim. Biophys. Acta **1057** (1991) 313.

36 Ö. Saygin and H.T. Witt, FEBS Lett. **176** (1984) 73.

37 K. Brettel, E. Schlodder and H.T. Witt, Biochim. Biophys. Acta **766** (1984) 403.

38 P. Jahns, A. Polle and W. Junge, EMBO J. **7** (1988) 589.

39 P. Jahns and W.Junge, FEBS Lett. **253** (1989) 33.

40 P. Jahns and W. Junge, (Eur. J. Biochem.), in press.

41 A.E. Senior, Physiol. Rev. **68** (1988) 177.

42 H. Tiedge, H. Lünsdorf, G. Schäfer and H.O.Schairer, Proc. Natl. Acad. Sci. USA **82** (1985) 7874.

43 J. Hoppe and W. Sebald, Biochim. Biophys. Acta **768** (1984) 1.

44 E. Schneider and K.Altendorf, Microbiol. Rev. **51** (1987) 477.

45 W. Junge, B. Rumberg and H.Schröder, Eur. J. Biochem. **14** (1970) 575.

46 U. Junesch and P. Gräber, Biochim. Biophys. Acta **809** (1985) 429.

47 G. Schmidt and P. Gräber, Z. Naturforsch. **42C** (1985) 231.

48 R.S. Negrin, B.L. Foster and R.H. Fillingame, J. Biol. Chem. **255** (1980) 5643.

49 E. Schneider and K. Altendorf, Eur. J. Biochem. **126** (1982) 149.

50 N. Sone, T. Hamamoto and Y. Kagawa, J. Biol. Chem. **256** (1981) 2873.

51 P. Mitchell, FEBS Lett. **78** (1977) 1.

52 H. Lill, G. Althoff and W. Junge, J. Membr. Biol. **98** (1987) 69.

53 G. Althoff, H. Lill and W. Junge, J. Membr. Biol. **108** (1989) 263.

54 W. Laubinger and P. Dimroth, Biochemistry **27** (1988) 7531.

55 H. Lill, S. Engelbrecht, G. Schönknecht and W. Junge, Eur. J. Biochem. **160** (1986) 627.

56 G. Althoff, G. Schönknecht and W. Junge, Eur. J. Biophys **19** (1991) 213.

57 A. Peskoff and D.M. Bers, Biophys. J. **53** (1988) 863.

58 A. Brünger, Z. Schulten and K. Schulten, Z. Phys. Chem. **136** (1983) 1.

59 J.D. Lear, Z.R. Wassermann and W.F. DeGrado, Science **240** (1988) 1177.

60 B. Hille, *Ionic Channels of Excitable Membranes* Sinauer Ass. Inc., Sunderland, Massachussets, 1984

61 E.R. Decker and D.G.Levitt, Biophys. J. **53** (1988) 25.

62 K. Nunogaki and M. Kasai, J. Theor. Biol. **134** (1988) 403.

63 R. Wagner, A.C. Apley and W. Hanke, EMBO J. **8** (1989) 2827.

64 G. Schönknecht, G. Althoff, A.C. Apley, R. Wagner and W. Junge, FEBS Lett. **258** (1989) 190.

65 M.A. Weiss and R.E. McCarty, J. Biol. Chem. **252** (1977) 8007.

66 C.S. Andreo, W.J. Patrie and R.E. Mc Carty, J. Biol. Chem. **257** (1982) 9968.

67 M.L. Richter and R.E.McCarty, J.Biol.Chem. **262** (1987) 15037.

68 W. Junge, Y.Q. Hong, L.P. Qian and A. Viale, Proc. Acad. Sci. USA **81** (1984) 3078.

69 S. Engelbrecht and W.Junge, FEBS Lett. **219** (1987a) 321.

70 H. Lill, S. Engelbrecht and W. Junge, J. Biol. Chem. **263** (1988) 14518.

71 S. Engelbrecht and W.Junge, Eur.J.Biochem. **172** (1987b) 213.

72 R. Wagner, E.C. Apley, S. Engelbrecht and W. Junge, FEBS Lett. **230** (1988) 109.

73 S. Engelbrecht and W.Junge, Biochim. Biophys. Acta **1015** (1989) 379.

74 S.G. Engelbrecht, K. Deckers–Hebestreit, Altendorf and W. Junge, Eur. J. Biochem. **181** (1989) 485.

75 G. Schmidt, A.J.W. Rodgers, S.M. Howitt, A.L. Munn, G.S. Hudson, T.A. Holten, P.R. Whitfeld, W. Bottmoley, F. Gibson and G.B. Cox, Biochim. Biophys. Acta **1015** (1990) 195.

76 M.J. Miller, M. Oldenburg and R.H. Fillingame, Proc. Natl. Acad. Sci. USA **87** (1990) 4900.

77 P. Mitchell, FEBS Lett. **182** (1985) 1.

78 P. Boyer, Biochem. **26** (1987) 8503.

79 W. Junge, Ann. N.Y. Acad. Sci. **574** (1989) 268.

80 M. Miyao and N. Murata, Photosynth. Res. **10** (1986) 489.

81 W. Junge, G. Schönknecht and V. Förster, Biochim. Biophys. Acta **852** (1986) 93.

82 M. Tester and M.K. Blatt, Plant Physiol. **91** (1989) 249.

Electron and Proton Transfer in Chemistry and Biology, edited by A. Müller et al.
Studies in Physical and Theoretical Chemistry, Vol. 78

Diffusion of Proton in Microscopic Space: Effect of Geometric Constraints and Dielectric Discontinuities

Menachem Gutman, Eyal Shimoni and Yossi Tsfadia

Laser Laboratory for Fast Reactions in Biology
Department of Biochemistry
Tel Aviv University
Ramat Aviv 69978 (Israel)

Summary

Diffusion of proton is a random walk process along hydrogen bonded water molecules. In the presence of electric field the randomness is biased by the gradient of the electrochemical potential of the proton. As a result the analysis of proton diffusion is a direct mode for evaluating these terms.

Measurements of proton diffusion in a small, mesoscopic space, are attainable by time resolved fluorescence measurements of a compound which in the excited state releases-recaptures a proton. Placement of the compound at the site of study allows the experimentalist to probe exclusively the immediate environment with no background information coming from the bulk.

In this study we exhibit the capability of the methodology by investigating the interior of the heme binding site of Apomyoglobin, a site containing 30 water molecules or less.

The electric field gauged by the proton confirms the estimation based on the model of Gilson *et al.* (J. Mol. Biol. **183** (1985) 505-515) for the calculation of electrostatic potentials in heterogenic mesoscopic systems.

This observation confirms the applicability of time resolved proton diffusion measurements for the study of the general dependence of the physical properties of a small site on its geometric features (shape and size).

Introduction

Biologists and biochemists measure the size of a cell in μm units. Another way to characterize its dimension is the time domain: How long will it take for a free diffusing particle to propagate from one site in the cell to another? This quantitation provides an estimate of the tightness of coordination between various activities proceeding simultaneously within the cell. For a small solute , diffusing by random walk in a homogeneus matrix, the time (t), needed to propagate a distance (l) is given by $t = l^2/2D$ where D is the diffusion coefficient.

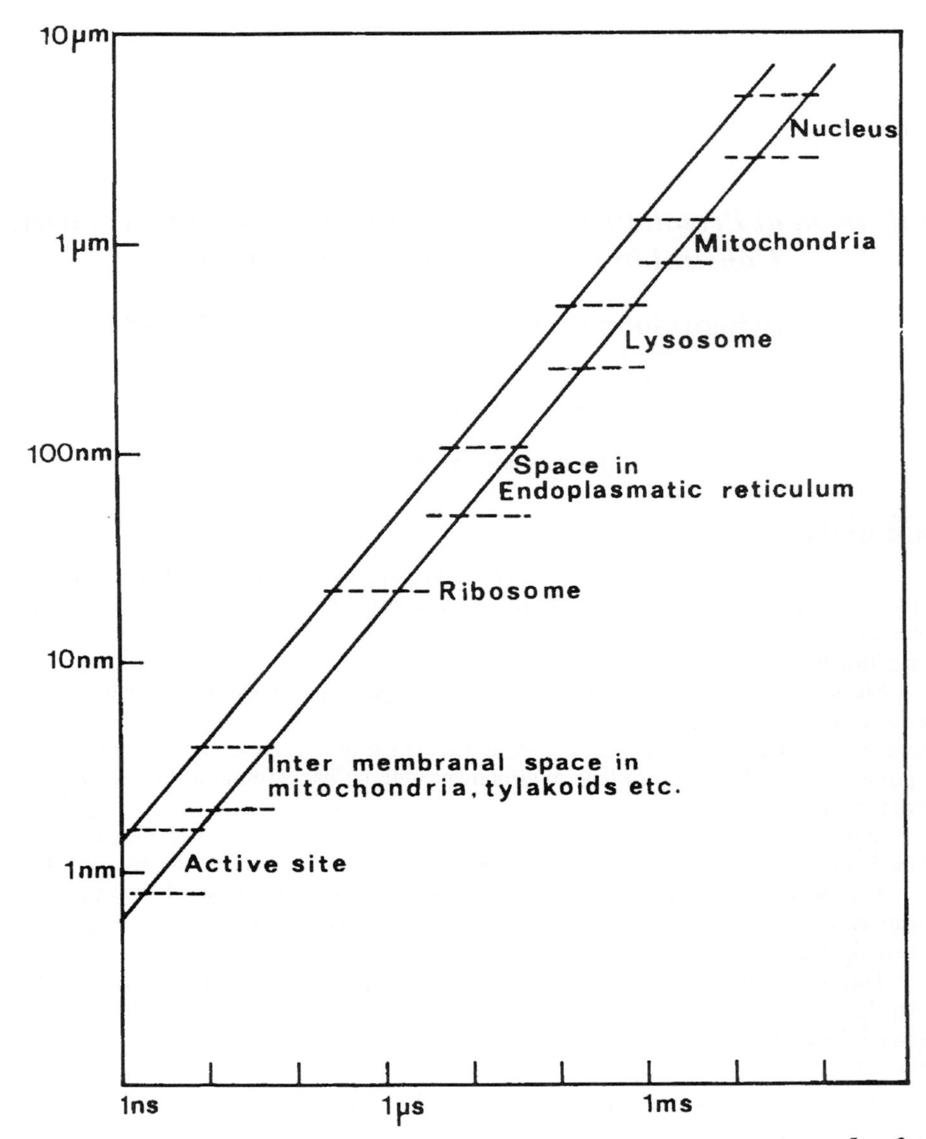

Fig.1. Correlation between time and diffusion distance of a small ion $D \cong 1.5 \cdot 10^{-5}$ cm^2/sec or a molecule like ATP or glucose $D \cong 3 \cdot 10^{-6}$ cm^2/sec. The dimension of structural elements which act as restrictive boundaries is indicated in the graph.

Figure 1 depicts this correlation for a small ion ($D \cong 1.5 \cdot 10^{-5}$ cm^2/sec) or solute like ATP or glucose ($D=3 \times 10^{-6}$ cm^2/sec). We also indicated in this figure the repeating distance of various intracellular structures. As evident solutes cannot propagate freely for an appreciable time frame without interacting with some impermeable intracellular structures. If generated within mitochondria or thylakoid, the particle will encounter the membranal structure within a few nanoseconds. A compound in the endoplasmatic reticulum will collide with the structural elements in the microsecond time scale or ~10 μs if inside a lysosome. As matter of fact the whole length of bacteria or mitochondria can be probed by a free moving solute within 1 millisecond or so. As all these time frames are significantly shorter than the turnover number of most enzymes (1-10 msec), we must recognize that the reactivity of a substrate, or its availability to the active site, is modulated by the packing of the structural elements in the environment. It will be affected by these structures even before its first collision with its target site.

The diffusion of particle inside the cytoplasm is the sum of two elements: diffusion in the immediate vicinity of a nonpermeable boundary and a random walk far from structural elements. The latter can be treated as diffusion in bulk water. It is the former which calls for experimental investigation. In this manuscript we shall describe how the geometric constraints affect the diffusion of a small gauge particle generated in the immediate vicinity of biological boundaries of known structure and composition.

Methodology

The experimental algorithm for these measurements is very ancient and was first introduced by Noah in his Ark. Noah despatched an unbiased messenger, the dove, and measured the time length of the round trip. From that parameter Noah could deduce certain facts about the environment beyond his observation horizon. In our laboratory we substitute a proton for the dove and an excited fluorophore for the Ark. The time scale is shortened to subnanosecond, the period suitable for measuring distances of only a few nanometers from the window of the Ark.

The technique used for these measurements is the laser induced proton pulse (refs. 1 and 2). A very short laser pulse is used to excite a molecular proton emitter to its first excited electronic singlet state. These compounds, such a pyranine (8-hydroxypyrene 1,3,6 trisulfonate) are weak acids which upon excitation become very acidic. The pK shift for pyranine is from $pK_0 = 7.7$ to $pK^* = 1.4$ (ref. 3) (for comprehensive tabulation, see ref. 4).

The excitation of pyranine is followed by extremely fast dissociation of the hydroxy proton ($\tau \cong 100$ psec) leaving a charged anion ΦO^{*-} (Z = -4). This compound decays radiatively ($\tau = 5.5$ nsec) emitting at $\lambda_{max} = 515$ nm. If within its excited period the anion is reprotonated (the homecoming dove) ΦOH^* is reformed, emitting light at $\lambda_{max} = 450$nm. Thus by monitoring the time profile of the ΦOH^* emission we can observe both the initial ejection of the proton and its return to the site of origin. An environment that restricts the diffusion of the proton will increase the probability of recapturing the proton with subsequent ennrichment of the ΦOH^* population at the expense of the ΦO^{*-} one.

Experimental Results

Demonstration of the capacity of the observation to reflect the characteristics of the environment is given in Figure 2.

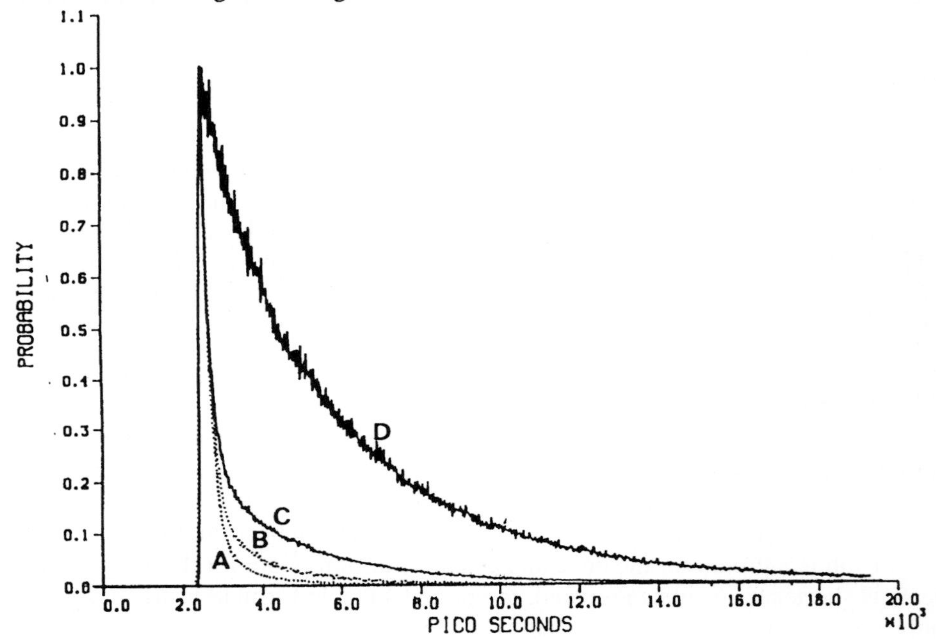

Fig. 2. Time resolved dynamics of the excited protonated form of pyranine as measured in various aqueous environments. The pyranine (8 hydroxypyrene 1,3,6 trisulfonate) was excited by a 50 psec laser pulse and the emission of ΦOH^* was monitored ($\lambda = 450$ nm) using a single photon counting system over a 20 ns period. Curve A was measured with pyranine dissolved in water (pH = 5.5). Curve B was measured with pyranine trapped in the thin water layer of multilamellar vesicles made of Di Palmitoylphosphatidyl–Choline (pH = 5.5). Curve C was measured with a pyranine-PhoE protein complex 1:1 (pH = 5.5). Curve D was measured with pyranine entrapped in the heme binding site of Apomyoglobin (pH=5.5).

Curve A in Figure 2 was measured with pyranine dissolved in water. After excitation there is a rapid decay of ΦOH^* emission ($\tau = 110$ psec), representing the dissociation to ΦO^{*-} and H^+, yet the relaxation does not proceed all the way down to the baseline. There is a persistent "tail" lasting for many nanoseconds with a non exponential relaxation dynamics (refs. 3, 5 and 6). Curve B in figure 2 was measured for the same fluorophore, but enclosed in a thin water layer (30Å deep) sandwiched between phospholipid membrane (refs. 7 and 8). In this case the initial dissociation is very similar to the process in bulk water, but the "tail" gains size and becomes a prominent feature. Considering the larger diffusion coefficient of proton ($D_{H+} = 9.3 \cdot 10^{-5}$ cm^2/sec), it can cross the whole width of the aqueous layer in 0.5 nsec. Thus ~1 nsec after dissociation the ΦO^{*-} may encounter protons whose diffusion trajectory was distorted by the proton impermeable phospholipid membrane.

Curve C in Figure 2 depicts a more intimate environment. In this case the pyranine was enclosed in the anion specific pore of the PhoE protein (ref. 9). This micropore has an opening of 18x27Å and bore 30Å into the membrane (for detailed structure see ref. 10). The inner side of the pore, lined with positive and negative charges ($Z_{tot} = +3$) (ref. 11), serves as a high affinity anchor site for the -3 pyranine anion ($K_{as} = 1.8 \cdot 10^7 M^{-1}$ (ref. 9)). Excitation of the dye initiates a rapid dissociation but the reflective environment herds the proton back to its origin with a respectable swelling of the tail.

Curve D was measured in a close tigh environment – the heme binding site of Apomyoglobin (ref. 12). In this cavity, comparable in size to the heme molecule, the pyranine is deeply buried and its discharged proton can escape through a narrow opening (~3x15Å) (ref. 13). In such small box the ejected proton is subjected to multiple reflections from the walls, recombination with ΦO^{*-} and redissociation of the just formed ΦOH^*.The high frequency of these events merges the tail with the head to form an almost smooth relaxation curve.

Mode of Analysis

Figure 2 demonstrated that proton is a faithful messenger, reporting the features of the surroundind environment. The interpretation of its message necessitates the utilization of Agmon's formalism for treating a diffusion of a charged particle within the electric field of the pyranine anion (refs. 3, 5, 6 and 14). The process is based on the iterated solution of the Debye–Smoluchowski differential operator which states that the flux of an ion between two concentric shells around a central counterion will be a function of the electrochemical potential gradient arround the central ion and the temporal population of each shell by the mobile ion. The frequency at which the particle jumps between the adjacent shells is a function of its diffusion coefficient. The iterative procedure is carried out using adjustable parameters with the aim of reproducing a theoretical curve superpositioned over the experimental tracing.

The parameters used for reconstruction consist of two categories: those characterizing events within the reaction sphere and those representing environment out of it. The reaction radius in this formalism is the distance where the OH bond of the pyranine is replaced reversibly by the ion pair ΦO^{*-} and H^+. Its size is comparable to the width of ~2 hydration shells. The reaction inside the reaction sphere, which is too fast to be resolved by the subnanosecond dynamics, is characterized by rate constant representing the events as detected by an outside observer. These are the rate of H^+ appearance on the perimeter of the reaction sphere and the rate it is reabsorbed.

The second class of parameters quantitates the properties of the medium where the proton is diffusing:

- The diffusion coefficient of proton (D_{H+}).
- The intensity of the electrostatic force (given by the distance where the electrostatic potential is reduced to the level of thermal energy) ($R_D = Z_1 Z_2 e_0^2/\varepsilon RT$) and
- The distance to boundaries which interfere with the random walk (R_{max}). For more comprehensive discussion see refs. 3, 5, 6 and 14.

This procedure is applicable for any symmetric environment, be it a proton discharged in bulk water or within the cavity of the heme binding site, only that the parameters will vary with the geometry of the reaction space. This variance is informative by itself, indicating to what extent water in microcavity differs from bulk water.

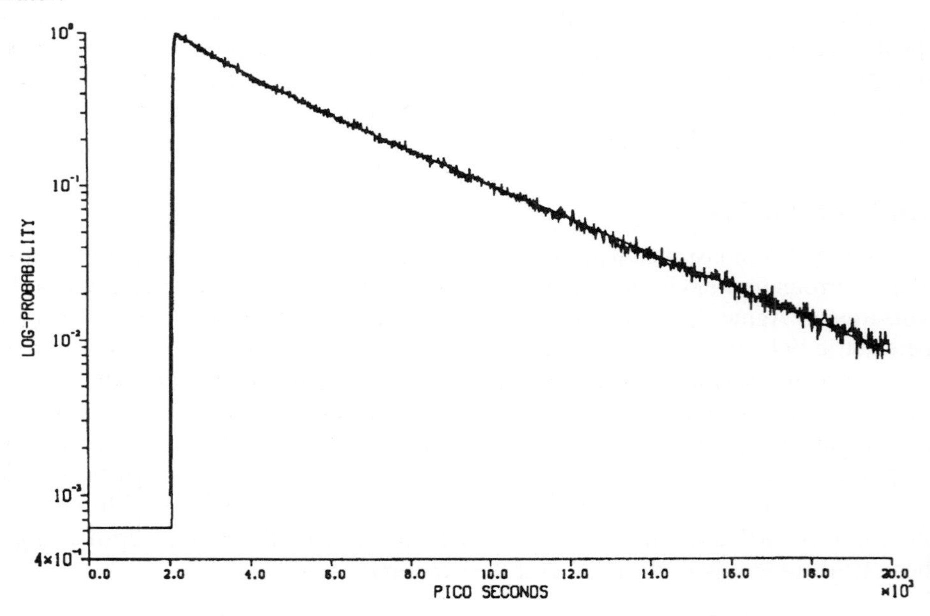

Fig. 3. Reconstruction of the fluorescence decay curve of protonated pyranine in Apo-myoglobin. The curve, the same one as presented in Figure 2, curve D, is drawn on a logarithmic Y axis. The reconstructed curve, a continuous line superpositioned over the experimental one, is a numerical reconstruction using Agmon's computer program (ref. 3) with the following microscopic parameters: reaction radius: 4Å; rate of proton ejection out of the reaction sphere: $2.2\cdot10^8$ sec^{-1}; rate of proton capture by the reaction sphere: $2.2\cdot10^{-8}$ sec^{-1}; proton's diffusion coefficient: $D_{H+} = 4\cdot10^{-5}$ cm^2/sec; Debye radius: 65 Å, compatible with average dielectric constant $\varepsilon_{eff} = 8.5$; outer diameter of diffusion space: $R_{max} = 10$Å; absorption of external boundary at $R_{max} = 10\%$.

Figure 3 depicts a reconstructed dynamics of a proton dissociating inside the Apomyoglobin cavity. For these computations we approximated the site as a spherical space, 10Å in radius (R_{max} = 10Å). The pore is represented by a 10% probability of a proton at $R_i = R_{max}$ to be absorbed by the boundary, corresponding with diffusion out of the cavity.

The fit of the curve, as shown in Figure 3, was achieved with two, non interchangeable parameters: The diffusion coefficient of proton in the microscopic space and the intensity of the electrostatic attraction.

Discussion

The diffusion coefficient of proton in the cavity is $D = 4.0 \cdot 10^{-5}$ cm^2/sec. This value is substantially smaller than that measured in bulk water ($9.3 \cdot 10^{-5}$ cm^2/sec). Similar slow diffusion was determined for proton in other microscopic enclosures like the phoE anion channel (ref. 9) or the surface of lysosome (ref. 15). The slower diffusion coefficient in the immediate proximity of hydrophilic surface is explained in terms of higher ordering of water at the interface (for more comprehensive discussion, see ref. 16).

The most crucial parameter for reconstruction of the experimental dynamics is the intensity of the electrostatic attraction between proton and the anion. In bulk water, where the pyranine anion bears 4 negative charges, the Debye radius (*vide supra*) is 28.3 Å (ref. 3). The binding of the dye to the Apomyoglobin is stabilized by 3 salt bridges between the sulfono groups with Arg-45 His-95 and His-97 (refs. 12 and 17). Thus the distance where the water dielectrics will reduce the electrostatic attraction between H^+ and the oxy anion of the dye is expected to be 7Å. The simulated curve shown in Figure 3 was calculated with R_D = 65Å or an effective dielectric constant of $\varepsilon_{eff} = 8.5$. This value may be the most prominent outcome of these measurements; the capacity to measure the dielectric constant of the aqueous matrix inside small cavities of biological structures.

The unusually small dielectric constant calls for theoretical evaluation (for comprehensive review see Hoing *et al.*, 18). A charge located near a dielectric discontinuity interacts with molecules on both sides of the boundary. If placed in a low dielectric matrix, such as protein the charge is stabilized by interaction with the dipole of water molecules out of the protein (refs. 18 and 19). On the other hand , a charge in aqueous phase near a low dielectric medium (protein, phospholipid membrane) is destabilized by the unequal polarity of the two media . This destabilization is quantitated by introduction of image charges (method best suited for charge near a planar discontinuity). For non–planar geometries, an effective dielectric constant is a more convenient procedure (ref. 19).

The effective dielectric constant ε_{eff} is the correlation factor between the electrostatics potential (G_{ij}) of charges and the distance between them (R_{ij}):

$\varepsilon_{eff} = (Q_iQ_j)/(G_{ij}R_{ij})$

Its magnitude is a function of the coordinates of the charges (with respect to the boundaries) and it varies over the whole reaction space.

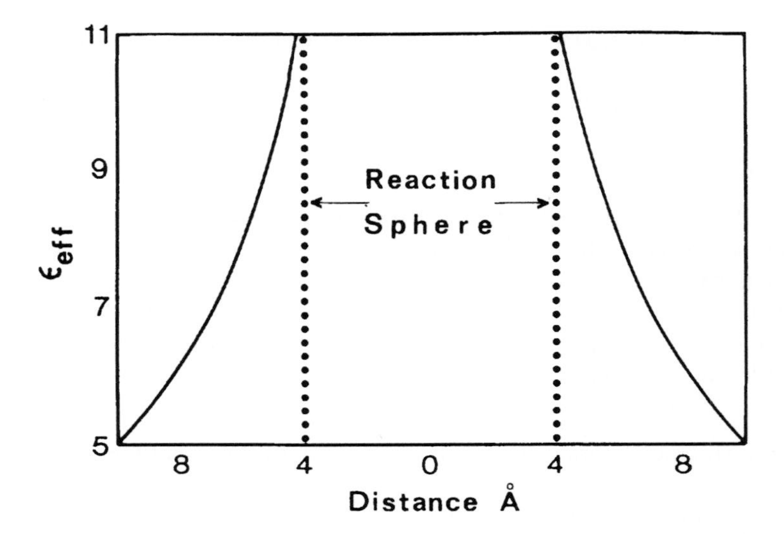

Fig. 4. Mapping of the effective dielectric constant for the diffusion space described in the legend to figure 3. The dielectric constant of the proteinaceous surface is taken as $\varepsilon_{prot} = 5$. The dielectric constant within the reaction sphere is not presented. The averaged dielectric constant over the diffusion space 4.5Å < R_i < 9Å is $\overline{\varepsilon_{eff}} = 8.5$. This value is compatible with the experimentally determined value (Figure 3). The curve was calculated according to Honig's equation (ref. 19) applied for aqueous bodies (ref. 20).

Figure 4 maps the effective dielectric constant for the reaction space compatible with the reconstructed dynamics shown in Figure 3. The values were calculated according to the analytic solution of Honig and his workers (ref. 19) as adapted by Tsfadia for an aqueous sphere surrounded by low dielectric medium (ref. 20).

In this Figure we draw the effective dielectric constant as it varies with the distance, in the range $R_0 \leq R_i \leq R_{max}$ (see legend for details).

At $R_i = R_{max}$ the dielectric constant of the aqueous phase equals that of the protein and rises towards the center. At the surface of reaction sphere $\varepsilon_{eff} = 11$. The averaged dielectric constant, obtained by integration of over the reaction space is $\overline{\varepsilon_{eff}} = 8.5$. This value is compatible with the experimentally measured one.

The potential energy of a proton (G_{ij}) in the model space is drawn in Figure 5.

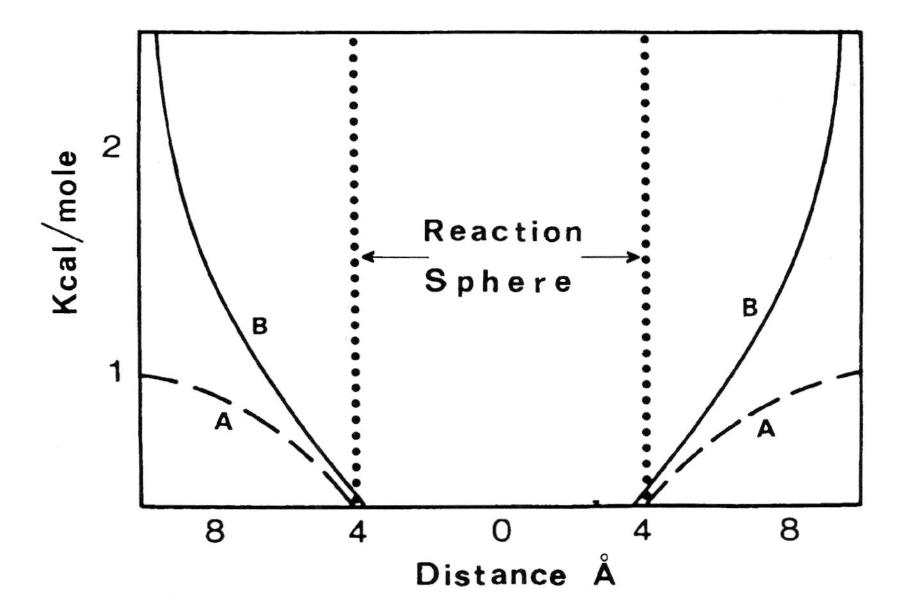

Fig. 5. The electrostatic potential of a proton in the diffusion space defined in the legend to Figure 3. The potential energy is given as the sum of Coulombic interaction with the central pyranine anion and the self energy of the proton (19). Curve A was calculated for a case of no dielectric discontinuity, *i.e.* $\varepsilon_{pro} = \varepsilon_{water} = 78$ and represents potential of the proton pyranine pair in bulk water. Curve B was calculated with the inclusion of a dielectric discontinuity $\varepsilon_{prot} = 5$, $\varepsilon_{water} = 78$. The potential within the reaction sphere is not draw.

Line A was calculated for a proton in a cavity without a dielectric discontinuity ($\varepsilon_{prot} = \varepsilon_{water}$). The thermodynamic stable position of the proton is a rather shallow through of ~1Kcal/mol near the centrally located anion. As R increases the slope of the curve gets smaller approaching the situation where the potential is independent of R.

This is not the case if R_{max} is also a dielectric boundary (see curve B in Figure 5). In this case, due to the incremental self energy of the proton, the potential rises as the proton approaches the dielectric discontinuity.

As described above, and fully elaborated by Agmon (refs. 3, 5, 6 and 14) the gradient of the potential determines the probability of proton transfer between two sites in the reaction space. The map of transition probability in our model space is given in Figure 6.

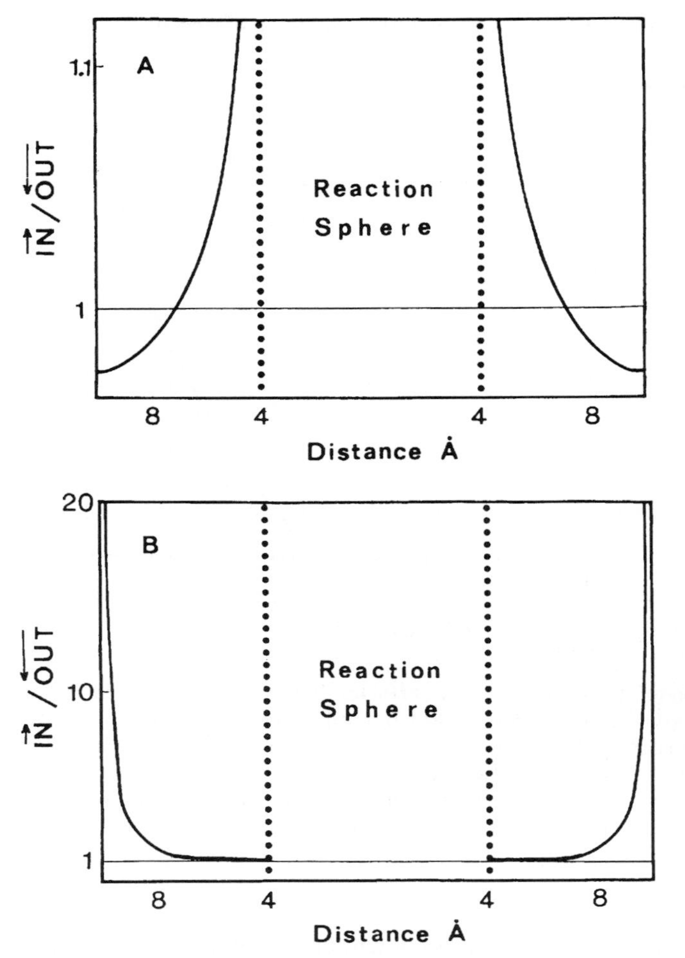

Fig. 6. The mapping of the transition probability for a proton in the diffusion space defined in the legend to Figure 3. The ordinate is the ratio of the transition probability of stepping towards the central ion over that stepping outwards.

Frame A was calculated in the range 4Å $\leq R_i \leq$ 10Å for the proton–pyranine pair in water. Note that 7Å from the center the ratio equals 1 and from that point onwards the proton will diffuse spontaneously away. Frame B was calculated for the same space as above but with a dielectric discontinuity, $\varepsilon_{prot} = 5$. Note the change in scale of the ordinate. Over the whole volume there is no place where the proton will be driven away from the central anion.

The function draw in this figure is the ratio between the probability to diffuse away from the central anion *vs*. the probability to diffuse towards the anion. Frame A was calculated for a proton in a cavity without dielectric discontinuity. Within the first few Å from the reaction sphere, the proton will prefer to step inside (by a factor of 110%) but a little bit further, at R ~7Å it reverses the thermodynamic favored direction. Figure 6B depicts the same parameter as calculated for $\varepsilon_{prot} = 5$, $\varepsilon_{water} = 78$. All over the volume the probability of stepping towards the central anion is much higher than walking outwards. What is more, the reflection is effective even before $R_i = R_{max}$.

In spite of the energy barrier at the interface, protons leak out the cavity with a time constant of ~14nsec (ref. 18). The force driving the proton, against the unfavored potential, is their probability density gradient.

The variation of the proton's radial probability is shown in Figure 7.

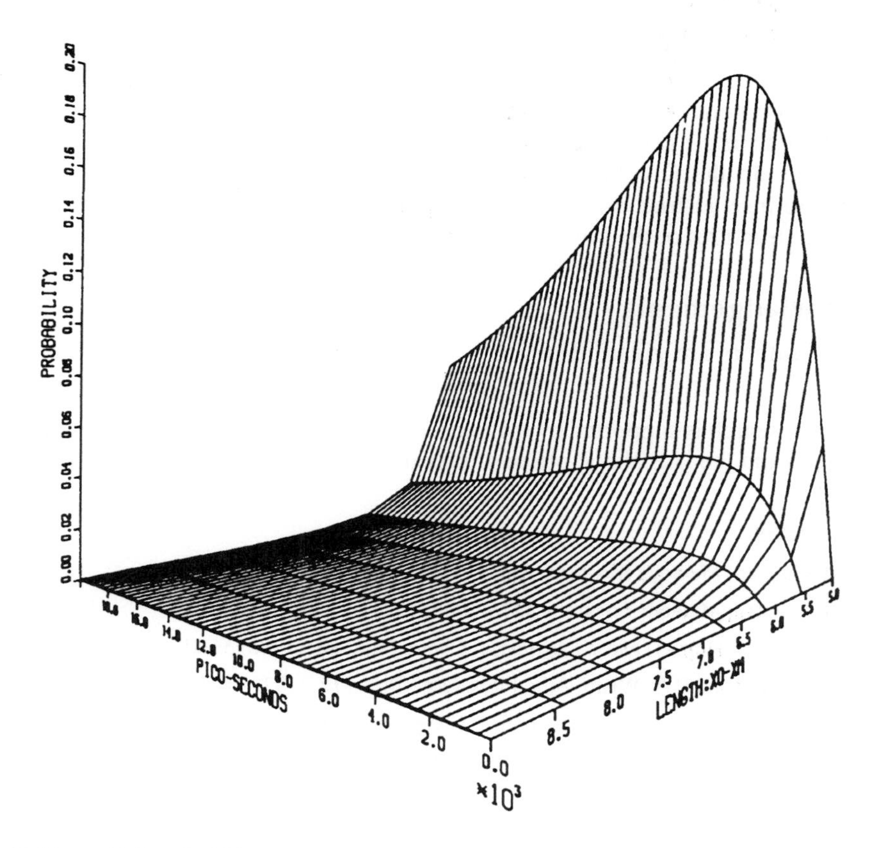

Fig. 7. The radial probability function of proton in the heme binding site,calculated with the parameters used for reconstructing the dynamics shown in Figure 3. The vertical axis is the radial probability of protons — the fraction of whole proton population included in spherical shell around the inner pyranine anion. The right side axis denotes the distance from the center of the sphere while the left side axis depicts the time. Protons reaching the 9Å line are either lost to the bulk (10% probability) or being reflected towards the center. The loss of proton to bulk, not shown in the figure, has a time constant of $\tau = 14$ nsec. (ref. 17).

The vertical axis quantitates the proton population (Σ p = 1) at each concentric shell (right side axis) with the progression of time (left side axis). These calculation, (based on the numerical reconstruction given in Figure 3), demonstrate how the innermost shell is populated by protons and how they leave it to more external shells.

The outermost ones behave almost as if "proton free". In this region the steep potential barrier restricts proton entry, while the few which do enter leak out through the pore and are lost from the cavity.

Measurements of return time of a synchronously delivered signal is a common technique. Photons are used by radar and laser range finders for long distance measurements in the gas phase. Condensed media are probed by sonic and ultrasonic radiation as do the Dolphins, echo sounders and ultrasonic scanners. Apparently protons are also instrumental for this purpose and their realm is the microscopic molecular structures of the biological world.

Acknowledgements

The authors are greatfull to Dr. J. Tommassen for the generous gift of PhoE protein.

This research is supported by the American Israel Binational Science Foundation (87-00235) and the U.S.Navy, office of Naval Research (N00014-89-J-1622).

References

1 M. Gutman, Methods Biochem. Analysis **30** (1984) 1.

2 M. Gutman, Methods Enzymol. **127** (1986) 522.

3 E. Pines, D. Huppert and N. Agmon, J. Chem. Phys. **88** (1988) 5620.

4 J.F. Ireland and P.A. Wyatt, Adv. Phys. Org. Chem. **12** (1976) 122.

5 N. Agmon, E. Pines and D. Huppert, J. Chem. Phys. **88** (1988) 5631.

6 N. Agmon, J. Chem. Phys. **88** (1988) 5639.

7 M. Gutman, E. Nachlier and S. Moshiach, Biochemistry **28** (1989) 2936.

8 S. Rochel, E. Nachliel, D. Huppert and M. Gutman, J. Mem. Biol. **118** (1990) 225.

9 Y. Tsfadia and M. Gutman *Diffusion of Proton in the Microscopic Space of the PhoE Channel* in: T. Bountis (Ed.), *Proton Transfer in Hydrogen-Bonded Systems* Plenum Publishing Corporation, New York, in press.

10 B.K. Jap, J. Mol. Biol. **205** (1989) 407.

11 P. Ley and J. Tommassen, In: Torriani-Gorini, F.G. Rothman, S. Silver, A. Wright and E. Yagil (Eds.) *Phosphate Metabolism and Cellular Regulation in Microorganisms* American Society for Microbiology, 1987, pp. 159-163.

12 M. Gutman, E. Nachlier and D. Huppert, Eur. J. Biochem. **125** (1982) 175.

13 P.A. Adams, Biochem. J. **159** (1976) 371.

14 N. Agmon and A. Szabo, J. Chem. Phys. **92** (1990) 5270.

15 R. Yam, Ph.D. Thesis, Tel Aviv University, 1991.

16 M. Gutman and E. Nachiel, Biochim. Biophys. Acta **1015** (1990) 391.

17 E. Shimoni, M.Sc. Thesis, Tel Aviv University, 1991.

18 B.H. Honig, W.L. Hubbell and R.F. Flewelling, Ann. Rev. Biophys. Chem. **15** (1989) 163.

19 M.K. Gilson, A. Rashin, R. Fine and B. Honig, J. Mol. Biol. **183** (1985) 503.

20 Y. Tsfadia and M. Gutman, personal communication.

Electron and Proton Transfer in Chemistry and Biology, edited by A. Müller et al.
Studies in Physical and Theoretical Chemistry, Vol. 78

Protonation of the Schiff Base Chromophore in Rhodopsins

C. Sandorfy

Département de Chimie
Université de Montréal
Montréal, Québec, H3C 3J7 (Canada)

Summary

In a neutral medium the weak acids present in rhodopsins would not be able to protonate the Schiff base 100%. A supporting mechanism is needed. Indications to the effect that water molecules play a role in this are reviewed.

Introduction

The aim of this paper is to draw attention to the probable role of water molecules in making the protonation of the Schiff base (SB) possible in rhodopsins.

As it is well known, the SB chromophore would have its absorption maximum at about 370 nm (or about 27000 cm^{-1}) in a neutral solution. Protonation takes this down to about 440 nm (or about 22700 cm^{-1}). Actual wavelengths for the band maximum observed in rhodopsins vary from roughly 350 to about 600 nm. This so called "protein-shift" (which can be positive or negative!) will not concern us at this point, however. Instead we shall ask the question: how can the weak acids present in the pigments protonate the SB? In a symbolic notation

$$\mathrm{C{=}N{:}} + \mathrm{HOOC} \leftrightarrow \mathrm{C{=}N^{+}{-}H}\cdots \left[\begin{smallmatrix} \mathrm{O} \\ \mathrm{O} \end{smallmatrix} \!\!>\!\mathrm{C}{-} \right]^{-}$$

What makes the balance favorable to the right hand side in the pigments? We do not know the answer at the present stage, although a number of proposals have been put forward. However, we are going to recall a few facts that should help us in approaching the problem.

Another well known fact is that the resonance Raman and differential infrared spectra of rhodopsins exhibit $\nu(C{=}NH^+)$ stretching bands at frequencies higher than the unprotonated species. It is at 1655 cm^{-1} in bovine rhodopsin, 1641 cm^{-1} in

Abbreviations: CPR, 3–chloropropanoic acid; FTIR, Fourier transform infrared; PR,propionic acid; SB, Schiff base; TRSB, trans–retinylidene–tert–butylamine.

bacteriorhodopsin and at about 1620 cm^{-1} in the nonprotonated SB. It is generally assumed that mechanical coupling between the ($C=NH^+$) stretching and the NH^+ bending motions pushes the former to higher frequencies. This is not obvious but it is supported by normal coordinate calculations.

How can complete protonation be achieved ?

In our laboratory we recorded the FITR spectra of a model compound, trans–ret–inylidene–tert–butylamine (TRSB) (ref. 11) and of its salts formed with HCl, HBr, HI and a series of carboxylic acids, whose pK_a values varied from 0.66 to 4.87 at room temperature in chloroform solution. Moreover, we recorded the FTIR spectra of TRSB with propionic acid (PR) and 3–chloropropionic acid (CPR) at temperatures ranging from room temperature to about -150°C in a mixed solvent formed by 1:1 mixtures of $CFCl_3$ (freon–11) and CF_2BrCF_2Br (freon 114–B–2) or $CFCl_3$ and methylcyclohexane. Both mixtures set to solid glasses at low temperature. The pK_a values for PR and CPR are 4.87 and 3.99, while for aspartic and glutamic acids they are 3.87 and 4.25, respectively. The relative concentrations (imine/acid) were 1:1 or 1:2 or 2:1.

The spectra have shown that the hydrohalides were protonated at all temperatures and that in $>C=N^+–H\cdots X^-$ the hydrogen bond is the strongest in the chloride and weakest in the iodide. Prominent $\nu(NH^+)$ stretching bands could be identified. For the imine/carboxylic acid mixtures, however, these prominent $\nu(NH^+)$ bands are replaced by a region of continuous absorption; this, according to Zundel and his co–workers (ref. 20), indicates systems in which, on the average, the proton is at about midway between the proton donor and the proton acceptor, and its motions are governed by a double well potential. This is in keeping with the simultaneous presence of the $\nu(C=N)$ and $\nu(C=NH^+)$ bands, near 1619 and 1657 cm^{-1}, respectively. Furthermore, it was shown (ref. 11), that above pK_a~2 at room temperature, the acids protonate the SB only partially: PR only to about 15% and CPR to about 60-70%. The latter has a pK_a value close to that of aspartic or glutamic acids.

The low temperature spectra have shown that while in 1:1 TRSB/PR mixtures (concentrations 0.06 or 0.1M) at 25°C the nonprotonated $\nu(C=N)$ band at 1621 cm^{-1} is preponderant and the $\nu(C=NH^+)$ at 1663 cm^{-1} is weak, upon lowering the temperature the latter becomes gradually stronger, at about -100°C the intensity relationship of the two bands is reversed. In an 1:2 imine/acid mixture the reversal occurs before, at about -75°C. The $\nu(C=O)$ and $\nu(C=C)$ bands follow the trend. For TRSB/CPR, where the acid is stronger, the nonprotonated band disappears at about -50°C showing that the SB is almost entirely protonated.

Two conclusions can be drawn from the above mentioned works. 1) The extent of protonation is temperature dependent in SB/carboxylic acid systems in neutral solutions; 2) acids having pK_a values (in water) as high as aspartic or glutamic acid could not achieve 100% protonation of a retinal SB at physiological temperatures. Hence, a supporting mechanism is needed to make protonation complete.

Several possibilities can be invoked. It is perhaps good to remember in this respect that liquid or dissolved alcohols, phenols, water, amines, *ect.* are almost entirely self–associated, except at great dilution. The example of self–associated alcohols is typical in this respect: the "free" R–OH band has its stretching frequency between 3640 and 3610 cm^{-1}, according to cases; the hydrogen bonded dimer R\O–H···: R\O–H··· at about 3485 cm^{-1}, whereas in the "multimer" or "polymer" it is typically near 3350 cm^{-1}:

R\O–H··· : R\O–H··· : R\O–H··· : R\O–H···

This frequency is due to those O–H groups which serve both as proton donors and proton acceptors: as an oxygen is involved with hydrogen bonding at its lone pair, it becomes more acid and a better proton donor. Blatz (ref. 4) found many years ago that retinal SB with weak acids are usually fully protonated in methanol while in a "neutral" solvent they are not. There is an analogy between this "leveling effect" and the case of self–associated alcohols. In the actual pigments, a second, third, ... carboxylic acid or amino acid could play this role. Suggestions for such hydrogen bond chains (proton relay networks) have been made by several authors; these have been reviewed by Sandorfy and Vocelle (refs. 16-17).

The possible role of water

We intend to concentrate on the mechanism which we consider most probable. It involves water. Rafferty and Shichi, (refs. 13-14), studied the effect of extensive dehydration on air dried films of bovine rod outer segment under nitrogen atmosphere. They found that instead of the normal 498 nm band, the dehydrated pigment had two bands in the visible-ultraviolet region, at 390 and at 476 nm. Upon irradiation into the 390 nm band this band disappeared and was replaced by a band at 478 nm. (Resembling the one of metarhodopsin I). This result clearly showed that under dry conditions rhodopsin is at least partly deprotonated. In (ref. 7) it has been suggested that there must be a water molecule between the Schiff base and the proton donating amino–acid and/or other water molecules around the chromophore. Only a few structural water molecules could be involved, since bulk water would hydrolyze the protonated Schiff base. Having one (or more) water molecules between the base and the acid is not unknown in biological systems (refs. 10 and 18).

Hildebrandt and Stockburger (1984) and Massig *et al.* (1985), (refs. 9 and 12), provided resonance Raman evidence for the involvement of water molecules in bacteriorhodopsin put under high vacuum. The $\nu(C{=}NH^+)$ stretching band narrowed considerably when H_2O was replaced by D_2O and upon dehydration the visible band at 570 nm shifted to 530 nm. They suggested that the positive charge on the Schiff base and the negative charge on the counterion are stabilized by surrounding water molecules.

Ganter *et al.* (ref. 8) studied the photoreaction of vacuum dried rhodopsin at low temperature and found evidence for charge stabilization by water (by visible ultraviolet spectroscopy). In dry rhodopsin a thermal equilibrium exists for protonation; upon hydration water molecules stabilize the protonated form. They suggested that since large conformational changes are unlikely at low temperature, the most probable interpretation is in terms of a hydrogen bond with a double well potential.

Birge *et al.*, (ref. 3), suggested that the hydrogen bond to the Schiff base does not involve the counterion, so that water or secondary interactions with the protein are responsible for the stabilization of the ion pair.

Deng and Callender, (ref. 6), put forward the idea that hydrogen bonds may be broken and re–formed with a new group when bathorhodopsin is formed, alternatively there may be only a simple hydrogen bond involved which follows the movements of the Schiff base. This would be conceivable if there are water molecules at the active site.

The protonation of a model retinal Schiff base by aspartic or glutamic acid, or by tyrosine using FTIR spectroscopy combined with the attenuated total reflection (ATR) technique has been studied in (ref. 1). It has been found that the extent of protonation is highly dependent on environmental conditions, among others on the order of deposition of the base and the acid on the surface. With weak acids, the adsorbed Schiff base can only be partially protonated by the addition of these acids, but increasing the water content (humidity) increases the extent of protonation significantly.

Dencher *et al.*, (ref. 5), examined the topography of water molecules in the purple membrane by neutron diffraction and magnetic birefringence. According to their results, there are four water molecules in the vicinity of the the Schiff base of which three are tightly bound, even at zero degree of humidity, and can only be removed under high vacuum. They might well stabilize the ion pair.

The most recent, ponderous event in this respect is the identification (refs. 15 and 19) of glutamic acid–113 as the counterion of the retinylidene Schiff base in bovine rhodopsin. They succeeded in doing this by site–directed mutagenesis, replacing systematically the charged aminoacids in bovine rhodopsin transmembrane helix C.

Replacing Glu^{113} by aspartic acid gave a slightly redshifted pigment (505 nm), replacement by glutamine yielded a pigment that was "strikingly blue–shifted" with λ_{max} at 380 nm. The 380 species existed in a pH dependent equilibrium with a 490 nm species; at acidic pH all of the pigment was converted to the 490 species. Glu^{122} may also interact with the chromophore and may contribute to wavelength regulation in rhodopsin.

Thus, Glu[113] appears to be the counterion although it is not excluded that it is only a booster that is, a second charged aminoacid that makes the counterion more acid and protonation possible (ref. 19). According to the latter authors the counterion in drosophila rhodopsin is probably Tyr[126].

Now, unfortunately, the exact distance between the nitrogen of the Schiff base and the oxygen (or rather carboxylate group) of the counterion is not known. The knowledge of this distance (and angular geometry) would make it possible to see if the expectation, that water molecules are located between the Schiff base and the counterion, is realistic or not. Glu[113] is too weak an acid to achieve 100% protonation without a supporting mechanism. It is expected that our suggestions made in this respect will be vindicated (refs. 7, 16 and 17).

References

1 S. Badilescu, L.S. Lussier, C. Sandorfy, H.Le–Thanh and D. Vocelle, In: E.D. Schmid, F.W. Schneider and F. Siebert (Eds.) *Spectroscopy of Biological Molecules New Advances* Wiley and Sons, New York, 1988, pp. 191-208.

2 S. Badilescu and C. Sandorfy, J. Mol. Struct. **177** (1988) 69.

3 R.R. Birge, C.M. Einterz, H.M. Knapp and L.P. Murray, Biophys. J. **53** (1988) 376.

4 P.E. Blatz, J.H. Mohler and H.V. Navangul, Biochemistry **11** (1972) 848.

5 N.A. Dencher, D. Dresselhaus, G. Maret, G. Papadopoulos, G. Zaccai and G. Büldt, in: T. Hara (Ed.) *Light Induced Structural Changes in Bacteriorhodopsin and Topography of Water Molecules in the Purple Membrane Studied by Neutron Diffraction and Magnetic Birefringence* Proc. XXI Yamada Conf. on Mol. Physiology of Retinal Proteins, Osaka, Japan, Yamada Science Foundation, 1988, pp. 109-115.

6 H. Deng and R.H. Callender, Biochemistry **26** (1987) 7418.

7 P. Dupuis, F.I. Harosi, C. Sandorfy, J.M. Leclercq and D. Vocelle, Rev. Can. Biol. **39** (1980) 247.

8 U.M. Ganter, E.D. Schmid and F. Siebert, J. Photochem. Photobiol. B. Biol. **2** (1988) 417.

9 P. Hildebrandt and M. Stockburger, Biochemistry **23** (1984) 5539.

10 H. Kleeberg, G. Heinje and W.A.P. Luck, J. Phys. Chem. **90** (1986) 4427.

11 L.S. Lussier, C. Sandorfy, H. Le–Thanh and D. Vocelle, J. Phys. Chem. **91** (1987) 2282.

12 G. Massig, M. Stockburger and Th. Alshuth, Can. J. Chem. **63** (1985) 2012.

13 C.N. Rafferty and H. Shichi, Spectral changes induced by dehydration of rhodopsin suggest reversible deprotonation of the chromophore. Communication at the meeting of the American Society of Biological Chemists and the Biophysical Society, 1980.

14 C.N. Rafferty and H. Shichi, Photochem. Photobiol. **33** (1981) 229.

15 T.P. Sakmar, R.R. Franke and H.G. Khorana, Proc. Natl. Acad. Sci. USA **86** (1989) 8309.

16 C. Sandorfy and D. Vocelle, Can. J. Chem. **64** (1986) 2251.

17 C. Sandorfy and D. Vocelle *The Problem of Protonation in Rhodopsin and Model Schiff Bases* in: J. Maruani (Ed.) *Molecules in Physics, Chemistry and Biology* Kluwer Academic Publishers, 1989, Vol. 4, pp. 195-211.

18 G. Varo and L. Keszthelyi *The Role of Water in the Pumping Activity of Bacteriorhodopsin* in: T.G. Ebrey, H. Frauenfelder, B. Honig and K. Nakanishi (Eds.) *Biophysical Studies of Retinal Proteins* University of Illinois Press, 1987, pp. 199-205.

19 E.A. Zhukovsky and D.D. Oprian, Science **246** (1989) 928.

20 G. Zundel, J. Mol. Structure **177** (1988) 43.

Electron and Proton Transfer in Chemistry and Biology, edited by A. Müller et al.
Studies in Physical and Theoretical Chemistry, Vol. 78

Proton Transfer along the Hydrogen Bridge in Some Hydrogen-Bonded Molecular Complexes

Henryk Ratajczak

Institute of Chemistry, University of Wrocław
and
Institute of Low Temperature and Structure Research
of the Polish Academy of Sciences, Wrocław (Poland)

Summary

It is demonstrated that two types of proton transfer mechanisms along the hydrogen bond, A-H···B, can appear. In the first one the hydrogen bond is described by the potential energy surface with a single minimum. The equilibrium position of the proton shifts smoothly from the acid (AH) toward the base (B) as the acid becomes gradually more acidic or the base becomes more basic. This mechanism of the proton transfer takes place in the hydrogen bonded complexes formed between hydrogen halides with ammonia and aliphatic amines in the gas phase and in low temperature matrices (argon, nitrogen). Second one is generated by the potential energy surface of the hydrogen bond with a double minima. Here one would expect the reversible proton transfer phenomenon along the hydrogen bridge inside the complex. This mechanism of the proton transfer is observed in the hydrogen-bonded complexes formed by some derivated phenol with triethylamine in aprotic solvents.

Introduction

Proton transfer process is one of the most interesting phenomena in molecular science. It plays an important role in chemistry, physics and biology (refs.1-3). So far, the proton transfer has been widely studied in aqueous solutions and solids by various physical experimental methods and different theoretical approaches. However, much less is known on proton transfer in nonaqueous (aprotic) solutions and isolated systems (gas phase). This is important since aprotic solvents are of some interest for biochemical model studies; they simulate to some extent the interior environment in proteins, membranes and the central parts of the double helices of DNA. In this respect at least two important questions should be answered: what is the mechanism of a proton transfer along the hydrogen bridge in hydrogen-bonded complexes? What is the influence of solvent (environment) on a proton transfer process? The mechanism of a proton transfer along the hydrogen bond is fundamentally related to the potential energy surface of the hydrogen bridge. In principle, the hydrogen bond can be described at least by two different types of potentials: a single minimum potential or a double minimum one (Fig.1).

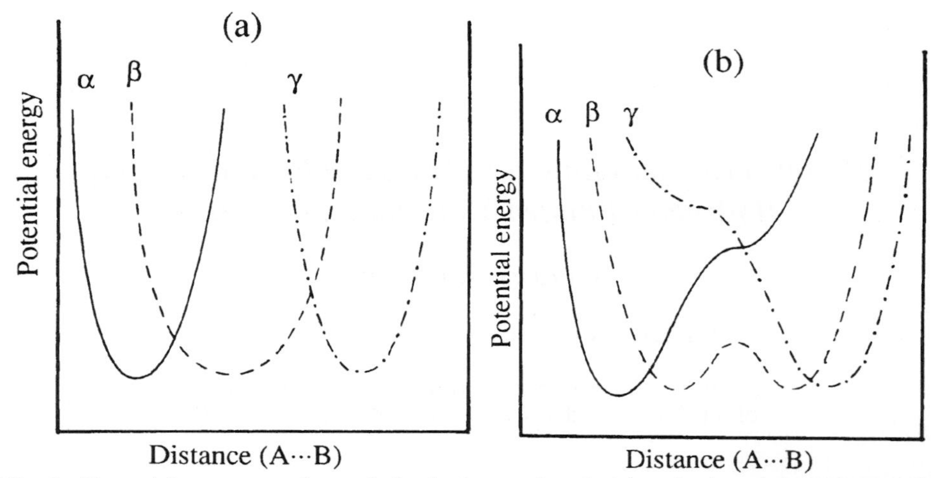

Fig. 1. Potential energy surface of the hydrogen bond; (a) a single minimum potential: α) A-H···B, β) A···H···B, γ) A^{-}···H-B^{+}, (b) a double minimum potential: α) A-H···B, β) A-H···B $\rightleftarrows$ A^{-}···H-B^{+}, γ) A^{-}···H-B^{+}.

In the first case the equilibrium position of a proton in the hydrogen bridge, A-H···B, depends strongly upon proton-acceptor and proton-donor properties (which might be expressed by proton affinity of the base PA(B), and conjugate base PA(A^{-}) respectively) of the two components of the complex. The equilibrium position of the proton shifts smoothly from the acid (AH) toward the base (B) as the acid becomes gradually more acidic or the base becomes more basic (see Fig. 1a). One can distinguish three limited situations: in the first one, the proton is located much closer to the conjugate base A^{-}, this corresponds to a "normal" A-H···B hydrogen bond (Fig. 1a, α). In the second case, the proton is "equally shared" by two bounded subunits, and a symmetric or pseudosymmetric A···H···B hydrogen bond appears (Fig. 1a, β). In the last one, the proton is localized at the base subunit A^{-}···H-B^{+}; a hydrogen-bonded ion-pair is created (Fig. 1a, γ).

In the case of a double minimum potential of the hydrogen bond (Fig. 1b) one can also distinguish three limited situations: the first one can be described by the potential energy which is presented in Fig. 1b, α, the normal hydrogen bond of the A-H···B type occurs. The proton is located at the acid A atom. The second minimum on the potential energy surface is placed much higher than the first one. By gradual increasing proton-donor-acceptor properties of the interacting molecules one can change the nature of the potential energy surface in the way presented in Fig. 1b. It means one can reach the second limited situation which is described by Fig. 1b, β of the A-H···B $\rightleftarrows$ A^{-}···H-B^{+} type. Thus one would expect the reverse proton transfer phenomenon along the hydrogen bridge inside the complex which would result in the tautomeric equilibrium of the type:

$$\text{A-H}\cdots\text{B} \rightleftarrows \text{A}^{-}\cdots\text{H-B}^{+}$$

The position of the tautomeric equilibrium (interconversion of tautomers involving movement of the proton along the hydrogen bond) should depend upon the proton donating power of the acid, $PA(A^-)$, the basicity of the base PA(B), the temperature, and the nature of the solvent. The proton in a double minimum potential can move along the hydrogen bond "classically" crossing a potential energy barrier by activation energy process or can "move" quantum-mechanically by tunneling effect (ref.4). The proton tunneling phenomenon plays probably an important role in some chemical, physical and biological processes (refs. 1 and 4). The third situation can be described by a strongly asymmetric double energy potential with the lower energy minimum localized at the base subunit (Fig. 1b, γ). The hydrogen-bonded ion-pair $A^-\cdots H\text{-}B^+$ appears.

In isolated complexes the shape of the hydrogen bond potential is generated by the electronic structure of the complex and by coupling between the protonic motion and other intramolecular vibrations (ref. 67). This is only valid for complexes in the gas phase. In condensed systems the potential depends also strongly upon intermolecular interactions with environment (coupling with the bath) (refs. 63 and 67).

Recently developed computational quantum chemical methods permit to evaluate the potential energy surface of the hydrogen bond but they have to be used with care since the nature of the calculated potential is extremely sensitive to the basis sets (in particular to the exponents of the polarization functions) as well as to the effects of electron correlation (ref. 5).

For methodological reason it is convenient to classify molecular proton-transfer hydrogen-bonded systems into four subgroups:

- proton transfer between neutral molecules which produces an ion-pair, $A\text{-}H\cdots B \rightleftharpoons A^-\cdots H\text{-}B^+$
- proton transfer in the hydrogen-bonded ion, *i.e.* $A^+\text{-}H\cdots B \rightleftharpoons A\cdots H\text{-}B^+$
- intramolecular proton transfer, *e.g.* malonaldehyde
- simultaneous proton transfer in acid dimers

A group of prime candidates to study the proton transfer phenomenon in more systematic way are complexes formed between phenols and aliphatic amines (refs. 6-29), between carboxylic acids and strong organic bases (ref. 30) or between hydrogen halides (HCl, HBr, HI) and aliphatic amines (refs. 31 and 32-54).

In this article we will review proton transfer processes in two types of neutral hydrogen bonded complexes: hydrogen halides (HF, HCl, HBr, and HI) with ammonia and aliphatic amines and phenols with aliphatic amines.

Proton transfer in the hydrogen bonded complexes formed between hydrogen halides with ammonia and aliphatic amines

Hydrogen halides with ammonia and aliphatic amines form ionic solids (salts) with complete proton transfer from the acid to the base. Those salts are not soluble in aprotic solvents. They are soluble in water where dissociation and hydration processes appear.

Until recently experimental studies of these complexes in gas phase faced many difficulties. Pimentel *et al.* (ref. 31) first used low temperature matrix isolation technique in order to minimize the influence of environment on bimolecular complex between ammonia or aliphatic amines and hydrogen halides.

Since that those complexes have been extensively studied by Barnes *et al.* (refs. 32-34) and Perchard *et al.* (ref. 35) who demonstrated that in low temperature matrices (argon, nitrogen) stable 1:1 strongly hydrogen-bonded molecular or ionic complexes are formed. The frequencies of the proton stretching vibrations of some hydrogen halides-ammonia, aliphatic amines complexes in argon matrix are given in Table 1.

Recently Legon *et al.* (ref. 36) successfully applied the pulsed nozzle FT microwave spectroscopy to study strongly hydrogen-bonded complexes in the gas phase (so that the interactions of interest are unperturbed by lattice, matrix or solvent effect). The following hydrogen halides - ammonia, aliphatic amine complexes have been studied: $(CH_3)_3N$-HF, H_3N-HCl, CH_3H_2N-HCl, $(CH_3)_3N$-HCl, H_3N-HBr and $(CH_3)_3N$-HBr. Some experimental data are summarized in Table 1.

Since the pioneering SCF calculations of Clementi (ref. 43) carried out for the H_3N-HCl system quantum chemical methods have been extensively used to study properties, structure and potential energy surface of hydrogen halides-ammonia, aliphatic amine complexes. Some results of those calculations are given in Table 1.

Obtained results are consistent with a single minimum potential energy surface for the hydrogen bond in those complexes in which the equilibrium position of the proton depends upon the proton-donor-acceptor properties of the interacting molecules. The mechanism of a proton transfer along the hydrogen bond might be generated by potentials presented in Fig. 1a as described in the Introduction.

In weaker complexes (H_3N, CH_3H_2N, $(CH_3)_2HN$, $(CH_3)_3N$ with HF; H_3N-HCl, H_3N-HBr) the "normal" hydrogen bond of the A-H···B type appears. The calculated two-dimensional ($R_{A \cdots B}$, $R_{A \cdots H}$) potential energy surface has one energy minimum located at the acid subunit. One observes a small elongation of the A-H bond due to the complex formation. The change of distribution of electron density after complexation is relatively small. The evaluated amount of charge transfer from the base to the acid is no more than about 0.1 e.

Table 1.
Some experimental and calculated data for the hydrogen-bonded complexes formed between hydrogen halides with ammonia and aliphatic amines.

Complex	Experimental data					Calculated data									Type
	$R_{N...A}$ [Å]	Δr [Å]	k_σ [Nm^{-1}]	Δq [e]	ν(X-H)[a) X = A, B [cm^{-1}]	Basis set	Method	$R_{N...A}$ [Å]	$R_{H...N}$ [Å]	r_{A-H} [Å]	Δr [Å]	-ΔE [kcal/mol]	Δμ [D]	ref.	
H_3N-HF					3041	4-31G	SCF	2.684	1.735	0.949	0.027	16.74	1.49	[55]	MOL
CH_3H_2N-HF						4-31G	SCF	2.640	1.690	0.954	0.032	17.30	1.57	[55]	MOL
$(CH_3)_2HN$-HF						4-31G	SCF	2.627	1.679	0.956	0.034	17.36	1.59	[55]	MOL
$(CH_3)_3N$-HF	2.586 [37]	0.041 [38]	38.3 [38]		2589	4-31G	SCF	2.606	1.674	0.959	0.037	17.32	1.60	[55]	MOL
H_3N-HCl	3.1367 [39]		17.6 [39]		1371	(10,6/8,4/4) → [6,4/5,3/3]	SCF	2.892	1.145	1.747	0.442	10.96		[40]	
						(12,8/10,6/6) → [7,5/6,4/4]	SCF	2.880	1.123	1.757	0.461	12.52		[40]	
						(10,6,1/8,4,1/4,1) → [6,4,1/5,3,1/3,1]	SCF	3.259	1.964	1.295	0.024	8.58	1.248	[40]	
							CEPA	3.105	1.782	1.323	0.047	10.58		[40]	
						(12,8,2/10,6,2/6,1) → [7,5,2/6,4,2/4,1]	SCF	3.240	1.947	1.293	0.021	6.28		[40]	
						4-31G	SCF	3.13	1.856	1.274		10.8		[41]	
						6-31G (2d, p)	SCF	3.297	2.004	1.293	0.023	9.29	1.13	[42]	
							MP2	3.144	1.827	1.317	0.040	11.03	1.52	[42]	
						(11,7/11,7/6) → [5,4/5,3/3]	SCF	2.867	1.245	1.622	0.322	19.0		[43]	
						(12,9,1/9,5,1/5,1)	SCF	3.284				7.89		[44]	
							CI	3.228	1.905	1.323		9.03		[44]	MOL
						ECP-DZP[b)	SCF	3.263	1.966	1.297			1.36	[45]	MOL
						ECP-DZP	ACCD	3.225						[45]	MOL

Table 1. continued

Complex	Experimental data					Calculated data									Type
	$R_{N...A}$ [Å]	Δr [Å]	k_σ [Nm^{-1}]	Δq [e]	$\nu(X\text{-}H)$[a] X = A, B [cm^{-1}]	Basis set	Method	$R_{N...A}$ [Å]	$R_{H...N}$ [Å]	$r_{A\text{-}H}$ [Å]	Δr [Å]	$-\Delta E$ [kcal/mol]	$\Delta\mu$ [D]	ref.	
$CH_3H_2N\text{-}HCl$						(10,6,1/8,4,1/4,1) → [6,4,1/5,3,1/3,1]	SCF	3.212		1.299	0.028	8.91	1.385	[40]	
						6-31G (d, p)	MP2	2.96	1.61	1.36		13.7		[46]	
							MP3	3.06	1.75	1.32		11.7		[46]	
$(CH_3)_2HN\text{-}HCl$						6-31G (d, p)	MP2	2.99	1.45	1.55		15.3		[46]	
							MP3	3.00	1.55	1.46		12.7		[46]	
$(CH_3)_3N\text{-}HCl$	2.8166 [47]		84.6 [47]		1486	6-31G (d, p)	MP2	2.91	1.29	1.62		17.0		[46]	IP
							MP3	2.92	1.23	1.69		14.8		[46]	IP
$H_3N\text{-}HBr$	3.255 [48]		13.4 [48]	0.1 [48]	729	(13,9,4/8,4,1/4,1) → [6,5,3/5,3,1/3,1]	SCF	3.380	1.936	1.444	0.032	8.12	1.648	[40]	MOL
						ECP-DZP	SCF	3.468	2.042	1.426	0.023	5.8	1.51	[45]	MOL
							ACCD	3.406				7.5		[45]	
						(15,12,5/9,5,1/4,1) → [8,6,2/3,2,1/2,1]	SCF	3.512	2.082	1.430	0.02	5.53		[49]	MOL
							MP2					7.82		[49]	MOL
						ECP-DZP	MP2	3.234	1.762	1.472	0.054	10.37	2.12	[50]	MOL
$CH_3H_2N\text{-}HBr$						(13,9,4/8,4,1/4,1) → [6,5,3/5,3,1/3,1]	SCF	3.311		1.452	0.040	8.49	1.88	[40]	MOL
								2.997		1.891	0.479	9.63	7.71	[40]	IP
						(15,12,5/9,5,1/4,1) → [8,6,2/3,2,1/2,1]	MP2					15.79		[49]	
						ECP-DZP	MP2	2.952	1.244	1.716	0.298	13.14	5.43	[50]	

Table 1. continued

Complex	Experimental data					Calculated data									Type
	$R_{N...A}$ [Å]	Δr [Å]	k_σ [Nm^{-1}]	Δq [e]	$\nu(X\text{-}H)$[a] X = A, B [cm^{-1}]	Basis set	Method	$R_{N...A}$ [Å]	$R_{H...N}$ [Å]	$r_{A\text{-}H}$ [Å]	Δr [Å]	$-\Delta E$ [kcal/mol]	$\Delta\mu$ [D]	ref.	
$(CH_3)_2HN\text{-}HBr$						ECP-DZP	MP2	2.961	1.158	1.810	0.392	17.47	6.85	[50]	IP
$(CH_3)_3N\text{-}HBr$	2.9607 [51]		93.7 [51]		1660	ECP-DZP	MP2	2.972	1.161	1.811	0.393	20.13	7.04	[50]	IP
$H_3N\text{-}HI$					1256	4-31G	SCF	3.4	1.1	2.3	0.7	15.4	10.35	[52]	
						ECP-DZP	SCF	3.822	2.208	1.614	0.016	4.0	2.04	[45]	MOL
								3.294	1.076	2.221		8.8	7.69	[45]	IP
							ACCD	3.968				5.7		[45]	MOL
												6.6		[45]	IP
							MP2	3.384	1.695	1.689	0.089	8.28	3.13	[50]	MOL
								3.205	1.165	2.040	0.440	9.45	7.71	[50]	IP
							MP4	3.579	1.933	1.646	0.039	6.77		[50]	MOL
								3.206	1.168	2.038	0.431	5.93		[50]	IP
						I: ECP-(3,3) 6-31G	SCF	3.30				7.7		[53]	IP
												14.8		[53]	MOL
$CH_3H_2N\text{-}HI$						ECP-DZP	MP2	3.210	1.122	2.097	0.497	16.89	8.35	[50]	IP
$(CH_3)_2HN\text{-}HI$						ECP-DZP	MP2	3.212	1.097	2.112	0.562	22.76	8.99	[50]	IP
$(CH_3)_3N\text{-}HI$					1889 1882	ECP-DZP	MP2	3.219	1.085	2.134	0.534	25.91	9.59	[50]	IP

[a] ref. 34

[b] Core electrons are described by the effective core potential (ECP) and valence electrons by the double zeta set plus polarisation functions.

It is interesting to notice that for two of the complexes studied, namely H_3N-HCl and H_3N-HBr, there is a disagreement between IR matrix isolation studies and theoretical calculations or gas phase rotational spectroscopy concerning the proton position in the hydrogen bond.

IR spectra of the complex isolated in argon matrix suggest that the proton is shared more or less equally between the nitrogen and chlorine or bromine atoms respectively. This "apparent" disagreement clearly shows how sensitive the strongly hydrogen bonded systems are to even weak interactions with their environment. Barnes *et al.* (ref. 34) nicely demonstrated strong dependence of the infrared spectrum, in particular the dependence of proton stretching vibrations, in hydrogen halides (HCl, HBr) - ammonia, aliphatic amine complexes upon the matrix material. The strong frequency shift of the proton stretching vibration from argon to nitrogen matrix was interpreted by the increase of degree of proton transfer (which is equivalent to the shift of proton from the acid to the base) in nitrogen matrix.

Both theoretical calculations and gas phase or matrix isolation studies demonstrate that in very strong complexes [$(CH_3)_3N$-HCl, $(CH_3)_3N$-HBr, $(CH_3)_3N$-HI] the hydrogen-bonded ion-pair $A^-\cdots H\text{-}B^+$ occurs and one observes a complete proton transfer from the acid to the base. The hydrogen bond is described by one minimum potential energy surface. The energy minimum is localized at nitrogen atom of the base. One observes significant increase of polarity of the hydrogen bond, $\Delta\vec{\mu}$ = 7-10 D, which reflects the large change of distribution of electron density due to transfer of proton from the acid to the base. The careful analysis of the ^{35}Cl and ^{14}N-nuclear quadrupole coupling constants carried out for $(CH_3)_3N$-HCl is in agreement with the picture of partial proton transfer in the direction of $(CH_3)_3N^+\text{-}H\cdots Cl^-$(ref. 70). NMR measurements carried out by Golubev and Denisov (ref. 54) also nicely demonstrated that $(CH_3)_3N$-HBr complex appears in the hydrogen-bonded ion-pair form in the gas phase. In the $(CH_3)_3N$-HCl, HBr, HI complexes the extent of proton transfer steadily increases from HCl to HBr and further to HI, see also Table 1.

An interesting class of complexes is that which is between the two above mentioned extrema. Analysis of IR spectra based on the Pimentel vibrational correlation diagram strongly suggests that in these complexes the proton is "completely" or "partly" shared between two heavy subunits (A···B). The degree of proton transfer continuously increases coming to the more basic base or to the more acidic acid. The potential energy curve of the hydrogen bond has a single minimum. The position of energy minimum depends systematically upon the difference between the proton affinities of base (B) and conjugate base (A^-). More advanced quantum chemical calculations recently carried out for those systems are in general agreement with conclusions concerning the mechanism of proton transfer drawn from IR spectra. The calculated potential energy of the BrH-ammonia, aliphatic amines and the IH-aliphatic amine complexes are given in Fig. 2.

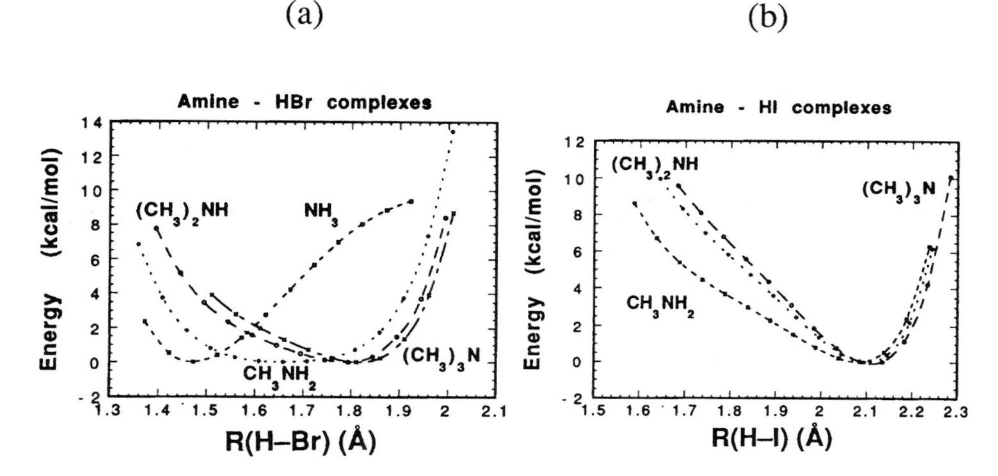

Fig. 2. Proton potential energy curves calculated at the MP2 level with MP2 optimized geometry: a) amine-HBr complexes, b) amine-HI complexes.

As one can see from Fig. 2a the equilibrium position of the proton in the Br···H···N bond shifts gradually from the bromine to nitrogen atoms. In the H_3N-HBr complex the proton is localized at the bromine atom and the normal hydrogen bond of the Br-H···N type appears. Coming to the stronger $(CH_3)H_2N$-HBr complex, one observes a very broad single minimum in which the proton lies approximately midway between the N and Br atoms (ref. 50). This result, received on more advanced calculation level, is in contrary to that previously obtained by Brciz *et al.* (ref. 40). Their SCF potential energy surface has two minima, the ion pair being more stable of the two. In the case of the $(CH_3)_2HN$, $(CH_3)_3N$-HBr complexes a single minimum is located at the nitrogen atom and the hydrogen-bonded ion-pair occurs. These conclusions are supported by the analysis of the change of electron density due to the complexation (see Table 1).

As one can see from Fig. 2b in the amine-HI complexes the proton is completely transferred from the I to N atoms. The hydrogen-bonded ion-pair is formed. Small differences in the shape of potentials of the N^+···H-I^- bonds are observed (see Fig. 2b).

An interesting situation appears in the H_3N-HI complex. Our SCF and electron correlation on Møller-Plesset level calculations (ref. 50) as well as Jasien and Stevens (ref. 45) and Seel *et al.* (ref. 53) strongly suggest that two minima appear, the ion pair being less stable of the two (the energy difference between two protonic states is in the range 0.9 (ref. 50), 1.17 (ref. 45) and 7.1 (ref. 53.) kcal/mol, respectively.

Schuster *et al.* (ref. 68) have proposed on the basis of simple consideration of the proton transfer process a method to estimate the appearance of the hydrogen-bonded ion pair in the gas phase. According to Schuster the energy of ion pair (ΔE_{NI}) consists of two major contributions of opposite sign: the proton transfer energy

$\Delta E_{PT} = \Delta E_{PA}(B) - \Delta E_{PA}(A^-)$ corresponding to proton transfer at infinite separation and the electrostatic contribution to the energy of ion pair formation, $\Delta E_{coul}(IP) = -\frac{e^2}{R_{AB}}$; thus $\Delta E_{NI} = \Delta E_{PT} + \Delta E_{coul}(IP) + ...$ The corresponding data for the series of complexes formed by hydrogen halides (HF, HCl, HBr, HI) with ammonia and aliphatic amines are given in Table 2. As one can see from Table 2 there is a quite good agreement between the predicted appearance of the hydrogen-bonded ion pair in the gas phase with experimental evidence obtained by spectroscopic methods and quantum chemical calculations analised previously.

Table 2.

Stability of ion-pair complexes in associations of hydrogen halides with ammonia and aliphatic amines.

Complexes	PA(B) [kcal/mol]	ΔE_{PT} [kcal/mol]	$\frac{e^2}{R_{NA}}$ [kcal/mol]	R_{NA} [Å]	$\Delta E_{PT} - \frac{e^2}{R_{NA}}$ [kcal/mol]
HF complexes, PA(F^-) = 372 [kcal/mol]					
NH_3	205	167	123.7	2.684	43.3
NH_2CH_3	214.1	157.9	125.8	2.640	32.1
$NH(CH_3)_2$	220.5	151.5	126.4	2.627	25.1
$N(CH_3)_3$	224.3	147.7	127.4	2.606	20.3
HCl complexes, PA(Cl^-) = 333 [kcal/mol]					
NH_3	205	128	100.7	3.297	27.3
NH_2CH_3	214.1	118.9	112.2	2.96	6.7
$NH(CH_3)_2$	220.5	112.5	111.1	2.99	1.4
$N(CH_3)_3$	224.3	108.7	114.1	2.91	-5.4
HBr complexes, PA(Br^-) = 324 [kcal/mol]					
NH_3	205	119	102.7	3.234	16.3
NH_2CH_3	214.1	109.9	112.5	2.952	-2.6
$NH(CH_3)_2$	220.5	103.5	112.1	2.961	-8.6
$N(CH_3)_3$	224.3	99.7	111.7	2.972	-12.0
HI complexes, PA(I^-) = 314 [kcal/mol]					
NH_3	205	109	100.6	3.3	8.4
NH_2CH_3	214.1	99.9	103.4	3.210	-3.5
$NH(CH_3)_2$	220.5	93.5	103.4	3.212	-9.9
$N(CH_3)_3$	224.3	89.7	103.2	3.219	-13.5

Influence of environmental effects on the structure of hydrogen halide-amine complexes has been studied by Kurnig and Scheiner (ref. 69) by *ab initio* quantum chemical methods. The effects of a polarisable medium was included *via* the SCRF formalism. It has been noticed that the shift of the proton toward the nitrogen is enhanced by increasing of the solute solvent interaction such that relatively modest coupling leads to complexes of ion-pair type. In the series of complexes studied (HF, HCl, HBr - amines) the HBr systems are the most sensitive to either the basicity of the amine or the influence of the medium, whereas the HF analogs are affected very little.

Proton transfer in the hydrogen-bonded complexes formed between derivated phenol and amines in aprotic solvents.

The hydrogen-bonded complexes formed by phenols and amines can be used as model systems to study properties of the hydrogen bond in a very systematic way. So far they have been widely studied in aprotic solvents and solids by using various physico-chemical methods: IR (refs. 18, 21-23 and 56), UV-visible (refs. 9, 15 and 24), ^{1}H NMR (refs. 19, 26 and 27) and ^{13}C NMR (refs. 19 and 20) spectroscopy, dielectric (refs. 7, 8 and 12), and calorimetric measurements (ref. 13) and quantum chemical calculations (ref. 55). The experimental data have been interpreted on the basis of the following thermodynamic equilibria which might appear in those systems:

$$\begin{array}{ccccc} & & (AH\cdots B)_{n,s} \overset{k_6}{\rightleftharpoons} (A^-\cdots H\overset{+}{B})_{n,s} & & \\ & & \text{Aggregates} & & \\ & & k_4 \updownarrow \qquad\qquad k_5 \updownarrow & & \end{array}$$

$$(AH)_s + (B)_s \overset{k_1}{\rightleftharpoons} (AH\cdots B)_s \underset{k_{rev}}{\overset{k_{PT}}{\rightleftharpoons}} (A^-\cdots H\overset{+}{B})_s \overset{k_2}{\rightleftharpoons} (A^- || H\overset{+}{B})_s \overset{k_3}{\rightleftharpoons} (A^-)_s + (H\overset{+}{B})_s$$

Acid	Base	H-bonded complex	Ion-pair hydrogen-bonded	Solvent separated ion-pair	Free ions

[s] - solvated

Much attention has been paid to the study of the tautomeric equilibrium,

$$(AH\cdots B)_s \underset{k_{rev}}{\overset{k_{PT}}{\rightleftharpoons}} (A^-\cdots HB^+)_s \qquad (1)$$

It has been demonstrated (refs. 7 and 57) that 1:1 complex formed between substituted phenols and triethylamine are ideally suited for studies of this proton transfer process since their formation constants are large enough so the dissociation into free phenol and amine is relatively small in solution in aprotic solvents and furthermore that the position of the tautomeric equlibrium can be adjusted by a judicious choice of substituents in the aromatic ring of the phenol molecule. It has been further demonstrated, that in solvents with low dielectric permittivity, k_3 (dissociation of a hydrogen-bonded ion pair into free ions) is very small (ref. 58). Recently the equilibrium constants k_4, k_5 (aggregation phenomenon) have been evaluated by Dega-Szafran and Szafran (ref. 59) for some phenols-triethylamine complexes in benzene.

A study of the increment of the dipole moment, $\Delta\vec{\mu}$, due to the hydrogen bond formation in the derivated phenol-triethylamine complexes in benzene as a function of the pK_A of the phenolic acid (Fig. 3) clearly shows that a distinct transition in the nature of the acid-base complex occurs in the pK_A range 4-8. This is also reflected in the NMR chemical shift and linewidth of the hydrogen-bonded proton signal (Figs. 4 and 5).

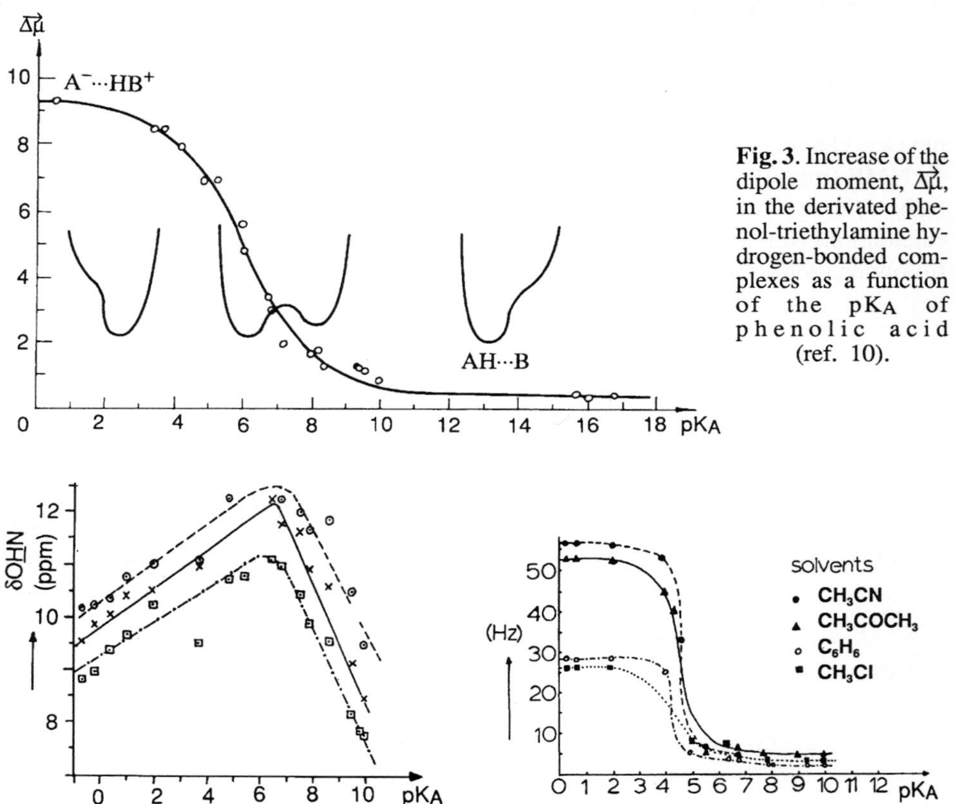

Fig. 3. Increase of the dipole moment, $\Delta\vec{\mu}$, in the derivated phenol-triethylamine hydrogen-bonded complexes as a function of the pK_A of phenolic acid (ref. 10).

Fig. 4. Variation of the chemical shift of the hydrogen bridge proton of phenol-triethylamine complexes in different solvents as a function of phenol pK_A at room temperature; circles: $CDCl_3$ (0.25 molal), crosses: C_6H_6 (0.15 molal), squares: $(CD_3)_2CO$ (0.25 molal).

Fig. 5. Variations of the linewidth of the hydrogen bridged proton signal of phenol-triethylamine complexes in different solvents as a function of phenol pK_A at room temperature.

Table 3.

Thermodynamic quantities of the proton transfer process in some derivated phenol-amines complexes.

Phenol derivative	amine	solvent	$-\Delta H^0_{PT}$ kcal/mol	$-\Delta S^0_{PT}$ cal/(mol·K)	ref.
2,3-$(Cl)_2$-	triethylamine	$CDCl_3$	1.5	6.9	[20]
2,5-$(Cl)_2$-	triethylamine	$CDCl_3$	2.3	11.2	[20]
2,5-$(Cl)_2$-	triethylamine	CD_2Cl_2	1.8	9.1	[20]
2,6-$(Cl)_2$-	triethylamine	CD_2Cl_2	4.8	16.3	[20]
2,3,5-$(Cl)_3$-	triethylamine	$CDCl_3$	3.4	12.0	[20]
2,3,5-$(Cl)_3$-	triethylamine	CD_2Cl_2	2.8	9.3	[20]
2,4,6-$(Cl)_3$-	triethylamine	CD_2Cl_2	6.7	22.9	[20]
2,3-$(Cl)_2$-	octylamine	octylamine	3.3	11.0	[64]
2,3,4-$(Cl)_3$-	octylamine	octylamine	3.8	12.4	[64]
2,3,5-$(Cl)_3$-	octylamine	octylamine	6.5	20.1	[64]
4-(NO_2)-	n-butylamine	n-butanol	2.5		[65]
4-(NO_2)-	diisopropylamine	dichloroethane	2.1		[65]
4-(NO_2)-	diisopropylamine	n-butanol	3.7		[65]
4-(NO_2)-	tri-n-butylamine	dichloroethane	3.3		[65]
4-(NO_2)-	tri-n-butylamine	n-butanol	2.5		[65]
4-(NO_2)-	piperazine	n-butanol	3.8		[65]
4-(NO_2)-	piperazine	dichloroethane	3.4		[65]
4-(NO_2)-	N-methylpiperazine	n-butanol	3.3		[65]
4-(NO_2)-	2,5-dimethylpiperazine	n-butanol	2.3		[65]
4-(NO_2)-	piperidine	dichloroethane	2.5		[65]
4-(NO_2)-	piperidine	n-butanol	3.1		[65]
4-(NO_2)-	N-methylpiperidine	n-butanol	2.3		[65]
4-(NO_2)-	4-methylpiperidine	dichloroethane	2.4		[65]
4-(NO_2)-	4-methylpiperidine	n-butanol	3.3		[65]
4-(NO_2)-	2,6-dimethylpiperidine	n-butanol	2.1		[65]
2,6-$(NO_2)_2$-	imidazole	dioxane	2.1	4.9	[58]
2,6-$(NO_2)_2$-	triethylamine	dioxane	2.8	4.4	[58]
2,6-$(NO_2)_2$-	n-butylamine	dioxane	2.6	7.8	[58]
2,4-$(NO_2)_2$-	n-butylamine	dioxane	2.5	4.3	[58]

These results are consistent with the following interpretation: by gradual changes the acidic properties of phenols, expressed by the pK_A of phenol, one can obtain the complexes with a "normal" hydrogen bond AH···B, a tautomeric equilibrium of the AH···B $\rightleftharpoons$ A^-···HB^+ type and a hydrogen-bonded ion-pair A^-···HB^+.

Thus the potential energy surface of the hydrogen bond in the complexes formed between phenols and triethylamine in aprotic solvents has a double minima; the shape of the potential strongly depends upon the acidity of phenol. This situation is presented in Fig. 1b and is described in Introduction.

In the pK_A range 4-8 one expects the existence of a tautomeric equilibrium between an ion-pair and a molecular complex of the type (1) where the rate of proton transfer is rapid on the NMR time scale at ambient temperatures.

By UV-visible, IR and NMR spectroscopy methods it has been possible to determine the thermodynamic properties of the proton transfer process. The results obtained are summarized in Table 3.

Table 4.

Relations between ΔpK_A[a] values and appearance of the splitting of the OHN proton signal in derivated phenol-amine complexes in C_2H_5Cl[b].

derivated phenol (pK_A)	**amine (pK_A)**		
	triethylamine (10.8)	N-methylpiperidine (10.1)	N,N-dimethylaniline (5.1)
2.4 $(Cl)_2$ – (7.9)	$\Delta pK_A = 2.9$	—	—
2.3 $(Cl)_2$ – (7.7)	$\Delta pK_A = 3.1$	—	—
2.5 $(Cl)_2$ – (7.5)	—	$\Delta pK_A = 2.6$	—
2.3.4.5 $(Cl)_4$ – (5.6)	—	—	$\Delta pK_A = -0.5$
2.3.4.6 $(Cl)_4$ – (5.2)	—	—	$\Delta pK_A = -0.1$
2.3.5.6 $(Cl)_4$ – (5.0)	—	—	$\Delta pK_A = 0.1$
$(Cl)_5$ – (4.7)	—	—	$\Delta pK_A = 0.4$

[a] $\Delta pK_A = pK_A$ (protonated amine) - pK_A (phenol)

[b] ref. 66

Table 5.

Thermodynamic and activation parameters for the proton transfer process in some derivated phenol-amine complexes.

phenol	amine	solvent	$-\Delta H^0_{PT}$ kcal/mol	$-\Delta S^0_{PT}$ cal/(mol·K)	$\Delta H^{\neq}_{PT}$ kcal/mol	$\Delta H^{\neq}_{rev}$ kcal/mol	$-\Delta S^{\neq}_{PT}$ cal/(mol·K)	$-\Delta S^{\neq}_{rev}$ cal/(mol·K)	T K	$\Delta G^{\neq}_{PT}$ kcal/mol	$\Delta G^{\neq}_{rev}$ kcal/mol	k_{PT} sec^{-1}	k_{rev} sec^{-1}	ref.
2.3.5.6-(Cl_4)-	N,N-dimethylaniline	C_2H_5Cl	0.8	6.6	3.1	3.8	24.8	18.4	162.1			650.0	1538.3	[27]
2.4-(Cl_2)-	triethylamine	C_2H_5Cl	1.1	6.9					195	9.0	8.7	338	599	[26]

However, due to experimental difficulties, there is a paucity of kinetic data (ref. 60). Recently, Denisov *at al.* (refs. 60 and 61) and independently Ratajczak *et al.* (refs. 19, 26 and 27) have shown that low-temperature ^1H NMR spectroscopy is an eminently suitable method for studying this proton transfer phenomenon in solution since, in favourable circumstances, one can observe separate signals for the molecular and ion-pair tautomeric forms. In principle, by studying the temperature dependence of these signals, one may evaluate the thermodynamics of the process from their intensities and the kinetics from their line shapes.

Table 4 shows the hydrogen-bonded complexes in which, in low temperature, the splitting of the OHN signal is observed due to the existence of the two forms of the complexes in the tautomeric equilibrium of the type (1). The thermodynamic quantities ($\Delta G^{\neq}$, $\Delta H^{\neq}$, $\Delta S^{\neq}$) for the proton transfer process in some hydrogen-bonded complexes have been obtained and are given in Table 5. The obtained results permit to construct the potential energy surface of the hydrogen bond.

It is interesting to notice that in the series of complexes which lie in region of existence of a tautomeric equilibria the IR spectra show a large continuous absorption called the Zundel type absorption, (refs. 18 and 62).

Conclusions

It has been shown that two different mechanisms of a proton transfer along the hydrogen bond can appear. It depends upon the nature of the potential energy profile of the hydrogen bond: 1) the proton occupies two states in a double minimum potential well which results in the tautomeric equilibrium, 2) the proton occupies only one state in a single minimum potential surface. It has been demonstrated that both these mechanisms are possible. The tautomeric equilibrium has been observed in the hydrogen bonded systems of the O-H···N type in a series of complexes formed between the derivated phenol with triethylamine in aprotic solvents as well as in the systems with the hydrogen bond of the C-H···Y type (ref. 60). The second proton transfer mechanism has only been observed so far in the hydrogen-bonded complexes formed by hydrogen halides with ammonia and aliphatic amines in the gas phase and in low temperature matrices (argon, nitrogen). However, it is not clear yet to what extent the electronic structure of the hydrogen-bonded complex and to what extent the nature of the environment influence the mechanism of a proton transfer along the hydrogen bridge.

In this respect it would be useful to mention the recently published paper on the formulation of new theoretical approach developed by Juanós i Timoneda and Hynes to study proton transfer phenomenon in solutions (ref. 63). The theory is presented for the electronic structure and multidimensional free energy surfaces for hydrogen-bonded complexes AH···B capable of proton transfer to form an ion-pair $A^{-}\cdots HB^{+}$ in solution.

Acknowledgements

I am very indebted to my co-workers: Professor Z. Mielke and Drs. Z. Latajka, M. Wierzejewska-Hnat, M. Ilczyszyn and M.K. Marchewka for very fruitful discussions and comments on the problems described in this article.

References

1 H. Ratajczak and W.J. Orville-Thomas (Eds.) *Molecular Interactions* Vol. 2, J. Wiley, Chichester, 1981.

2 E.F. Caldin and V. Gold (Eds.) *Proton-Transfer Reactions* Chapman and Hall, London, 1975.

3 R.P. Bell *The Proton in Chemistry* Chapman and Hall, London, 1973.

4 R.P. Bell *The Tunnel Effect in Chemistry* Chapman and Hall, London, 1980.

5 H. Ratajczak and Z. Latajka, to be published.

6 C.L. Bell and G.M. Barrow, J. Chem. Phys. **31** (1959) 1158.

7 H. Ratajczak and L. Sobczyk, Zhur. Strukt. Khim. **6** (1965) 262.

8 H. Ratajczak and L. Sobczyk, J. Chem. Phys. **50** (1969) 556.

9 H. Baba, A. Matsuyama and H. Kokubun, Spectrochim. Acta A**25** (1969) 1709.

10 H. Ratajczak and L. Sobczyk, Bull. Acad. Polon. Sci., ser. sci. chim. **18** (1970) 93.

11 H. Ratajczak, J. Phys. Chem. **76** (1972) 3000.

12 J. Jadzyn and J. Małecki, Acta Phys. Polon. A**41** (1972) 599.

13 D. Neerinck, A. Van Audenhaege, L. Lamberts and P. Huyskens, Nature **218** (1968) 461.

14 B.H. Robinson in: E.F. Caldin and V. Gold (Eds.) *Proton-Transfer Reactions* Chapman and Hall, London, 1975, p. 121.

15 R. Scott, D. De Palma and S. Vinogradov. J. Phys. Chem. **72** (1972) 3192.

16 J. Oszust and H. Ratajczak, J. Chem. Soc., Faraday Trans. I **77** (1981) 1209.

17 J. Oszust and H. Ratajczak, J. Chem. Soc., Faraday Trans. I **77** (1981) 1215.

18 G. Albrech and G. Zundel, J. Chem. Soc., Faraday Trans. I **80** (1984) 553.

19 M. Ilczyszyn, L. Le-Van, H. Ratajczak and J.A. Ladd in: P. Laszlo (Ed.) *Protons and Ions Involved in Fast Dynamic Phenomena* Elsevier, Amsterdam, 1978, p. 257.

20 M. Ilczyszyn, H. Ratajczak and K. Skowronek, Magn. Res. Chem. **26** (1988) 445.

21 G. Zundel and A. Nagyrevi, J. Phys. Chem. **82** (1978) 685.

22 J. Fritsch and G. Zundel, J. Phys. Chem. **85** (1981) 556.

23 G. Zundel and J. Fritsch, J. Phys. Chem. **88** (1984) 6295.

24 H. Romanowski and L. Sobczyk, J. Phys. Chem. **79** (1975) 2535.

25 J. Kraft, S. Walker and M.D. Magee, J. Phys. Chem. **79** (1975) 881.

26 M. Ilczyszyn, H. Ratajczak and J.A. Ladd, Chem. Phys. Lett. **153** (1988) 385.

27 M. Ilczyszyn, H. Ratajczak and J.A. Ladd, J. Mol. Structure **198** (1989) 499.

28 J. Kalenik, I. Majerz, Z. Malarski and L. Sobczyk, Chem. Phys. Lett. **165** (1990) 15.

29 M. Maćkowiak, P. Kozioł and J. Stankowski, Z. Naturforsch. **41**a (1986) 225.

30 Th. Zeegers-Huyskens and P. Huyskens in: H. Ratajczak and W.J. Orville-Thomas (Eds.) *Molecular Interactions* Vol. 2, J. Wiley, Chichester, 1981, p. 1 and references therein.

31 B.S. Ault, E. Steinback and G.C. Pimentel, J. Phys. Chem. **79** (1975) 615.

32 A.J. Barnes, T.R. Beech and Z. Mielke, J. Chem. Soc., Faraday Trans. 2 **80** (1984) 455.

33 A.J. Barnes, J.N.S. Kuzniarski and Z. Mielke, J. Chem. Soc., Faraday Trans. 2 **80** (1984) 465.

34 A.J. Barnes and M.P. Wright, J. Chem. Soc., Faraday Trans. 2 **82** (1986) 153.

35 L. Schriver, A. Schriver and J.P. Perchard, J. Am. Chem. Soc. **105** (1983) 3843.

36 A.C. Legon, Chem. Brit. **26** (1990) 562.

37 C.A. Rego, R.C. Batten and A.C. Legon, J. Chem. Phys. **89** (1988) 696.

38 A.C. Legon and C.A. Rego, Chem. Phys. Lett. **154** (1989) 468.

39 N.W. Howard and A.C. Legon, J. Chem. Phys. **88** (1988) 4694.

40 A. Brciz, A. Karpfen, H. Lischka and P. Schuster, Chem. Phys. **89** (1984) 337.

41 P. Kollman, J. Mc Kelvey, A. Johansson and S. Rothenberg, J. Am. Chem. Soc. **97** (1975) 955.

42 Z. Latajka and S. Scheiner, J. Chem. Phys. **81** (1984) 4014.

43 E. Clementi, J. Chem. Phys. **46** (1967) 3851.

44 R.C. Raffenetti and D.H. Phillips, J. Chem. Phys. **71** (1979) 5434.

45 P.G. Jasien and W.J. Stevens, Chem. Phys. Lett. **130** (1986) 127.

46 Z. Latajka, S. Sakai, K. Morokuma and H. Ratajczak, Chem. Phys. Lett. **110** (1984) 464.

47 A.C. Legon and C.A. Rego, J. Chem. Phys. **90** (1989) 6867.

48 N.W. Howard and A.C. Legon, J. Chem. Phys. **86** (1987) 6722.

49 Z. Latajka, S. Scheiner and H. Ratajczak, Chem. Phys. Lett. **135** (1987) 367.

50 Z. Latajka, S. Scheiner and H. Ratajczak, Chem. Phys., submitted.

51 A.C. Legon, A.L. Wallwork and C.A. Rego, J. Chem. Phys. **92** (1990) 6397.

52 P. Kollman, A. Dearlang and E. Kochanski, J. Phys. Chem. **86** (1982) 1607.

53 M. Seel and H. Ratajczak, unpublished data.

54 N.S. Golubev and G.S. Denisov, Khim. Fiz. **5** (1982) 563.

55 Z. Latajka and H. Ratajczak, unpublished data.

56 A. Sucharda, L. Sobczyk and H. Ratajczak, Wiadomości Chemiczne **22** (1968) 339.

57 H. Ratajczak, Z. phys. Chem., (Leipzig) **231** (1966) 33.

58 S.N. Vinogradov, R.A. Hudson and R.M. Scott, Biochim. Biophys. Acta **214** (1970) 6.

59 Z. Dega-Szafran and M. Szafran, J. Chem. Soc. Perkin Trans. II (1987) 897.

60 G.S. Denisov, S.F. Bureiko, N.S. Golubev and K.G. Tokhadze, in: H. Ratajczak and W.J. Orville-Thomas (Eds.) *Molecular Interactions* Vol. 2, J. Wiley, Chichester, p. 107.

61 G.S. Denisov and N.S. Golubev, J. Mol. Structure **75** (1981) 311.

62 G. Zundel in: P. Schuster, G. Zundel and C. Sandorfy (Eds.) *The Hydrogen Bond* Vol. II, North-Holland, Amsterdam, p. 683.

63 J. Juanós i Timoneda and J.T. Hynes, J. Phys. Chem. **95** (1991) 10431.

64 R. Krämer and G. Zundel, Z. Phys. Chem. (Neue Folge) **14, S** (1985) 265.

65 A.S.N. Murthy and A.R. Reddy, Adv. Mol. Rel. Int. Processes **22** (1982) 199.

66 M. Ilczyszyn and H. Ratajczak, to be published.

67 D.C. Borgis, S. Lee and J.T. Hynes, Chem. Phys. Lett. **162** (1989) 19.

68 P. Schuster, P. Wolschann and K. Tortschanoff, in: I. Pecht and R. Rigler (Eds.) *Molecular Biology, Biochemistry and Biophysics* Vol. 24, Springer-Verlag, Berlin, 1977, p. 107.

69 I.J. Kurnig and S. Scheiner, Int. J. Quant. Chem., Quant. Biol. Sym. **14** (1987) 47.

70 P.W. Fowler, A.C. Legon, C.A. Rego and P. Tole, Chem. Phys. **134** (1989) 297.

Electron and Proton Transfer in Chemistry and Biology, edited by A. Müller et al.
Studies in Physical and Theoretical Chemistry, Vol. 78

Hydrogen–Bonded Systems with Large Proton Polarizability due to Collective Proton Motion as Pathways of Protons in Biological Systems

Georg Zundel

Institute of Physical Chemistry, University of Munich,

Theresienstr. 41, D-8000 München 2 (Germany)

Summary

Homoconjugated, as well as heteroconjugated hydrogen bonds show polarizabilities which are 1-2 orders of magnitude larger than polarizabilities caused by distortion of electron systems. It is shown experimentally as well theoretically that hydrogen–bonded chains may show particularly large proton polarizability. Thus, such systems are very effective proton pathways. It is demonstrated that a large number of hydrogen bonds and especially hydrogen–bonded systems, which may be present in biological systems, show large proton polarizability. Finally, it is postulated that hydrogen bonds with large proton polarizability are of importance for proton shifts in active centers of enzymes. Furthermore, it is postulated that hydrogen–bonded chains with large proton polarizability due to collective motion are of large importance for the proton transport, especially in biological membranes.

Proton Polarizability of Hydrogen Bonds

Homoconjugated hydrogen bonds, *i.e.* bonds of the types $B^+H\cdots B \rightleftarrows B\cdots H^+B$ or $AH\cdots A^- \rightleftarrows A^-\cdots HA$, respectively, as well as heteroconjugated $AH\cdots B \rightleftarrows A^-\cdots H^+B$ bonds show polarizabilities so–called proton polarizabilities which are about 1-2 orders of magnitude larger than polarizabilities caused by distortion of electron systems (refs.1-4). These polarizabilities occur since the proton can easily be shifted in a double minimum proton potential or in broad flat potential wells. In the case of the homoconjugated hydrogen bonds the proton potentials are symmetrical if these hydrogen bonds are considered without environment. They become, however, more or less asymmetrical caused by the interaction with the environments (ref. 5). In the case of the heteroconjugated hydrogen bonds, however, more or less symmetrical double minimum proton potentials are usually only created by their environments since if they are isolated the well at the AH group is much deeper, but to this potential, in liquid state, the potential of the reaction field is added (ref. 6), and in this way double minima arise. In addition specific interactions of the groups with the hydrogen bond as, for instance, the interaction with acidic CH groups act in the same way (ref. 6). Thus, in $AH\cdots B \rightleftarrows A^-\cdots H^+B$ bonds the double minimum proton potentials are created by their environments. Then they show large proton polarizability, and small changes of the environments may shift the equilibria. Caused by the strong interaction with their environments – resulting from their large proton polarizability – such hydrogen bonds cause continua in the infrared spectra instead of bands. Besides, the IR continua such bonds with large proton polarizability are indicated by intense Rayleigh wings (elastic scattering) at the excitation line when Raman spectra are taken (ref. 7). For summary of the behavior and importance of hydrogen bonds with large proton polarizability see (ref. 8).

Hydrogen–Bonded Chains Formed by Side Chains of Proteins and Phosphates

Poly–α–amino acid films have been studied by IR spectroscopy. With these studies phosphates were added to these films (refs. 9-14).

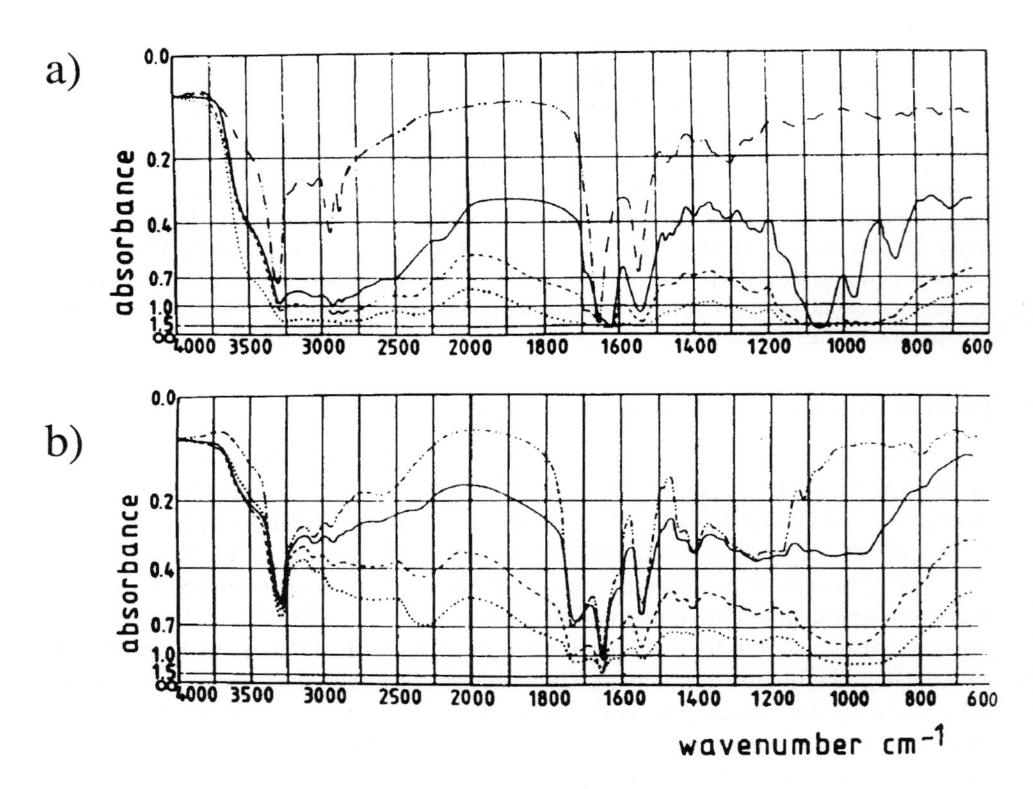

Fig. 1. a) IR spectra of (L-Lys)$_n$–KH_2PO_4 systems: (–··–), pure (Lys)$_n$; (———), Lys : KH_2PO_4 = 3 : 1; (- - - -), Lys : KH_2PO_4 = 1 : 1; (····), Lys : KH_2PO_4 = 1 : 2. b) IR spectra of (L–Glu)$_n$ – KH_2PO_4 systems : (–··–), pure (Glu)$_n$; (———), Glu : KH_2PO_4 = 2 : 1; (----), Glu : KH_2PO_4 = 1 : 1; (····), Glu : KH_2PO_4 = 1 : 3.

Fig. 1 (a) shows polylysine + dihydrogen phosphate systems. With increasing phosphate content a very intense continuum arises. The same is true with the polyglutamic acid dihydrogen phosphate systems as shown in Fig. 1 (b).

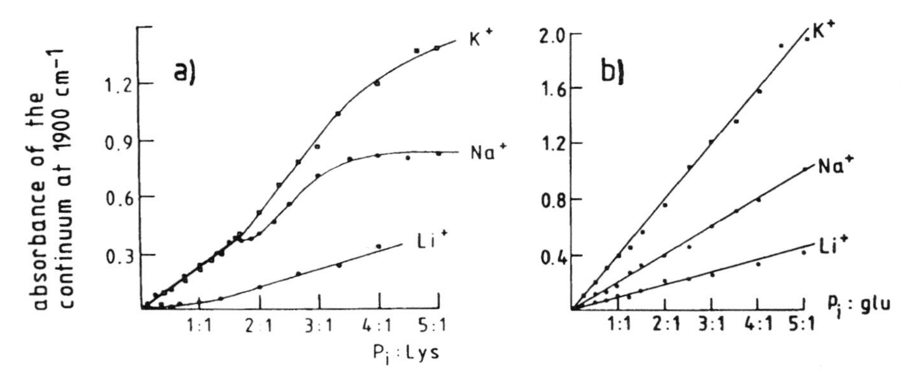

Fig. 2. Absorbance of the continuum at 1900 cm^{-1} as a function of the P_i: residue ratio; a) (L–Lys)n–KH_2PO_4 systems, b) $(L–Glu)_n$–KH_2PO_4 systems.

In Fig. 2 the intensities of these continua are shown as a function of the phosphate to side chain ratio. Fig 2 (a) shows that with the polylysine systems the intensity of the continuum increases up to three phosphates per lysine residue if sodium, or up to five if potassium ions are present. Fig. 2 (b) shows that in the case of the polyglutamic acid dihydrogen phosphate systems the intensity of the continuum increases up to phosphate – glutamic acid residue ratios of 5:1, the highest phosphate content which could be measured.

These results demonstrate that side chains form with phosphates hydrogen–bonded chains with large proton polarizabilities which are caused by collective proton motion within these chains. Probably these chains are structurally symmetrical. Then with the polylysine + phosphate systems in the case of the Na^+ system 7 and with the K^+ system 11 hydrogen bonds are built up, whereas in the polyglutamic acid + dihydrogen phosphate systems extended hydrogen–bonded networks are formed.

In Table 1 all results obtained with poly–α–amino acid phosphate systems are summarized. This table shows that also in $(his)_n$ + phosphate and $(tyr)_n$ + phosphate systems hydrogen–bonded chains may form which show large proton polarizability caused by collective proton motion. Additionally, Table 1 shows that the behaviour of such protein side chain–phosphate systems depends very sensitively from the cations present. **1.** They influence the position of the proton transfer equilibria in the protein side chain–phosphate hydrogen bonds. **2.** They influence the size of the proton polarizability of these hydrogen bonds. This result is obtained if one compares the systems with various cations under comparable conditions. **3.** They determine the length of the hydrogen–bonded chain which are built up. All these results demonstrate

Table 1
Hydrogen bonds with large proton polarizability formed between side chains of proteins and phosphates.

System	Hydrogen bond	Cation	% Proton transfer	Chains formed[x)]	Phosphates in the chain
Imidazole (R-substituted, HN, N) + $H_2PO_4^-$	$N \cdots HOP \rightleftharpoons N^+H \cdots {}^-OP$	Li^+ Na^+ K^+	55 41 } to His 32	$=N^+H \cdots OPO \cdots HOPO \cdots$ (with H, O above; O_{2-}, O_- below)	No chain No chain 2
$R(CH_2)_4NH_2 + H_2PO_4^-$	$N \cdots HOP \rightleftharpoons N^+H \cdots {}^-OP$	Li^+ Na^+ K^+	100 85-95 } to Lys 75-85	$-N^+H \cdots OPO \cdots HOPO \cdots$ (with H, H, O, H, O above; H, O_{2-}, O_- below)	>4[xx)] 3 5
Phenol (OH, R-substituted) + HPO_3^{2-}	$OH \cdots {}^-OP \rightleftharpoons O^- \cdots HOP$	Li^+ Na^+ K^+	0 5 } to P_i 10	${}^-O \cdots HOPOH \cdots OPOH \cdots$ (with O, O above; O_-, O_{2-} below)	No chain 2 4
$R(CH_2)_2CO_2H + HPO_4^{2-}$	$OH \cdots {}^-OP \rightleftharpoons O^- \cdots HOP$	Li^+ Na^+ K^+	<5 75[xxx)] } to P_i 95	No chains	No chains
$R(CH_2)_2CO_2H + H_2PO_4^{2-}$	-	Li+ Na^+ K^+	>0 to P_i	$-C(=O)OH \cdots OPOH \cdots OPOH \cdots$ (with H, O, H, O above; O_-, O_- below)	>5

x) Only one of the proton limiting structure given.
xx) Charge fluctuation only within the phosphate–phosphate hydrogen bonded part of the chain.
xxx) If the hydrogen bonds are completely formed at a P_i:Glu ratio of 1.6:1. At small P_i:Glu ratios, the % proton transfer is only slight; it increases with increasing P_i:Glu ratio.

that such systems can easily be controlled by the cations present and also by other local electrical fields since they show large proton polarizability. This fact is probably of large importance for regulation processes in biological systems.

Studies of the Proton Polarizability using Intramolecular Hydrogen-Bonded Chains

Fig. 3 shows the spectrum of a solution of the monosalt of 1,2,3-benzenetricarboxylic acid (ref. 15).

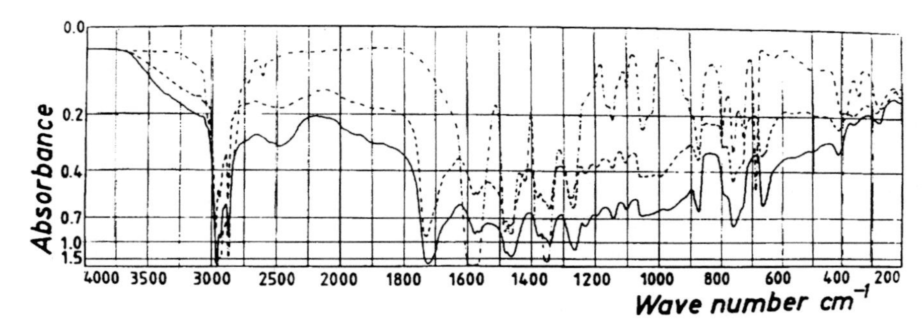

Fig. 3. IR spectra of acetonitrile-d_3 solutions (sample thickness 0.096 mm) of: 1,2,3-benzenetricarboxylic acid tetrabutylammonium monosalt: 0.50 mole/dm^3 (——) and 0.25 mole/dm^3 (– – –), and of the tetrabutylammonium salt (– · – · –)

With increasing concentration an intense continuum arises, demonstrating that this system formed by two intramolecular hydrogen bonds shows large proton polarizability due to collective proton motion and must be represented by these three proton limiting structures

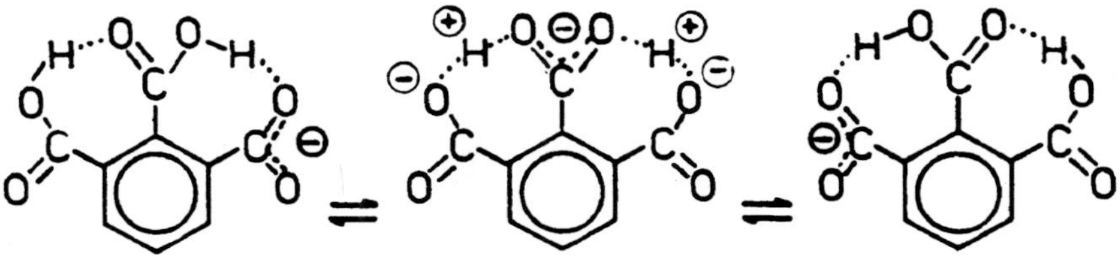

I ⇌ II ⇌ III

Another system which we studied was a solution of the monosalt of 1,11,12,13,14–pentahydroximethylpentacene (ref. 16). Also with this system an intense continuum is observed (Fig. 4), indicating the large proton polarizability due to collective proton motion in the hydrogen–bonded chain

$$OH\cdots OH\cdots OH\cdots OH\cdots O^- \rightleftarrows O^-\cdots HO\cdots HO\cdots HO\cdots HO$$

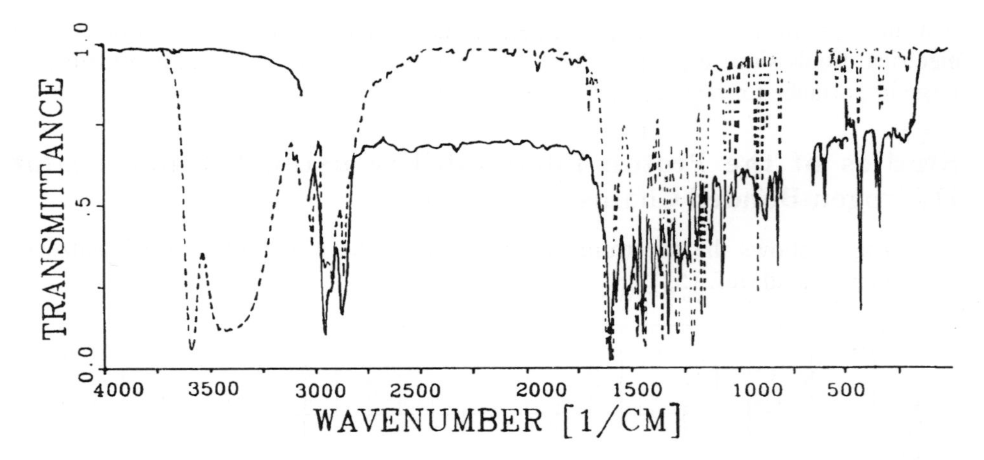

Fig. 4. FTIR spectra of 0.1 mol dm^{-3} CH_2Cl_2 solutions of: (– – – –) ,1,11,12,13,14–pentahydroxymethylpentacene; and (———), of its tetrabutylammonium salt.

The next studied systems contain carboxylic acid as well as phenol groups (ref. 17)

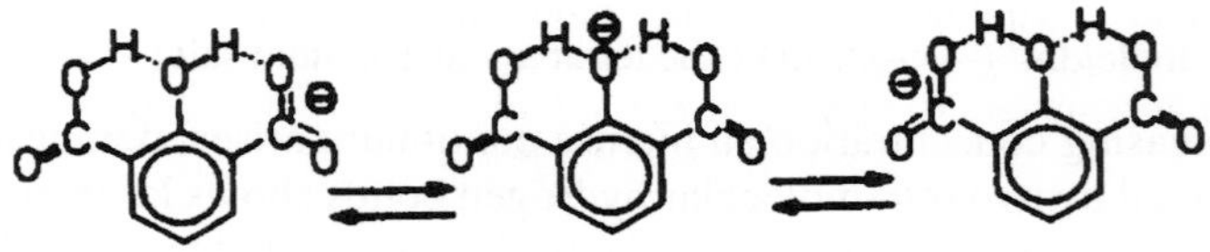

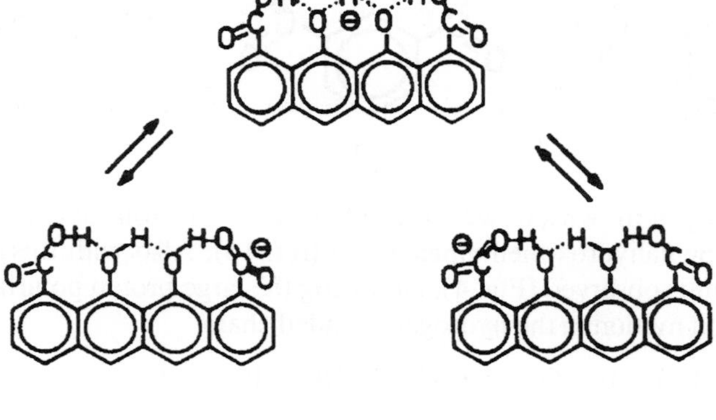

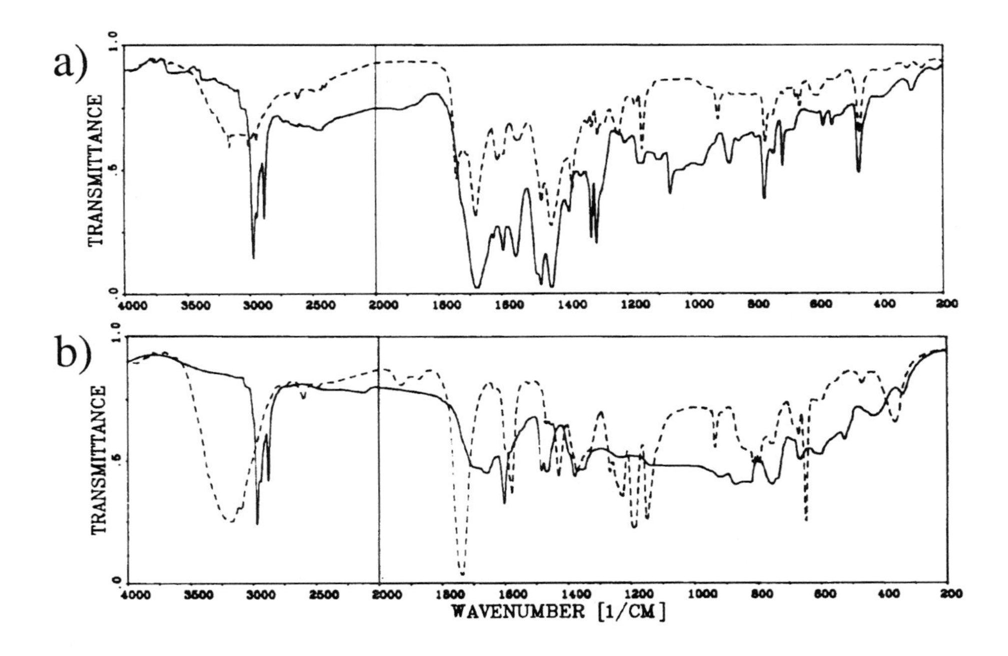

Fig. 5. IR spectra of acetonitrile-d_3 solutions (– – –), of the substance and (——), of the tetrabutylammonium monosalt: a) 2–hydroxyisophtalic acid, b) 11,12–dihydroxy 1,10 naphtacene dicarbonic acid.

Fig. 5 shows that in the solutions of both molecules intense continua are observed, demonstrating that also these systems show large proton polarizability due to collective proton motion and must be represented by the proton limiting structures shown above.

Particularly interesting are monoperchlorates of 2,6–disubstituted MANNICH bases (ref. 18). With these substances the acidity of the phenolic group can be changed by the substituents R_1 and R_2.

In Table 2, in the last column, the absorbance of the continuum of the monoperchlorates is shown as a function of the acidity of the phenolic group. With increasing acidity the intensity of the continuum increases. If the acidity increases further, the intensity decreases again and finally, instead of the continuum, NH^+ stretching vibration bands are observed. Thus, the proton polarizability, due to collective proton motion in these two hydrogen bonds, first increases and, if the acidity becomes still larger, it decreases again, and vanishes finally.

TABLE 2

Intensity of the IR continuum with disubstituted protonated MANNICH bases.

Compound		Intensity of the Continuum	
R_1	R_2	Monobase $\ln\frac{I_0}{I}$ continuum	Mono Perchlorate of the di-base $\ln\frac{I_0}{I}$ continuum
H	OBu	0.000	0.000
H	F	0.011	0.213
H	Ph	0.04	0.264
H	Cl	0.047	0.284
Cl	Cl	0.065	0.261
H	$COOCH_3$	0.085	0.178
H	$COOC_2H_5$	0.089	0.088
H	NO_2	0.0105	0.037
NO_2	NO_2	-	bands at 3283 and 3120 cm^{-1}

The results can be understood if one considers the weights of the three proton limiting structures

I III II

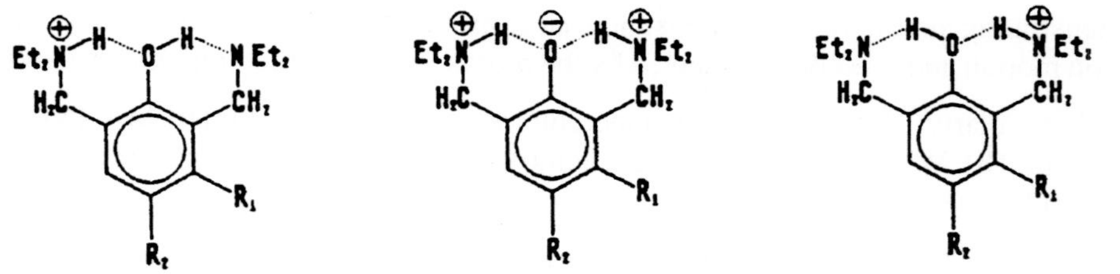

With increasing acidity of the phenolic group, the collective proton fluctuation between the two proton limiting structures (I) $N^+H\cdots OH\cdots N \rightleftharpoons N\cdots HO\cdots H^+N$ (II) of the system is favored, and thus, the proton polarizability increases. With further increasing acidity a third proton limiting structure, the structure $N^+H\cdots O^-\cdots H^+N$ (III)

obtains weight. The system must now be represented by these three proton limiting structures $N^+H\cdots OH\cdots N \rightleftharpoons N^+H\cdots O^-\cdots H^+N \rightleftharpoons N\cdots HO\cdots H^+N$. If the acidity increases still further, the weight of the limiting structures I and II decrease and thus, the proton polarizability of the system becomes smaller. Finally, only limiting structure III remains. The protons are now localized at the nitrogens and instead of a continuum, bands are observed.

Similar results were obtained as a function of the acidity of the phenolic group (ref. 19) with the monoperchlorates of 2,6–bis((diethylamino)methyl)4–R–phenol di–N–oxides.

All these results, with regard to the proton polarizability of hydrogen–bonded chains, can be justified by theoretical treatments.

With *ab initio* SCF calculations one is limited on small systems. Therefore, we calculated (ref. 20) energy and dipole moment surfaces of the systems formic acid-water–formate and formic acid–water–water–formate

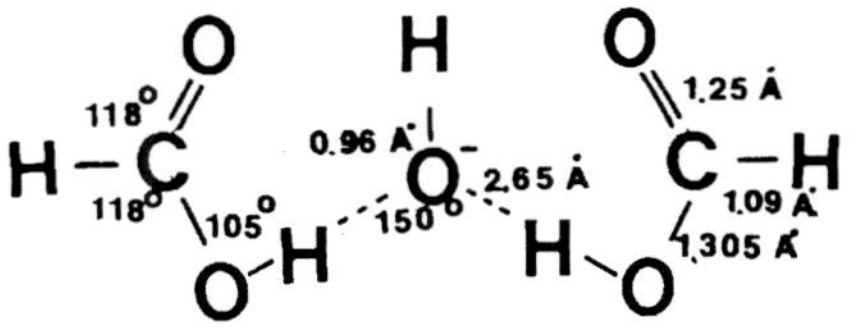

1

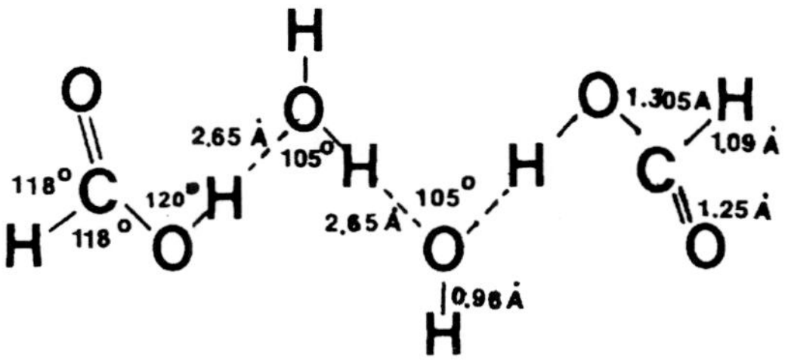

2

With the first system, one observes an energy surface with three minima (Fig. 6a). If one considers the reaction path, a three minima proton potential is found (Fig. 6b). Each minimum corresponds to one of these three proton limiting structures.

With the second system the energy surface is a four–dimensional hypersurface. Fig. 7 shows the proton potential along the reaction path. A four minima proton potential is present. Each minimum corresponds to one of these four proton limiting structures.

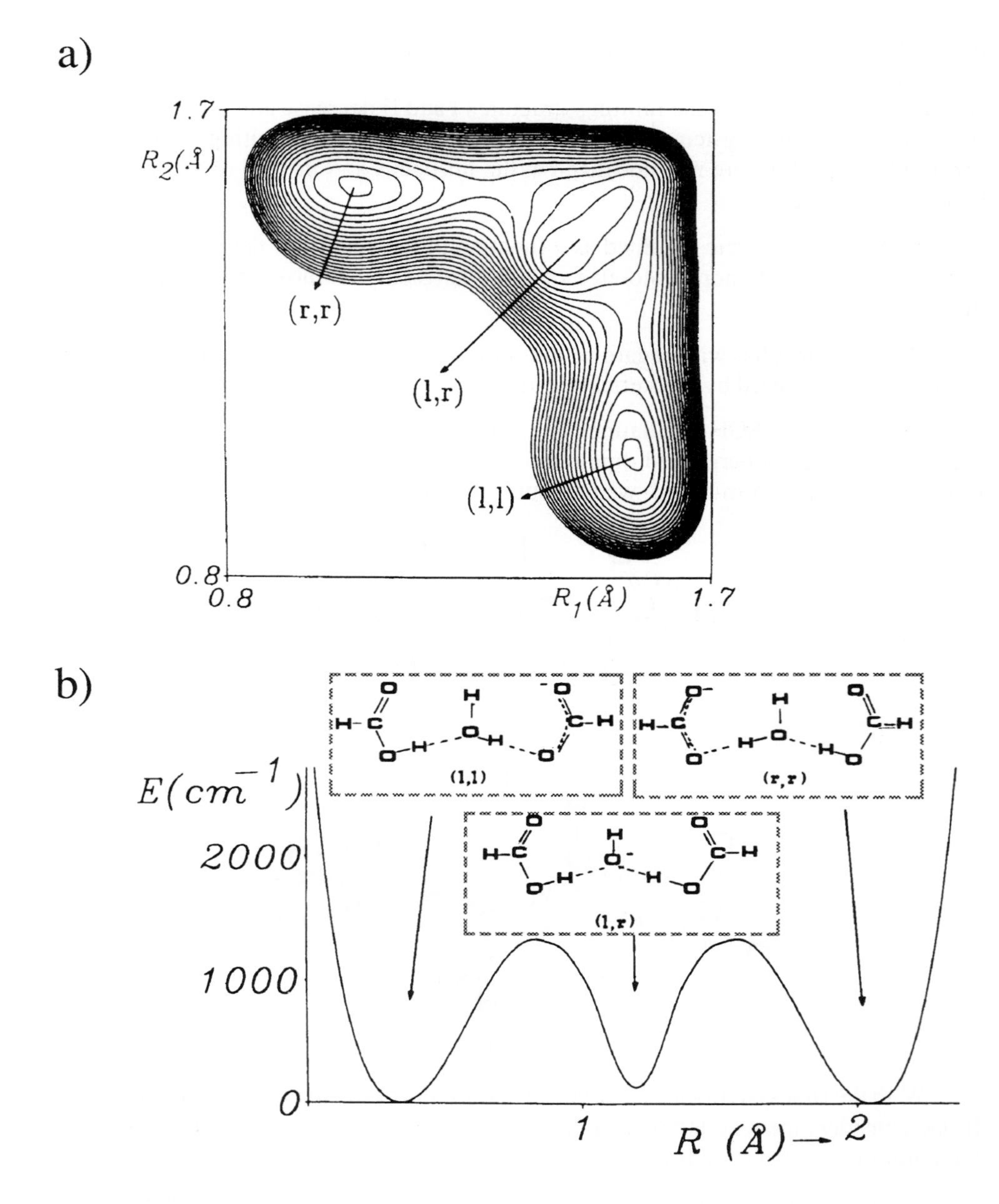

Fig. 6. The system formic acid–water–formate (definition of R_1, R_2 and R see ref. (20)); a) Energy surface. b) Proton potential of this system along the reaction path and proton limiting structures.

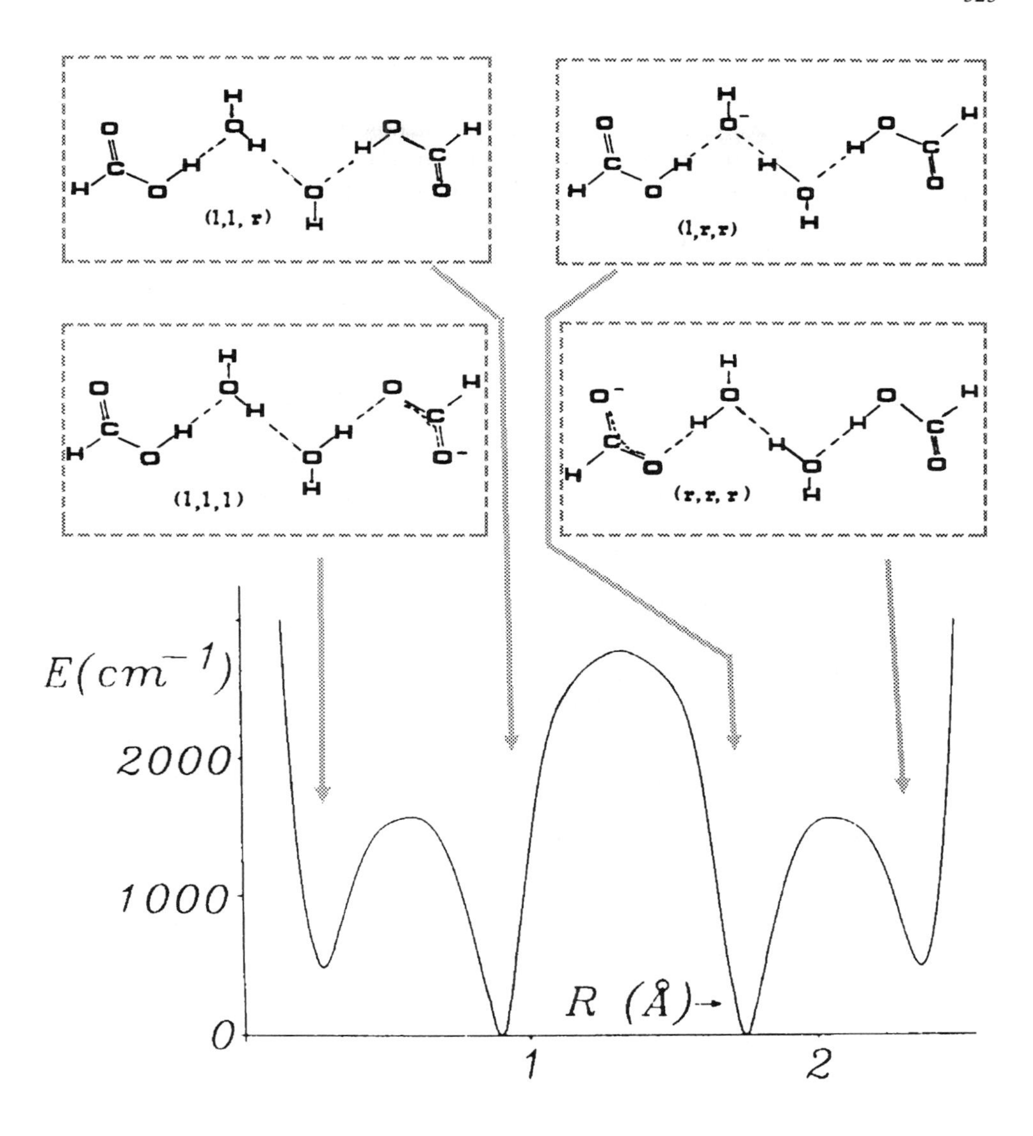

Fig. 7. The system formic acid–water–water–formate, proton potential along the reaction path and proton limiting structures of the system (definition of R see ref. (20)).

Fig. 8a shows the proton polarizabilities of the formic acid–water–formate system as a function of the electrical field strength. In the case of the formic acid–water–formate system, maxima of the proton polarizability are observed with electrical fields of $\pm 0.6 \cdot 10^7$ V/cm. The proton polarizability amounts to $(150\text{-}200) \cdot 10^{-24}$ cm^3 as a function of the temperature. Thus, the proton polarizability of this system is 2 orders of magnitude larger than polarizabilities due to distortion of the electron systems.

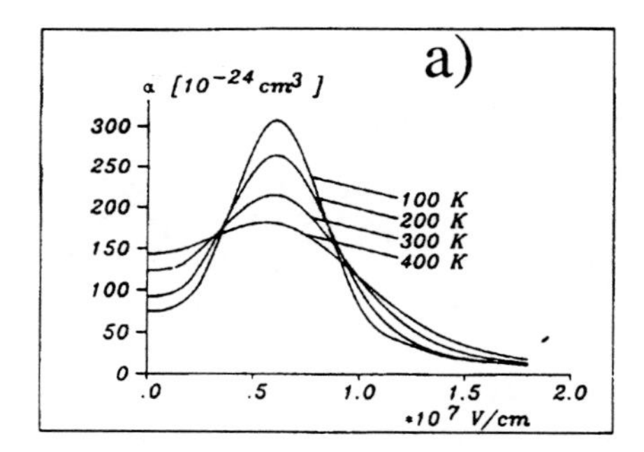

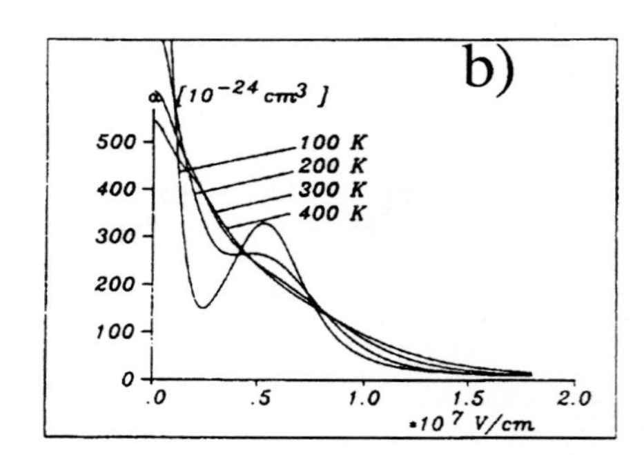

Fig. 8. Proton polarizability in direction of the hydrogen–bonded systems as a function of the electrical field. The parameter is the temperature T. Only the proton polarizabilities in the case of positive electrical fields are shown. For negative fields these qualities are $\alpha(-F)=\alpha(+F)$: (a) system 1 (formic acid–water–formate; (b) system 2 (formic acid–water–water–formate).

Fig. 8b shows the proton polarizability of the formic acid–water–water–formate system. At higher temperatures only one maximum of the polarizability is observed. It is found if no electrical field is present at the hydrogen–bonded system. If no electrical field is present the proton polarizability amounts to $(500\text{-}900) \cdot 10^{-24}$ cm^3 as a function of temperature.

No longer chains can be calculated in this way *ab initio*. Therefore, we proceeded our calculations with model proton potentials (ref. 21).

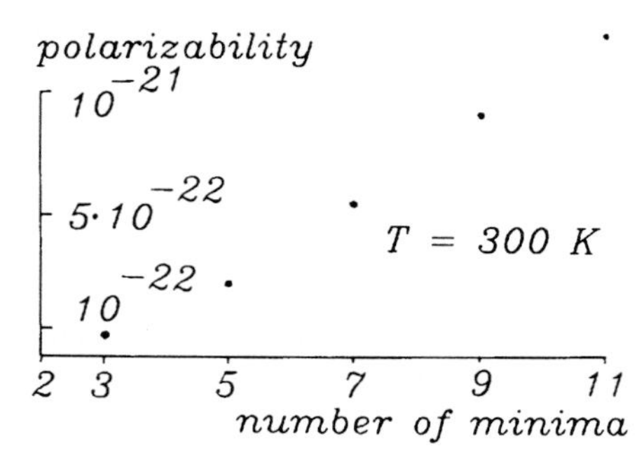

Fig. 9. Proton polarizability of hydrogen–bonded chains as a function of the number of the minima of the proton potential. Results of calculations with model potentials.

Fig. 9 shows the proton polarizability as a function of the number of minima. With increasing number of minima it becomes 3-4 orders of magnitude larger than usual polarizabilities.

Hydrogen Bonds and Hydrogen–Bonded Systems with Large Proton Polarizability in Biological Systems

We have already seen that hydrogen bonds formed between side chains of proteins and phosphates and especially also hydrogen–bonded chains formed between side chains of proteins and several phosphates may show large proton polarizability.

Studies with model systems have shown that also a very large number of homoconjugated as well as of heteroconjugated hydrogen bonds which may form between side chains of proteins may show large proton polarizability (refs. 22-30), (for summary see (ref. 14)). All these hydrogen bonds are summarized in Table 3.

TABLE 3
Hydrogen bonds with large proton polarizability formed between side chains of proteins.

System	Hydrogen Bond
$(L\text{-}Cys)_n + (L\text{-}Cys^-NEt_4^+)_n$	$SH\cdots S^- \rightleftarrows {}^-S\cdots HS$
$(L\text{-}Lys)_n + (L\text{-}Lys^+ClO_4^-)_n$	$N^+H\cdots N \rightleftarrows N\cdots H^+N$
$(L\text{-}Tyr)_n + (L\text{-}Tyr^-NEt_4^+)_n$	$OH\cdots O^- \rightleftarrows {}^-O\cdots HO$
$(L\text{-}His)_n + (L\text{-}His^+Cl^-)_n$ $(Boc\text{-}His\text{-}OMe)_2 + HClO_4$	$N^+H\cdots N \rightleftarrows N\cdots H^+N$
$(L\text{-}Glu)_n + (L\text{-}Glu^-NEt_4^+)_n$ $(Z\text{-}Glu\text{-}OBzl)_2 + HClO_4$	$OH\cdots O^- \rightleftarrows {}^-O\cdots HO$
$(Z\text{-}Asp\text{-}OMe)_2 + HClO_4$	$OH\cdots O^- \rightleftarrows O^-\cdots HO$
$(L\text{-}Tyr)_n + (L\text{-}Arg)_n$	$OH\cdots N \rightleftarrows O^-\cdots H^+N$
$(L\text{-}Cys)_n + (L\text{-}Lys)_n$	$SH\cdots N \rightleftarrows S^-\cdots H^+N$
$(L\text{-}Tyr)_n + (L\text{-}Lys)_n$	$OH\cdots N \rightleftarrows O^-\cdots H^+N$
$(L\text{-}Glu)_n + (L\text{-}His)_n$ Z-Glu-OBzl + Boc-His-OMe	$OH\cdots N \rightleftarrows O^-\cdots H^+N$
Z-Asp-OBzl + Boc-His-OMe	$OH\cdots N \rightleftarrows O^-\cdots H^+N$

It is of particular interest that all hydrogen bonds which may form in the active center of the bacteriorhodopsin molecule, show also large proton polarizability (refs. 31-33).

The proton motion in neighboring hydrogen bonds with large proton polarizability is coupled *via* proton dispersion forces, *i.e.* *via* electromagnetic coupling. Thus, hydrogen–bonded chains which are built up from the hydrogen bonds in Table 3 may show large proton polarizability, too.

Via such hydrogen bonds with large proton polarizability protons can easily be shifted in active centers of enzymes as in trypsin, in alcoholdehydrogenases, in aspartate proteases and in others. But of particular interest are the hydrogen–bonded chains with large proton polarizability due to collective proton motion. Such chains are very effective pathways for the transport of protons in biological systems, for instance in the cristae mitochondriales or in the thylakoid membranes of chloroplasts.

References

1 E.G. Weidemann and G.Zundel, Z. Naturforschung **25a** (1970) 627.

2 R. Janoschek, E.G. Weidemann, H. Pfeiffer and G. Zundel, J. Amer. Chem. Soc. **94** (1972) 2387.

3 R. Janoschek, E.G. Weidemann and G. Zundel, J. Chem. Soc. Faraday Trans. II **69** (1973) 505.

4 M. Eckert and G. Zundel, J. Phys. Chem. **91** (1987) 5170.

5 G. Zundel and Th. Ruhland, J. Chem. Soc. Faraday Trans. **86** (1990) 3557.

6 R. Krämer and G. Zundel, J. Chem. Soc. Faraday Trans. **86** (1990) 301.

7 W. Danninger and G. Zundel, J. Chem. Phys. **74** (1981) 2769.

8 G. Zundel and M. Eckert, J. Mol. Struct. **200** (1989) 73.

9 U. Burget and G. Zundel, J. Mol. Struct. **145** (1986) 93.

10 U. Burget and G. Zundel, Biopolymers **26** (1987) 95.

11 U. Burget and G. Zundel, Biochem. (Life Science Adv.) **7** (1988) 35.

12 U. Burget and G. Zundel, J. Chem. Faraday Trans. I **84** (1988) 885.

13 U. Burget and G. Zundel, Biophys. J. **52** (1987) 1065.

14 G. Zundel, in: A. Pullman *et al.* (Eds.) *Carriers, Channels and Pumps* Kluver Academic Publishers, Dordrecht, 1988, pp. 409.

15 B. Brzezinski, G. Zundel and R. Krämer, Chem. Phys. Letters **124** (1986) 395.

16 B. Brzezinski and G. Zundel, Chem. Phys. Letters **178** (1991) 138.

17 B. Brzezinski, G. Zundel and R Krämer, J. Phys. Chem. **91** (1987) 3077.

18 B. Brzezinski, H. Maciejewska, G. Zundel and R. Krämer, J. Phys. Chem. **94** (1990) 528.

19 B. Brzezinski, H. Maciejewska and G. Zundel, J. Phys. Chem. **94** (1990) 6983.

20 M. Eckert and G. Zundel, J. Phys. Chem. **92** (1988) 7016, see also **93** (1989) 5324.

21 M. Eckert and G. Zundel, J. Mol. Struct. (Theochem) **181** (1988) 141.

22 G. Zundel and J. Mühlinghaus, Z. Naturforschung **26B** (1971) 546.

23 R. Lindemann and G. Zundel, Biopolymers **16** (1977) 2407.

24 R. Lindemann and G. Zundel, Biopolymers **17** (1978) 1285.

25 W. Kristof and G. Zundel, Biophys. Struct. Mech. **6** (1980) 209.

26 W. Kristof and G. Zundel, Biopolymers **19** (1980) 1753.

27 P.P. Rastogi, W. Kristof and G. Zundel, Internat. J. Biol. Macromol. **3** (1981) 154.

28 P.P. Rastogi and G. Zundel, Biochem. Biophys. Res. Commun. **99** (1981) 804.

29 P.P. Rastogi and G. Zundel, Z. Naturforschung **36c** (1981) 961.

30 W. Kristof and G. Zundel, Biopolymers **21** (1982) 25.

31 H. Merz, U. Tangermann and G. Zundel, J. Phys. Chem. **90** (1986) 6535.

32 H. Merz and G. Zundel, Biochem. Biophys. Res. Comm. **138** (1986) 819.

33 H. Schmideder, O. Kasende, H. Merz, P.P. Rastogi and G. Zundel, J. Mol. Struct. **161** (1987) 87.

Electron and Proton Transfer in Chemistry and Biology, edited by A. Müller et al.
Studies in Physical and Theoretical Chemistry, Vol. 78

NMR Studies of Multiple Proton and Deuteron Transfers in Liquids, Crystals and Organic Glasses

Hans-Heinrich Limbach

Institut für Physikalische Chemie der Universität Freiburg i.Br.,
Albertstr. 21, D-7800 Freiburg (Germany)
and
Freie Universität Berlin, Fachbereich Chemie
Takustr. 3, W-1000 Berlin 33 (Germany)

Summary

Using dynamic high resolution NMR spectroscopy of liquids and solids we have found some novel intra–and intermolecular multiple degenerate proton transfer reactions of organic compounds. The kinetic hydrogen/deuterium isotope and solid state effects on these reactions are studied in different liquid and solid environments. In the case of intramolecular double proton transfer reactions the kinetic isotope and solvent effects indicate stepwise reaction pathways. By contrast, intermolecular double proton transfer reactions behave in a different way. Here both protons are in flight in the rate determining reaction step. The origin of the different behavior of both types of reactions is discussed.

Some of the reactions studied in the liquid are also subject to proton transfer in the solid state. The dynamics of the latter are studied by variable temperature ^{15}N CPMAS NMR spectroscopy. It is found that the gas phase degeneracy of the reactions studied is, generally, lost in the solid state due to intermolecular interactions. Whereas in the crystalline state all molecular reaction systems are influenced in the same way a broad distribution of equilibrium and rate constants is found for the amorphous solid state. Recent progress in this field is reported.

Introduction

Degenerate neutral multiple proton transfer reactions including their kinetic hydrogen/deuterium isotope effects have been of special interest in the past years (refs. 1 and 2). The simplest reaction of this kind is a double proton transfer shown in eqn. (1):

$$AH^* + BH \rightleftharpoons A\begin{smallmatrix} \diagup H^* \cdots \\ \cdots H \diagup \end{smallmatrix}B \rightleftharpoons A\begin{smallmatrix} \cdots H^* \diagdown \\ \diagdown H \cdots \end{smallmatrix}B \rightleftharpoons AH + BH^* \qquad (1)$$

These processes represent interesting reaction systems for the study of the elementary steps of bond breaking and bond formation in different liquid and solid environments. They can most conveniently be studied by dynamic NMR spectroscopy. Using this method, rate constants of some multiple inter– and intramolecular double proton transfers have been determined not only in liquid solutions (refs. 1-14) but also in the solid state (refs. 15-26). Neutral multiple proton transfers can also be induced at cryogenic temperatures by visible light (refs. 27 and 28). Recently, also an IR–induced proton transfer has been found (ref. 29). Multiple proton transfer reactions are related in organic and biochemical systems to bifunctional catalysis and biological activity (refs. 30 and 31). On the other hand, these reactions have also be studied theoretically (refs. 32-39); they are of special interest for the development of the theory of primary kinetic hydrogen/deuterium isotope effects (refs. 2-6, 11, 14, 40 and 41).

Here, an overview of the problems which are currently studied in our laboratory by dynamic high resolution liquid and solid state NMR spectroscopy is presented. Our interest has focused on NH···N proton transfer systems. In order to avoid undesired complications arising from nuclear quadruple moments the compounds studied were labeled with the stable ^{15}N isotope which has the spin 1/2.

Results and Discussion

Liquid State NMR Studies of Kinetic HH/HD/DD Isotope Effects on Double Proton Transfer Reactions

Kinetic hydrogen/deuterium isotope effects on chemical reactions are, generally, studied using conventional kinetic methods by performing so called "proton inventories". In such experiments reaction rates are measured as a function of the deuterium fraction D in the mobile proton sites (ref. 30). Not only the size of kinetic isotope effects but also the number of protons *m* in flight in the rate limiting step can be obtained from such experiments. In order to extract *m* from a proton inventory the validity of the "rule of the geometric mean" has, generally, needs to be assumed. This rule states for a double proton transfer that

$$k^{HD} = (k^{HH}k^{DD})^{1/2}, \textit{ i.e. } k^{HH}/k^{HD} = k^{HD}/k^{DD} \quad (2)$$

In order to test the validity of eqn. (2) experimentally, NMR proton inventory techniques (refs. 2-6) have been proposed for the direct determination of k^{HH}, k^{HD}, and k^{DD} of inter– and intramolecular reactions. Thus, by a combination of ^1H and ^2H NMR measurements it was possible to measure complete sets of kinetic isotope effects for the 1:1 and the 2:1 proton exchange between acetic acid and methanol in tetrahydrofuran (Figure 1) (refs. 3 and 4).

Deviations from eqn. (2) were observed for the 1:1 exchange in the sense that replacement of the first H atom by D resulted in a stronger decrease of the rate constants than replacement of the second H atom by D, *i.e.* $k^{HH}/k^{HD} = 5.1$ and $k^{HD}/k^{DD} = 3$ at 298K. Larger deviations, in the sense that

$$k^{HH}/k^{DD} >> 1 \text{ and } k^{HD}/k^{DD} \cong 1 \quad (3)$$

were observed for the tautomerism of meso–tetraphenylporphyrin (TPP, Figure 2).

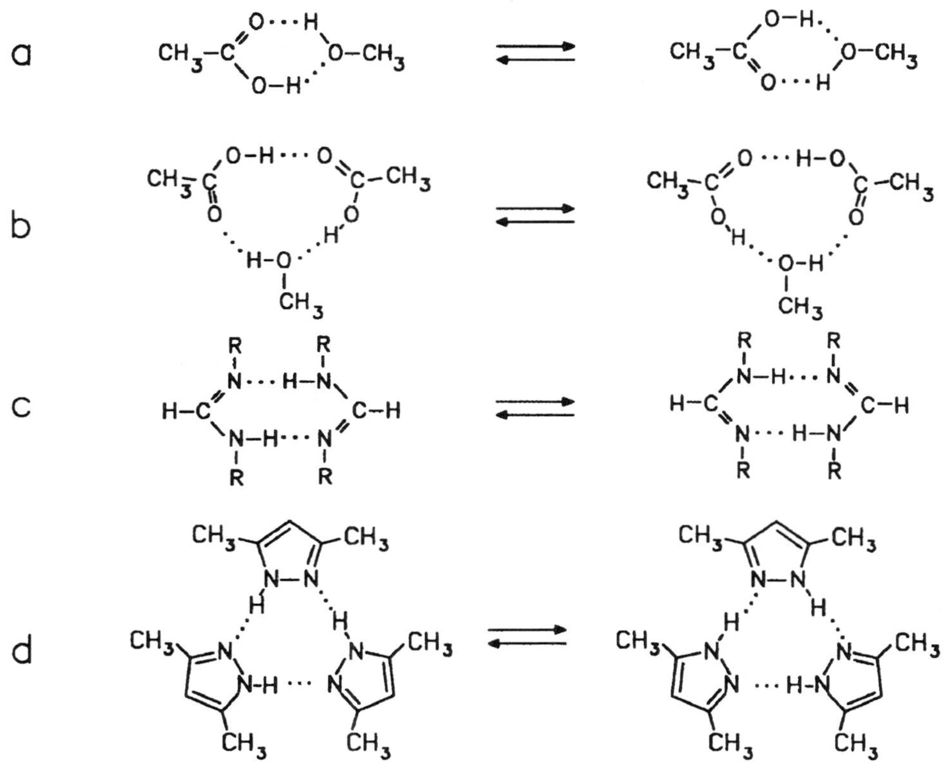

a

b

c

d

Fig. 1. Intermolecular multiple proton transfer reactions for which kinetic hydrogen/deuterium isotope effects were studied by dynamic liquid state NMR spectroscopy. a and b: Double and triple proton transfer between acetic and methanol according to (refs. 3 and 4); c: double proton transfer in formamidine dimers according to (refs. 5 and 6); d: triple proton transfer in cyclic trimers of 3,5–dimethylpyrazole according to (refs. 25 and 26).

These deviations from the RGM were first interpreted in terms of a concerted proton transfer pathway involving tunneling (refs. 3, 4 and 32).

After the completion of these initial studies it seemed desirable to know whether these deviations are of a general nature. Therefore, we have studied the dynamics of different inter-and intramolecular proton and deuteron transfer reactions shown in Figures 1 and 2. The results of these studies will be shortly reviewed in this section.

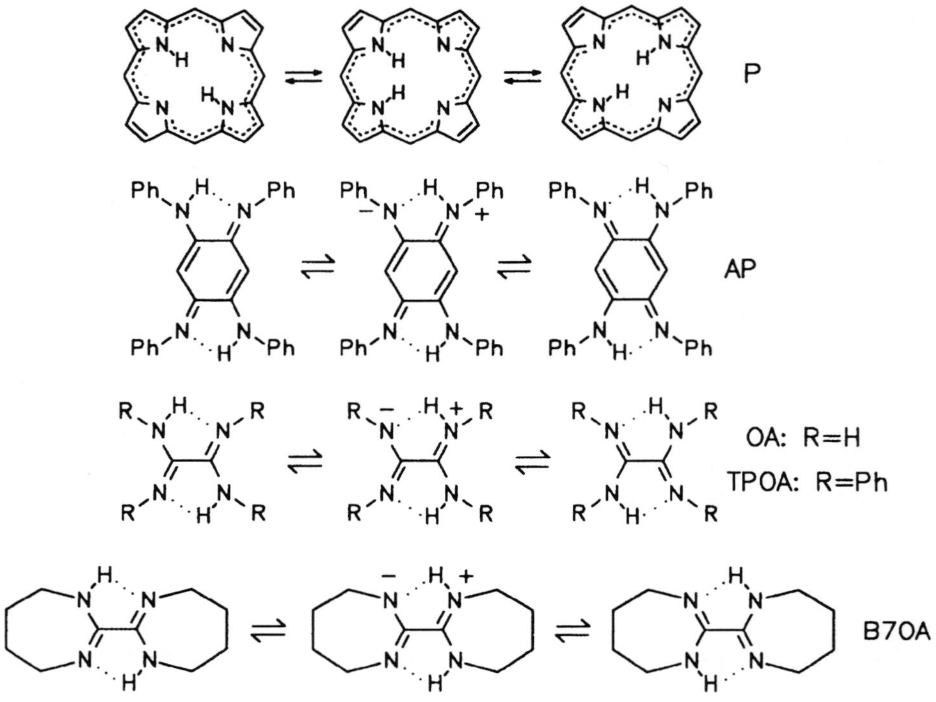

Fig. 2. Intramolecular proton transfer reactions for which kinetic hydrogen/deuterium isotope effects were studied by dynamic liquid state NMR spectroscopy. P: porphyrin and derivatives (ref. 3); AP: azophenine (refs. 3,11 and 12); OA: oxalamidine and derivatives (ref. 13), B7OA: bicyclic oxalamidine (ref. 14). The dynamics of the tautomerism of P and of AP have also be studied by NMR in the solid state (refs. 15-17) and (ref. 12).

Let us first consider the intramolecular double proton transfer reactions for which full kinetic HH/HD/DD isotope effects have been measured. Let us start with the example of azophenine (AP). This molecule is subject in liquid solution to a fast intramolecular double proton transfer or tautomerism involving two degenerate tautomers as shown in Figure 2 (ref. 3). Theoretical calculations (ref. 36) proposed a stepwise proton transfer pathway for this reaction involving a zwitterion as intermediate. In order to obtain experimental information on the mechanism of the tautomerism of AP reaction rate constants were measured as a function of temperature by dynamic NMR spectroscopy of various isotopically labeled AP species dissolved in different organic solvents (ref. 12). The rate constants were almost independent of the dielectric constant of the solvent, varied between 2 (toluene) and 25 (benzonitrile). For $C_2D_2Cl_4$ as solvent, the full kinetic HH/HD/DD isotope effects could be obtained at different temperatures. The observed kinetic isotope effects of $k^{HH}/k^{HD} = 4.1$ and $k^{HD}/k^{DD} = 1.4$ at 298 K indicate that eqn. (3) is fulfilled as in the case of tetraphenylporphyrin. These results could be modeled using formal kinetic theory assuming a stepwise double proton transfer involving a zwitterionic intermediate. The problem with this interpretation was that the formation of the highly polar zwitterion should lead to a substantial kinetic solvent effects on the reaction, by contrast to the experiment.

This problem could be solved for the tautomerism of the related oxalamidine (OA) system (Figure 2). The existence of this reaction in liquid solution was first demonstrated for the phenylsubstituted derivative TPOA (ref. 13). Unfortunately, TPOA is also subject in the liquid state to conformational isomerism and to intermolecular proton exchange (ref. 13). Therefore, the ^{15}N and ^{2}H labeled bicyclic derivative B7OA (Figure 2) was studied in detail where only the intramolecular hydrogen bonded species exists. Rate constants of the tautomerism of B7OA including the kinetic HH/HD/DD isotope effects could be obtained at 362 K by ^{1}H NMR lineshape analysis (ref. 14), using methylcyclohexane–d_{14} (MCY) and acetonitrile–d_3 (AN) as solvents. The kinetic isotope effects were found to be similar in both solvents, and similar to those obtained for AP. *E.g.*, $k_{MCY}{}^{HH}/k_{MCY}{}^{HD} = 2.4$, $k_{MCY}{}^{HD}/k_{MCY}{}^{DD} = 1.2$, $k_{MCY}{}^{HH}/k_{MCY}{}^{DD} = 3$, $k_{AN}{}^{HH}/k_{AN}{}^{HD} = 2.6$, $k_{AN}{}^{HD}/k_{AN}{}^{DD} = 1.3$, $k_{AN}{}^{HH}/k_{AN}{}^{DD} = 3.5$ at 362 K. These results are again consistent with a stepwise double proton transfer mechanism as shown in Figure 2. This interpretation could further be supported by the observation of a substantial kinetic solvent effect of $k_{AN}{}^{HH}/k_{MCY}{}^{HH} = 4.5$ at 362 K, indicating a polar transition state as expected. Thus, both the observed kinetic HH/HD/DD isotope and solvent effects point in the same direction. *I.e., the observation that eqn. (3) is fulfilled in a degenerate double proton transfer reaction can be taken as a criterion for a stepwise reaction mechanism.* This result is of importance in the interpretation of kinetic HH/HD/DD isotope effects of degenerate proton transfer reactions like the porphyrin or the azophenine tautomerism where kinetic solvent effects are absent. In these cases the study of the kinetic HH/HD/DD isotope effects represents the only way to experimentally establish a stepwise double proton transfer mechanism.

The absence of kinetic solvent effects on the azophenine tautomerism could have two reasons: either the phenyl groups effectively shield the reaction center from the solvent in contrast to B7OA, or the intermediate of the AP tautomerism is an apolar singlet biradical as discussed in (ref. 12).

The knowledge of these results is also of help in the case of the proton tautomerism of porphyrins, where recent theoretical studies gave evidence for a stepwise proton transfer in as shown in Figure 2 (refs. 33, 35 and 37). The previous finding (ref. 3) that eqn. (3) is well fulfilled can now be taken as evidence for the stepwise reaction mechanism. Recent measurements of the low–temperature rate constants of the porphyrin tautomerism as well as of hydroporphyrins (ref. 10) corroborates this interpretation and supports the idea of proton tunneling in this reaction (ref. 29).

Let us discuss now the intermolecular proton exchange between diphenylformamidine (DPFA) molecules dissolved in tetrahydrofuran (Figure 1c). DPFA forms in THF an s–trans and an s–cis conformer which interconvert slowly on the NMR time scale below 250 K. According to recent NMR results, only the s–trans–conformer can form cyclic dimers in which a double proton transfer takes place according to Figure 1c (ref. 5). The energy of activation of this reaction is about 17 $kJmol^{-1}$. For the derivative ^{15}N, ^{15}N'–di–p–fluorphenylformamidine molecule (DFFA) the full kinetic HH/HD/DD isotope effects could be obtained at 189.2 K (ref. 6). By ^{1}H NMR spectroscopy, a linear dependence of the inverse proton lifetimes on the deuterium fraction *D* in the mobile proton site was established. From this dependence the number of protons transported in the rate limiting step of the proton exchange was shown to be m=2, as expected for a double proton transfer in a cyclic s–trans dimer. The complete kinetic HH/HD/DD isotope effects of 233 : 11 : 1 at 189 K were determined by dynamic ^{19}F NMR spectroscopy. These results indicate a substantial deviation from the rule of geometric mean, of the same order as in the case of acetic acid/methanol/tetrahydrofuran (ref. 4).

It is clear that these kinetic HH/HD/DD isotope effects of the formamidine tautomerism and of 1:1 proton exchange in the system acetic acid/methanol are not consistent with a stepwise double proton transfer because eqn. (3) is not fulfilled. Stated in another way, both transferred protons contribute substantially to the observed kinetic isotope effects by contrast to the intramolecular case. This is in agreement with the finding that the overall kinetic HH/DD isotope effects are larger in intermolecular than in intramolecular double proton transfer reactions.

How can one then explain the small deviations from the RGM in the case of intermolecular exchange reactions? In the case of the system acetic acid/methanol/THF these deviations could be interpreted in terms of incoherent tunneling from excited vibrational states (refs. 3 and 4). The deviations arise because tunneling enhances the reaction rates especially of the light hydrogen isotopes. Further temperature dependent kinetic studies on the formamidine tautomerism are currently performed in order to confirm this interpretation also for this system.

A qualitative explanation for the different reaction pathways of intra– and intermolecular double proton transfer systems has been proposed recently (ref. 6). This explanation is based on the idea that the molecular frame of intramolecular proton transfer systems such as porphyrin and azophenine is relatively rigid. Thus, a high energy would be required to reduce the hydrogen bond distance in such systems by contrast to intermolecular hydrogen bonded systems which exhibit low frequency hydrogen bond stretching vibrations (ref. 39). The rigidity of the molecular frame of intramolecular proton transfer systems is schematically expressed in Figure 3a by an outer square.

It is plausible that a high energy is required to break the bonds of both protons to their neighboring heavy atoms at the same time. Therefore, the proton transfer will be stepwise. Note that proton tunneling in this case will always require a minimum energy of activation which corresponds to the energy difference between the intermediate and the initial state.

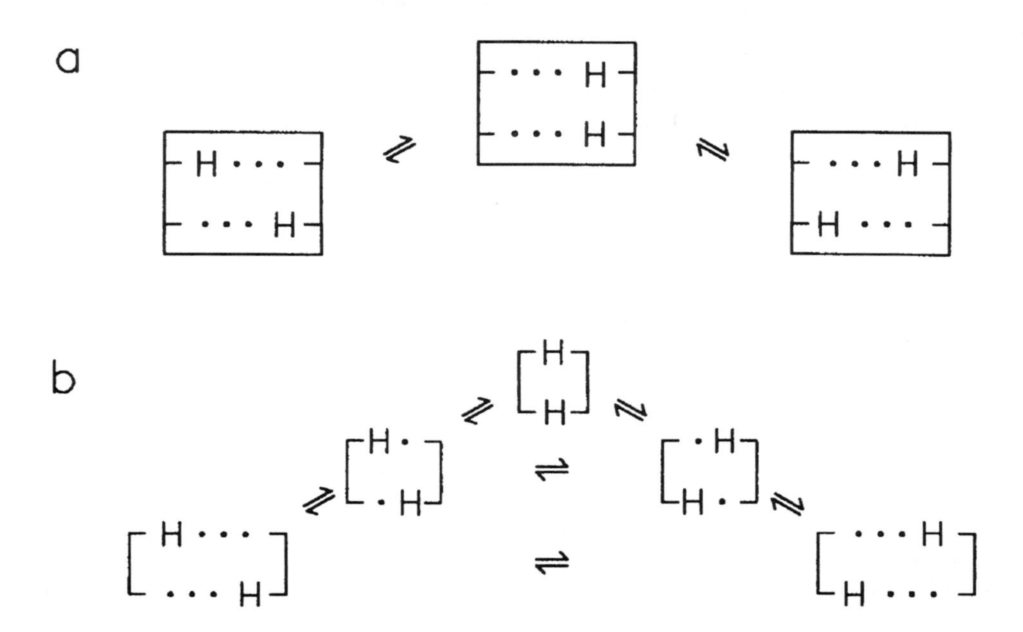

Fig. 3. a: Stepwise double proton transfer in the case of a fixed molecular frame of heavy atoms; b: double proton transfer in the case of variable hydrogen bond lengths according to a model proposed in (ref. 4). Reproduced with permission from (ref. 6).

By contrast, the presence of low frequency hydrogen bond stretching vibrations in the flexible intermolecular proton transfer systems allows a relatively easy compression of the hydrogen bond as schematically shown in the reaction model of Figure 3b. This model has been used for the interpretation of the Arrhenius curves of the proton exchange in the cyclic 1:1 dimers of acetic acid and methanol (refs. 3 and 4). It was assumed that the hydrogen bond lengths are variable, *i.e.* that the energy of activation of the proton transfer is used to excite the hydrogen bond stretching vibration. Upon excitation the hydrogen bond lengths are shortened resulting in a reduction of the barrier for the proton transfer. At very short hydrogen bond lengths the barrier vanishes. Therefore, also the difference between a stepwise and a concerted proton transfer mechanism vanishes. The imaginary frequency characterizing the transition state corresponds then to the hydrogen bond stretching and not to a AH–stretching vibration. It is well known that the latter are shifted to lower frequencies when the hydrogen bond distance is shortened (ref. 42). Therefore, a considerable loss of zero point energy of both vibrations is expected for the highly compressed transition state. The RGM will then be fulfilled at high temperatures. At lower temperatures the transfer could occur by tunneling leading to the deviations from the RGM mentioned above. Thus, for the intermolecular proton transfer systems a reaction mechanism according to Figure 3b is proposed. Note that there might be intermolecular proton transfer systems with rigid hydrogen bond distances and intramolecular proton transfer systems with flexible hydrogen bonds leading to an inverse behavior of kinetic isotope effects.

^{15}N CPMAS NMR Spectroscopy of Proton Transfer Systems in the Crystalline State.

A major problem arising in the theoretical interpretation of kinetic isotope effects of reactions in condensed matter is the question of how the reaction energy surface is affected by intermolecular interactions in the liquid, crystalline or amorphous glassy state. Information on this problem can be obtained by using variable temperature high resolution solid state NMR spectroscopy (ref. 43). This method works for spin 1/2 nuclei and uses line narrowing techniques such as proton decoupling and magic angle spinning (MAS) in order to remove the effects of dipolar coupling to protons and of the chemical shift anisotropy on the NMR spectra (ref. 44). Cross–polarization (CP) from protons to the nucleus studied enhances the signal to noise ratio (ref. 44). Using this method, several solid state hydrogen transfer systems have been studied by natural abundance ^{13}C CPMAS NMR (refs. 20,25,45 and 46). However, with exception of hydride transfers in carbonium ions (ref. 46), carbon atoms are not directly involved in proton transfers and their isotropic chemical shifts are not always modulated enough by these processes in order to obtain rate constants by lineshape analysis. Therefore, ^{15}N CPMAS NMR spectroscopy of ^{15}N labeled NH···N or NH···X proton transfer systems is more suitable. The ^{15}N labeling is necessary because of the quadruple moment of the ^{14}N nucleus. By ^{15}N CPMAS NMR spectroscopy fast proton transfer processes have been monitored in solid porphyrins (Fig. 2) (refs. 15-17), phthalocyanine (Fig. 4) (refs. 18 and 19), porphycene (Fig. 4) (ref. 16), azophenine (Fig. 2) (ref. 12) and tetraazaannulenes (Fig. 5) (refs. 21-24) as a function of temperature.

The ^{15}N CPMAS spectra contain a much information about solid state effects on the NH···N proton tautomerism, as expressed by Fig. 6.

Consider a reaction which is degenerate in the absence of intermolecular interactions. The reaction energy profile is then symmetric as indicated schematically in Fig. 6a, where possible intermediates have been omitted for simplicity. In the ordered crystalline state the molecular symmetry is, generally, removed by intermolecular interactions leading to an energy difference ΔE between the two stable wells of the potential energy curve (Fig. 6b). In first approximation the potential of the proton motion in a particular molecule does not depend on the tautomeric state of the

Pc

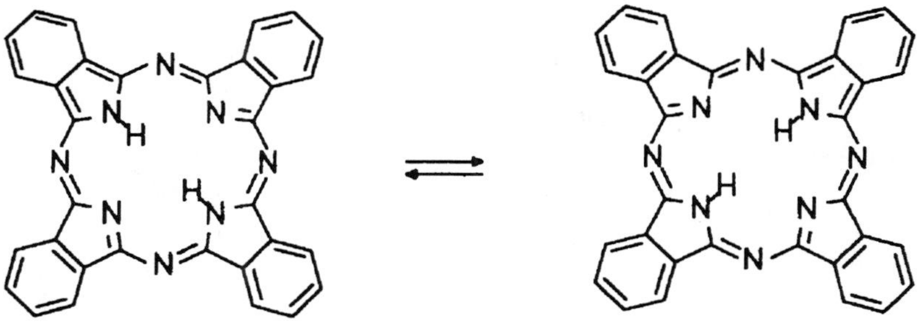

PHYC

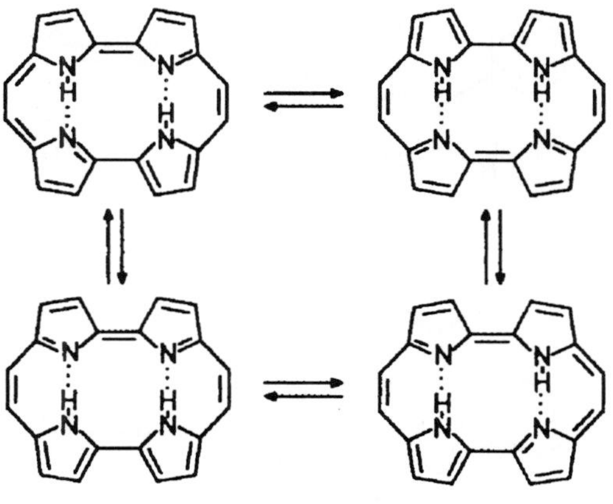

Fig. 4. Solid state tautomerism of phthalocyanine (Pc) (refs. 18-20) and of porphycene (PHYC), (ref. 16).

neighboring molecules. Using ^{15}N CPMAS NMR spectroscopy, ΔE could be measured for different porphyrins, azophenine, and for the reaction systems shown in Fig. 4 and 5. The ΔE values depend on the chemical structure and the crystal structure and were found to vary between 0 to 12 kJmol^{-1}. Note that the case $\Delta E \cong 0$ was realized within the error limits for the parent compound porphyrin (ref. 16), mesotetratolylporphyrin (ref. 15) and for 3,5–dimethylpyrazole (Fig. 1) which is subject to a triple proton transfer in a cyclic trimer (refs. 25 and 26). For phthalocyanine different solid state perturbations were observed depending on the crystal modification (ref. 19). By contrast, no substantial entropy difference between the different tautomeric states was observed.

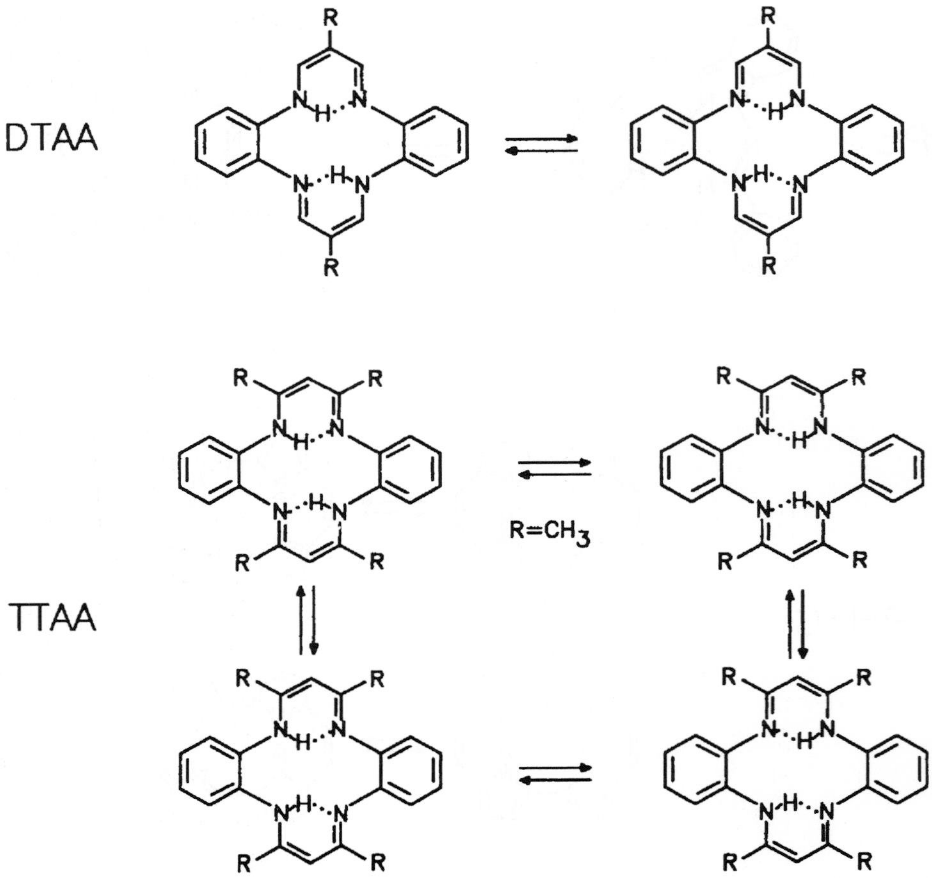

Fig. 5. Solid state tautomerism of dimethyltetraaza (ref. 14) annulene (DTAA) (ref. 22) and of tetramethyltetraaza (ref. 14) annulene (TTAA) (refs. 21, 23 and 24).

Solid state NMR experiments did not allow us only to determine thermodynamic but also kinetic information on the reaction rates in the solid state. Among the systems where rate constants could be obtained so far are porphyrins, phthalocyanine, dimethyl-tetra–azaannulene, and pyrazole derivatives (refs. 15-26). It is interesting to note that the rate constants of the tautomerism in crystalline porphyrin coincide with the rate constants obtained for liquid solutions (refs. 15-17). Thus, solid state effects on the thermodynamics of proton tautomerism seem to be more important than kinetic

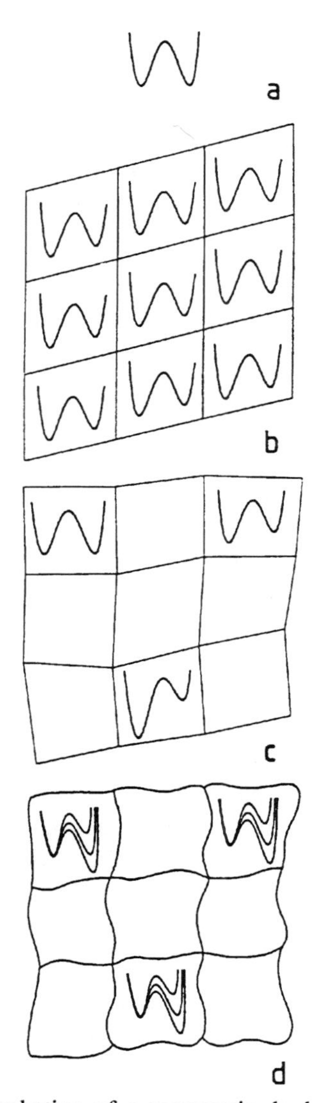

Fig. 6. Perturbation of a symmetric double minimum potential of a bistable molecule by intermolecular interactions. a: symmetric double minimum potential in the gas phase. b: Perturbation of the potential in the ordered crystalline state by intermolecular interactions which are the same for all molecules. c: Perturbation of the potential in the disordered solid state by intermolecular interactions which are the different for all molecules. d: Motional averaged symmetric potentials, (ref. 24).

solution solid state effects. Note that such effects are only meaningful when the hydrogen bond network is the same in the liquid and the solid.

In the case of strongly hydrogen bonded proton transfer systems, such as porphycen (ref. 16) (Fig. 4) or TTAA (ref. 21) (Fig. 5), the energy of activation of proton transfer is too small in order to be detectable by NMR lineshape analysis. Nevertheless, it is possible to distinguish whether the protons move in a slightly asymmetric double minimum potential or in a slightly asymmetric single minimum potential. In the first case a strong, in the second case a weak dependence of ^{15}N chemical shifts on temperature is expected. Thus, for TTAA and porphycen double minimum potentials could be established (refs. 16 and 21). Furthermore, it was also found in both cases that the two NH···N hydrogen bonds of each molecule are perturbed in a different way. This finding establishes an independent proton motion and the presence of four different tautomeric states shown in Fig. 4 and 5. Note that these NMR results are of use answer the problem of proton localization arising in the x–ray crystallographic analysis of compounds with labile protons (refs. 47-49).

^{15}N CPMAS NMR Spectroscopy of Proton Transfer Systems in the Disordered Solid State

For the crystalline solids discussed so far all molecules of a sample experience the same solid state perturbation ΔE as shown in Fig. 6b. This observation is not surprising because of the translational symmetry of crystals. It is now interesting to ask how the reaction energy surface of a particular proton transfer system is altered when the molecule is placed in a amorphous environment such as an organic glass? Information about this problem is desirable in order to obtain a better understanding of the structure and the dynamics of bistable molecules embedded in glasses.

Such information can be obtained by CPMAS NMR lineshape analysis. In order to extract thermodynamic and kinetic data from the spectra, first, an appropriate line shape theory is necessary as has been described recently (refs. 19 and 24). The model on which this theory is based is expressed schematically in Fig. 6c. The model assumes a continuous distribution of different sites characterized by different rate and equilibrium constants of proton tautomerism. The possibility of exchange between the different sites can be taken into account (refs. 19 and 24). Actually, bigaussian distributions of the reaction enthalpies of the and of the enthalpies of activation of proton exchange were employed. In order to reduce the number of adjustable parameters different simplifications of reducing the two dimensional site distribution function to a one dimensional distribution were discussed (ref. 19).

The theory was then used to simulate the ^{15}N CPMAS NMR spectra of ^{15}N labeled amorphous phthalocyanine (Pc) (ref. 19) which show both static as well as dynamic line broadening due to the solid state tautomerism. No evidence for exchange between the different sites was obtained. The amorphous modification was characterized by a broad distribution of differently perturbed asymmetric double minimum potentials, as expected for a disordered environment. In other words, in a disordered environment not only a distribution of equilibrium constants but also of rate constants is realized (ref. 19). This makes lineshape calculations difficult.

The problem is less severe in the case of fast proton transfer reaction systems where dynamic NMR line broadening is absent. Therefore, ^{15}N CPMAS NMR measurements were performed on TTAA (Fig. 5) dissolved in polystyrene (ref. 24). Since the matrix does not contain nitrogen atoms the signals of TTAA do not interfere with solvent signals, problem which arises in ^{13}C CPMAS studies. We find that the TTAA tautomerism in polystyrene is indeed very fast as compared to the NMR time scale, *i.e.* much faster than the reorientation of the solvent molecules; thus, no line broadening due to slow proton transfer complicates the spectra. However, a distribution of TTAA molecules with different equilibrium constants of tautomerism was found, as expressed schematically by Fig. 6c. Since the latter depend on temperature the distribution induces temperature dependent inhomogeneously broadened line shapes. Exchange between the different types of TTAA molecules *via* rotational and translational diffusion is very slow below the glass transition. From the simulation of the NMR spectra the parameters of a bigaussian distribution of free reaction energies of the tautomerism were obtained as function of temperature (ref. 24). The inhomogeneous line broadening was independently confirmed by two dimensional NMR experiments. In the glass transition region the phenomenon of motional averaging of the differently perturbed molecular sites into averaged site with an effective symmetric double minimum potential for the proton motion within the NMR time scale was observed (ref. 24). Thus, the different sites interconvert rapidly within the NMR time scale above the glass transition temperature as shown schematically in Fig. 6d. Thus, proton transfer systems such as TTAA are sensitive molecular probes for microscopic order in glasses.

Conclusions

It has been shown that the dynamics of neutral double and triple proton transfer reactions can conveniently be studied by dynamic liquid and solid state NMR spectroscopy. If the rate constants are not too fast it is possible to obtain possible to obtain information on the multiple kinetic isotope effects and the effects of intermolecular interactions on the reaction dynamics. It is found that intramolecular and intermolecular double proton transfer reaction systems behave in different ways. In the intramolecular cases only a single proton contributes significantly to the kinetic isotope effect indicating a stepwise reaction mechanism. Highly polar intermediates can lead to kinetic solvent effects if the reaction center is not shielded from the solvent. By contrast, in the intermolecular reactions both protons contribute to the overall kinetic isotope effects, thus increasing their size. The origins of the different behavior is discussed. Tunneling is found to play an essential role in these reactions.

Using novel solid state NMR techniques the dynamics of solid state NH···N tautomerism of organic dyes can be studied in a time scale of slow molecular motion, in the crystalline ordered state as well as in the disordered — amorphous and glassy — solid state. Evidence has been found that the proton transfer systems studied are subject to static perturbations of the reaction energy profiles of the proton motion, even in cases where kinetic solvent effects are absent. Molecular motions lead to motionally averaged effective reaction surfaces in the liquid state.

Acknowledgements

I would like to thank all collaborators which have contributed to the work described above: J. Hennig, D. Gerritzen, H. Rumpel, G. Otting, L. Meschede, B. Wehrle, M. Schlabach, G. Scherer, F. Aguilar–Parilla and J. Braun. I further thank C.S. Yannoni, IBM, San Jose, C. Djerassi, Stanford, E. Vogel, Köln and H. Zimmermann, Heidelberg for their contribution to this work.

The financial aid of the Deutsche Forschungsgemeinschaft, Bonn–Bad Godesberg, the Stiftung Volkswagenwerk, and the Fonds der Chemischen Industrie, Frankfurt, is also gratefully acknowledged.

References

1 H.H. Limbach *The Use of NMR Spectroscopy in the Study of Hydrogen Bonding in Solution* in: J. Gormally and E. Wyn-Jons (Eds.), *Aggregation Processes* Elsevier, Amsterdam ,1983, Chapter 16 and references cited therein.

2 A more detailed account of the work presented here can be found in: H.H. Limbach *Dynamic NMR Spectroscopy in the Presence of Kinetic Hydrogen/Deuterium Isotope Effects* in: *NMR Basic Principles and Progress* Vol. 23, Springer, Heidelberg, 1990.

3 H.H. Limbach, J. Hennig, D. Gerritzen and H. Rumpel, Faraday. Discuss. Chem. Soc. **74** (1982) 229.

4 D. Gerritzen and H.H. Limbach, J. Am. Chem. Soc. **106** (1984) 869.

5 L. Meschede, D. Gerritzen and H.H. Limbach, Ber. Bunsenges. Phys. Chem **92** (1988) 469.

6 H.H. Limbach, L. Meschede and G. Scherer, Z. Naturforsch., Teil A **44** (1989) 459.

7 C. B. Storm and Y. Teklu, J. Am. Chem. Soc. **94** (1974) 1745; Ann. N. Y. Acad. Sci. **206** (1973) 631.

8 J. Hennig and H.H. Limbach, J. Am. Chem. Soc. **106** (1984) 292; J. Hennig and H.H. Limbach, J. Magn. Reson. **49** (1982) 322.

9 M. Schlabach, B. Wehrle, H.H. Limbach, E. Bunnenberg, A. Knierzinger, A. Shu, B.R. Tolf and C. Djerassi, J. Am. Chem. Soc., **108** (1986) 3856.

10 M. Schlabach, H. Rumpel and H.H. Limbach, Angew. Chem. **101** (1989) 84; Int. Ed. Engl. **28** (1989) 76.

11 H. Rumpel, G. Zachmann and H.H. Limbach, J. Phys. Chem. **93** (1989) 1812.

12 H. Rumpel and H.H. Limbach, J. Am. Chem. Soc. **111** (1989) 5429.

13 G. Otting, H. Rumpel, L. Meschede, G. Scherer and H.H. Limbach, Ber. Bunsenges. Phys. Chem. **90** (1986) 1122.

14 G. Scherer and H.H. Limbach, J. Am. Chem. Soc. **111** (1989) 5946.

15 H.H. Limbach, J. Hennig, R.D. Kendrick and C.S. Yannoni, J. Am. Chem. Soc. **106** (1984) 4059.

16 B. Wehrle, H.H. Limbach, M. Kocher and E. Vogel, Angew. Chem. **99** (1987) 914; Int. Ed. Engl. **26** (1987) 934.

17 H.H. Limbach, B. Wehrle, M. Schlabach, R. Kendrick and C.S. Yannoni, J. Magn. Reson. **77** (1988) 84.

18 R.D. Kendrick, S. Friedrich, B. Wehrle, H.H. Limbach and C.S. Yannoni, J. Magn. Reson. **65** (1985) 159.

19 B. Wehrle and H.H. Limbach, Chem. Phys. **136** (1989) 223.

20 B.H. Meier, C.B. Storm and W.L. Earl, J. Am. Chem. Soc. **108** (1986) 6072.

21 H.H. Limbach, B. Wehrle, H. Zimmermann, R.D. Kendrick and C.S. Yannoni, J. Am. Chem. Soc. **109** (1987) 929.

22 H.H. Limbach, B. Wehrle, H. Zimmermann, R.D. Kendrick and C.S. Yannoni, Angew. Chem. **99** (1987) 241;
Angew. Chem. Int. Ed. **26** (1987) 247.

23 H.H. Limbach, H. Zimmermann and B. Wehrle, Ber. Bunsenges. Phys. Chem. **91** (1987) 941.

24 B. Wehrle, H. Zimmermann and H.H. Limbach, J. Am. Chem. Soc. **11** (1988) 7014.

25 A. Baldy, J. Elguero, R. Faure, M. Pierrot and E.J. Vincent, J. Am. Chem. Soc. **107** (1985) 5290.

26 J.A.S. Smith, B. Wehrle, F. Aguilar-Parrilla, H.H. Limbach, M.C. Foces-Foces, F.H. Cano, J. Elguero, A. Baldy, M. Pierrot, M.M.T. Khurshid and J.B. Larcombe–McDouall, J. Am. Chem. Soc. **111** (1989) 7304.

27 S. Volker and J.H. van der Waals, Mol. Phys. **32** (1976) 1703.

28 J. Friedrich and D. Haarer, Angew. Chem. Int. Ed. **23** (1984) 113.

29 T. Butenhoff and C.B. Moore, J. Am. Chem. Soc. **110** (1988) 8336;
T. Butenhoff, R. Chuck, H.H. Limbach and C.B. Moore, (J. Phys. Chem.), in press.

30 R.D. Gandour and R.L. Schowen *Transition States of Biochemical Processes* Plenum Press, New York, 1978.

31 J.D. Hermes and W.W. Cleland, J. Am. Chem. Soc. **106** (1984) 7263.

32 H.H. Limbach and J. Hennig, J. Chem. Phys. **71** (1979) 3120;
H.H. Limbach, J. Hennig and J. Stulz, J. Chem. Phys. **78** (1983) 5432;
H.H. Limbach, J. Chem. Phys. **80** (1984) 5343.

33 A. Sarai, Chem. Phys. Lett. **83** (1981) 50;
J. Chem. Phys. **76** (1982) 5554;
ibid., **80** (1984) 5341.

34 G.I. Bersuker and V.Z. Polinger, Chem. Phys. **86** (1984) 57.

35 Z. Smedarchina, W. Siebrand and F. Zerbetto, Chem. Phys. **136** (1989) 285.

36 M.J.S. Dewar and K.M. Merz, J. Mol. Struct. (Theochem) **124** (1985) 183;
K.M. Holloway, C.H. Reynolds and K.M. Merz, J. Am. Chem. Soc. **111** (1989) 3466.

37 K.M. Merz and C.H. Reynolds, J. Chem. Soc. Chem. Comm. (1988) 90.

38 R.P. Bell *The Tunnel Effect in Chemistry* Chapman and Hall, London 1980.

39 P. Schuster, G. Zundel, C. Sandorfy, (Eds.) *The Hydrogen Bond* North Holland, Amsterdam, 1976.

40 J. Bigeleisen, J. Chem. Phys. **23** (1955) 2264.

41 W.J. Albery, J. Phys. Chem. **90** (1986) 3773.

42 A. Novak, Struct. Bonding (Berlin) **14** (1974) 177.

43 J.R. Lyerla, C.S. Yannoni and C.A. Fyfe, Acc. Chem. Res. **15** (1982) 208;
C. A. Fyfe *Solid State NMR for Chemists* C.F.C. Press, Guelph, Ontario, 1983.

44 J. Schaeffer and E.O. Steijskal, J. Am. Chem. Soc. **98** (1976) 1031.

45 N.M. Szeverenyi, A. Bax and G.E. Maciel, J. Am. Chem. Soc. **105** (1983) 2579.

46 P.C. Myrrhe, J.D. Kruger, B.L. Hammond, S.M. Lok, C.S. Yannoni, V. Macho, H.H. Limbach and H.M. Vieth, J. Am. Chem. Soc. **106** (1984) 6079.

47 J. M. Robertson, J. Chem. Soc. (1936) 7719;
F. W. Karasek and J.C. Decius, J. Am. Chem. Soc. **74** (1952) 7716;
B.F. Hoskins, S.A. Mason and J.C.B. White, J. Chem. Soc. Chem. Comm. (1969) 554.

48 L.E. Webb and E.B. Fleischer, J. Chem. Phys. **43** (1965) 3100;
B.M.L. Chen and A. Tulinsky, J. Am. Chem. Soc. **94** (1972) 4144;
A. Tulinsky, Ann. N. Y. Acad. Sci. **206** (1973) 47;
M.J. Hamor, T.A. Hamor and J.L. Hoard, J. Am. Chem. Soc. **86** (1964) 1938;
S.J. Silvers and A. Tulinsky, J. Am. Chem. Soc. **89** (1967) 3331;
R.J. Butcher, G.B. Jameson and C.B. Storm, J. Am. Chem. Soc. **107** (1985) 2978.

49 V.L. Goedken, J.J. Pluth, S.M. Peng and B.Bursten, J. Am. Chem. Soc. **98** (1976) 8014.

Electron and Proton Transfer in Chemistry and Biology, edited by A. Müller et al.
Studies in Physical and Theoretical Chemistry, Vol. 78

Proton Transfer Reactions in Solutions: A Molecular Approach

D. Borgis

Laboratoire de Physique Theorique des Liquides
Universite P. et M. Curie, T16, 5 eme etage
75252 PARIS Cedex (France)

Summary

We review some recent attempts to interpret unimolecular proton transfer reactions at a molecular level. The modern formulation of chemical rate constants in condensed phases is based on the reactive flux correlation function formula of Yamamoto. We show how this formulation can be applied to proton transfer reactions in different limits. The ideas are illustrated by the results of some molecular dynamics simulations of a model proton transfer system in a polar solvent.

I. Introduction

The transfer of a proton is an ubiquitous elementary reaction in chemistry, which has been extensively studied and described. Some selected books and reviews are given in refs. 1-4. Two main physical pictures are commonly adopted in order to describe proton transfer reactions in solution. The widely used theory of Bell (ref. 1) is derived from the general Transition State picture of atom transfer reactions (ref. 5), with quantum corrections accounting for the possible tunneling of the proton before it has reached the classical saddle point of the potential energy surface. This theory has been extremely successful in predicting experimentally observed kinetic isotope effects (ref. 2). The Marcus view of proton transfer reactions (ref. 6), which has been extensively developed by the Russian school (refs. 7 and 8), emphasizes the electrostatic coupling of the reaction system to the surrounding polar solvent, and the associated "solvent activation energy". Both pictures, the one of "à la Bell" and the one "à la Marcus"—although never fully and consistently reconciled—are important and provide, from two complementary viewpoints, a sound phenomenological understanding of proton transfers. Nevertheless, a fully microscopic description of those processes is still lacking. In this paper, we review a few recent attempts to interpret proton transfer reactions within the framework of equilibrium statistical mechanics.

In Section II, we give a brief qualitative introduction to unimolecular proton transfers in solution, with a special attention to the fact that, even in the simple case of a unimolecular process, there are different classes of proton transfers. An important distinctive factor is the strength of the H–bond along which the proton is transfered. The specificity of "weak" or "strong" H–bonds has been stressed for many years by spectroscopists; the distinction is important in a reaction context also (refs. 9, 10). In Section III, we turn to a molecular description. We introduce the Yamamoto flux correlation formula for computing the rate of a general chemical reaction in a condensed medium and discuss its application to the different limiting types of proton transfers. In Section IV, some of the ideas are illustrated by a molecular dynamics simulation of a model of proton transfer reaction in a polar solvent.

II. General Description

A. System.

The present paper is devoted to the study of a proton transfer reaction in an inter– or intramolecular H–bonded complex AH–B in polar solvents. The process is supposed to be unimolecular, with well–defined forward and reverse rate constants:

$$\mathrm{AH{-}{-}B} \quad \underset{k_f}{\overset{k_r}{\rightleftharpoons}} \quad \mathrm{A{-}{-}HB} \tag{1}$$

Intramolecular proton transfer reactions are unimolecular by nature. For reactions in intermolecular complexes, we focus on the transfer step per se which can be treated as unimolecular; we do not consider the diffusion–controlled part although it may be, in the case of acid–base reactions between oxygen atoms for instance, the rate limiting step of the process (ref. 2).

The simplest description of proton transfers relies on a one dimensional picture, in which the proton is transfered along a linear AH–B bond. The reaction system is characterized by two coordinates: the proton position q, measured for example from the center of mass of the A–B bond, and the A–B distance Q. For a given Q, the proton potential V(q,Q) is generally a doublewell, each well corresponding to a chemical bond formed between H and either A or B. The shape of the potential and particularly the barrier height are strongly modulated by the A–B distance. The barrier is lower and the proton transfer is easier at shorter A–B distance; on the other hand, there is an energy price to pay in order to overcome the short range repulsion of the chemical groups A and B. The balance between those two antagonistic effects is a key feature of proton transfer reactions. Another important effect comes from the dependence of the dipole moment of the complex on the position of the proton. The dipole variation, which may be important for a reaction involving a charge separation for example, is at the origin of the coupling between the reaction process and the dynamics of the surrounding polar medium: the proton potential is dynamically modulated by the interaction with the solvent. In contrast with the influence of Q, the polar solvent affects the relative asymmetry of the wells rather than the barrier height. Because of the solvation free energy, the average proton potential in solution is different from that in the gas phase. A more specific and detailed description will be given in the two cases in which the equilibrium A–B distance, Q_0, is long ($Q_0 > 2.7$ Å for a typical OH–O transfer) or short ($Q_0 < 2.5$ Å). Those limits corresponds to the partition between weak and strong H–bonds (refs. 9-10). For the most part,the discussion is sketched in terms of a symmetric proton transfer system; the extension to asymmetric systems is indicated.

B. Proton Transfer along Weak H–Bonds

In this case, the proton potential presents a high barrier; this latter is modulated by the A-B distance Q. See Fig. 1a.

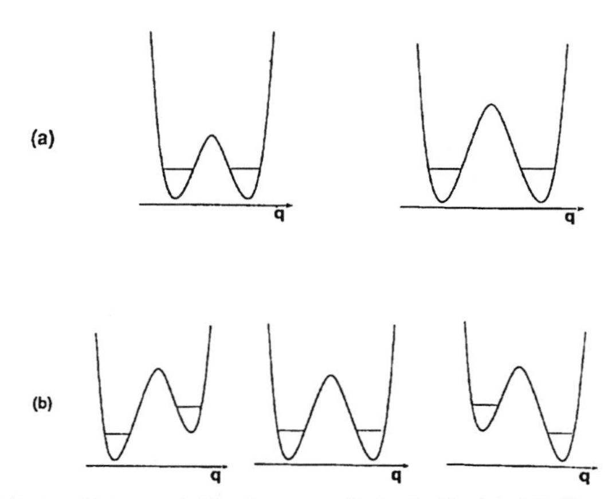

Fig. 1. Proton double–well potential in the nonadiabatic limit. a) Influence of Q; right: $Q = Q_0$; left: $Q<Q_0$. b) Illustration of the solvent induced asymmetry. The straight lines indicate the "diabatic" energy levels in each well.

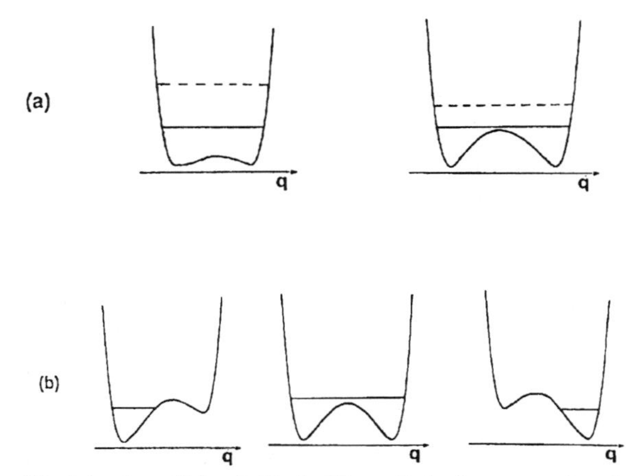

Fig. 2. Same than Fig.1 in the adiabatic limit. The adiabatic energy levels are shown instead of the diabatic ones.

In the gas phase, there is a coherent tunneling between the two wells. The tunneling frequency is $2\pi E_{01}/h$, where E_{01} represents the splitting between the ground and the first excited energy levels. In solution, however, the proton potential is asymmetrized by the interaction with the solvent and the tunneling is quenched. In a Golden Rule perspective, the quantum transition occurs when a symmetric proton potential is restaured through solvent fluctuations, as illustrated in Fig. 1b. In this view, which has many similarities with the Marcus standard interpretation of charge transfer processes in condensed media (ref. 11), the activation energy of the reaction is found in the solvent degrees of freedom rather than in the proton coordinate. For a more detailed description, see (refs. 12-14). Illustrative experimental example of the type of reactions discussed here are the proton exchange in the malonaldehyde molecule or in carboxylic acid dimers (refs. 15-16). The key notion for describing those reactions is the MODULATION OF PROTON TUNNELING by solvent fluctuations.

C. Proton Transfer along Strong H–Bonds

In this limit, the proton potential has a low barrier — or is broad and flat with no barrier — such that there is a large gap between the ground and first excited energy states (E_{01}>kT). The picture applying here is illustrated in Fig.2. The proton may be considered as confined to its ground state. In the gas phase and for a symmetric transfer, the associated wave function is delocalized over the entire potential range. However, in solution, the proton potential is asymmetrized by the electric field of the solvent molecules, and the proton wave function becomes localized on one side or the other (See Fig 2b). In this picture, the electric charge distribution associated with the proton follows exactly the external electric field, playing the role of an apparent "electronic" polarizability. This "protonic", rather than "electronic" polarizability may be very large compared to the standard ones; this feature has been described and discussed in many details by Zundel *et al.* (ref. 17) (The protonic polarizability in strong H–bonds is often referred to as the "Zundel polarizability"; it is familiar in a spectroscopic context). Different physical situations may arise, depending on the magnitude of the "localization" energy in Fig 2b., *i.e.*, of the difference in the solvent/solute interaction energy when going from a "localized" to a "delocalized" proton. If this energy is small, it does not cost much energy to the solvent to invert back and forth the electric field acting on the complex. Doing so, it drags the proton charge along. The proton transfer occurs in this case on a typical solvent time scale (in fact, a solvent time scale in the absence of any protonic charge redistribution, as if the complex had a fixed averaged dipole). In the limit considered here, the proton transfer is a spectroscopic problem rather than a reaction one: there is no activation energy.

If the "localization" energy is large, that is, if there is a substantial charge redistribution accompanying the proton transfer, it is more difficult for the solvent to go from a configuration with a localized proton and a finite dipole to a configuration with a symmetrically delocalized proton and a vanishing dipole. The proton transfer becomes in this case an activated process; the activation free energy appears along the solvent coordinates rather than the proton coordinate (We have seen that, in this limit, the proton barrier is small or even inexistent).

Complexes of the form $[R_2OH^+\cdots OR_2]$, $[R_3NH^+\cdots NR_3]$, where R stands for any carboxylic group (ref. 18), or the maleate and malonate ions (ref. 19), are experimental examples of strong, symmetric H–bonded systems, with a weak to moderate degree of dipole reorganization associated with the proton transfer. Proton transfers with large charge reorganizations can be found in asymmetric complexes of the type OH···N (*e.g.*, phenol–amine complexes (ref. 20)). Note that the physical picture depicted in Fig. 2b is still valid if the starting gas phase potential is asymmetric (left potential). If the product is more polar and is energetically favoured by the interaction with the solvent, solvent fluctuations may symmetrize the proton potential (middle) and induce eventually a reaction to the product (right). The symmetric double well potential in Fig. 2b may be understood as an effective one, including solvation effects. Keywords for describing the proton transfer in the systems considered here would be SOLVENT INDUCED PROTON LOCALIZATION.

III. Microscopic Approach

A. Hamiltonian

We now turn to a microscopic description of proton transfers. The physical system under consideration is constituted by one reactive complex immersed in an ensemble of N polar, non polarizable molecules. The hamiltonian of the system is written in the following form:

$$H = T_q + T_Q + T_S + V(q,Q) + V_S(S) + V_{int}(q,Q,S) \tag{2}$$

T designates the kinetic energies. V(q,Q) is, as before, the gas phase internal potential energy of the complex. $S=(R_{N+1},\Omega_{N+1})$ stands for the positions and orientations of the complex and of the N solvent molecules and $V_S(S)$ is the associated pairwise interaction potential (including the short range interaction of the solvent molecules with the complex). Finally, V_{int} represents the complex–solvent electrostatic interaction, and may be written in the form:

$$V_{int}(q,Q,S) = -\mu(q,Q)\cdot E(S) \tag{3}$$

where μ is the dipole moment of the complex and depends strongly on the proton position. E is the electric field created by the solvent dipoles at the center of mass of the complex. Note that we consider here that the dipole moment — or charge distribution — of the complex is independent of the solvent; for a given system, it may be deduced from an *ab initio* calculation in the gas phase. Such an approximation is valid as long as the interaction with the solvent is weak compared to typical intramolecular energies. It has been used several times in the molecular study of chemical reactions in solution (ref. 20). An alternative approach may be found in the Empirical Valence Bond (EVB) theory developed, in a microscopic context, by Warshel, (ref. 21). This author has applied the EVB concepts to the simulation of a model proton transfer reaction in solution (ref. 22). In his work, the proton was treated classically. This has to be contrasted with the quantum description presented in this paper.

B. Yamamoto's Rate Coefficient Formula

Once the Hamiltonian of the system is defined, we need a proper microscopic definition of the relevant experimental quantity characterizing the process, here the proton transfer rate constant. Such a definition was proposed by Yamamoto (ref. 23). For a general first order reaction, the first requirement is to partition properly the phase space into a Reactant and a Product region. For proton transfer, a natural definition of the species is realized by the operator (refs. 25-27):

$$N_R(q,Q) = 1 - \theta(q - q^{\#}(Q)) \tag{4}$$

where θ is the Heaviside function and $q^{\#}(Q)$ is, *e.g.*, the maximum of the gas phase double–well potential $V(q,Q)$ for a given Q. Other choices of N_R are possible as will be seen later. With a proper definition of N_R, the Yamamoto formula can be derived from the linear response theory. The result reads:

$$k = \int_0^{t_p} dt\ [2\pi/\beta h \int_0^{\beta\hbar} d\tau <\dot{N}_R \dot{N}_R\ (t + i\tau)>] \tag{5}$$

k is the rate coefficient, from which the forward and reverse rate constants can be derived according to:

$$k_f = [Z/Z_R]k, \quad k_r = [Z/Z_P]k, \quad Z_R = \mathrm{Tr}\ N_R\ e^{-\beta H}, \quad Z_P = Z - Z_R \tag{6}$$

In Eqn. (5), $\dot{N}_R$ represents the time derivative of N_R, *i.e.*:

$$\dot{N}_R = -\ 2\pi i/h[H,N_R] \tag{7}$$

t_p represents a "plateau time", long with respect to molecular time scales but short with respect to k^{-1}. In fact, t_p can be taken equal to infinity as far as only the short time dynamics ($t<<k^{-1}$) is considered in the time evolution of N_R. In this case, using the expression of the Fourier transform of a quantum correlation function at zero frequency (ref. 23), Eqn. (5) can be shown to be equivalent to:

$$k = \frac{1}{2} \int_{-\infty}^{+\infty} dt \, <\dot{N}_R \dot{N}_R^+(t)> \tag{8}$$

where $\dot{N}_R^+$ designates the complex conjugate. Different variants of Eqn. (8), which may be useful for practical purposes, can be found by performing explicitly the integration over t. For a clear and complete discussion, see (refs. 24 and 25) in the classical case and (ref. 26) in the general quantum case. Gillan has recently proposed a simplification of Eqn. (8), equivalent to a Quantum Transition State Theory approximation (refs. 27-28), which avoids the need of computing the entire quantum time correlation function. In his formalism, the proton position is replaced by "the centroid of the proton quantum path". Purely static properties, such as the free energy of the centroid, may be computed using a Quantum Monte–Carlo sampling (ref. 28). A method for incorporating quantum dynamical effects in Gillan's formulation remains to be designed.

C. Born-Oppenheimer Approximation

In this paper, we restrict the theory to the case where a Born–Oppenheimer separation can be applied between the protonic motion and the motion of the heavy particles in the system, *i.e.* the heavy atoms A, B of the complex and the solvent molecules. In this approximation, for a fixed coordinate Q and any fixed position and orientation of the solute and of the solvent molecules (characterized by the general coordinate S), the motion of the proton is described by a fixed electronically adiabatic potential V(q;Q,S), for which the Schrödinger equation for the proton can be solved. This procedure gives an infinite set of wave functions $|\psi_n(q;Q,S)>$ and of associated energy levels $E_n(Q,S)$, parameterized by the intra–complex vibration Q and the solvent. The slow motion of the heavy coordinates is then described by a set of adiabatic hamiltonians $H_n(Q,S)$, given by:

$$\begin{aligned} H_n(Q,S) &= T_Q + T_S + E_n(Q,S) \\ E_n(Q,S) &= <\psi_n|T_q + V(q,Q) + V_S + V_{int}|\psi_n> \end{aligned} \tag{9}$$

The theory is restricted here to the consideration of only two levels, the ground state and the first excited one. This restriction is sufficient to treat proton transfer reactions in weak H–bonded systems with moderate intrinsic and/or solvent-induced asymmetry. In this case, instead of $|\psi_{0,1}>$, it is more convenient to adopt a "localized" wave function representation $|\psi_{R,P}>$ defined from $|\psi_{0,1}>$ by:

$$\begin{aligned} |\psi_R> &= a_R\,|\psi_0> + a_P\,|\psi_1> \\ |\psi_P> &= a_P\,|\psi_0> - a_R\,|\psi_1> \\ a_R^2(Q,S) &= \int_{-\infty}^{q^{\#}} dq\, \psi_0^2(q;Q,S); \quad a_P^2 = 1 - a_R^2 \end{aligned} \tag{10}$$

$a_R^2(Q,S)$ represents the weight of the proton ground state wave function in the Reactant region for a given configuration of the heavy atoms. The Hamiltonian (3) becomes in this representation:

$$H = |\psi_R\rangle H \langle\psi_R| + |\psi_P\rangle H \langle\psi_P| + C(|\psi_P\rangle\langle\psi_R| + |\psi_R\rangle\langle\psi_P|) \quad (11)$$

$H_{R,P}(Q,S)$ and $C(Q,S)$ are defined by:

$$H_R = a_R^2 H_0 + a_P^2 H_1$$
$$H_P = a_P^2 H_0 + a_R^2 H_1 \quad (12)$$
$$C = a_R a_P (H_1 - H_0)$$

In the adiabatic limit, the original representation $|\psi_{0,1}\rangle$ is more adequate ; the dynamics is restricted to H_0 alone. In any case, the projector

$$N_R = |\psi_R\rangle\langle\psi_R| \quad (13)$$

provides a sensible specification of the species. It is different from the definition (4) ; however, it presents the same characteristics that the observable $\langle\psi|N_R|\psi\rangle$ is one when the proton is localized in Reactant side and zero when it is localized in the Product side, with a sharp transition for the solvent configurations that produce a symmetric proton double–well. Eqs.(4) or (13) can be used alternatively, the value of the rate constant being independent of the initial choice (for a discussion of this property in the classical case, see (ref. 29)).

D. Non–Adiabatic Limit

This limit applies to the case B of section II, *i.e.* for proton transfer reactions along weak H–bonds for which, for any accessible value of the intra–complex coordinate Q, the proton potential presents a high barrier. In this limit, the coupling term C in the hamiltonian (11) is small with respect to $\Delta E = H_P - H_R$ and the "localized" wave functions $|\psi_{R,P}\rangle$ provide an adequate eigen–representation of the hamiltonian. Using the species definition (13), one has:

$$\dot{N}_R = - 2\pi i/h \, C(|\psi_R\rangle\langle\psi_P| + |\psi_P\rangle\langle\psi_R|) \quad (14)$$

Using this expression in the quantum rate formula (8), we get at dominant order in the coupling C (ref. 30)

$$k = [4\pi^2 Z_R/Zh^2] \int_{-\infty}^{+\infty} dt \, \langle C(0) \exp_{(-)} [2\pi i/h \int_0^t d\tau \Delta E(\tau)] \, C(t)\rangle_R \quad (15)$$

The average and the dynamics are here defined in terms of the REACTANT hamiltonian, *i.e.*:

$$<A>_R = 1/Z_R\ T_r\ (A\ e^{-\beta H})$$

(16)

$$A(t) = \exp(2\pi i H_R t/h)\ A\ \exp(-\ 2\pi i H_R t/h)$$

In Eqn.(15), $\exp_{(-)}$ designates the right time–ordered exponential. A careful ordering of the different operators has to be kept when the Q–coordinate is treated quantum–mechanically (the solvent is always considered classically). $\Delta E(t)$ represents the time–dependent splitting between the protonic diabatic states, $\Delta E = <\psi_P|H|\psi_P> - <\psi_R|H|\psi_R>$. This quantity is mostly solvent–dependent. As we have seen before, C(t) is the time–dependent coupling, $<\psi_P|H|\psi_R>$. This quantity depends on the proton potential barrier height, which is strongly modulated by the A–B distance Q(t).

Eqn.(15) is a time correlation function formula which provides a way to compute the rate of a proton transfer reaction at a microscopic level using, *e.g.*, molecular dynamics simulations. However, the evaluation of k using Eqn. (15) in a straight way is not easy, because of the presence of a highly oscillating function in the integrand. A route permitting one to circumvent this difficulty has been proposed in (refs. 12-14). It can be verified numerically that the time–dependent coupling C(t) is in fact solvent–independent and varies with Q according to :

$$C(Q) = C_0 \exp(-\ \alpha\ \delta Q - \gamma^2\ \delta Q^2 + ...)$$

(17)

$$\delta Q = Q - Q_0$$

Taking advantage of the functional form of C, we can carry out a cumulant expansion of the right hand side of Eqn.(15) and obtain at dominant order:

$$k = (8\pi^2 C_0^2/h^2)\ \mathrm{Re}\int_0^\infty dt\ \exp\ [(2\pi i/h)<\Delta E>t - (4\pi^2/h^2)\int_0^t d\tau \int_0^\tau d\tau' <\delta\Delta E \delta\Delta E(\tau')> +$$
$$+\ \alpha^2\ (<\delta Q \delta Q(t)> + <\delta Q^2>) + ...\ ] \qquad (18)$$

where $\delta\Delta E = \Delta E - <\Delta E>$. It can be seen that the problem has been reduced to the evaluation of the correlation functions $<\delta Q \delta Q(t)>$ and $<\delta\Delta E \delta\Delta E(t)>$. Those can be easily determined from a CLASSICAL molecular dynamics simulation, as will be seen below.

E. Adiabatic Limit

This limit applies to proton transfers in strong H–bonded systems for which the gap between the two adiabatic hamiltonian H_0 and H_1 is large compared to kT and the dynamics of the system can be practically restricted to the lowest adiabatic surface associated with H_0. See section II–C. In this case, the value of any quantum observable is given by the average over the ground state wave function $|\psi_0\rangle$. In particular, taking either the definition (4) or (13), the species specification is determined by:

$$\langle\psi_0|N_R|\psi_0\rangle = a_R^2(Q,S) \tag{19}$$

$a_R^2(Q,S)$ is the weight of the ground state wave function in the Reactant side. The rate coefficient k is given by:

$$k = \frac{1}{2}\int_{-\infty}^{+\infty} dt < \dot{a}_R^2 \, \dot{a}_R^2(t) > \tag{20}$$

Note that a_R^2 is one in Reactant side, zero in Product side and varies strongly in the phase space region corresponding to a symmetric double well potential for the proton, *i.e.* around $\Delta E = H_P - H_R = 0$. We can therefore make the approximation

$$a_R^2(Q,S) = \theta(\Delta E(Q,S)) \tag{21}$$

In the limit in which $\dot{Q}$ is classical and "recrossing" trajectories are neglected - *i.e.* $a_R^2(\infty)$ is replaced by $\theta(\Delta \dot{E})$: trajectories coming at the barrier top from the Product region with positive velocities $\Delta \dot{E}$ are reactive and only those – Eqn. (18) may be integrated in time, yielding a familiar transition state formula:

$$k = <\Delta\dot{E}\theta(\Delta\dot{E}) \; \delta(\Delta E)> \tag{22}$$

An important part of the evaluation comes from the computation of the activation free energy of the reaction, *i.e.*:

$$<\delta(\Delta E)> = \exp(-\,\beta G(\Delta E{=}0))$$

$$G(\Delta E) = -\,kT\,\ln(P(\Delta E)) \tag{23}$$

where P(E) is the normalized probability distribution of the collective variable ΔE.

IV. Illustration: Molecular Dynamics Simulation of a Model System

A. Model

The purpose here is to apply the ideas developed in the previous sections to a model of proton transfer reaction in a polar solvent. Once the model has been defined, in terms of pairwise additive interactions acting on each atomic site, the system is monitored in time by letting the particles evolve according to the proper dynamical law: the Schrödinger equation for the proton and the Newton equations for all the heavy entities.

Regarding the solute, the A,B groups are modeled by two Lennard–Jones spheres with an equilibrium distance Q_0. The proton is a point charge moving on the line A — B. The chemical bonding of H with either A or B is modeled by two Morse potentials V_R and V_P. The resulting interaction is additive, *i.e.*, the potential of the complex is given by:

$$\begin{aligned} V(q,Q) &= U_{AB}(Q) + V_R(q,Q) + V_P(q,Q) \\ V_R(q,Q) &= D_R[\exp(-2\alpha_R(q + Q/2 - d_R)) - 2\exp(-\alpha_R(q + Q/2 - d_R))] \\ V_P(q,Q) &= D_P[\exp(2\alpha_P(q - Q/2 + d_P)) - 2\exp(\alpha_P(q - Q/2 + d_P))] \end{aligned} \tag{24}$$

Here, q is measured from the center of the A—B bond and $d_{R,P}$ denote the AH and BH bond length, respectively. U_{AB} denotes the A–B Lennard–Jones interaction. The variation of the dipole moment of the complex along the proton coordinate is modeled by assigning some q–dependent point charges on A,H and B, in the form:

$$\begin{aligned} e^{\alpha}(q) &= (1 - f(q))\, e_R^{\alpha} + f(q)\, e_P^{\alpha}, \quad \alpha = A, H, B \\ f(q) &= \frac{1}{2}[1 + \tanh(\gamma(q - q_0))] \end{aligned} \tag{25}$$

where e_R^{α}, e_P^{α} describe the charge distribution in the Reactant and Product configurations; q_0 locates the inflexion point of the charge switching function.

The solvent chosen for the simulation is a prototype of polar aprotic solvent, which has been used in previous simulations of chemical reactions involving some charge dynamics (refs. 31-32). It is constituted of diatomic molecules, with point charges on the sites, creating a dipole of either 2.4 or 3 Debye. The solvent molecules interact with each other *via* Lennard–Jones and Coulomb potentials. They have a Lennard-Jones interaction with A and B. The electrostatic interaction with the complex has to be averaged over the proton wave function, *i.e.*, according to the form (3) of the Hamiltonian:

$$V_{C/S} = \int dq\, \psi^2(q;Q,S)\, V_{int}(q,Q,S) \tag{26}$$

According to the discussion of Section III, and depending on the limit considered, the proton wave function $|\psi\rangle$ has to be understood as the Reactant wave function $|\psi_R\rangle$ in the case of a nonadiabatic reaction and as the ground state one $|\psi_0\rangle$ in the case of an adiabatic reaction.

B. Simulation

a. Nonadiabatic Limit

The reaction system is typical of a molecule or complex with a weak H–bond interaction (like malonaldehyde or acid dimers), for which the proton tunneling is the dominant effect. The proton transfer distance is 3 Å and there is a high potential barrier: $\Delta V = 25$ kcal/mole. We have verified that in this case, C(t) is effectively solvent independent, and that the expansion (16) is well justified. The determination of the rate coefficient k then rests on the evaluation of the time correlation functions $\langle\delta Q\delta Q(t)\rangle$ and $\langle\delta\Delta E\Delta E(t)\rangle$. Expanding all quantities around Q_0, those correlation functions can be computed from the classical molecular dynamics simulation of a RIGID complex. In the case of a quantum Q–vibration, the dephasing theory of molecular vibration can be used to compute the vibrational correlation function. The details of the simulation have been presented in (refs. 13-14).

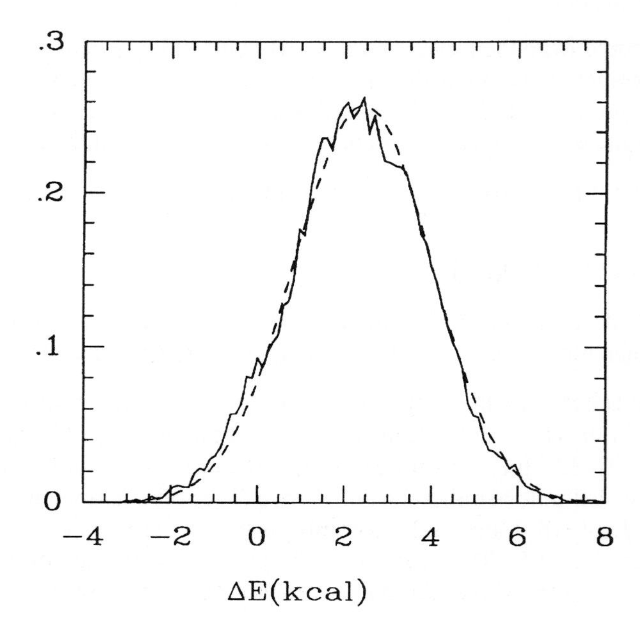

Fig. 3. Probability distribution function of ΔE, energy splitting between the weels, in the nonadiabatic limit. ——: simulation results; ---: gaussian approximation.

The following important conclusions could be drawn from the simulation results.

- The collective variable ΔE possesses a marked gaussian character, which is illustrated in Fig. 3. This quantity is a remarkable microscopic analog of the solvent coordinate introduced by Marcus in the case of outer–sphere electron transfers (ref. 11). The gaussian property of ΔE is a strong support to the validity of the truncation of the cumulant expansion used in the theory.
- The proton transfer reaction occurs in the slow modulation limit of the solvent fluctuations, in which the solvent acts as a source of static statistical disorder, with no chance, on the time scales involved in the transfer step, to have a real dynamical influence.
- The modulation of the reaction by the donor–acceptor distance Q contributes in a major may to the value of the rate constant. In this contribution, however, the dynamical coupling of Q to the solvent has a negligible influence.

The above features are expected to be general for any nonadiabatic proton transfer reactions in polar aprotic solvents. They are more questionable in the case of protic solvents, in which the interference of the reaction with the high frequency librational and vibrational modes may be important.

b. Adiabatic Limit

We present some preliminary results concerning the adiabatic proton transfer in a strong, symmetrical $[OH\text{-}O]^+$ bond. The A—A distance is 2.5 Å. For this distance, and for the set of parameters that we have chosen ($D_{R,P}$ = 53.5 kcal, $\alpha_{R,P}$ =3.5 Å, $d_{R,P}$= = 0.95 Å; these parameters are able to reproduce the spectroscopic properties of OH–O bonds (ref. 33)), the proton ground state is still below, but close to the potential barrier top. The energy splitting E_{01} between the ground and first excited state is 1 kcal/mole (*i.e.* 2 kT at the temperature of the simulation : T = 250 K) (ref. 34). The simulations are carried out for a fixed A—B distance. In the limit described here, the role of Q is to modulate the zero point energy of the proton (see Fig. 2b). This modulation is far less important than the one of the energy splitting considered in the last section. It will be ignored in favor of the solvent effects.

The procedure employed for the simulation is the following. For a given solvent configuration, the proton potential is computed and the Schrödinger equation is solved. We use the Numerov method, which provides a very convenient and accurate algorithm for solving a one–dimensional Schrödinger equation. The algorithm is an iterative one, requiring a first guess for the energy states. At each time step, the converged energy levels of the previous step are used as an input; the convergence for the new states is obtained after only 4-5 iterations. Once the eigenstates are accurately known — we monitor the first four lowest levels — we use the ground state wave function to get the forces on the heavy atoms; according to the Hellmann–Feynman theorem (ref. 35), for a general coordinate Y, the Y–component of the force coming from the solvent–solute electrostatic interaction can be computed as:

$$F_Y = \langle \psi_0 | \partial H / \partial Y | \psi_0 \rangle \qquad (27)$$

All one–dimensional integrals over q are obtained using an accurate gaussian quadrature; the gaussian weighting functions are fitted on the positions and curvatures of the wells of the gas phase potential. We have verified that the total energy of the molecular dynamics box was very well conserved on the time scale of our runs. This provides an indication that the algorithm is correct.

Note that the algorithm that we use bears some simularities with but is different from the Car–Parinello (CP) algorithm (refs. 36-37). Here, we just solve iteratively the Schrödinger equation for the proton at each time step — a sort of fictitious discrete dynamics — and we use subsequently the Hellmann–Feynman theorem to integrate the equations of motions for the heavy particles; we do not carry out a coupled integration as in the CP method. The Numerov algorithm is only designed for a one–dimensional system. If the proton were considered as 3–dimensional, it would be more appropriate to use a localized basis set (ref. 38) or a floating gaussian (ref. 39) expansion of the proton wave function, coupled to a CP integration. All simulations were carried out with a smooth truncation of the interactions at half box length, in order to avoid the complication of Ewald summations when dealing with non–neutral systems (ref. 40).

We have studied three different systems, having the same proton gas phase potential but different variations of the dipole moment when the proton is transfered; we thus modulate the strength of the electrostatic coupling to the solvent. In the first system, there is no charge on the A's and the proton carries a full positive charge ; the magnitude of the dipole variation, $\Delta\mu$, is 1.5 D. In the two other systems, there is some charge reorganization on the flanking groups, resulting in a larger dipole variation of 5 D and 7 D respectively. The second value is somehow realistic (ref. 18), whereas the third one is certainly too high for a symmetrical system and would be more typical of an asymmetric one. However, the conclusions drawn here remain valid if the mean solvation effects tend to reduce the asymmetry of the double–well potential: the symmetric double–well must then be interpreted as an effective potential rather than a gas phase one.

In Fig. 4, the proton wave function dynamics is displayed for two typical runs involving the systems with the lowest or highest dipole variation. The first one, characterized by a small coupling to the solvent, may be considered as an illustration of the Zundel polarizability phenomenon: the proton charge follows the solvent field back and forth, mimicking and effective electronic polarizability. The transfer occurs on a picosecond time scale, which is a typical time scale for the dynamics of the solvent dipoles. In the second case, in which there is a larger coupling, the proton spends much more time on one side or the other: the interaction with the solvent favors a more localized charge distribution. Once the proton is localized, the entire system is "trapped" in an energetically favorable configuration,until a sufficiently large fluctuation is able to "flip" the proton on the other side.

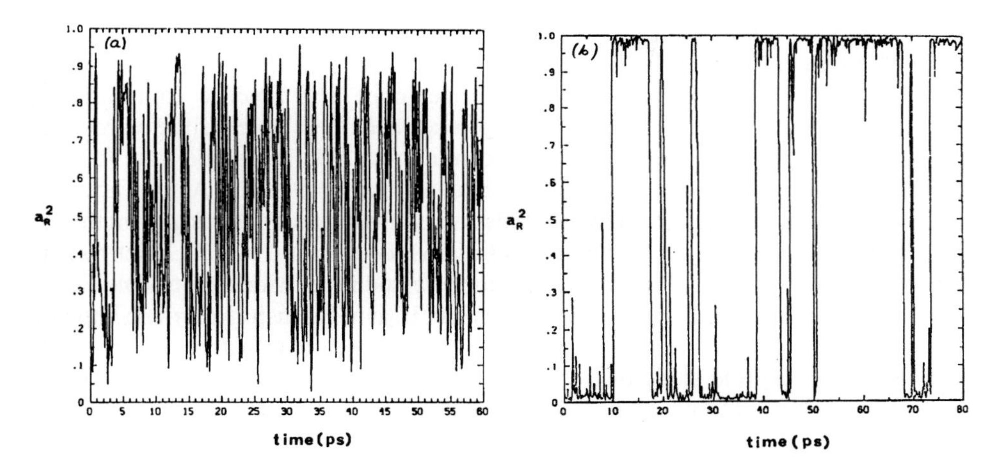

Fig. 4. Time–dependent probability of presence of the proton in Reactant well: $a_R^2(t) = \langle\psi_0(t)|(1 - \theta)|\psi_0(t)\rangle$. Two typical runs for $\Delta\mu = 1.5$ D (a) and $\Delta\mu = 7$ D (b).

The transition between a "spectroscopic" process and a "reactive" one is more clearly illustrated by plotting the free energy curves along the solvent coordinate ΔE, as suggested by the considerations of section III–E; see Fig. 5.

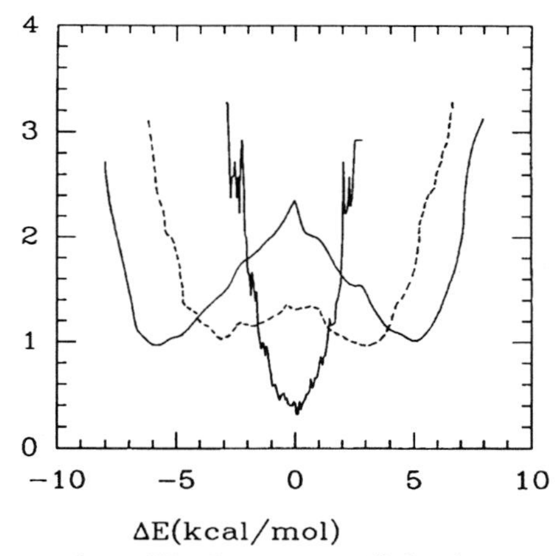

Fig. 5. Free energy curve along ΔE, the energy splitting between the wells. Inner curve: $\Delta\mu = 1.5$ D; intermediate: $\Delta\mu = 5$ D; outer: $\Delta\mu = 7$ D. The free energy is deduced from the measured probability distribution $P(\Delta E)$.

Recall that $\Delta E = H_R - H_P$ is a measure of the relative energy splitting between the two proton wells. In the first system ($\Delta\mu = 1.5$ D), the potential in ΔE is confined and relatively harmonic, with no energy barrier. In the second case ($\Delta\mu = 5$ D), the potential is much broader, with a slight barrier. A more substantial barrier is coming up with the highest coupling ($\Delta\mu = 7$ D). The corresponding potential is typical of an activated chemical process; the activation energy is about 2.5 kT. It can be seen from the analysis of the MD trajectory given in Fig. 4b that a Transition State Theory approximation should provides a reasonable value of the rate: once a symmetric configuration, characterized by $a_R{}^2 = 1/2$ (or $\Delta E = 0$), has been reached, there is very little recrossing; the system is immediately trapped in the incoming well. The transition state formula of Section III–C can be used to estimate the rate; we find $k \approx 9.9$ ps^{-1}. A more accurate computation would require to analyze an ensemble of trajectories starting from the top of the barrier, in order to get exactly the transmission coefficient (ref. 20). Such a study goes beyond the purpose of simple illustration adopted here. Another possible extension is to include nonadiabatic transitions between H_0 and H_1, using a surface hopping procedure (ref. 41). A more complete analysis of proton transfers along strong H–bonds, including the study of asymmetric complexes and possible corrections to adiabaticity will be presented in a near future.

V. Conclusions

We have shown in this paper that the modern formulation of chemical rate constants in condensed phases, based on the flux correlation function formula of Yamamoto, can be used to study proton transfer reactions in solution. The specificity and most difficult aspect of proton transfers lie in the quantum character of the proton which precludes the use of standard trajectory methods for determining the rate (refs. 25 and 26). Two routes can be followed in order to avoid a direct quantum dynamical simulation –which, in fact, is still beyond the present capabilities. The centroid approach of Gillan gives a method for computing the rate at the level of quantum transition state theory. His formulation is quite general; it is independent of the type of proton transfer under consideration. However, there are still difficulties in order to incorporate dynamical effects. The approach presented in this paper may be referred to as a semiclassical one (ref. 42). The basic idea is to eliminate adiabatically the protonic degrees of freedom and reduce the problem to the definition of a discrete set of CLASSICAL hamiltonians for the remaining heavy particles, with possible Franck–Condon transitions between them (in this paper, we have restricted the description to $H_{0,1}$ or, equivalently, to the two "diabatic" hamiltonians $H_{R,P}$; this approximation should be sufficient for most cases). We have presented two limits in which the formalism for computing the rate is well set. In the intermediate regime, it is possible to generalize the adiabatic approach of Section III–E by using a surface hopping algorithm (ref. 41) that describes the possible quantum transitions between H_0 and H_1. The semiclassical formulation has the advantage of treating consistently the dynamical solvent effects. On the other hand, it is only exact in the nonadiabatic regime for which the Golden Rule — or the equivalent time–dependent formula (15) — is valid or in the fully adiabatic regime in which there is no quantum transitions. In between, the surface hopping picture is only approximate: no attention is paid to the quantum phase and the possible interference effects.

The microscopic study of proton transfer reactions in solution is in a rapidly expanding stage. Many new methodological ideas concerning the simulation of quantum systems have emerged recently. Hopefully, they will help us in deepening our vision of those very important and interesting processes.

References

1 R.P. Bell *The Proton in Chemistry* Chapman and Hall, London, 1973.

2 E.F. Caldin and V. Gold (Eds.) *Proton Transfer reactions* Chapman and Hall, London, 1975.

3 Special issue, Chem. Phys. **136** (1989).

4 F. Hibbert, Adv. Phys. Chem. **22** (1986) 113.

5 S. Glasstone, K. Laidler and H. Eyring *The Theory of Rate Processes* Mc Graw-Hill, N.Y., 1941.

6 R.A. Marcus, J. Chem. Phys. **72** (1968) 892.

7 R.R. Dogonadze, A.M. Kuznetsov and V.G. Levich, Electrochim. Acta **13** (1968) 1025;
E.D. German, A.M. Kuznetzov and R.R. Dogonadze, J.C.S. Faraday II **76** (1980) 1128;
N. Brünische–Olsen and J. Ulstrup, J.C.S. Faraday I **75** (1979) 205;
E.D. German and A.M. Kuznetsov, J.C.S. Faraday II **77** (1981) 2203;
A.M. Kuznetsov, Mod. Aspects Electrochem. **20** (1989) 95.

8 J. Ulstrup *Charge Transfer Processes in Condensed Media* Springer Verlag, N.Y., 1980.

9 A. Nowak, Struct. Bonding **18** (1974) 177.

10 S. Bratos, J. Chem. Phys. **63** (1975) 3499.

11 R.A. Marcus, J. Chem. Phys. **24** (1956) 966, 979.

12 D. Borgis, S. Lee and J.T. Hynes, Chem. Phys. Let. **162** (1989) 19.

13 D. Borgis and J.T. Hynes, J. Chim. Phys. (Paris) **87** (1990) 819.

14 D. Borgis and J.T. Hynes, J. Chem. Phys., **94** (1991) 3619.

15 N. Shida, P.F. Barbara and J. Almlöf, J. Chem. Phys. **92** (1989) 4061.

16 S. Hayashi, J. Umenura, S. Kato and K. Morokuma, J. Phys. Chem. **88** (1984) 1330.

17 G. Zundel and J.Mühlinghaus, Z. Naturforsch. **26b** (1971) 546;
G. Zundel and J. Fritsh, J. Phys. Chem **88** (1984) 6295;
G. Zundel, in: P. Schuster, G. Zundel and C. Sandorfy (Eds.)
The Hydrogen Bond – Recent Developments in Theory and Experiments
North Holland, Amsterdam, 1976, Vol. II, Chp. 15.

18 S. Scheiner, Acc. Chem. Res. **18** (1985) 174.

19 J. Mauri, W. van Gunsteren and H. Berendsen, (J. Phys. Chem.), in press.

20 H. Ratajczak, J. Phys. Chem. **76** (1972) 3000; *Ibid*, **76** (1972) 3991.

21 A. Warshel and R. Weiss, J. Am. Chem. Soc. **102** (1980) 6218.

22 A. Warshel, J. Phys. Chem. **86** (1982) 2218.

23 T. Yamamoto, J. Chem. Phys. **33** (1960) 281.

24 C.H. Wang *Spectroscopy of Condensed Media* Academic Press, London, 1985.

25 D. Chandler, J. Chem. Phys. **68** (1978) 2959.

26 S.H. Northrup and J.T. Hynes, J. Chem. Phys. **73** (1980) 2700.

27 G. Voth, D. Chandler and W.H. Miller, J. Phys. Chem. **93** (1989) 7009;
J. Chem. Phys. **91** (1989) 7749.

28 M.J. Gillan, J. Phys. C **20** (1987) 3621;
Phys. Rev. Lett. **58** (1987) 563.

29 D. Borgis and M. Moreau, Mol. Phys. **57** (1986) 33.

30 This formula can also be derived from the Golden Rule, see refs. 12-14.
An alternate derivation from Yamamoto's formula is given in ref. 27.

31 G. Ciccotti, M. Ferrario, J.T. Hynes and R. Kapral, J. Chem. Phys. (Paris) **85** (1988) 925;

32 D. Zichi, G. Ciccotti, J.T. Hynes and M. Ferrario, J. Phys. Chem. **93** (1989) 6261.

33 E. Matsushita and T. Matsubara, Prog. Theor. Phys. **67** (1982) 1.

34 The appropriate condition for adiabaticity is that the Landau-Zener Parameter is larger than one, *i.e.*, $\gamma = \pi E_{01}^2/h(\langle \Delta E_{01}^2 \rangle_{\Delta E=0})^{1/2} > 1$, where $\langle\ \rangle_{\Delta E=0}$ is the conditional average computed at $\Delta E = 0$. See, *e.g.*, (ref. 42).

35 R. Feynman, Phys. Rev. **56** (1979) 340;
H. Hellmann *Einführung in die Quantenchemie* Deutick, 1937.

36 R. Car and M. Parrinello, Phys. Lett., **55** (1985) 2471.

37 D. Remler and P. Madden, Mol. Phys. **70** (1990) 921.

38 G. Ciccotti and R. Kapral, personal communication.

39 M. Sprik and M. Klein, J. Chem. Phys. **89** (1988) 1592; *Ibid*, **90** (1989) 7614.

40 M.P. Allen and D.J. Tildesley *Computer Simulation of Liquids*
Clarendon Press, Oxford, 1987.

41 J.C. Tully , J. Chem. Phys. **93** (1990) 1061.

42 E.E. Nikitin *Theory of Elementary Atomic and Molecular Processes in Gases*
Clarendon Press, Oxford, 1974.

Electron and Proton Transfer in Chemistry and Biology, edited by A. Müller et al.
Studies in Physical and Theoretical Chemistry, Vol. 78

Recent Developments in Solitonic Model of Proton Transfer in Quasi–One–Dimensional Infinite Hydrogen–Bonded Systems[+]

Eugene Kryachko[*]

Institute of Physical Chemistry
University of Munich
Theresienstr. 41, D-8000 (Germany)

Summary

The solitonic models of orientational and ionic defects in the hydrogen-bonded network of liquid water are outlined. Properties of these kink defects are discussed and compared with experimental data. The interpretation of the central bifurcated H–bond in the orientational kink defect as a "free" OH–group is proposed. A relaxation of heavier subsystem under formation of protonic kink defect in the chain $(A–H\cdots)_\infty$ with the symmetric double–Morse potential is under study. The energy splitting due to tunneling of proton in a homoconjugated H–bond with the symmetric double–Morse barrier is explicitly derived to first order in $h/2\pi$ within the instanton approach of the Feynman's path integral formalism.

Solitonic Background for Liquid Water

The main aim of my contribution is to penetrate the "solitonic dinosaure" concept in a more extent into the physico–chemical media and to demonstrate that physical chemistry appears to be very rich area for them, that is why they are still alive, and I hope a richness of the physical chemistry "land" allows to avoid a degeneration of them, comparing with their natural prototypes.

One can continue further this comparison by noticing that our "solitonic dinosaures" are very young. The first one visited us only about 160 years ago, that was thoroughly documented (ref. 1) under observing very strange things occurring on the surface of liquid water in the narrow Union Canal which linked Edinburgh and Glasgow. The Scottish civil engineer John Scott Russel investigated a possible relation of the shapes of the hulls of ships with the speed and forces required to propel the ships, and revealed an odd wave on the surface which was formed when a boat moving through the channel was suddenly stopped. He describes this wave as follows (refs. 1 and 2):

It accumulated round the prow of the vessel in the state of violent agitation, then suddenly leaving it behind, rolled forward with great velocity, assuming the form of a

[*]Permanent address: Institute for Theoretical Physics, Kiev 252130 (Ukraine)
[+]Paper II in the series ***"Collective Proton Transfer in the $(A–H\cdots)_\infty$ System with Double Morse Symmetric Potential"*** see also Paper I (ref. 33).

large solitary elevation, a rounded, smooth, and well–defined heap of water, which continued its course along the channel, apparently without change of form or diminution of speed. I followed it on horseback, and overtook it still rolling on at a rate of some eight or nine miles per hour, preserving its original figure.

The second "dinosaure" came to our observations from water, too. This is a well–known tsunami.

I guess everyone should be surprised by the fact that both "solitonic dinosaures", the first ones observed among their "relatives" were originated on water surfaces, and I stress that in both these cases, the surfaces of water are quite different to take the extreme part in their origin, and no one still observed them on surfaces of ammonia or alcohols, for instance. Hence, making indeed a somewhat illogical jump, one could assume that it is not the phenomena of surface, but related to microscopic properties inherent to liquid water. Claiming further, no one denies that liquid water is not a subject of physical chemistry. And that is the reason of my starting statement on physical chemistry as a rich "land" for "solitonic dinosaures" to be still alive.

In my present contribution I would like to make an unexpectable and, maybe, non–existing linkage between the Russel soliton with those which are now under the way in science to describe a proton migration occurring again in the same liquid water, via the "channels" of hydrogen (or H—) bonds, and emphasize what the new insight in liquid water the concept of proton–transfer solitons brings us, at least by the formula "What you see is what you get".

The human knowledge, especially the scientific knowledge, embarks from the observation accompanied by the observer's state of surprise as source of thinking and inventing models or theories. In the epistemology, we have to postulate the existence of a zero-point of view, or an priori view as the beginning in the evolution of theories. The state of surprise starts by observing a something "strange" or "unusual". The term "strange" means that we cannot match our "old dressed" theory, which we had just before the observation, to that strangeness, or appearance (phenomenon). As long as that phenomenon remains a weak spot of the before–observation theory, or in other words, incompatible with it, we refer it to the "anomalous" one. This term, applied to the phenomenon under surprise, is still vital epistemologically until the before–observation theory is evolutionarized towards the new, after–observation theory enabling to understand it, though maybe, it is not yet deficient of errors, that can be established only in the way of "error elimination", (ref. 3).

Liquid water still continues to surprise observers as that with a plenty of "unusual" properties. That is why it is called sometimes as "schizophrenic". Notwithstanding the fact that many properties of that kind, being "unusual" only in the sense that the concepts of simple liquids are transferred commonly to liquid water, have already theoretically grounded (how these theories are rigorous is the other question; that is the state of complete surprise has already changed by that of error elimination accompanied sometimes by a certain surprise), the term "anomalous", applied to liquid water, still "roves" from one textbook on physical chemistry to another. If water knew how much it is "anomalous", maybe it would prefer to be "usual"!

One of the most striking "anomalous" properties of liquid water is its high electric conductivity arising, as emphasized commonly in textbooks on physical chemistry, due to large mobility of the hydrogen ion H^+, or proton because there still exists the considerably discrepancy in terminology relative to the question how much a "quasi-atom" of H is atom or proton. Speaking rigorously, the entity, that remains the hydrogen atom, is neither the latter nor the proton, although we call it hereafter as the "proton", understanding that our "proton" is dressed by the electronic "fur–coat", but not completely to look like the hydrogen atom.

Well, and how large the proton mobility in liquid water, and to what is it compared? In the context of the diffusion theory working rather well for simple liquids for instance, comparing the van der Waals ionic radii of the H_3O^+ ion and K^+,

$$r(H_3O^+) = 1.38Å, \qquad r(K^+) = 1.33Å,$$

one concludes that their mobility should be approximately equal, about the value of the K^+ mobility, 7.6 $10^{-4}cm^2V^{-1}s^{-1}$. It is commonplace to suggest that the "proton" in liquid water exists as the aggregate called H_3O^+ — once more term to call the real entity! But, it is astonishing, the mobility of this entity which we convinced to call "proton", is 36.2 $10^{-4}cm^2V^{-1}$ at temperature 25°C?, that is about in 5 times greater, (ref. 4).

Hydrogen Bonding and Liquid Water

The origin of "schizophrenia" of water is in a specificity of interaction between water molecules — the so–called hydrogen bonding — which organizes them in the hydrogen–bonded network whose structure (speaking rigorously, V–structure (ref. 4) with the definite lifetime during of which the term "structure" for liquid water is appropriate) obeys the Bernal–Fowler–Pauling rules, (refs. 5 and 6).

A. Each linear hydrogen–bonded bridge between two neighboring molecules of water involves a single hydrogen atom ("proton"), and this bridging hydrogen is well located off the H–bond midpoint, *i.e.* it can be viewed to "belong" to one of the participating oxygens.

B. Each oxygen atom forms four H-bridges with neighboring oxygens in such a manner that only two hydrogens really "belong" to it.

Any configuration of water molecules obeying the aforementioned rules is called a Bernal–Fowler configuration. Therefore, from this viewpoint, water represents itself a random packing of Bernal-Fowler configurations. Unfortunately, such Bernal–Fowler–like water would not be the really existing one, still causing the observer's surprise, if it no longer contains non–Bernal–Fowler configurations, or bond defects. The latter can be defined as breakdown of any Bernal–Fowler–Pauling rules:

DA. A number of protons, occupying a chosen H–bridge, is equal to zero ("empty" bond) or two ("double" bond). A case of three or more protons is inconceivable.

DB. Oxygen possesses single or three neighboring protons.

Taking a look at these breakdowns' definitions, one can immediately give the examples of bond defects. A bond defect, corresponding to the DA definition, is called an orientational defect, because one can organize the "empty" of "double" H–bond by rotating a single OH–bond in the Bernal–Fowler configuration out or to a given H–bond, respectively. The simplest model of orientational defect was suggested by Bjerrum (ref. 7): L–defect corresponds to a single "empty" ("leer") H–bond without proton, O···O; D–defect to a single "double" ("doppelt") H–bond, O–H···H–O. The orientational defects of both types originate from rotating a single OH–bond in the Bernal–Fowler configuration around the bisector of $O_1 \cdots O \cdots O_2$, through the formation of DL–pair. It is clear that L–defect is characterized by a deficiency of positive charge, that is why it carries a certain negative charge, say $-e_B$, while the D–defect is positively charged, for its charge being $+e_B$.

Another bond defect, of the DB–type, refers to the ionic one. As in the above case, one can fairly easy give simplest candidate for the ionic defect. The H_3O^+ ion, *i.e.* the oxygen atom with three neighboring protons, represents itself a positive ionic defect (call it P–defect) with charge $+e_I$, and OH^- a negative one (or N–defect) granted by the charge $-e_I$.

Now we are quite prepared to describe schematically the mechanism of proton conductivity in liquid water, in terms of given simple models of bond defects. Assuming the H_3O^+-OH^- pair is formed due to thermal fluctuations in the H–bonded network and applying an external electric field, one can break it down, that results in migrating P– and N–defects separately, in the opposite directions along hydrogen bonds, until they reach the electrodes. One can provide a conductivity process to be continuous by suggesting that at the moment of arrival of ionic defects at electrodes, the H–bonded pathway, or "channel" is turned by means of generating orientational defects. Migrating and reaching the edges of the protonic channel, the latter give the start of the subsequent step of proton transfer, and so on. Therefore, the proton–transfer process consists of two stages, for each stage being responsible for carrying the definite charge which unify at the edges, results in the total charge of proton, by the absolute value, *i.e.* $e=e_B+e_I$, what was expected (ref. 4).

One should notice, concluding this section that the picture given above is rather simplified. First of all, due to the schematic model of bond defects applied, it neglects a cooperativity of H–bonding. In view of hydrogen bonding, cooperativity has different facets. The energetic one goes back to Frank and Wen (ref. 8) who used this terminology to explain the stability of the H–bonded structure of liquid water. The non–additivity of the binding energy of water oligomers (ref. 9) frames the energetic facet. The charge facet of cooperativity is determined by non–additivity of the total dipole moment of water clusters (ref. 10). The striking pattern of the other,the so–called spectroscopic facet, is the well–known drastic change in the harmonic force constant of stretching vibrations (ref. 11). For liquid water, under a passage from a gas phase to the liquid or amorphous, or ice, it consists of about -300 cm^{-1} (the red shift of OH–stretching vibration). This red shift cannot be explained in terms of dimer–like conformations with corresponding charge of the harmonic force constant and $\Delta\nu$ of about -60 ÷ -80 cm^{-1}, but becomes well understood by using the concept of cooperativity (ref. 12). The spectroscopic facet of H–bonding has been recently experimentally studied (refs. 13 and 14). For this reason, namely due to the sharp evidence of the really observed cooperativity, I disagree with the viewpoint of Schuster *et al.* (ref. 15) that a cooperativity of H–bonds is not an intrinsic physical phenomenon, or the property of H–bonding, but reflects only our level of description. In fact, if the cooperativity is not inherent for H–bonded associates, why do we need to go beyond the model of independent particles to describe "nothing"? It is observed, hence it does exist beyond us.

A considerable attention has been paid here to the structural facet of cooperativity. In application to the bond defects in the H–bonded network of liquid water, it was thoroughly investigated in (refs. 4, 16, 17 and 18). The aforementioned models of bond defects, especially orientational one, represent the local breakdown of the Bernal–Fowler–Pauling rules, neglecting a distortion of H–bonds in the vicinity of D– and L–defects. A "distortion" is meant here in terms of non–linear H–bonds O–H···O, for which the angle OHO differs from 180°, surrounding the defect centre. For ionic defects, it presumes the more extended aggregates like $H_5O_2^+$ (ref. 19), $H_9O_4^+$ (ref. 20), for example.

One of the more natural and fruitful way to define the bond defects accurately is to bring the concept of soliton. Clearly, from observing two "solitonic dinosaures", mentioned above, the terms "cooperativity" and "soliton" are very neighboring, that explains our use of the word "natural". A demonstration of fruitfulness of this way is the aim of the present paper.

Quasi–One–Dimensional Chain of Water Molecules: Discrete and Continuous Treatment

A choice of model is a crucial step in any theoretical study. A model's prerequisite is usually determined so that absorbs, as much as possible or as we can, the peculiarities of the phenomenon under study, and neglect, as much as possible too, all unessential details. We believe that a quasi–one–dimensional chain of water molecules (Fig. 1), taking off from the continuous H–bonded network, is a rather good "caricature" of real liquid water or ice (1D ice) purported to give the solitonic Ansatz for bond defects. Such a choice of the model system implies that in studying a proton transfer along H–bonded pathway, we isolate a certain zig–zag protonic channel (somewhat recalling the famous Edinburgh–Glasgow channel where Sir John Scott Russel "caught" his soliton) in a three dimensional network of liquid water or crystal of ice.

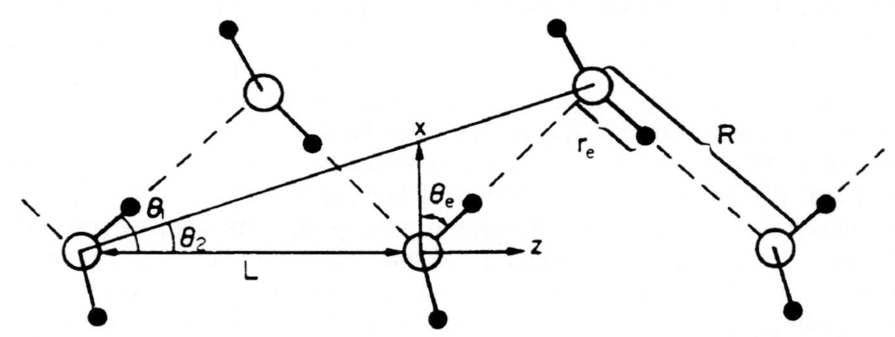

Fig. 1. Model of zig–zag infinite chain of water molecules. All plotted H–bonds, O–H···O, are in the plane of figure; O is the oxygen atom, and ● the hydrogen atom.

According to the definition of bond defects, two geometrical variables play the basic part. The first one is angle $\theta_i \equiv \frac{1}{2}(O_{i+1} - O_i - O_{i-1})$ describing a rotation of the proper OH–bond dipole belonging to the i–th water molecule in the interior of our zig–zag channel. Due to symmetry of the model system chosen, the own potential energy $U(\theta_i)$ of the proper i–th OH–bond dipole has two equivalent minima at $\theta_i \equiv \pm\, \theta_e$. The $2\theta_e$-th fluctuation jumping, classical over the barrier $U_B^{(0)} = U(\theta_i = 0))$, from minimum at $\theta_i = +\theta_e$ to the equivalent one, at $\theta_i = -\theta_e$, correlated with associated transition of neighboring OH-bond dipoles owing to the presence of the potential energy term like this:

$$æ^{(0)}(\theta_i - \theta_j)^2/2 \tag{1}$$

that describes a cooperativity between the jumps of i–th and j–th dipoles, generates the orientational defects. In eqn. (1) we assume that the (o)rientational coupling constant $æ^{(0)}$ is certainly non–vanishing. This term allows to overcome the aforementioned difficulties of the Bjerrum model corresponding to $æ^{(0)} = 0$. In the latter case, the DL–pair generated at the i–th site decays into quasi–isolated D– and L–defects as time is going on, due to the subsequent OH–bond reorientations, that finally results in the defected chain, comprising of two subchains. Each subchain involves its own

D– or L–defect, They differ from each other by the directions (topology) of the OH–bonds at their "edges". For the left one, involving the positive D–defect, its left–side–edge OH–bond occupies the minimum at $\theta = +\theta_e$, while the opposite one at $\theta = -\theta_e$.

The reverse arrangement of minima of the edge OH–bonds is peculiar for the right subchain with the negative L–defect. Generalizing that property on non–vanishing $æ^{(0)}$, one can define the orientational defects as those structures in our zig–zag chain which satisfy the following boundary conditions (refs. 21 and 22)

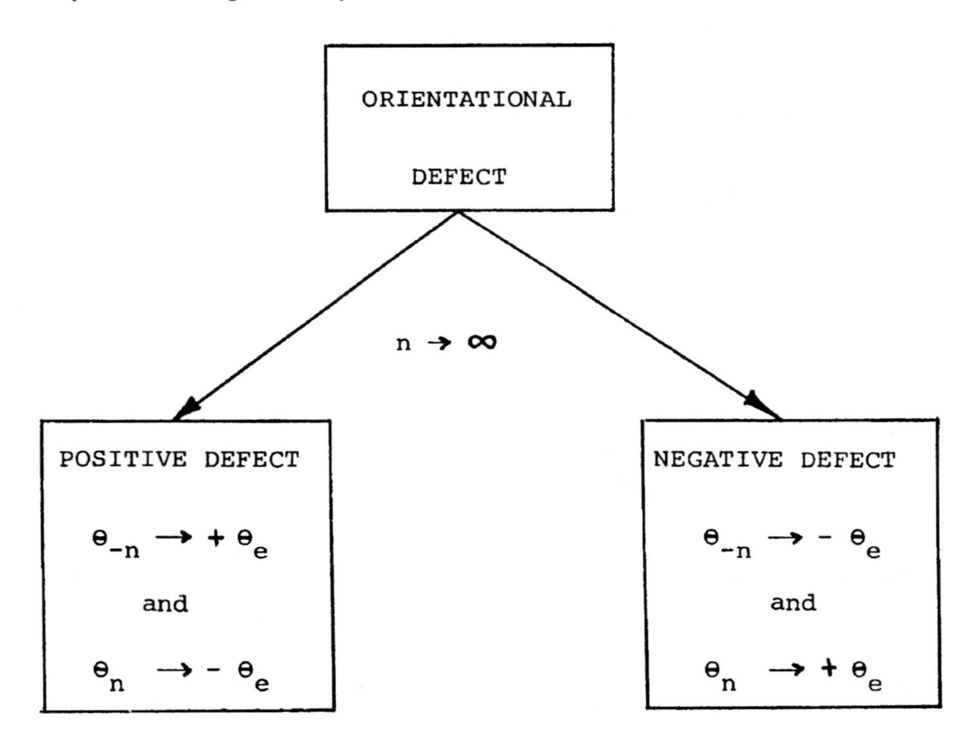

Therefore, the orientational defect represents itself a certain transition state between these limiting topologies of OH–bonds. For the word "transition" means a discrete set of OH–bond angles, $\{\theta_i\}$, realizing rotations of OH–bond dipoles from one limiting angle $\pm\ \theta_e$ to the opposite one $\mp\theta_e$.

The second variable, a co–length r of OH–bond is related to ionic defects. We assume that a proton can migrate one dimensionally, along a single H–bond, O–H···O, and that process is described by a unique reaction coordinate r directed along the O···O axis. We also assume that the i–th proton moves in the effective potential field $U(r_i)$, symmetrical relative to $r_i=0$, the centre of the O···O bond, and possessing two equivalent minima at $r_i = \pm r_0$. Here $r_0 = (R/2 - r_e)$, r_e is the equilibrium length of the OH–bond (the equilibrium co–length is r_0), and R that of the O···O bond. These minima are separated by the energy barrier with height $U_B^{(i)} = U(r_i = 0)$. A migration of the i–th proton is no doubt coupled with those of other protons in the chain, due to the following term:

$$æ^{(i)}(r_k - r_e)^2/2, \tag{2}$$

for instance, with the non–vanishing ionic coupling constant $æ^{(i)}$. Generalizing the schematic model of the ionic defect such as H_3O^+ and OH^-, on a domain of non–vanishing coupling constants, one can very similarly to the case of the orientational defects, give the definition of ionic defect as the definite structures in our quasi–one–dimensional chain, obeying the following "edge" conditions (refs. 23-25)

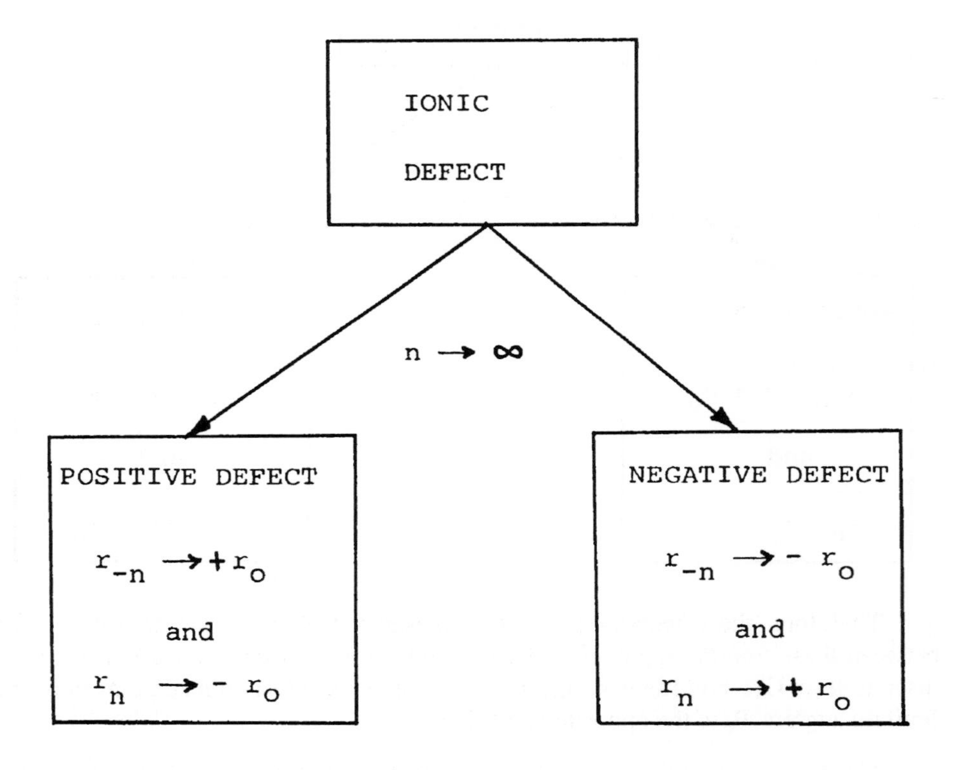

Speaking pictorially, the ionic defect is such one-dimensional structure of water molecules whose OH–bonds are "lengthened" discretely (that is described by the set $\{r_i\}$) between two "edge" OH–bonds with the equilibrium lengths but the different oxygen atom property.

Therefore, we suggest the existence of collective excitations (or structures) of the orientational and ionic defect types described by the sets of corresponding variable, $\{\theta_i\}$ or $\{r_i\}$. The way to describe such sets is to obtain them from the Euler–Lagrange equations of motion for the appropriate discretely representable Hamiltonians. Unfortunately, these equations of motion can be solved only numerically that forbids a thorough theoretical analysis to be carried out. However, and fortunately, it could be done by taking some Ansatz, the so–called continuum approximation. The latter means the replacement of an infinite set of discrete variables $\{X_i(t)\}$, where $X=\theta$ or r in our case, by the continuous function $X=X(z,t)$, with the z–th axis directed along the chain under study. In other words, if $z=z_i$ corresponds to the i–th molecule in the chain, $X(z_i,t)$ becomes just $X_i(t)$. It is clear that the continuum approximation is valid if a width of the excitation region $2\Delta z$ around $z=z_0$ (z_0 is a centre), where $X(z,t)$ sharply differs from the equilibrium values $\pm X_e$, much larger a lattice constant L (relative to the z–th axis).

A passage from the discrete Hamiltonian to its continuum analogue can be made by using, for example, the following relation,

$$X_{i+1}(t) \equiv X(z_{i+1},t) = X(z_i+L,t) = X_i(t) + L(\partial X/\partial z) + o[(\partial X/\partial z)^2] \tag{3}$$

and neglecting the last term in r.h.s. of eqn.(3). For this reason, the potential energy term

$$æ(X_{i+1} - X_i)^2/2 \tag{4}$$

[compare with eqns. (1) and (2)] transforms to:

$$æL^2(\partial X/\partial z)^2/2 \tag{5}$$

Let assume for a while that we have already got the solution $X(z,t)$ of the Euler–Lagrange equation within the continuum approximation. How should we come back to discrete case? Or, in other words, what is the inverse procedure to construct $\{X_i(t)\}$ from $X(z,t)$? At the first glance, the answer seems to be very simple: substituting discrete values of z into $X(z,t)$ results in $\{X_i(t)\}$. However, it is not that case, because when the continuum approximation is chosen, the coupling term (4) is linearized, that is, it is suggested that the term under $o[(\partial X/\partial z)^2]$ in r.h.s. of eqn. (3) are considerably smaller that $L(\partial X/\partial z)$. But the main difficulty still remains if we even take into account the above terms in a somewhat way. It consists in different symmetries inherent for solutions. Within the continuum approach, neglecting the terms under $o[(\partial X/\partial z)^2]$ together with the common kinetic energy term leads to a Lorentz continuum translation invariance of the solution X(z,t), while in a lattice, the translational symmetry becomes discrete and forbids to take values of the displacement of soliton which are not integral multipliers of the lattice constant. So, the caution is of a great necessity when we speak about the defect in a discrete chain in terms of its continuum analogue, because the reverse step is not yet clear so far.

Orientational Defect

Solitonic model

We start by assuming that the heavier subsystem of the model displayed in Fig. 1, which consists of the improper OH–bonds realizing the linkage of our channel with the rest of bulk water or 3D ice crystal, is frozen. Hence, the total Hamiltonian of the subsystem of proper OH–bonds is (refs. 21, 22 and 26):

$$H = \sum_{i} \left\{ \frac{1}{2} mr^2\dot{\theta}_i^2 + U_B^{(0)} \left[1 - \left(\frac{\theta_i}{\theta_e} \right)^2 \right]^2 + \sum_{n \neq i} \kappa_n^{(0)} \left(\theta_i - \theta_{i+n} \right)^2 \right\} \tag{6}$$

In eqn. (6), m is the reduced mass of the OH–bond dipole, and the dot means the time derivative. The second term under summation in eqn. (6) represents the potential energy of a single OH–bond rotator taking in the simplest form of "θ–four" potential. The third term there has been already explained. Assuming the dipole–dipole type of interaction between OH–dipoles in the chain $(H_2O)_\infty$ with the dipole moment $\mu_0 = 2.85$ D for a single OH–bond that corresponds to the effective charge 0.6e of the proton (ref. 27), one can evaluate

$U_B^{(0)} = 7.68$ kcal/mol.

Within the continuum approximation, the total Hamiltonian (eqn. 6) transforms to the following:

$$H = \frac{1}{2L} \int_{-\infty}^{+\infty} dz \left\{ mr^2\dot{\theta}^2 + mr^2c_0^2 \left(\frac{\partial \theta}{\partial z} \right)^2 + 2U_B^{(0)} \left[1 - \left(\frac{\theta}{\theta_e} \right)^2 \right]^2 \right\} \tag{7}$$

where $c_0 = 4.24 \cdot 10^6$ cm$\cdot$s^{-1}, and the total coupling constant $æ^{(0)}$=19.6 kcal mol^{-1}. c_0 is of dimension of velocity, and for this reason it can be interpreted as a characteristic velocity of the OH–dipole system. Comparison of $U_B^{(0)}$ with $æ^{(0)}$ justifies our passage from the discrete to the continuum representation.

Deriving the Euler–Lagrange equation of motion from eq. (7), and taking the boundary conditions into account, one writes the solution:

$$\theta_D(z,t) = \quad \tanh[(\gamma/\theta_e)(z - z_0 - Vt)] \tag{8}$$

Here

$$\gamma^2 = [2U_B/mr^2c_0^2][1-(V/c_0)^2]$$

and z_0 is a centre of orientational defect defined by $\theta_D(z_0, t = 0) = 0$, which owing to the Lagrange theorem, is the corollary of the boundary conditions. The upper sign in r.h.s. of eqn. (8) corresponds to a positive, and the lower one to negative defect of the orientational type. It should be understood that this solution gives us only the continuum analogue of the real 1D–orientational defect.

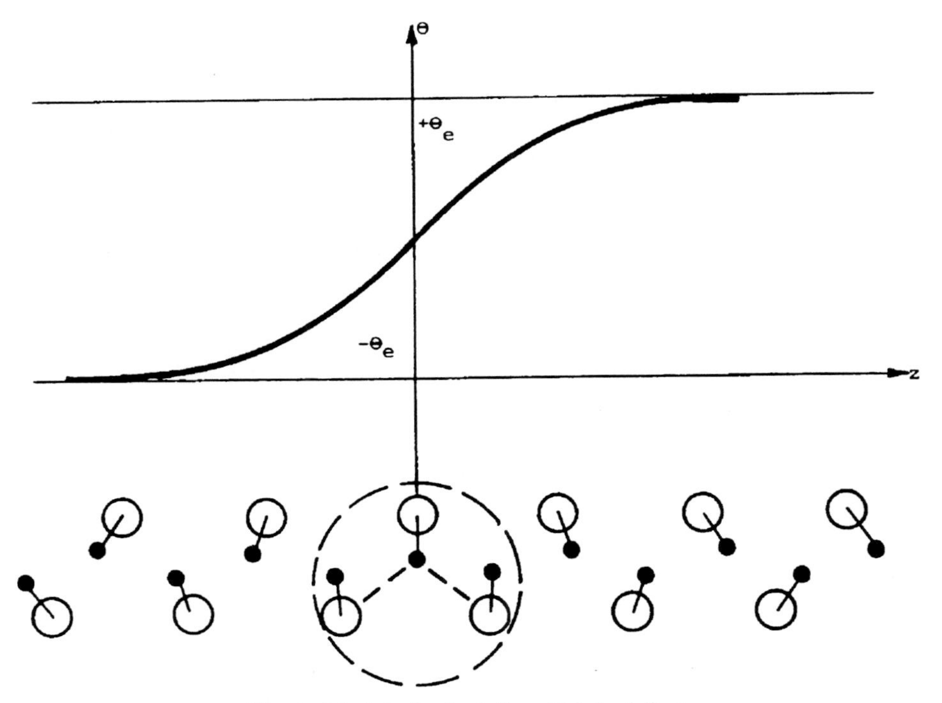

Fig. 2. Model of orientational kink–defect.

Now let us see what we get. Firstly we plot in Fig. 2 a positive orientational defect, which has the form of kink smoothly linking two angle–boundary structures. It is called a kink–soliton. It is characterized by the following energy density:

$$\mathcal{E}^{(0)}(z,0) \equiv \frac{1}{4} mr^2c_0{}^2[\partial\theta_D(z,0)/\partial z]^2 + \frac{1}{2} U_B{}^{(0)} [1 - (\theta_D(z,0)/\theta_e)^2]^2 \qquad (9)$$

integrating which over the whole z-th space, one gets the total energy of the static (V=0) kink,

$$E_D^{(0)} = \frac{4r\theta_e C_o}{3L}\sqrt{2mU_B^{(c)}} \simeq 15.7 \text{ kcal/mol.} \qquad (10)$$

In this context, the energy (eqn. 10) is interpreted as that of formation of the orientational kink–defect in our chain. The discrepancy between the theoretical value and the experimental one, $E_D{}^{(0)} = (15.6 \pm 0.9)$ kcal/mol (ref. 4) is rather small.

One can estimate how well the orientational kink–defeat is localized in two ways (ref. 22). The first one consists in inspecting the energy distribution around water molecules in the interval of length L/2. Integrating the energy density over the intervals $(z_0 - L/4, z_0 + L/4)$, $(z_0 - L/2, z_0 - L/4)U(z_0 + L/4, z_0 + L/2)$,$(z_0\text{-}L,z_0\text{-}L/2)U(z_0\text{+}L/2,z_0\text{+}L)$ *etc.* results in that the contribution to $E_D{}^{(0)}$ of the central molecule attains 43.9% of $E_D{}^{(0)}$, of two next molecules 2x21.6% = 43.2% of $E_D{}^{(0)}$, further 2x2.7% = 7.4%, and so on, that is why one concludes that energetically the orientational kink–defect is localized in the z-th space over region of about 2L ≅10Å, enveloping five water molecules. Introduction of a deviation parameter $d = \alpha\theta_e$, where α is 0.9, say one can say that the orientational kink–defect representing the transition state between two Bernal–Fowler configurations, is spread over a region with dimension of about 15Å (refs. 21 and 22). Evidently, the blurred character of the orientational kink–defect is the consequence of cooperativity of hydrogen bonding. This estimate of the kink-dimension also confirms the validity of the continuum approximation.

Effective Charge

The orientational kink–defect is formed fluctuationally by cooperative rotation of OH–bond dipoles of water molecules. A quantitative measure of such reorientational process is a change of projections of dipole moments of each OH bond onto the z–the axis. It is convenient to evaluate that charge in terms of the effective charge, e_B, of the orientational defect (ref. 22)

$$e_B \equiv -\sum_n \frac{1}{\Delta Z_{n,n-1}} (\mu_z^n - \mu_z^{n-1}) =$$

$$= -\frac{2}{L} \int_{-\infty}^{+\infty} dz \frac{\partial \mu_z}{\partial z} = \frac{2}{L} [\mu_z(+\infty) - \mu_z(-\infty)] = \pm \frac{2\mu_0}{R}$$

$$= \pm 0.43e$$

which exactly coincides with the topological charge of soliton (ref. 28). Here $\mu_z{}^n = \mu_0 \cos\theta_n$ and $\Delta z_{n,n-1} = L/2$. The experimental magnitude of e_B ranges from 0.36e to 0.44e (ref. 27).

Model of "Free" OH–Groups and Concept of States in Liquid Water

Let us now take a look at the discrete distribution of angles of OH–bonds contributed to the orientational kink–defect (Fig. 3).

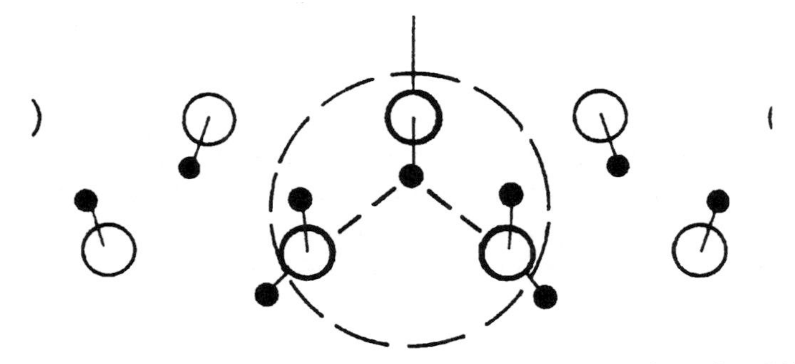

Fig. 3. "Free" OH–group as the central bifurcated H–bond in the orientational kink–defect.

According to the geometrical definition of the hydrogen bond, which belongs to the family of on–off definitions, the OH–bond of the central water molecule and two bonds of the neighboring ones, are "weak" or "broken" H–bonds though one can differ them saying that the former is the bifurcated H–bond, and the two latter are partially H–bonded bridges. The rest H–bonds participating in kink–defect formation, refer to "nice" or "strong" H–bonds in liquid water. However, in the context of the unique collective structure of the orientational kink–defect, especially within the continuum approximation, it is impossible to isolate one H–bond there from other bonds. In other words, there exists a semi–continuum (or semi–discrete) transition from "weak" to "strong" H–bonds which is intractable in terms of the ill–defined single threshold value of the "yes–no" property.

Consider the bifurcated OH–bond in the orientational kink–defect. Applying the model of OH–stretching vibrations in liquid water and ice (ref. 12), one can evaluate the harmonic force constant of this bond in the stretching regime (ref. 22), and obtain that its difference from that of monomer consists only 0.3% of the latter, or the red shift is -50-60 cm^{-1} only. That is rather small comparing with the ordinary shift about -300 cm^{-1}. For this reason, one can call the bifurcated OH–bond as "monomeric", or "free" OH–group, of course, well understanding a relativity of this definition. Thus, a "free" OH–group is counterpart of the orientational kink–defect. The completely similar structure of "free" OH–group was proposed also by Giguére (ref. 29) from the experimental "peroxide"–bearing point of view.

So, roughly speaking, we have two extremal H–bonds: "nice", or "strong" H–bond and the bifurcated, or "free" H–bond. In other words, that overcrude picture appears to belong to the family of "white–black" pictures. But, even for the model of orientational kink–defect, one can conclude that this two–state demarcation line cannot reflect the near semi–continuum variety of H–bonds participating in its formation, that is why the caution is strongly recommended with interpreting some properties of liquid water in terms of few threshold quantities.

Ionic Defect

Solitonic Model

We have already frozen the heavier subsystem of the total system depicted in Fig.1, but in the present case the variable r play the fundamental role. Now everything is clear that allows us to make the jump immediately to the continuum approach which will be justified somewhat later. So, that Hamiltonian of the protonic system of $(H_2O)_\infty$ becomes (refs. 23-26):

$$H=\frac{1}{2L}\int_{-\infty}^{+\infty} dz\left\{m(\dot{r})^2+mc_0^2\left(\frac{\partial r}{\partial z}\right)^2+2U(r)\right\} \tag{11}$$

Here m is the mass of proton, and c_0, interpreted as a "proton sound", is determined by the following formula (ref. 25)

$$c_0^2=2m^{-1}[\omega'(0)]^2\sum\left\{2\frac{3\cos^2\theta_1-1}{nL}+\frac{(nR)^2\cos(\theta_1+\theta_n)[3\cos(\theta_1-\theta_n)-1]}{[n^2L^2-nL^2+R^2]^{3/2}}\right\} \tag{11a}$$

based on assuming the dipole–dipole interaction between OH–dipoles and the sine–approximation of OH–bond dipole moment $\mu(r)$ verified for symmetric ices in (ref. 30). U(r) is a certain symmetric potential, for example, of a "r–four" form

$$U(r) = U_B^{(i)} [1 - (r/r_0)^2]^2 \qquad (12)$$

The classical field r=r(z,t) of collective protonic displacement is then defined by the Euler–Lagrange equation

$$m(r_{tt} - c_0^2 r_{zz}) + [4U_B^{(i)}/r_0^4]r^3 - [4U_B^{(i)}/r_0^2]r = 0 \qquad (13)$$

Introducing a new independent variable $\xi = z - Vt$, one can rewrite eqn. (13) as follows

$$m(V^2 - c_0^2)[d^2r/d\xi^2 = - 4U_B^{(i)}/r_0[(r/r_0)^3 - r/r_0] \qquad (14)$$

that, assuming $V < c_0$ and integrating, results in the formula

$$\int_0^r \frac{dr}{\sqrt{U(r)}} = \pm\sqrt{2}\gamma(\xi - \xi_0) \qquad (15)$$

where $\xi_0 = z_0 - Vt_0$. Substituting eqn. (12) into l.h.s. of eqn. (15) and integrating again, one obtains the soliton solution for the ionic defect

$$r_D(z,t) = \pm\, r_0 \tanh[\gamma/r_0(z - z_0 - Vt)] \qquad (16)$$

where

$$\gamma^2 = 2U_B^{(i)}/mc_0^2[1 - (V/c_0)^2].$$

Equation (16) tells us that the solutions of eqn. (13), which satisfy the ionic–defect type boundary conditions, are of two kinds. The first one starts from $r = -r_0$ at $z = -\infty$ and "arrives" at $r = +r_0$ at $z = +\infty$. This is positive ionic kink–defect. It is plotted in Fig. 4 (ref. 25).

The other solution is antikink, or negative ionic kink–defect. In the context of the positive kink–defect, the collective protonic jump from $-r_0$ to $+r_0$ occurs in the z–th region with the effective dimension Δz covering five molecules of water (ref. 25).

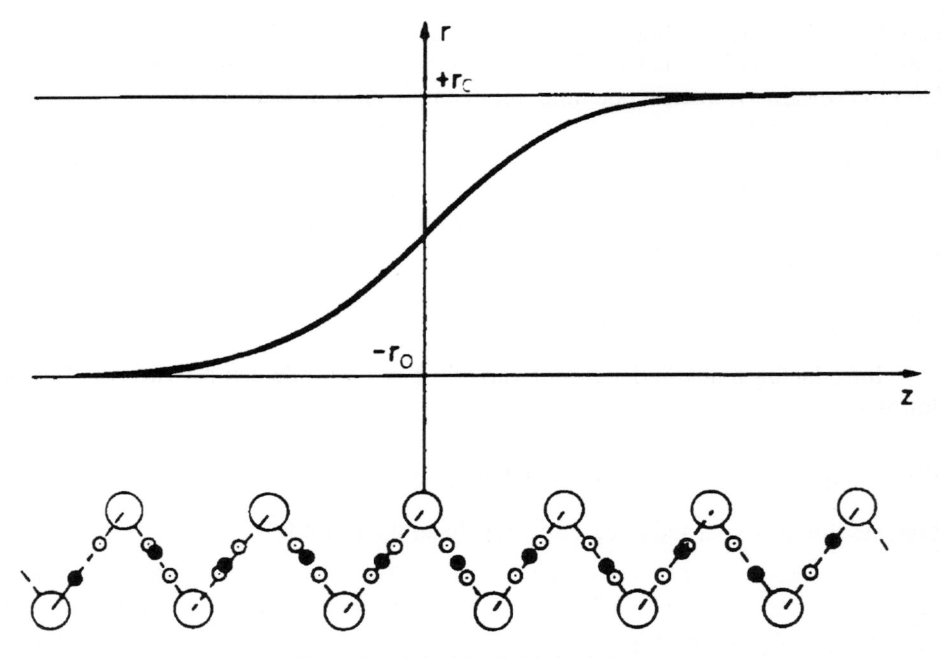

Fig. 4. Model of ionic kink–defect.

Therefore, the continuum approximation is valid. In the limit of small velocities V ($<c_0$), the total energy of the migration kink–defect is

$$E_D^{(i)}(V) = E_D^{(i)} + m_{eff}V^2 \tag{17}$$

where $E_D^{(i)}$ is the energy of the static (V = 0) defect (ref. 25).

$$E_D^{(i)} = \frac{8r_0c_0}{3L}\sqrt{2U_B^{(i)}m} \simeq 23.74 \text{ kcal/mol} \tag{18}$$

and m_{eff} is the effective mass of the ionic kink–defect (ref. 25).

$$m_{eff} = \frac{4r_0}{3Lc_0}\sqrt{2U_B^{(i)}m} \simeq 0.11m. \tag{19}$$

The quantity $E_D^{(i)}$ can be interpreted, similarly to the case of the orientational defect, as the energy of formation of the ionic defect. Comparison of the theoretical estimate with the experimental one, $E_D^{(i)expt} = 22\pm3$ kcal/mol (ref. 4) seems to be rather promising. The estimate of m_{eff}, comparing with m, elucidates the collective origin and high mobility of "protons" in liquid water and ice.

In the discrete picture, the ionic defect corresponds to a lengthy, topologically H–bonded region with the coordinates $\{r_i\}$ of all protons ranging quasi continuously in the interval $(-r_0, +r_0)$. The proposed model no doubt sounds in unison with the experimentally grounded conclusion by Hertz (ref. 31).

Few Comments on Isotopic Substitution

Let us come back to eqns. (18) and (19). Neglecting the second–order effects related to change of all lengths and angles when H is substituted by D, and taking into account eqn. (11a), one concludes that the energy of formation of the ionic kink–defect and its effective mass are isotopically invariant.

However, on the other side, the mass difference and geometrical isotopic effect could play the dominant role in such the nonlinear phenomena as the dependence of conductivity of liquid water on fractional isotopic substitution (ref. 32). In fact, in terms of the ionic and orientational collective solitonic structures, the fractional isotopic substitution implies a mass difference and geometrical inhomogeneity incorporated in our kink defects. In the limits of isotopically pure liquid water of H_2O and D_2O type, there is no that inhomogeneity. Between these limits, the interplay of the mass difference and geometrical disorder and nonlinearity (soliton) takes the part which determines certainly the origin of aforementioned phenomena (ref. 32). Not so time––consuming numerical simulations are needed to support this idea.

Relaxation of Heavier Subsystem

The "X–four" ($X = \theta$ or r) potential, applied before to outline the solitonic models of orientational and ionic defects in liquid water and ice (refs. 21, 22, 25 and 26) is insensitive, like the sine–Gordon potential also used, to change of the length R of the homoconjugated A–H···A hydrogen bridge in the system $(A\text{–}H\cdots)_\infty$ (for generality we use here A instead of OH). In other words, these potentials are always of double–well character irrespective on R. Such behaviour contradicts to some ab initio quantum chemical calculations of strong homoconjugated H–bonds which demonstrate an existence of a threshold value of R separating single- and double–minimum form of proton potential. For this reason we need to choose more realistic potential. A more suitable realistic potential of double–Morse type has been used firstly in ref. 33 to obtain the explicit solitonic solution.

A symmetric double–Morse potential is an equally weighted back–to–back linear combination of the three–parameter Morse potentials

$$U(r) = D\{1 - \exp[-\alpha(r + \tilde{R}/2)]\}^2 + D\{1 - \exp[\alpha(r - \tilde{R}/2)]\}^2 =$$

$$D\{[1 - 2\exp(-\alpha\tilde{R}/2)\mathrm{ch}\alpha r]^2 + [1 - 2\exp(-\alpha\tilde{R})]\} \tag{20}$$

Here D is the single–potential well depth, r_e is the equilibrium length of the isolated AH–bond, the parameter related to the AH–stretching harmonic constant, and $R = \tilde{R} + 2r_e$ is A···A distance. The threshold value of R for the symmetric double–Morse potential is expressed as follows (ref. 33).

$$R_{thr} = 2(r_e + \ln 2/\alpha) \quad \text{or} \quad \alpha\tilde{R}_{thr} = \ln 4 \cong 1.3863. \tag{21}$$

If nonzero R is smaller R_{thr}, the potential (20) has a single minimum at $r = 0$. In the opposite case, it possesses two equivalent minima at $r = \pm r_0$ where

$$r_0 = (1/\alpha)\cosh^{-1}[\tfrac{1}{2}\exp(\alpha\tilde{R}/2)] \tag{22}$$

separated by the barrier

$$U_B = D[1 - 2\exp(-\alpha\tilde{R}/2)]^2 \tag{23}$$

The explicit solution of solitonic type for the symmetric double–Morse potential ($R > R_{thr}$) takes the following form (ref. 33):

$$r_{\pm}(\xi;\xi_0) = \frac{1}{\alpha}\tanh^{-1}\left(\frac{\sinh[\pm\delta(\xi-\xi_0)]\sinh\alpha r_0}{\cosh\alpha r_0\cdot\cosh[\delta(\xi-\xi_0)]+1}\right) \tag{24}$$

where

$$\delta = \{2D/[m(c_0^2 - V^2)]\}^{1/2}\,\alpha\tanh\alpha r_0.$$

One can demonstrate by evaluating the protonic force acting upon the heavier subsystem that the formation of the proton kink (24) in the system $(A–H\cdots)_\infty$ results in contracting A···A distances around the soliton, that in turn, leads to the barrier lowering, or even it disappears. All these cases are studied analytically in ref. 33.

And the last comment. In this general case of the system $(A–H\cdots)_\infty$ when we can vary A and the parameters D, r_e, and α, one can observe a large freedom in dimension of the region where the proton kink is localized. Choosing A = OH and the appropriate values of the parameters one can get the proton kink which effectively reminds the well–known $H_5O_2^+$ complex (ref. 19).

Soliton–Instanton Approach for Proton Tunneling

The present Section considerably differ from the preceding ones, where, firstly we have considered the classical mechanism of migration between two equivalent minima, and secondly, the chain consisting of rather large number of H–bridges. At the first glance it is astonishing, however it is true that the concept of solitons can be very fruitful to study a proton tunneling and, to first order in $h/2\pi$, evaluated the explicitly the tunneling energy splitting for some double–well potentials.

Consider a proton of mass m in which tunnels across the barrier of the symmetric double–Morse potential, starting at time $\tau = -\mathbf{t}/2$ from the left well at $r = -r_0$ and arriving at $\tau = +\mathbf{t}/2$ at the right one, with $r = +r_0$. Within the Feynman's path integral formation (ref.34), the transition probability is defined as follows

$$(-r_0,-\tau/2\,|\,+r_0,+\tau/2)=\int \mathscr{D}[r(\tau)]\exp\left(\frac{i}{\hbar}S[r(\tau)]\right) \tag{25}$$

where

$$S[r(\tau)]=\int_{-\tau/2}^{+\tau/2} d\tau\,\mathscr{L}[r(\tau)] \tag{26}$$

is the Hamiltonian action for the path $r(\tau)$ expressed in terms of the classical Lagrangian

$$\mathscr{L}[r(\tau)] = (m/2)(dr/d\tau)^2 - U(r(\tau)) \tag{27}$$

and the path integral implies a summation over all periodic path $r(\tau)$ starting in the left well and ending in the right one.

Passing from the real time τ to the imaginary time $t = i\tau$ in Eq. (25), one obtains:

$$(-r_0,-\tau/2\,|\,+r_0,+\tau/2)=\int \mathscr{D}[r(t)]\exp\left\{-\frac{1}{\hbar}\int_{-\tau/2}^{+\tau/2} dt\left[\frac{m}{2}\left(\frac{dr}{dt}\right)^2+U(r(t))\right]\right\}. \tag{28}$$

Now we have two different formulas for the transition amplitude. In the small $h/2\pi$ limit the stationary phase method(ref. 34) tells us that the classical path, $r_{cl}(t)$ or $r_{cl}(\tau)$, gives the dominant contribution to the transition amplitude. The classical path is determined as follows:

Real Time

$$m[d^2r_{cl}(\tau)/d\tau^2] + d/dr[U(r_{cl}(\tau))] = 0 \quad (29a)$$

$$r(\mp\tau/2) = \mp r_0 \quad (29b)$$

Imaginary Time

$$m[d^2r_{cl}(t)/dt^2 + d/dr[-U(r_{cl}(t))] = 0 \quad (30a)$$

$$r(\mp T/2) = \mp r_0 \quad (T = i\tau). \quad (30b)$$

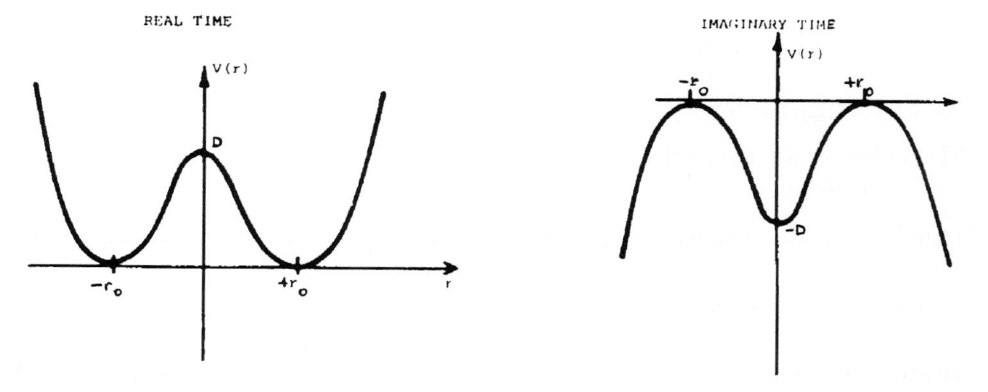

Fig. 5. Inversion of barrier with a passage from the real to the imaginary time.

Let us compare firstly r.h.s. of Eqs. (29a) and (30a). Mathematically the continuation of the real time to the imaginary time half–plane implies an inversion of a potential barrier or the transition from the symmetric double–well potential to the symmetric double–hump well shown in Fig. 5. Hence the concept of imaginary time (ref. 35) ensures the rigorous ground for the procedure of inverting of potential barriers was referred to "more intuitive rather than straight logical" ((ref. 36), p. 52).

The comparison of Eqs. (30a) with (13), taking at zeroth velocity V, with the simultaneous replacement of mc_0^2 by m and z by t in the latter, results in the explicit formula for the double–Morse classical path (compare with Eq. (16))

$$r_{cl}(t)=\frac{1}{\alpha}\tanh^{-1}\left[\frac{\sinh\left(\frac{\mu t}{\sqrt{2}}\right)\cdot\sinh\alpha r_0}{\cosh\left(\frac{\mu t}{\sqrt{2}}\right)\cdot\cosh\alpha r_0+1}\right],\ \mu=2\left(\frac{D}{M}\right)^{1/2}\alpha\tanh\alpha r_0. \tag{31}$$

Unlike the kink–soliton, given by Eq. (16), the solution determines a localized structure in time, that is why it is called "instanton", ref.28 .

Using the concept of instantons and applying the procedure outlined in ref. 28 , one can obtain the formula for the tunneling energy splitting in the case of the symmetric double–Morse potential (to first order in h/2π)

$$\Delta E=2\left(\frac{D}{M}\right)^{3/4}\sinh\alpha r_0\left(\frac{2\sqrt{2}}{\pi}\hbar\alpha M\tanh^3\alpha r_0\right)^{1/2}\exp\left[-\frac{2\sqrt{2DM}}{\alpha\hbar}(\alpha r_0-\tanh\alpha r_0)\right]. \tag{32}$$

In ref. 37 the formula (32) is employed to investigate analytically the isotopic effects in proton tunneling and the expression for proton polarizability derived in ref. 19 to base the concept of large proton polarizability.

I would like to conclude my contribution by the comment that now almost everything with a richness of physical chemistry as the new land for "solitonic dinosaures" is clear, and as the matter of fact we can believe that they will be alive as long as liquid water at least will surprise its own observers.

Acknowledgments

Professor Georg Zundel for the fruitful discussions and warm hospitality and the AvH Foundation for the Research Grant and Professor A. Müller for his kind invitation to take part in the International Workshop "Electron and Proton Transfer in Chemistry and Biology" (Bielefeld, September 1990) are greatly acknowledged.

References

1 J. Scott Rusell, Proc. Roy. Soc. Edinburgh (1844) 319.

2 A.C. Scott, Science, March/April, (1990) 29.

3 K.R. Popper, In: G. Tarozzi and A. van der Merwe (Eds.) *Open Questions in Quantum Physics* Reidel, Dordrecht, 1985, pp. 395-413.

4 D. Eisenberg and W. Kauzman *The Structure and Properties of Water* Oxford University Press, New York, 1969.

5 J.D. Bernal and R.H. Fowler, J. Chem. Phys. **1** (1933) 515.

6 L. Pauling, J. Am. Chem. Soc. **57** (1935) 2680.

7 N. Bjerrum, K. Dan. Vidensk. Selsk. Mat.–Fys. Medd., **27** (1953) 3; Science **115** (1952) 385.

8 H.S. Frank and W.–Y. Wen, Disc. Faraday Soc. **23** (1957) 72.

9 D. Hankins, J.W. Moskowitz and F.H. Stillinger, J. Chem. Phys. **53** (1970) 4544; J.E. Del Bene and J.A. Pople, *Ibid.* **52** (1970) 4858; B.R. Lentz and H.A. Scheraga, *Ibid.* **58** (1973) 5296; E. Clementi *Determination of Liquid Water Structure Coordination Numbers for Ions and Solvations for Biological Molecules. Lecture Notes in Chemistry* Springer–Verlag, Berlin, Vol. 2, 1976.

10 E.S. Cambell and M. Mezei, J. Chem. Phys., 67 (1977) 2338; Mol.Phys. **41** (1980) 883; E.S. Cambell and D.Belford, Theor. Chim. Acta **61** (1982) 295.

11 G.C. Pimentel and A.L. MacClellan *The Hydrogen Bond* Freeman, San Francisco, 1960.

12 E.S. Kryachko, Int. J. Quantum Chem., 30 (1986) 495.

13 H. Graener,T.Q. Ye and A. Laubereau, J. Chem. Phys. **90** (1989) 3413; *Ibid.* **91** (1989) 1043; Phys. Rev. **841** (1990) 2587.

14 H. Kleeberg and W.A.P. Luck, Z. Phys. Chem. **270** (1989) 270.

15 P. Schuster A. Karpfen and A. Beyer *Molecular Interactions* W.J. Orville–Thomas and H. Ratajczak (Eds.), Wiley, London, 1979 pp. 1-54.

16 J.D. Dunitz, Nature, 197 (1963) 860.

17 N.V. Cohan, M. Cotti, J.V. Iribarne and M. Weissmann, Trans. Faraday Soc. **58** (1962) 490.

18 D. Eisenberg and C.A. Coulson, Nature **199** (1963) 368.

19 G. Zundel *Hydration and Intermolecular Interactions* Academic Press, New York, 1969.

20 M. Eigen and L. DeMaeyer, Proc. Roy. Soc., London **A247** (1958) 505.

21 E.S. Kryachko, Chem. Phys. Lett. **141** (1987) 346.

22 O.E. Yanovitskii and E.S. Kryachko, Phys. stat. sol. (b), **147** (1988) 69; E.S. Kryachko, In: *Physics of Many–Particle Systems* vol.XI, Naukova Dumka, Kiev, 1987.

23 V.Ya. Antonchenko, A.S. Davydov and A.V. Zolotariuk, Phys. stat. sol.(b), **115** (1983) 631; A.S. Davydov *Solitons in Molecular Systems* Kluwer, Dordrecht, 1990.

24 Y. Kashimori, T. Kikuchi and K. Nishimoto, J. Chem. Phys., 77 (1982) 1904.

25 E.S. Kryachko, Solid State Commun. **65** (1988) 1609.

26 E.S. Kryachko, In: D. Mukherjee (Ed.) *Correlation Aspects in Extended Molecular Systems Lecture Notes in Chemistry* Vol. 50, Springer Verlag, Berlin, 1989, pp. 301-318.

27 S. Scheiner and J.F. Nagle, J.Phys.Chem. **87** (1983) 4267.

28 R. Rajaraman *Solitons and Instantons* North-Holland, Amsterdam, 1982.

29 P.A. Giguére, J. Chem. Phys. **87** (1987) 4835.

30 D.D. Klug and E. Whalley, J. Chem. Phys. **83** (1985) 925.

31 H.G. Hertz, Chem. Scr., **27** (1987) 479.

32 H. Weingärtner and C.A. Chatzidimitriou–Dreismann, Nature **346** (1990) 548.

33 E.S. Kryachko, Chem. Phys. **143** (1990) 359.

34 R. Feynman and A. Hibbs *Quantum Mechanics and Path Integrals* McGraw–Hill, New York, 1965.

35 K. Freed, J. Chem. Phys. **56** (1972) 692;
D.W. MacLaughlin, J. Math. Phys. **13** (1972) 1099.

36 R.P. Bell *The Tunnel Effect in Chemistry* Chapman and Hall, London, 1980.

37 E.S. Kryachko, M. Eckert and G. Zundel, (J. Mol. Structure), in press.

Index of Contributing Authors

Subject index